Lecture Notes in Mathematics 1552

Editors:
A. Dold, Heidelberg
B. Eckmann, Zürich
F. Takens, Groningen

Joachim Hilgert   Karl-Hermann Neeb

Lie Semigroups and their Applications

Springer-Verlag

Berlin Heidelberg New York
London Paris Tokyo
Hong Kong Barcelona
Budapest

Authors

Joachim Hilgert
Mathematisches Institut der Universität Erlangen
Bismarckstr. 1 1/2
D-91054 Erlangen, Germany

Karl-Hermann Neeb
Fachbereich Mathematik
Technische Hochschule Darmstadt
Schloßgartenstr. 7
D-64289 Darmstadt, Germany

Mathematics Subject Classification (1991): 22A15, 22A25, 22E46, 22E30, 53C30, 53C50, 53C75

ISBN 3-540-56954-5 Springer-Verlag Berlin Heidelberg New York
ISBN 0-387-56954-5 Springer-Verlag New York Berlin Heidelberg

© Springer-Verlag Berlin Heidelberg 1993
Printed in Germany

46/3140-543210 - Printed on acid-free paper

Table of Contents

3. Geometry and topology of Lie semigroups

4. Ordered homogeneous spaces

5. Applications of ordered spaces to Lie semigroups

6. Maximal semigroups in groups with cocompact radical

7. Invariant Cones and Ol'shanskiĭ semigroups

8. Compression semigroups

9. Representation theory

viii

Introduction

Although semigroups of transformations appear already in the original work of S. Lie as part of his efforts to find the right analogue of the theory of substitutions in the context of differential equations, it was Ch. Loewner who first studied such objects purposely [Loe88]. He considered semigroups of self-maps of the unit disc as a tool in geometric function theory. In the late seventies, subsemigroups of Lie groups were considered in relation to control systems with symmetries (cf. [JK81a,b], [Su72]). At about the same time Ol'shanskiĭ introduced such semigroups to in.order to study the representation theory of infinite dimensional classical groups ([Ols91]). Moreover causality questions led people in relativity theory to consider subsemigroups of Lie groups generated by one-parameter semigroups as well. Motivated by this evidence Hofmann and Lawson worked out, in [HoLa83], systematic groundwork for a Lie theory of semigroups. These efforts eventually resulted in the monograph [HHL89].

In the meantime it has become increasingly clear that certain subsemigroups of Lie groups play a vital role in the harmonic analysis of symmetric spaces and representation theory. The purpose of this book is to lead the reader up to these applications of Lie semigroup theory. It is intended for a reader familiar with basic Lie theory but not having any experience with semigroups. In order to keep the overlap with [HHL89] to a minimum we have occasionally quoted theorems without proof from this book – especially when the version there is still the best available. On the other hand the last few years have seen rapid development, and so we are able to present improved versions of many results from [HHL89] with completely new proofs.

This book is not meant to be comprehensive. We have left out various topics that belong to the theory but, at the time being, don't show close connections with the applications we have in mind. Also we have chosen to focus on closed subsemigroups of Lie groups and thereby avoid certain technical complications.

A Lie semigroup is a closed subsemigroup S of a Lie group G which, as a closed subsemigroup, is generated by the images of all the one-parameter semigroups

$$\gamma_X : \mathbb{R}^+ \to S, \qquad t \mapsto \exp(tX).$$

The set of all these one-parameter semigroups can be viewed as a set $\mathbf{L}(S)$ in the Lie algebra $\mathfrak{g}$ of G. It is a closed convex cone satisfying

$$e^{\operatorname{ad} X} \mathbf{L}(S) = \mathbf{L}(S) \qquad \forall X \in \mathbf{L}(S) \cap -\mathbf{L}(S),$$

an algebraic identity which reflects the fact that S is invariant under conjugation by elements from the unit group $S \cap S^{-1}$. Convex cones satisfying these properties are called Lie wedges and play the role of Lie algebras in the Lie theory of semigroups. Following the general scheme of Lie theory one wants to study the properties of Lie semigroups via their Lie wedges using the exponential function for the translation mechanism. In Chapter 1 we describe the essential features of this mechanism. In particular the topological and algebraic obstructions that arise when one tries to find a Lie semigroup with prescribed Lie wedge are pointed out. The problem, called the globality problem, has not been solved in a definitive way, but one has far-reaching results which essentially reduce the globality problem to finding the maximal subsemigroups of Lie groups. The topic of maximal subsemigroups is taken up again later in the book (Chapter 6 and Chapter 8).

The main result of Chapter 1 is Theorem 1.35 which characterizes the Lie wedges that occur as the tangent wedge of a Lie semigroup in terms of the existence of certain functions on the group G. In order to prove it we use a result about ordered homogeneous spaces which is presented only later, in Chapter 4 (Cor. 4.22). We chose this way of organizing things to be able to present the globality problem without too much technical ballast. Moreover the material about ordered homogeneous spaces presented in Chapter 4 is of independent interest even though the separation between semigroup and ordered space aspects may seem artificial to insiders.

Chapter 2 is completely devoted to a list of examples which either have model character or serve as counterexamples at some point. In Chapter 3 we present various geometric and topological properties. Most importantly, it is shown that the interior of a Lie semigroup S is dense if $\mathbf{L}(S)$ generates $\mathfrak{g}$ as a Lie algebra (cf. Theorem 3.8). Also important for later applications is the fact that Lie semigroups admit simply connected covering semigroups (cf. Theorem 3.14).

In Chapter 5 some more consequences of the theory of ordered homogeneous spaces are listed. Among other things it is shown that the unit group of a Lie semigroup is connected. Moreover we explain how the existence of a Lie semigroup with prescribed Lie wedge in a given connected Lie group G is related to the existence of such a Lie semigroup in a covering group of G.

Chapter 6 deals with the characterization of maximal subsemigroups with interior points in simply connected groups with cocompact radical. They all have half-spaces as tangent wedges and a closed subgroup of codimension one as unit group. Finally we show how one can use this result to solve some controllability questions on reductive groups.

The main result of Chapter 7 is Lawson's Theorem on Ol'shanskiĭ semigroups which in particular says that for a connected Lie group G sitting inside a complexification $G_{\mathbb{C}}$ with a Lie algebra $\mathfrak{g}$ which admits a pointed $\mathrm{Ad}(G)$-invariant cone W with interior points, then $(g, X) \to g \exp iX$ is a homeomorphism $G \times W \to G \exp iW$ onto a closed subsemigroup of $G_{\mathbb{C}}$. This semigroup is called a complex Ol'shanskiĭ semigroup. Before we get there we show what consequences the existence of invariant cones with interior points has for a Lie algebra $\mathfrak{g}$, give a characterization of those Lie algebras containing pointed generating invariant cones, and describe the complete classification of such cones.

Complex Ol'shanskiĭ semigroups and their real analogues appear in many different contexts. They consist of the elements of $G_{\mathbb{C}}$ which map the positive part of the ordered homogeneous space $G_{\mathbb{C}}/G$ into itself, where the ordering is induced by the invariant cone field associated to the invariant cone. In the semisimple case they (at least the ones coming with the maximal invariant cones) can also be viewed as semigroups of compressions

$$\mathrm{compr}(\mathcal{O}) = \{g \in G_{\mathbb{C}} : g.\mathcal{O} \subseteq \mathcal{O}\}$$

of certain open G-orbits $\mathcal{O}$ in suitable flag manifolds associated with $G_{\mathbb{C}}$. In order to show this we study the open G-orbits on complex flag manifolds via the symplectic (in fact, pseudo-Kähler) structure that is given on these orbits. Using the results obtained in this process one can eventually show that complex Ol'shanskiĭ semigroups for maximal invariant cones are maximal subsemigroups.

Apart from their different geometric realizations, complex Ol'shanskiĭ semigroups occur as the natural domains in $G_{\mathbb{C}}$ to which one can analytically continue highest weight representations of G. We show in Chapter 9 how this is done for general G. Moreover we give some examples of this continuation procedure such as the holomorphic discrete series representations and the metaplectic representation which gives rise to Howe's oscillator semigroup. The largest subrepresentation of $L^2(G)$ - for general G - which admits an analytic continuation to a complex Ol'shanskiĭ semigroup leads to a Hardy space of holomorphic functions on this semigroup satisfying an L^2-condition on G-cosets. This Hardy space coincides with the classical notions for tube domains and polydiscs if G is a vector group or a torus.

In Chapter 10 we collect the results presented in this book for semigroups related to $\mathrm{Sl}(2)$.

For the orientation of the reader we conclude this introduction with some comments on the overlap with [HHL89].

Apart from some elementary properties of Lie wedges and cones Chaper 1 is independent of [HHL89]. The idea of monotone functions is only briefly touched in [HHL89] and the corresponding results we present in Chapter 1 are stronger and the proofs less complicated.

Some of the characteristic examples such as 2.1, parts of 2.2, 2.9 − 2.11 described in Chapter 2 occur also in [HHL89]. We have included them for the convenience of the reader since they illuminate some specific features of the theory.

Chapter 3 is independent of [HHL89]. The results of Section 3.2 were known at that time, but the new proof of Hofmann and Ruppert is shorter and it offers some new insights.

Ordered homogeneous spaces do also occur in [HHL89], where they are used to obtain the results about the structure of Lie semigroups near their group of units (cf. Sections 4.2, 4.3). The results on the global structure of ordered homogeneous spaces concerning properties such as global hyperbolicity are new (cf. Sections 4.4, 4.7).

The *Unit Group Theorem* and the *Unit Neighborhood Theorem* (cf. Section 5.1) were already proved in [HHL89]. Here we obtain these results out of a general theory of ordered homogeneous spaces.

Sections 6.2 − 6.6 are more or less contained in [HHL89]. The results in Sections 6.1 and 6.7 are new and complement the existing results in an interesting way. Since the area of maximal subsemigroups, in particular of maximal subsemigroups in simple groups, still presents many open problems, we decided to include the whole state of the theory of maximal semigroups in Chapter 6. We note also that the Sections 8.1 and 8.6 contain recent results on maximal subsemigroups in semisimple Lie groups complementing the material in Chapter 6 which is mostly concerned with groups G, where $\mathbf{L}(G)$ contains a compact Levi algebra.

Even though the theory of invariant cones and their classification by intersections with compactly embedded Cartan algebras is contained in [HHL89], our

Section 7.1 does not significantly overlap with [HHL89]. Our approach to invariant cones is based on coadjoint orbits. It seems to be more fruitful and far-reaching than the direct approach. Those results on invariant cones which we need in the sequel are proved along these lines. This made it possible to shorten some of the proofs considerably.

The remainder of Sections 7 – 9 is absolutely independent of [HHL89]. For more recent results lying already beyond the scope of this book and concerning the material contained in these sections we refer the reader to [Ne93a-f].

User's Guide

Since many results in this book do not depend on every preceding chapter, we give a list containing for each section, the set of all other sections on which it depends. If, e.g., Section 4.5 depends on Section 4.4 and Section 4.4 depends on Section 4.3, then Section 4.3 appears only in the list of Section 4.4. So the reader has to trace back the whole tree of references by using the lists of several sections. Nevertheless we hope that this is helpful to those readers interested merely in some specific sections of the book.

Chapter 1:

1.2 [1.1], 1.3 [1.1], 1.4 [1.3], 1.8 [1.7], 1.9 [1.1, 1.8], 1.10 [1.4, 1.9]

Chapter 2:

2.3 [1.7], 2.6 [1.4, 1.5, 1.9], 2.7 [1.10]

Chapter 3:

3.1 [1.5, 1.10, 2.2, 2.7], 3.3 [1.10, 2.11], 3.4 [1.10, 3.2], 3.5 [3.4], 3.6 [3.5]

Chapter 4:

4.2 [1.9, 4.1], 4.3 [1.4, 1.9, 4.1, 4.2], 4.4 [1.7, 4.3], 4.5 [4.3], 4.6 [3.2, 4.3], 4.7 [4.4, 4.6]

Chapter 5:

5.1 [1.4, 1.8, 4.3], 5.2 [1.10, 3.2, 4.3, 5.1], 5.3 [3.2, 4.4, 5.1], 5.4 [3.1, 4.2, 5.3], 5.5 [2.5, 5.4]

Chapter 6:

6.2 [2.9], 6.3 [1.7, 6.2], 6.4 [6.3], 6.5 [6.3], 6.6 [6.5], 6.7 [1.2, 1.10, 3.2, 4.2, 4.3, 6.6]

Chapter 7:

7.1 [1.2], 7.2 [1.3, 7.1], 7.3 [4.2, 5.3, 7.2]

Chapter 8:

8.1 [1.7], 8.4 [1.3, 1.7, 7.2, 8.1, 8.3], 8.5 [2.6, 7.2], 8.6 [6.7, 8.4, 8.5]

Chapter 9:

9.3 [3.4, 7.3], 9.4 [8.4], 9.5 [7.1], 9.7 [7.3]

1. Lie semigroups and their tangent wedges

The basic feature of Lie theory is that of using the group structure to translate global geometric and analytic problems into local and infinitesimal ones. These questions are solved by Lie algebra techniques which are essentially linear algebra and then translated back into an answer to the original problem. Surprisingly enough it is possible to follow this strategy to a large extend also for semigroups, but things become more intricate. Because of the missing inverses one has not only to deal with linear algebra but also with convex geometry at the infinitesimal level. Similarly to the group case the Lie algebraic counterpart of a subsemigroup can either be defined as geometric (sub-)tangent vectors to the semigroup or a family of one-parameter semigroups contained in the semigroup (or at least in its closure). It turns out to be a convex cone in the Lie algebra, possibly containing non-trivial vector subspaces. For this reason we prefer the notion wedge. It is well known from Lie group theory that Lie subalgebras always correspond to analytic subgroups, but these analytic subgroups need not be closed, i.e., embedded manifolds. This difficulty of course does not disappear in the context of semigroups. But in order to avoid undue technical complications we will often simply restrict our attention to closed subsemigroups of Lie groups. Apart from the additional problems on the infinitesimal level caused by the replacement of vector spaces by wedges there is a new, even more serious, obstacle to a successful translation mechanism in the semigroup context: It is much harder to translate answers back to the global level since it turns out that the relation between Lie wedges and semigroups is quite complicated. In this chapter we study a class of semigroups which *can* be recovered from their Lie wedges, but before we do that, we collect in Section 1.1 a few facts about the geometry of wedges.

In Section 1.2 we consider wedges in vector space V which are invariant under the linear action of a compact group K. Of particular interest in this setting is the projection mapping which maps V onto the submodule of K-fixed points. These results will be used in Chapter 7 for invariant cones in Lie algebras and their duals.

Section 1.3 is a self-contained introduction into the characteristic function of a cone and its basic properties. This function is a basic tool in the study of groups acting on cones as automorphisms. It proves particularly valuable for non-compact groups.

In Section 1.4 we develop the notion of a Lie semigroup which will be fundamental throughout this book. To illustrate the condition that a Lie semigroup is determined by its infinitesimal data, a Lie wedge in the corresponding Lie algebra, we deal in some detail with the examples of compression semigroups of Lorentzian cones and Euclidean balls. As already mentioned above, the relation between Lie semigroups and Lie wedges incorporates some new difficulties which do not arise for Lie algebras and groups. These diffculties are described in a categorial framework in Sections 1.5 and 1.6.

In the remainder of Chapter 1 we are dealing with the globality problem, i.e., the problem to find for a given Lie wedge a Lie semigroup. The concept of a monotone function is a basic tool to cope with the globality problem. It is developed in Section 1.7 and some deeper insight into the existence of smooth and analytic monotone functions are gained in Section 1.8. The characterization of global Lie wedges by the existence of certain monotone functions is given in Section 1.9 but we postpone a technical part of the proof to Chapter 4, where we will use the theory of ordered homogeneous spaces to complete it. The last section contains various criteria for the globality of a Lie wedge which are deduced from the characterization from above and which will be useful throughout the other chapters.

1.1. Geometry of wedges

Let L be a finite dimensional vector space. A subset W is called a *wedge* if it is a closed convex cone. The vector space $H(W) := W \cap -W$ is called the *edge of the wedge*. We say that W is *pointed* if the edge of W is trivial and that W is *generating* if $W - W = L$. We denote the dual of L with L^*. The *dual wedge* $W^* \subseteq L^*$ is the set of all functionals which are non-negative on W. Furthermore we set $\operatorname{algint} W := \operatorname{int}_{W-W} W$ and $W^\perp := H(W^*)$. The following proposition is a collection of elementary facts about the relations between wedges and their duals.

Proposition 1.1.   *We identify the dual of L^* with L. Then the following assertions hold for a wedge $W \subseteq L$:*

(i) $(W^*)^* = W$.

(ii) *W is generating if and only if W^* is pointed and, conversely, W is pointed if and only if W^* is generating.*

(iii) *$\omega \in \operatorname{algint} W^*$ if and only if $\omega(x) > 0$ for all $x \in W \setminus H(W)$ and*

$$\operatorname{algint} W = \{x \in W : \omega(x) > 0 \ \text{ for all } \ \omega \in W^* \setminus H(W^*)\}.$$

(iv) *For a family $(W_i)_{i \in I}$ of wedges in L we have that*

$$\Big(\bigcap_{i \in I} W_i\Big)^* = \overline{\sum_{i \in I} W_i^*} \quad \text{and} \quad \Big(\sum_{i \in I} W_i\Big)^* = \bigcap_{i \in I} W_i^*.$$

(v) *If $V \subseteq L$ is a convex cone, then $\overline{V} = (V^*)^*$, $\operatorname{algint} V = \operatorname{algint} \overline{V}$, and $\overline{V} = \overline{\operatorname{algint} V}$.*

Proof.   (i) It is clear that $W \subseteq (W^*)^*$. Let $x \notin W$. Then, by the Theorem of Hahn-Banach, there exists a $\omega \in W^*$ with $\omega(x) < 0$. Therefore $x \notin (W^*)^*$.

(ii) Note first that
$$H(W^*) = W^\perp = (W - W)^\perp.$$

Therefore W^* is pointed if and only if $W - W = L$, i.e., if W is generating. The dual assertion follows from (i) by applying the first one to W^*.

(iii) In view of (i), it suffices to prove that

$$\text{algint}\, W = \{x \in W : \omega(x) > 0 \ \text{ for all }\ \omega \in W^* \setminus H(W^*)\}.$$

"$\subseteq$": Let $x \in \text{algint}\, W$. Then $0 \in \text{int}_{W-W}(W - x)$, so that $W - W = W - \mathbb{R}^+ x$. If $\omega \in W^*$ and $\omega(x) = 0$, then

$$\omega(W - W) = \omega(W) \subseteq \mathbb{R}^+.$$

Therefore $\omega(W - W) = 0$ since ω is linear, and hence $\omega \in H(W^*)$.

"$\supseteq$": Suppose that $x \notin \text{algint}\, W$. Then the Theorem of Hahn-Banach implies the existence of a linear functional $\omega' \neq 0$ on $W - W$ such that $\omega'(W) \subseteq \mathbb{R}^+$ and $\omega'(x) = 0$. Extending ω' to $\omega \in L^*$ we find that $\omega \in W^* \setminus H(W^*)$. This proves the inclusion $\supseteq$.

(iv) A linear functional is non-negative on $\sum_{i \in I} W_i$ if and only if it is non-negative on every wedge W_i, thus

$$\Big(\sum_{i \in I} W_i\Big)^* = \bigcap_{i \in I} W_i^*.$$

Replacing each wedge in this identity by its dual it follows with (i) that

$$\Big(\sum_{i \in I} W_i^*\Big)^* = \bigcap_{i \in I} W_i.$$

So the assertions follow from (i) and (v).

(v) Since every linear functional on L is continuous, it is clear that $V^* = (\overline{V})^*$. Therefore $(V^*)^* = \overline{V}$ follows from (i). To prove the rest of (v), in view of $\overline{V} \subseteq V - V$, we may assume that V is generating. First we note that $V + \text{int}\, V \subseteq \text{int}\, V$.

Let $x \in \text{int}\, V$ and $v \in V$. Then $v + tx \in \text{int}\, V$ for all $t > 0$ and therefore $v \in \overline{\text{int}\, V}$. To see that $\text{int}\, \overline{V} \subseteq \text{int}\, V$ (the other inclusion is trivial), let $x \in \text{int}\, \overline{V}$ and U a 0-neighborhood with $x - U \subseteq \overline{V}$. Set $W := U \cap \text{int}\, V$. Then $x - W$ is an open subset of $\overline{V}$ and therefore it contains an element of $v \in V$. Then $v = x - w$ holds with $w \in W$, so $x = v + w \in V + \text{int}\, V \subseteq \text{int}\, V$. ∎

The geometry of a closed convex set in a finite dimensional vector space is completely determined by the set of extremal points. But between extremal points and the whole set one has interesting sets, the faces, which share properties of convex sets *and* extremal points. We only give the definitions in the context of wedges: Let F, $W \subseteq L$ be wedges. Then we set

$$L_F(W) := \overline{W + F - F} \quad \text{and} \quad T_F(W) := H\big(L_F(W)\big) = L_F(W) \cap -L_F(W).$$

The fact that $W + F - F$ is convex and stable under multiplication with non-negative scalars shows that $L_F(W)$ is a wedge. Note that $L_F(W) = \overline{W - F}$ if $F \subseteq W$. We say that a wedge $F \subseteq W$ is an *exposed face* of W if

$$F = W \cap T_F(W)$$

and a *face* of W if its complement $W \setminus F$ is an ideal in the additive semigroup W. The geometric meaning of these concepts will be clarified by the following two propositions. We write $\mathcal{F}(W)$ for the set of faces of W and $\mathcal{F}_e(W)$ for the set of exposed faces of W.

The following proposition describes how the faces of W and its dual wedge are related.

Proposition 1.2.   *The set $\mathcal{F}_e(W)$ is stable under arbitrary intersections and therefore a complete lattice with $H(W)$ as minimal and W as maximal element. Moreover, the following assertions hold.*

(i) *The mappings*

$$\mathrm{op}^* : \mathcal{F}_e(W^*) \to \mathcal{F}_e(W), \; E \mapsto W \cap E^\perp$$

and

$$\mathrm{op} : \mathcal{F}_e(W) \to \mathcal{F}_e(W^*), \; F \mapsto W^* \cap F^\perp$$

are order reversing bijections mapping a face to its "opposite" face in the dual cone. Moreover, for every subset $E \subseteq W^*$ the set

$$\mathrm{op}^*(E) := E^\perp \cap W$$

is an exposed face of W and for every exposed face there exists $\omega \in W^*$ with $F = \ker \omega \cap W$.

(ii) *For a wedge $F \subseteq W$ we have that*

$$L_F(W)^* = W^* \cap F^\perp.$$

Proof.   Let $(F_i)_{i \in I}$ be a family of exposed faces of W and $F := \bigcap_{i \in I} F_i$. Then F is a wedge. The relation

$$F \subseteq W \cap T_F(W) \subseteq W \cap T_{F_i}(W) = F_i \qquad \forall i \in I$$

shows that $F = W \cap T_F(W) \in \mathcal{F}_e(W)$. We conclude that every non-empty subset in $\mathcal{F}_e(W)$ has an infimum. Thus this partially ordered set is a complete lattice. That $H(W)$ and W are the minimal and maximal elements, follows directly from the definition.

(ii) This follows from the definition of $L_F(W)$ and from Proposition 1.1(iv).

(i) Let $F \in \mathcal{F}_e(W)$. Then, in view of Proposition 1.1(ii) and (iv), we have

$$(1.1) \qquad F^* = (W \cap T_F(W))^* = \overline{W^* - \mathrm{op}(F)} = L_{\mathrm{op}(F)}(W^*).$$

Therefore

$$T_{\mathrm{op}(F)}(W^*) \cap W^* = W^* \cap F^\perp = \mathrm{op}(F)$$

shows that $\mathrm{op}(F) \in \mathcal{F}_e(W^*)$. Equation (1.1) and (ii) imply that

$$\mathrm{op}^* \circ \mathrm{op} = \mathrm{id}_{\mathcal{F}_e(W)}.$$

Replacing W by W^* we also find that

$$\mathrm{op} \circ \mathrm{op}^* = \mathrm{id}_{\mathcal{F}_e(W^*)}.$$

Let $E \subseteq W^*$ and $F := T_E(W^*) \cap W^*$ be the exposed face generated by E. Then

$$\mathrm{op}^*(E) = L_E(W^*)^* = L_F(W^*)^* = \mathrm{op}^*(F)$$

shows that $\mathrm{op}^*(E)$ is an exposed face of W. Finally, suppose that $F \in \mathcal{F}_e(W)$ and take $\omega \in \mathrm{algint}\,\mathrm{op}(F)$. Then, by Proposition 1.1(iii),

$$\ker \omega \cap W = \mathrm{op}(F)^\perp \cap W = \mathrm{op}^* \circ \mathrm{op}(F) = F.$$

■

Proposition 1.3.   *The set $\mathcal{F}(W)$ of faces of W is stable under arbitrary inter-sections and therefore a complete lattice with $H(W)$ as minimal and W as maximal element. Moreover, the following assertions hold.*

(i)
$$\mathcal{F}(V) = \{F \in \mathcal{F}(W) : F \subseteq V\} \quad \text{for every} \quad V \in \mathcal{F}(W),$$

i.e., the faces of a face V are exactly the faces of W which are contained in V.

(ii) *For every element $f \in \text{algint}\, W$ the whole wedge W is the only face containing f.*

(iii) *A subset $F \subseteq W$ is a face if and only if there exists a finite chain*

$$F_0 = F \subseteq F_1 \subseteq \ldots \subseteq F_n = W$$

of wedges such that $F_i \in \mathcal{F}_e(F_{i+1})$ for $i = 0, \ldots, n-1$. In particular, every face $F \neq W$ is contained in $\ker \omega \cap W$ for a suitable $\omega \in W^*$, and $\mathcal{F}_e(W) \subseteq \mathcal{F}(W)$.

(iv) *A subset $E \subseteq W^*$ is a face if and only if there exists a finite chain*

$$W = W_0 \subseteq W_1 \subseteq \ldots \subseteq W_n = E^*$$

of wedges such that $W_{i+1} = L_{F_i}(W_i)$ for an exposed face $F_i \in \mathcal{F}_e(W_i)$. In particular, $H(E^*) \cap W \neq H(W)$ if $E \neq W^*$.

(v) *For every face $E \in \mathcal{F}(W^*)$ we have that*

$$E = H(E^*)^{\perp} \cap W^* = (E - E) \cap W^* \quad \text{and} \quad E^* = L_{H(E^*)}(W).$$

(vi) *For faces $E_1, E_2 \in \mathcal{F}(W^*)$ the relations $E_1 \subseteq E_2$ and $E_2^{\perp} = H(E_2^*) \subseteq E_1^{\perp} = H(E_1^*)$ are equivalent.*

(vii) *If $F \subseteq W$ is a subsemigroup such that $W\backslash F$ is a semigroup ideal, then $F \in \mathcal{F}(W)$.*

Proof.   Let $(F_i)_{i \in I}$ be a family of faces of W and $F := \bigcap_{i \in I} F_i$. Then F is a wedge and $x + y \in F, x, y \in W$ implies that $x, y \in F_i$ for all $i \in I$, so $x, y \in F$. Hence $F \in \mathcal{F}(W)$.

(i) Let $F \in \mathcal{F}(V)$ and suppose that $x, y \in W$ with $x + y \in F \subseteq V$. Then $x, y \in V$ holds since V is a face of W, and therefore $x, y \in F$. Hence F is a face of W which is contained in V. Conversely, assume that F is a face of W which is contained in V. Then it is a face of V, because $x + y \in F$ and $x, y \in V$ implies that $x, y \in F$. We conclude that every non-empty subset in $\mathcal{F}(W)$ has an infimum. Thus this poset is a complete lattice.

(ii) Suppose that $f \in F \in \mathcal{F}(W)$ and let $x \in W$. Then, since $f \in \text{algint}\, W$, $f - W$ is an open neighborhood of 0 in $W - W$ (Proposition 1.1) and we find an $n \in \mathbb{N}$ such that $\frac{1}{n}x \in f - W$. Therefore we find $y \in W$ with $\frac{1}{n}x + y = f \in F$. Thus $\frac{1}{n}x \in F$ and this implies that $x \in F$.

(iii) The sufficiency of the condition follows from (i) by an easy induction which shows that F_i is a face of W for each subscript $i = n, n-1, \ldots, 0$. To see that this condition is necessary, we use induction on $\dim(W - W)$. If $F \cap \text{algint}\, W \neq \emptyset$ then, in view of (ii) above, we have that $F = W$. Suppose that $W \neq F$. Then we find a linear functional ω such that $\omega(\text{algint } W) \subseteq]0, \infty[$ and $\omega(F) = \{0\}$ (Hahn-Banach). Thus $\omega \in W^* \setminus H(W^*)$. Then $F' = \ker \omega \cap W$ is an exposed face of W containing F with $\dim(F' - F') < \dim(W - W)$. Now the induction hypothesis applies and shows that there are wedges

$$F_0 = F \subseteq F_1 \subseteq \ldots \subseteq F_{n-1} = F'$$

such that $F_i \in \mathcal{F}_e(F_{i+1})$. If we set $F_n := W$, the proof is complete.

(iv) Necessity: Let $E \in \mathcal{F}(W^*)$. Then, in view of (iii), we find a sequence of wedges

$$E = E_0 \subseteq E_1 \subseteq \ldots \subseteq E_n = W^*$$

such that $E_i \in \mathcal{F}_e(E_{i+1})$. We set $W_i := E^*_{n-i}$ for $i = 0, \ldots, n$. Then $W_0 = (W^*)^* = W \subseteq W_1 \subseteq \ldots \subseteq W_n = E^*$ and according to Proposition 1.2(i), we find exposed faces $F_i \subseteq W_i = E^*_{n-i}$ such that

$$E_{n-i-1} = E_{n-i} \cap F_i^\perp = L_{F_i}(W_i)^*.$$

Thus $W_{i+1} = E^*_{n-i-1} = L_{F_i}(W_i)$ with $F_i \in \mathcal{F}_e(W_i)$.

Sufficiency: For every sequence satisfying the conditions we know from Proposition 1.2 that

$$W^*_{i+1} = L_{F_i}(W_i)^* = F_i^\perp \cap W_i^* \in \mathcal{F}_e(W_i^*)$$

and that

$$E = W_n^* \subseteq W_{n-1}^* \subseteq \ldots \subseteq W_0^* = W^*.$$

Now (iii) implies that $E \in \mathcal{F}(W^*)$.

(v) The second statement follows from the first one by duality (Proposition 1.2(ii)). It is clear that $E - E = \text{span } E = H(E^*)^\perp = (E^\perp)^\perp$. Let $f = e - e' \in W^*$. Then $e = f + e' \in E$ and therefore $f, e' \in E$. This proves that $(E - E) \cap W^* = E$.

(vi) This is a direct consequence of (v).

(vii) If $f \in F$, then $f = tf + (1-t)f$ implies that $[0,1]f \subseteq F$, hence $\mathbb{R}^+ f \subseteq F$ so that F is a convex cone. Recall that F is a semigroup ideal in its closure $\overline{F}$ and algint $\overline{F} \subseteq F$ (cf. Proposition 1.1). Thus F is closed and hence a face.   ∎

To visualize the difference between faces and exposed faces, we give an example whith $\mathcal{F}_e(W) \neq \mathcal{F}(W)$.

$$W = \{(x, y, z) : x \leq 0, z \geq 0, x^2 + y^2 \leq z^2 \text{ or } x \geq 0, |y| \leq z\}.$$

Then $F := \mathbb{R}^+(0, 1, 1)$ is a face and the smallest exposed face containing F is $\widetilde{F} := \mathbb{R}^+(0, 1, 1) + \mathbb{R}^+(1, 0, 0)$ (cf. [Ru88a, 3.7]). The dual wedge is

$$W^* = \{(x, y, z) : x \geq 0, z \geq 0, x^2 + y^2 \leq z^2\}.$$

Using Proposition 1.3 or the geometric visualization (cf. Figure 1.1) it is easy to see that

$$\mathcal{F}_e(W^*) = \mathcal{F}(W^*)$$

but the above face F shows that $\mathcal{F}(W) \neq \mathcal{F}_e(W)$.

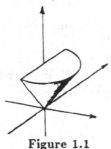

Figure 1.1

Since $\mathcal{F}(W^*)$ contains only two chains of length 3 but $\mathcal{F}(W)$ contains 4 chains of length 3, the lattices $\mathcal{F}(W)$ and $\mathcal{F}(W^*)$ are not antiisomorphic as it is the case for the lattices of exposed faces (Proposition 1.2).

Suppose that W is a Lorentzian cone and W' is a ray on the boundary of W. Then it is easy to see that $W - W'$ is no longer a wedge. In fact, it is the union of an open halfspace and the line $W' - W'$, hence not closed (cf. Figure 1.2).

Figure 1.2

If W' stays away from $W \setminus H(W)$, then this cannot happen, as the following proposition shows.

Proposition 1.4. *Let W, W' be wedges in the finite dimensional vector space L and suppose that*

$$W \cap -W' \subseteq H(W) \cap H(W'),$$

i.e., $W \cap -W'$ is a vector space. Then $V := W + W'$ is closed and hence a wedge. The edge of V is $H(V) = H(W) + H(W')$. If, in addition, W' is a vector space, then

$$\operatorname{algint}(W + W') = \operatorname{algint} W + W'.$$

Proof. It is clear that the above condition implies that $W \cap -W' = H(W) \cap H(W')$ is a vector space. Let $E := L/(W \cap -W')$, $C := W/(W \cap -W')$ and $C' := W'/(W \cap -W')$. Then C and C' are wedges in E and $C \cap -C' = \{0\}$. Let $v_n \in C$ and $v'_n \in C'$ with $v = \lim_{n \to \infty} v_n + v'_n$. Let $\| \cdot \|$ denote a norm on E. We have to show that $v \in C + C'$. Then we may assume that $\|v_n\| \to \infty$. If not, then there exists a subsequence v_{n_k} with $v_{n_k} \to v_1 \in C$ and $v'_{n_k} \to v - v_1 \in C'$ shows that

$v \in C + C'$. Passing to a subsequence we may assume that $e := \lim_{n \to \infty} \frac{1}{\|v_n\|} v_n \in C$. Then

$$0 = \lim_{n \to \infty} \frac{1}{\|v_n\|} v = \lim_{n \to \infty} \left(\frac{1}{\|v_n\|} v_n + \frac{1}{\|v_n\|} v'_n \right)$$

entails that $\frac{1}{\|v_n\|} v'_n \to -e$. Thus $-e \in C'$ and consequently $e \in C \cap -C'$, a contradiction. This shows that $C + C'$ and therefore $W + W'$ is closed, i.e., a wedge. Let $v = w + w' \in H(V)$. Then $-v = -w - w' = w_1 + w'_1 \in V$ with $w_1 \in W$ and $w'_1 \in W'$. Therefore

$$w + w_1 = -w' - w'_1 \in W \cap -W' = H(W) \cap H(W')$$

and consequently $w, w_1 \in H(W)$, $w', w'_1 \in H(W')$. Hence $v \in H(W) + H(W')$. The converse, that $H(W) + H(W') \subseteq H(V)$ holds, is clear. To prove the assertion on the algebraic interior of W, we assume that W' is a vector space. According to Proposition 1.1(iv), the dual wedge of V is

$$V^* = W^* \cap W'^* = W^* \cap W'^\perp.$$

Therefore

$$V^* \setminus H(V^*) = W^* \cap W'^\perp \setminus W^\perp.$$

Let $\omega \in V^* \setminus H(V^*)$ and $x \in \operatorname{algint} W + W'$. Then $\omega \in W^* \setminus H(W^*)$ and consequently $\omega(x) > 0$ because $\omega(W') = \{0\}$ (Proposition 1.1). From $V - V = W - W + W'$ we conclude that

$$\operatorname{algint} W + W' = \operatorname{int}_{W-W}(W) + W' \subseteq \operatorname{int}_{V-V}(V) = \operatorname{algint} V.$$

Let $x \in \operatorname{algint} V$. Suppose that $x \notin \operatorname{algint} W + W'$. Then, using the Separation Theorem of Hahn-Banach, we find $\nu \in (\operatorname{algint} W + W')^*$ such that $\nu(x) = 0$ and $\nu(V) \neq \{0\}$. This contradicts the above fact that $\omega(x) > 0$ for all $\omega \in V^* \setminus H(V^*)$. ∎

Now we can say what happens if we add two faces:

Corollary 1.5.   *Let $W \subseteq L$ be a wedge and $F_1, F_2 \in \mathcal{F}(W)$. Then*

(i) *$F_1 + F_2$ is closed and hence a wedge.*

(ii) *If $F_1, F_2 \in \mathcal{F}_e(W)$. Then their supremum in the lattice $\mathcal{F}_e(W)$ is*

$$F_1 \vee_e F_2 = T_{F_1 + F_2}(W) \cap W.$$

(iii) *The supremum of two faces $F_1, F_2 \in \mathcal{F}(W)$ may be obtained by setting*

$$W_0 := F_1 \vee_e F_2 \text{ in } \mathcal{F}_e(W) \quad and \quad W_{i+1} := F_1 \vee_e F_2 \text{ in } \mathcal{F}_e(W_i) \text{ for } i \in \mathbb{N}.$$

There exists an $i \in \mathbb{N}$ such that $W_{i+1} = W_i$ and

$$F_1 \vee F_2 = W_i = \bigcap_{j \in \mathbb{N}} W_j.$$

Proof. (i) Since every face contains the edge $H(W)$ (Proposition 1.2), we find that

$$H(W) \subseteq F_1 \cap -F_2 \subseteq W \cap -W = H(W)$$

is a vector space. An application of Proposition 1.4 proves (i).

(ii) It is clear that $F := T_{F_1+F_2}(W) \cap W$ is an exposed face of W because

$$F \subseteq T_F(W) \cap W \subseteq T_{F_1+F_2}(W) \cap W = F.$$

For every exposed face F' containing F_1 and F_2 we have

$$F = T_{F_1+F_2}(W) \cap W \subseteq T_{F'}(W) \cap W = F',$$

and therefore F is the supremum of F_1 and F_2 in $\mathcal{F}_e(W)$.

(iii) Let $F := F_1 \vee F_2$ in $\mathcal{F}(W)$. Since $W_{i+1} \in \mathcal{F}_e(W_i)$, it follows from Proposition 1.3 that the wedges F_1 and F_2 are faces of W_i for all $i \in \mathbb{N}$. Hence $F \subseteq \bigcap_{i \in \mathbb{N}} W_i$. If $F_1 + F_2$ does not intersect the algebraic interior of W_i, then $W_{i+1} \neq W_i$ has smaller dimension than W_i because it is contained in its boundary (Proposition 1.3). We conclude that there exists an index i such that $W_j = W_i$ for $j \geq i$. Then $F_1 + F_2$ intersects the interior of W_i and therefore $F_1 + F_2 \subseteq F \subseteq W_i$ implies that $F = W_i$ (Proposition 1.3). ∎

1.2. Wedges in K-modules

In this section we collect some material concerning the geometry of invariant wedges in finite dimensional modules of a compact Lie group K. The reason for this is that we will see later that invariant wedges in Lie algebras - the ones that correspond to invariant subsemigroups - are determined by their intersections with certain subalgebras which are the set of fixed points of an action of a compact group of automorphisms of the Lie algebra.

Let K be a compact group, m a normalized Haar measure on K, L a finite dimensional K-module and L^* the dual module. According to the real version of Weyl's unitary trick, L is isomorphic to an orthogonal K-module, hence it is semisimple as a K-module , i.e., every invariant subspace has an invariant complement. Consider $L_{\text{fix}} = \{x \in L : k.x = x \text{ for all } k \in K\}$ and $L_{\text{eff}} = \text{span}\{k.x - x : x \in L, k \in K\}$. Using the semisimplicity of the K-module L we find a complementary submodule L' for L_{eff}. Then it is clear that $L' \subseteq L_{\text{fix}}$ so that $L_{\text{fix}} + L_{\text{eff}} = L$. On the other hand L_{eff} contains a complement L'' for the intersection with L_{fix}. Then $L_{\text{eff}} \subseteq L''$ follows and we have a direct decomposition of L as a direct sum of the K-modules L_{fix} and L_{eff}. The averaging operator $p : L \to L, x \mapsto \int_K k.x \, dm(k)$ is a K-invariant projection onto L_{fix} and $\ker p = L_{\text{eff}}$ since Haar measure is a left and right invariant probability measure on K. The dual module has the direct decomposition

$$L^* = L_{\text{fix}}^* \oplus L_{\text{eff}}^* = L_{\text{eff}}^\perp \oplus L_{\text{fix}}^\perp$$

and the adjoint of p, $p^* : L^* \to L^*, \omega \mapsto \omega \circ p$, is the projection onto L_{fix}^* along L_{eff}^*. Similar statements are true for the module structure of L with respect to the Lie algebra $\mathbf{L}(K)$ of K.

Proposition 1.6.   *Suppose that $W \subseteq L$ is a K-invariant wedge, $W^* \subseteq L^*$ its dual, and $C := W \cap L_{\mathrm{fix}}$. Then the following assertions hold:*

(i) $p(W) = C$, $p^*(W^*) = W^* \cap L_{\mathrm{eff}}^\perp = C^* \cap L_{\mathrm{eff}}^\perp$.

(ii) $\mathrm{algint}(W \cap L_{\mathrm{fix}}) = p(\mathrm{algint}\, W) = \mathrm{algint}\, W \cap L_{\mathrm{fix}} \neq \varnothing$, $\mathrm{algint}\, W^* \cap L_{\mathrm{eff}}^\perp \neq \varnothing$.

(iii) $W \cap L_{\mathrm{eff}} \subseteq H(W)$, $W^* \cap L_{\mathrm{fix}}^\perp \subseteq H(W^*)$.

(iv) $p(H(W)) = H(C)$.

Suppose that $F \subseteq W$, W' are also K-invariant wedges. Then we have that:

(v) $p(W \cap W') = p(W) \cap p(W')$.

(vi) $p(L_F(W)) = L_{p(F)}(C)$.

(vii) $p(T_F(W)) = T_{p(F)}(C)$.

Proof.   The second assertion of (i) through (iii) follows from the first one by duality.

(i) It is clear that $W \cap L_{\mathrm{fix}} \subseteq p(W)$. But W is closed and convex and m is a probability measure on K. Hence $p(x)$ is contained in the closed convex hull of $K.x \subseteq W$. Thus $p(W) \subseteq W \cap L_{\mathrm{fix}}$.

(ii) Let $x \in \mathrm{algint}\, W$. First we show that $p(x) \in \mathrm{algint}\, W$. To see this, let $\omega \in W^* \setminus H(W^*)$. Then

$$\langle \omega, p(x) \rangle = \int_K \langle \omega, k.x \rangle \, dm(k) > 0$$

since $\langle \omega, 1.x \rangle = \langle \omega, x \rangle > 0$ and each open subset of K has non-zero Haar measure. Now Proposition 1.1 implies that $p(x) \in \mathrm{algint}\, W$. So $p(\mathrm{algint}\, W) = \mathrm{algint}\, W \cap L_{\mathrm{fix}}$. Before we prove the last assertion, we have to prove (iii).

(iii) We consider the K-module $L' := L_{\mathrm{eff}}$ and $W' := W \cap L'$. Then $L'_{\mathrm{fix}} = \{0\}$ and (ii) above shows that $0 \in \mathrm{algint}\, W'$. Hence W' is a vector space and therefore $W \cap L_{\mathrm{eff}} \subseteq H(W)$.

(ii) (continued) Using Proposition 1.4, we find that

$$\begin{aligned}
p(\mathrm{algint}\, W) &= p(\mathrm{algint}\, W + L_{\mathrm{eff}}) = p\big(\mathrm{algint}(W + L_{\mathrm{eff}})\big) \\
&= \mathrm{algint}(W + L_{\mathrm{eff}}) \cap L_{\mathrm{fix}} \\
&= \mathrm{algint}\big((W + L_{\mathrm{eff}}) \cap L_{\mathrm{fix}}\big) = \mathrm{algint}(W \cap L_{\mathrm{fix}}).
\end{aligned}$$

(iv) The inclusion $H(C) \subseteq p(H(W))$ is trivial. But $p(H(W))$ is a vector space and therefore contained in $H(C)$.

(v) This follows from

$$p(W \cap W') = W \cap W' \cap L_{\mathrm{fix}} = p(W) \cap p(W').$$

(vi) The inclusion $p(L_F(W)) \subseteq L_{p(F)}(C)$ follows from

$$p(L_F(W)) = p(\overline{W - F}) \subseteq \overline{p(W) - p(F)} = L_{p(F)}(p(W)).$$

But

$$p(W) - p(F) = W \cap L_{\mathrm{fix}} - F \cap L_{\mathrm{fix}} \subseteq L_F(W) \cap L_{\mathrm{fix}} = p(L_F(W))$$

and therefore $L_{p(F)}(p(W)) \subseteq p(L_F(W))$.

(vii) This is a consequence of (vi) and (iv).   ∎

1.3. The characteristic function of a cone

One of the most effective tools to study the geometry of cones in vector spaces and group actions on such cones is the *characteristic function* defined below. It was invented by E. B. Vinberg to study homogeneous and symmetric cones (cf. [Vin63]). Nevertheless, it also provides a useful tool in the general case.

In this section C denotes a wedge in the finite dimensional vector space V. We say that a measure μ_C is a *Lebesgue measure* on C if it is the restriction of a Lebesgue measure on $C - C$ to C.

Suppose that C is generating. We fix a Lebesgue measure on the dual cone C^*. The characteristic function $\varphi = \varphi_C$ of C is defined by

$$\varphi_C : \operatorname{int} C \to \mathbb{R}^+ \cup \{\infty\}, \quad x \mapsto \int_{C^*} e^{-\langle \omega, x \rangle} d\mu_{C^*}(\omega)$$

This is the usual definition of the characteristic function if the cone C is also pointed. For later applications to Lie semigroups we don't want to assume this. We note that it is not a priori clear that the values of φ are finite.

We choose a vector space complement E for $H(C)$ in V. Then

$$C = H(C) \oplus (C \cap E)$$

and the cone $C \cap E$ is pointed and generating in E. Moreover, the subspace E is in duality with $C^* - C^* = H(C)^{\perp} \subseteq V^*$. For $c = h + e$ with $h \in H(C)$ and $e \in E$ we have that $\omega(c) = \omega(e)$ for every $\omega \in C^*$. Whence

$$\varphi_C(c) = \varphi_{C \cap E}(e)$$

entails immediately that φ_C is constant on the affine subspaces $e + H(C)$, where $e \in \operatorname{int} C$. To prove the finitenes of φ, the following lemma is helpful.

Lemma 1.7. *Let $K \subseteq \operatorname{int} C$ be a compact subset and $\| \cdot \|$ a norm on V^*. Then there exists $\rho_K > 0$ such that*

$$\langle x, \omega \rangle \geq \rho \|\omega\| \qquad \forall \omega \in C^*, x \in K.$$

Proof. Since the inequality is positively homogeneous in ω, it suffices to obtain it for all ω contained in a compact base D of the pointed cone C^*. Suppose that the assertion is false. Then there exists a sequence $(x_n, \omega_n) \in K \times D$ such that $\frac{\langle x_n, \omega_n \rangle}{\|\omega_n\|} \to 0$. Since D and K are compact, we may assume that $(x_n, \omega_n) \to (x, \omega) \in K \times D$. Then $\|\omega\| > 0$ entails that $\langle x, \omega \rangle = 0$, contradicting the fact that $x \in \operatorname{int} C$. ∎

Theorem 1.8.   *Let C be a pointed generating cone in the n-dimensional real vector space V. Then the characteristic function φ of C has the following properties:*

(i) *φ is finite, positive and analytic on int C.*

(ii) *For $g \in \mathrm{Aut}(C) := \{g \in \mathrm{Gl}(V) : g(C) \subseteq C\}$ we have that*

$$\varphi(g.x) = |\det(g)|^{-1}\varphi(x).$$

(iii) *$\varphi(\lambda.x) = \lambda^{-n}\varphi(x)$ for $\lambda \in]0, \infty[$.*

(iv) *If $x_n \to x \in \partial C$, then $\varphi(x_n) \to \infty$.*

(v) *$\log \varphi$ is a strictly convex function on $\mathrm{int}(C)$.*

(vi) *The mapping*

$$* : \mathrm{int}(C) \to \mathrm{int}(C^*), \quad x \mapsto -d(\log \varphi)(x)$$

is a bijection of $\mathrm{int}(C)$ onto $\mathrm{int}(C^*)$, $x^*(x) = n$, and $x \in C$ is unique with the property that

$$\varphi(x) = \min\{\varphi(y) : x^*(y) = n, y \in \mathrm{int}(C)\}.$$

Proof.   (i) For every $\omega \in C^*$ we consider the holomorphic function e_ω on $V_{\mathbb{C}}$ obtained by holomorphic continuation of $v \mapsto e^{-\langle \omega, v \rangle}$, i.e.,

$$e_\omega(v_1 + iv_2) := e^{-\langle \omega, v_1 \rangle - i\langle \omega, v_2 \rangle}.$$

If $K \subseteq \mathrm{int}\,C$ is a compact subset, then, according to Lemma 1.7, there exists $\rho_K > 0$ such that

$$\langle \omega, v \rangle \geq \rho_K \|\omega\| \qquad \forall v \in K, \omega \in C^*.$$

Therefore

$$|e_\omega(v_1 + iv_2)| \leq e^{-\rho_K \|\omega\|}$$

holds for all $\omega \in C^*$ and $v \in K + iV$. We conclude that the integral defining φ converges uniformly on compact subsets of $\mathrm{int}\,C + iV$. Since, for every compact subset $D \subseteq C^*$, the functions

$$v \mapsto \int_D e_\omega(v)\, d\mu_{C^*}(\omega)$$

are holomorphic on $V_{\mathbb{C}}$, we see that the function

$$v \mapsto \int_{C^*} e_\omega(v)\, d\mu_{C^*}(\omega)$$

is locally on $\mathrm{int}\,C + iV$ a uniform limit of holomorphic function, hence holomorphic. The analyticity and finiteness of φ now follows by restricting to $\mathrm{int}\,C$.

(ii) Let $g \in \mathrm{Aut}(C)$ and $x \in \mathrm{int}(C)$. Then

$$\varphi(g.x) = \int_{C^*} e^{-\langle \omega, g.x \rangle}\, d\mu_{C^*}(\omega) = \int_{C^*} e^{-\langle g^*.\omega, x \rangle}\, d\mu_{C^*}(\omega)$$

$$= |\det g^*|^{-1} \int_{g^*(C^*)} e^{-\langle \omega, x \rangle}\, d\mu_{C^*}(\omega)$$

$$= |\det g|^{-1} \int_{C^*} e^{-\langle \omega, x \rangle}\, d\mu_{C^*}(\omega)$$

follows from the transformation formula for integrals and the invariance of C^* under g^*.

(iii) This is immediate from (ii) since $\lambda C = C$ holds for all $\lambda > 0$.

(iv) Let $c \in \partial C$. Then there exists $\omega_1 \in C^* \setminus \{0\}$ such that $\omega_1(c) = 0$. We extend $\{\omega_1\}$ to a basis $\{\omega_1, \ldots, \omega_n\}$ contained in C^* such that the volume of the cube spanned by these elements is 1. Set $D := \sum_{i=1}^n \mathbb{R}^+\omega_i$. Then $D^* \subseteq V$ is a pointed generating cone (Proposition 1.1) containing C, and

$$\varphi_{D^*}(x) \leq \varphi_C(x) \qquad \forall x \in \mathrm{int}\, C \subseteq \mathrm{int}\, D^*.$$

Now

$$\varphi_{D^*}(v) = \int_D e^{-\langle \omega, v \rangle}\, d\mu_D(\omega)$$

$$= \int_0^\infty e^{-u_1\omega_1(v)} du_1 \cdot \int_0^\infty e^{-u_2\omega_2(v)} du_2 \cdot \ldots \cdot \int_0^\infty e^{-u_n\omega_n(v)} du_n$$

$$= \prod_{i=1}^n \frac{1}{\omega_i(v)}.$$

If $c_n \to c$ with $c_n \in \mathrm{int}(C)$, then $\omega_1(c_n) \to 0$, so that

$$\varphi_{D^*}(c_n) \to \infty$$

yields that $\varphi(c_n) \to \infty$.

(v) It clearly suffices to show that the second derivative

$$d^2(\log \varphi)(c) : V \times V \to \mathbb{R}$$

is positive definite for all $c \in \mathrm{int}\, C$. Let $u \in V$. Then

$$d^2(\log \varphi)(c)(u, u) = \frac{d^2}{dt^2}\Big|_{t=0} \log \varphi(c + tu)$$

$$= \frac{1}{\varphi(c)^2}\Big(\varphi(c) d^2\varphi(c)(u, u) - \big(d\varphi(c)(u)\big)^2\Big)$$

and

$$d\varphi(c)(u) = -\int_{C^*} \langle \omega, u \rangle e^{-\langle \omega, c \rangle}\, d\mu_{C^*}(\omega)$$

and

$$d^2\varphi(c)(u, u) = \int_{C^*} (\langle \omega, u \rangle)^2 e^{-\langle \omega, c \rangle}\, d\mu_{C^*}(\omega).$$

We set

$$f(\omega) := e^{-\frac{1}{2}\langle \omega, c \rangle} \quad \text{and} \quad g(\omega) := f(\omega)\langle \omega, u \rangle.$$

Then

$$\varphi(c) d^2\varphi(c)(u, u) - \big(d\varphi(c)(u)\big)^2$$

$$= \int_{C^*} f^2(\omega) d\mu_{C^*}(\omega) \int_{C^*} g^2(\omega) d\mu_{C^*}(\omega) - \left(\int_{C^*} f(\omega)g(\omega) d\mu_{C^*}(\omega)\right)^2 > 0$$

follows from the Cauchy-Schwartz-inequality because f and g are not proportional.
(vi) If $c \in C \setminus \{0\}$ and $v \in \text{int}(C)$, then

$$v^*(c) = \frac{1}{\varphi(c)} \int_{C^*} \langle \omega, c \rangle e^{-\langle \omega, v \rangle} d\mu_{C^*}(\omega) > 0$$

since the integrand is non-negative and does not vanish identically. Whence $v^* \in \text{int}(C^*)$.

Since φ is positively homogeneous of degree $-n$, Euler's identity on homogeneous functions yields

$$d\varphi(x)(x) = -n\varphi(x),$$

hence $x^*(x) = -\frac{1}{\varphi(x)} d\varphi(x)(x) = n$.

Let $\omega \in \text{int}(C^*)$. We claim that $\omega = v^*$ for an element $v \in \text{int}(C)$. To see this, we consider the hyperplane $E := \omega^{-1}(n)$. The intersection $E \cap C$ is compact and the function φ tends to infinity at $E \cap \partial C$. Whence φ attains a minimum in a point $v \in E \cap \text{int}(C)$. At this point $d\varphi(v)$ vanishes on the hyperplane $\ker \omega$, i.e., v^* is a multiple of ω. Now the fact that $v^*(v) = \omega(v) = n$ proves equality. If $w \in \text{int}(C)$ is another element satisfying $w^* = \omega$, then $w \in E \cap \text{int}(C)$ and $d\varphi(w)$ vanishes on $\ker \omega$. Now the strict convexity of the function $\log \varphi$ entails that w is a local minimum of φ on E, hence also a global minimum. Thus $w = v$ is a consequence of the uniqueness of the minimum. ∎

We consider some typical examples.
If $C = (\mathbb{R}^+)^n \subseteq \mathbb{R}^n$, then we have already seen in the proof of Theorem 1.8 that

$$\varphi_C(x_1, \ldots, x_n) = \frac{1}{x_1 \cdot \ldots \cdot x_n}.$$

If $C = \{(x_0, x_1, \ldots, x_n) \in \mathbb{R}^{n+1} : x_0 \geq \sqrt{x_1^2 + \ldots + x_n^2}\}$ is the *Lorentz cone* in $\mathbb{R}^{n+1}$, then $SO(1, n)_0 \subseteq \text{Aut}(C)$ and therefore the element

$$(x_0, x_1, \ldots, x_n) \in \text{int}(C)$$

is conjugate under the group $\text{SAut}(C)$ of automorphism with determinant 1 to

$$\sqrt{x_0^2 - x_1^2 - \ldots - x_n^2}(1, 0, \ldots, 0).$$

Now Theorem 1.8(ii),(iii) yield

$$\varphi_C(x_0, \ldots, x_n) = (x_0^2 - x_1^2 - \ldots - x_n^2)^{-\frac{n+1}{2}} \varphi_C(1, 0, \ldots, 0).$$

If V is a vector space endowed with a non-degenerate bilinear form, then one gets immediately an isomorphism α of V onto its dual V^* such that

$$\alpha \circ g = (g^*)^{-1} \circ \alpha$$

holds for all $g \in \text{Gl}(V)$ which leave the bilinear form invariant. The characteristic function φ of a cone C, more precisely, the mapping $v \mapsto v^*$ provides a similar tool for convex cones. Here the automorphism group of the cone C plays a similar role as the group of isometries of a bilinear form and the duality mapping is only defined on the interior of the cone C.

Proposition 1.9.   *Let $g \in \text{Aut}(C)$ and $x \in \text{int}(C)$. Then*

$$(g.x)^* = (g^{-1})^*.x^*,$$

i.e., $\text{int}(C)$ and $\text{int}(C^*)$ are isomorphic as analytic $\text{Aut}(C)$-spaces.

Proof.   First we recall from Theorem 1.8(ii) that $\varphi(g.x) = |\det(g)|^{-1}\varphi(x)$ implies that

$$\log \varphi(g.x) = -\log |\det(g)| + \log \varphi(x).$$

Hence

$$d(\log \varphi)(g.x) \circ g = d(\log \varphi)(x)$$

follows from the chain rule. We conclude that

$$(g.x)^* = -d(\log \varphi)(g.x) = -d(\log \varphi)(x) \circ g^{-1} = x^* \circ g^{-1} = (g^{-1})^*.x^*.$$

∎

An interesting special case where the preceding proposition applies is the situation, where B is a Lorentzian form on V and C is one of the closed light cones associated to B. Then $\text{Aut}(C)$ is the connected component of $\mathbf{1}$ in the Lorentzian group $O(B)_0$, and the duality mappings corresponding to B and C coincide on the interior of C. In general it is not true that the duality mapping for a cone can be extended to the whole space. To see this, we consider the Lie algebra

$$\mathfrak{g} := \mathfrak{h}_1 \rtimes \mathfrak{sl}(2, \mathbb{R}),$$

where $\mathfrak{h}_1$ is the three dimensional Heisenberg algebra. This Lie algebra contains a pointed generating invariant cone C ([Ne92b]), so we have a duality mapping

$$\text{int}(C) \to \text{int}(C^*), \quad c \mapsto c^*.$$

But the $\mathfrak{g}$-modules $\mathfrak{g}$ and $\mathfrak{g}^*$ are not isomorphic. This follows from the fact that

$$[\mathfrak{g}, \mathfrak{g}] = \mathfrak{g}, \quad \text{and} \quad \dim Z(\mathfrak{g}) = 1$$

because

$$[\mathfrak{g}, \mathfrak{g}]^\perp = \{\omega \in \mathfrak{g}^* : \mathfrak{g}.\omega = \{0\}\}.$$

For later applications to ordered homogeneous spaces we need a slight modification of the characteristic function. We define $\psi = \psi_C$ on V by

$$\psi(c) = \begin{cases} \varphi(c)^{-\frac{1}{n}} & \text{for } c \in \text{int}(C) \\ 0 & \text{for } c \notin \text{int}(C). \end{cases}$$

We say that a function ψ' on V is a *length functional* of C if it is proportional to ψ. This does not depend on the choice of Haar measure on V^*.

Theorem 1.10.   *The function ψ has the following properties:*

(i) $\psi(c) > 0$ *for* $c \in \text{int}(C)$.

(ii) $\psi(g.c) = |\det(g)|^{\frac{1}{n}}\psi(c)$ *for* $g \in \text{Aut}(C)$.

(iii) $\psi(\lambda c) = \lambda\psi(c)$ *for* $\lambda \in]0, \infty[$.

(iv) ψ *is continuous on* V, *real analytic on* $\text{int}(C)$, *and there exists a norm* $\|\cdot\|$ *on* V *such that* $\psi \leq \|\cdot\|$.

(v) ψ *is concave on* C *and strictly concave on* $\text{int}(C)$.

Proof.   (i)-(iii) These properties follows immediately from the corresponding properties of the characteristic function (Theorem 1.8).

(iv) Let $c_n \to c$ in C. Since φ is continuous on $\text{int}(C)$ and ψ is constant on $V \setminus C$, we may assume that $c \in \partial C$. Then

$$\lim \psi(c_n) = \lim \varphi(c_n)^{-\frac{1}{n}} = 0 = \psi(c)$$

shows that ψ is continuous (Theorem 1.8(iv)). The analyticity on the interior of C follows from the analyticity of φ on this set (Theorem 1.8(i)).

Since ψ is continuous, the set $\psi^{-1}([1, \infty[)$ is closed and does not contain 0. We choose a norm on V such that the unit ball does not intersect this set. Now $\|x\| = 1$ clearly implies that $\psi(x) \leq 1 = \|x\|$ and the last assertion follows from (iii).

(v) Since ψ is continuous, it suffices to prove the concavity on the interior of C and since ψ is positively homogeneous of degree 1, it suffices to show that $\psi(x+y) \geq \psi(x)+\psi(y)$ for $x, y \in \text{int}(C)$. Let $x, y \in \text{int}(C)$, $s < \psi(x)$, and $t < \psi(y)$. Then $\psi(\frac{1}{s}x) > 1$ and $\psi(\frac{1}{t}y) > 1$, together with the concavity of $\log\psi = -\frac{1}{n}\log\varphi$ (Theorem 1.8(v)), entails that

$$1 < \psi\left(\frac{s}{s+t}\frac{1}{s}x + \frac{t}{s+t}\frac{1}{t}y\right) = \frac{1}{s+t}\psi(x+y).$$

Whence $\psi(x+y) > s+t$. Now the arbitrariness of t and s shows that

$$\psi(x+y) \geq \psi(x) + \psi(y).$$

■

Note that ψ is not concave on V. To see this, let $x \in V \setminus C$ and $y \in \text{int}(C)$ such that $x+y \notin C$. Then

$$\psi\left(\frac{1}{2}(x+y)\right) = 0 \not> \frac{1}{2}\psi(x) + \frac{1}{2}\psi(y) = \frac{1}{2}\psi(y) > 0.$$

In the following $\text{SAut}(C)$ denotes the group of all automorphisms of C with determinant 1 and $\text{EAut}(C)$ the closure of $\text{Aut}(C)$ in the algebra $\text{End}(V)$.

Lemma 1.11.   *The following assertions hold:*

(i) *If* $g \in \text{EAut}(C) \setminus \text{Aut}(C)$, *then* $g(C) \subseteq \partial C$.

(ii) *Let* $g_n \in \text{Aut}(C)$ *and* $c \in \text{int}(C)$ *such that the sequence* $g_n.c$ *converges to an element in* $\text{int}(C)$. *Then there exists a subsequence* g_{n_k} *converging in* $\text{Aut}(C)$.

(iii) *Let $c \in \text{int}(C)$. Then the isotropy group $\text{Aut}(C)^c$ of c is compact.*

Proof. (i) Let $g = \lim_{n \to \infty} g_n$ with $g_n \in \text{Aut}(C)$. Then the fact that $g \notin \text{Aut}(C)$ implies that $g \notin \text{Gl}(V)$, i.e., that $\det g = 0$. We conclude that $\det g_n \to 0$.

We consider a length functional ψ of C. Let $c \in \text{int}(C)$. Then

$$\psi(g.c) = \lim_{n \to \infty} \psi(g_n.c) = \lim_{n \to \infty} |\det(g_n)|^{\frac{1}{n}} \psi(c)$$
$$= \psi(c) \lim_{n \to \infty} |\det(g_n)|^{\frac{1}{n}} = 0.$$

Whence $g.c \in \partial C$. Since $\text{int}(C)$ is dense in C, it follows that $g(C) \subseteq \partial C$.

(ii) Let $y := \lim_{n \to \infty} g_n.c$, $\leq$ denote the partial order on V defined by $v \leq w$ if $w - v \in C$, and note that the linear action of $\text{Aut}(C)$ on V preserves this order. Pick $z \in \text{int}\, C + y$. Then $y \in z - \text{int}\, C$ and therefore we find $n_0 \in \mathbb{N}$ such that $g_n.c \in z - \text{int}\, C$ for all $n \geq n_0$. This means in particular that

$$0 \leq g_n.c \leq z.$$

Let $[0, c] = C \cap (c - C)$ denote the order interval between 0 and c. This set is compact because C is pointed. Since $c \in \text{int}\, C$, the interval $[0, c]$ has non-empty interior, hence contains a basis B of V. Now

$$g_n(B) \subseteq g_n([0, c]) = [0, g_n.c] \subseteq [0, z] \qquad \forall n \geq n_0,$$

and the set $[0, z]$ is also compact. This proves that the sequence g_n is bounded in $\text{End}(V)$. Let $g_0 \in \text{End}(V)$ be a cluster point of this sequence and $g_0 = \lim_{k \to \infty} g_{n_k}$. Then

$$y = \lim_{k \to \infty} g_{n_k}.c = g_0.c \in \text{int}(C).$$

Thus $g_0 \in \text{Aut}(C)$ by (i).

(iii) Let g_n be a sequence in $\text{Aut}(C)^c$. Then $g_n.c = c$ is constant. In view of (ii), there exists a subsequence converging in $\text{Aut}(C)$. Hence $\text{Aut}(C)^c$ is compact because it is closed in $\text{Aut}(C)$. ∎

The following result gives very precise information on the orbits of the special automorphim group of a cone. Again, a similar statement is true for the orbits of the group of isometries of a symmetric bilinear form on a vector space.

Proposition 1.12. *Let C be a pointed generating cone in a finite dimensional vector space V, $x \in \text{int}(C)$, and $G \subseteq \text{SAut}(C)$ a closed subgroup. Then the orbit $G.x$ is closed and the orbit mapping*

$$G \to V, \quad g \mapsto g.x$$

is proper.

Proof. Let ψ denote a length functional of C. Since $G \subseteq \text{SAut}(C)$, this function is invariant under the action of G (Theorem 1.10(ii)). Let $y = \lim_{n \to \infty} g_n.x$. Then the fact that the level surfaces of ψ are closed in C (Theorem 1.10(ii)) implies that

$\psi(y) = \psi(x) > 0$, hence that $y \in \text{int}(C)$ (Theorem I.10(i)). Using Lemma I.11(ii), we find a convergent subsequence $g_{n_k} \to g_0 \in G$. Whence

$$y = \lim_{k \to \infty} g_{n_k}.x = g_0.x \in G.x.$$

Let $K \subseteq V$ be a compact set. Then $K \cap (G.x)$ is compact since $G.x$ is closed and the topology on $G.x$ coincides with the quotient topology of G/G^x because $G.x$ is locally compact ([Ho65, p.7]). Now the compactness of $G^x = G \cap \text{Aut}(C)^x$ implies the assertion (Lemma 1.11(iii)). ∎

Endomorphisms of a cone

Let C be a pointed generating invariant cone in the finite dimensional vector space V and ψ a length functional on C. We have seen in the preceding section that the elements of $\text{SAut}(C)$ preserve the function ψ on C. In this section we prove a refinement of this result for invertible endomorphisms of the cone C. We write $\text{SEnd}(C)$ for the semigroup of all elements in $\text{Sl}(V)$ mapping to cone C into itself.

Proposition 1.13.   *Let* $g \in \text{SEnd}(C)$. *Then*

$$\psi(g.c) \geq \psi(c) \qquad \forall c \in C.$$

Proof.   The assertion is trivial for $c \in \partial C$ since $\psi(c) = 0$ and $g.c \in C$. So we may assume that $c \in \text{int}(C)$. We have to show that

$$\varphi(g.c) \leq \varphi(c)$$

holds for the characteristic function φ of C.

We note that $g(C) \subseteq C$ entails that $g^*(C^*) \subseteq C^*$. Using the fact that $\det(g) = \det(g^*) = 1$, we calculate as follows:

$$\varphi(g.c) = \int_{C^*} e^{-\langle \omega, g.c \rangle} d\mu(\omega) = \int_{C^*} e^{-\langle g^*.\omega, c \rangle} d\mu(\omega)$$

$$= \int_{g^*(C^*)} e^{-\langle \omega, c \rangle} d\mu(\omega) \leq \int_{C^*} e^{-\langle \omega, c \rangle} d\mu(\omega) = \varphi(c).$$

∎

This proposition shows that, if one considers only the action of the semigroup $\text{SEnd}(C)$ on C, the orbits of elements in C are mapped towards the interior of the cone C. This contrasts the observation that non-invertible endomorphisms of C map the whole cone into its boundary (Lemma 1.11) whenever they are contained in the closure of the group $\text{Aut}(C)$.

1.4. Lie wedges and Lie semigroups

The tangent space of a Lie subgroup of a Lie group G is a vector space and, in addition, a Lie subalgebra of the Lie algebra $\mathbf{L}(G)$. Similarly one expects the semigroup analog to satisfy additional algebraic conditions. It will turn out that the tangent object of a sufficiently well behaved semigroup is a *Lie wedge*. If we want to stress the ambient Lie algebra we say a Lie wedge is a pair $(W, \mathfrak{g})$, where W is a wedge in a finite dimensional Lie algebra $\mathfrak{g}$ satisfying the condition

$$(1.2) \qquad e^{\operatorname{ad} X} W = W \quad \text{for all} \quad X \in H(W).$$

But by abuse of notation we will also write simply W for the Lie wedge $(W, \mathfrak{g})$. A Lie wedge $(W, \mathfrak{g})$ is said to be *Lie generating* if $\mathfrak{g}$ is the smallest subalgebra of $\mathfrak{g}$ containing W. A wedge $W \subseteq \mathfrak{g}$ is said to be *invariant* if it is invariant under the adjoint action, i.e.,

$$(1.3) \qquad e^{\operatorname{ad} X} W = W \quad \text{for all} \quad X \in \mathfrak{g}.$$

It follows immediately from (1.2) upon differentiation that $H(W)$ is a subalgebra of $\mathfrak{g}$ if $W \subseteq \mathfrak{g}$ is a Lie wedge. Moreover, writing $e^{\operatorname{ad} X}$ as a power series, one sees that a vector subspace $W \subseteq \mathfrak{g}$ is a Lie wedge if and only if it is a subalgebra. Now suppose that $W \subseteq \mathfrak{g}$ is a half space. Then it is a Lie wedge if and only if $H(W)$ is a subalgebra. In fact, while the one direction follows from above the other can be seen as follows. If $H(W)$ is a subalgebra and $X \in H(W)$, we choose $\omega \in \mathfrak{g}^*$ such that $H(W) = \ker \omega$ and $W = \omega^{-1}(\mathbb{R}^+)$. Then the invariance of $H(W)$ under $\operatorname{ad} X$ shows that $\omega \circ \operatorname{ad} X = \lambda \omega$. Thus

$$\omega \circ e^{\operatorname{ad} X} W = e^\lambda \omega(W) = e^\lambda \mathbb{R}^+ = \mathbb{R}^+$$

implies that $e^{\operatorname{ad} X} W = W$. Note also that if $W \subseteq \mathfrak{g}$ is invariant, then W is a Lie wedge, $H(W)$ and $W - W$ are ideals of $\mathfrak{g}$, and W is Lie generating if and only if $W - W = \mathfrak{g}$. Finally we note that any pointed cone in the Lie algebra $\mathfrak{g}$ is a Lie wedge.

Now we are ready to introduce the central objects of this book. But first we fix some notation.

In the following G denotes a connected finite dimensional Lie group, $\mathbf{L}(G)$ its Lie algebra and $\exp \colon \mathbf{L}(G) \to G$ its exponential function. We write $\lambda_g \colon G \to G, x \mapsto gx$ for left multiplication with g, $\rho_g \colon G \to G, x \mapsto xg$ for right multiplication with g, and $I_g := \lambda_g \circ \rho_{g^{-1}} \colon x \mapsto gxg^{-1}$ for conjugation with g.

Let $S \subseteq G$ be a closed subsemigroup. We define the *tangent wedge* of S by

$$\mathbf{L}(S) := \{ X \in \mathbf{L}(G) : \exp(\mathbb{R}^+ X) \subseteq S \}.$$

The largest group $H(S) := S \cap S^{-1}$ contained in S is called the *group of units* of S. Moreover, we denote the subgroup of G generated by S with $G_S := \langle S \cup S^{-1} \rangle$. The name tangent wedge for $\mathbf{L}(S)$ is justified by the following proposition.

Recall that a semigroup S with an identity element $\mathbf{1}$ is called a *monoid*.

Proposition 1.14.   *Let S be a closed submonoid of the Lie group G. Then the following assertions hold:*

(i) $L(S)$ *is a Lie wedge.*

(ii) $L\big(H(S)\big) = H\big(L(S)\big)$.

Proof.   (i) It is clear that $\mathbb{R}^+ L(S) = L(S)$. The closedness of $L(S)$ follows from

$$L(S) = \bigcap_{t>0} \frac{1}{t} \exp^{-1}(S)$$

and the continuity of $\exp$. Let $X, Y \in L(S)$. To see that $L(S)$ is a wedge we have to show that $X + Y \in L(S)$. This follows from

$$\exp\big(t(X+Y)\big) = \lim_{n \to \infty} \big(\exp(\tfrac{t}{n}X)\exp(\tfrac{t}{n}Y)\big)^n \in S \qquad \forall t \in \mathbb{R}^+.$$

If $X \in H\big(L(S)\big)$, then $\exp X, \exp -X \in S$ and therefore

$$L\big(I_{\exp X}(S)\big) = \mathrm{Ad}(\exp X)\, L(S) = e^{\mathrm{ad}\,X}\, L(S) = L(S).$$

We conclude that $L(S)$ is a Lie wedge.

(ii) If $X \in H\big(L(S)\big)$, then $\exp \mathbb{R}^+ X \subseteq S$ and $\exp \mathbb{R}^- X \subseteq S$. Therefore $\exp \mathbb{R} X \subseteq H(S)$ and $X \in H\big(L(S)\big)$. The inclusion $L\big(H(S)\big) \subseteq H\big(L(S)\big)$ is trivial.   ∎

To go the other way, we set

$$S_W := S_{W,G} := \overline{\langle \exp W \rangle}$$

for a wedge $W \subseteq L(G)$. This is the smallest closed subsemigroup of G whose tangent wedge contains W. A *Lie semigroup* is a pair (S,G), where G is a connected Lie group and $S \subseteq G$ a closed subsemigroup with

$$S = S_{L(S)} = \overline{\langle \exp L(S) \rangle}.$$

As in the case of Lie wedges we also use the notion Lie semigroup for the semigroup S itself if it is clear in which group it sits. Note that our definition is consistent with the one given in [HoLa83]. A Lie semigroup is said to be *generating* if $L(S)$ is Lie generating in $L(G)$.

If V is a finite dimensional real vector space endowed with its Lie group structure, then the exponential function is a diffeomorphism and we may identify V with $L(V)$. Then it is obvious that the Lie subsemigroups of V are the wedges W in the vector space V. The Lie semigroup (W,V) is generating if and only if W is a generating wedge in V. Note at this point that it is easy to construct closed subsemigroups of the plane which cannot be recovered from their tangent wedges. For instance consider

$$S := \{(x,y) \in \mathbb{R}^2 : y^2 \leq x^3\}$$

whose tangent wedge simply is $\{(x,0) \in \mathbb{R}^2 : x \geq 0\}$ (cf. Figure 1.3).

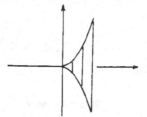

Figure 1.3

If G is a connected Lie group, then Proposition 1.14(ii) shows that the Lie subsemigroups $S \subseteq G$ with $H(S) = S$ are exactly the closed connected subgroups of G.

If (S, G) is a Lie semigroup and G is a vector space or $S = S^{-1}$, then $S = \exp L(S)$ or at least $S = \langle \exp L(S) \rangle$. In general $S \neq \langle \exp L(S) \rangle$. A simple example for this behaviour is the so called "parking ramp" which is described in Chapter 2.

Before we go on with the general theory we consider some examples for Lie semigroups and demonstrate some methodes to show that a given semigroup is a Lie semigroup.

The ordered space of Lorentzian cones

A quadratic form q on $\mathbb{R}^{n+1}$ is called *Lorentzian* if there exists $A \in \mathrm{Gl}(n + 1, \mathbb{R})$ such that

$$q(Ax) = q_0(x) := x_0^2 - x_1^2 - \ldots - x_n^2.$$

Since the stabilizer of the standard Lorentzian form q_0 is the group $\mathrm{O}(1, n)$, and the group $\mathrm{Gl}(n + 1, \mathbb{R})^+$ of invertible matrices with positive determinant acts transitively on the set of all Lorentzian forms, this set can be identified with the homogeneous space

$$\mathrm{Gl}(n + 1, \mathbb{R}) / \mathrm{O}(1, n) \cong \mathrm{Gl}(n + 1, \mathbb{R})^+ / \mathrm{SO}(1, n).$$

For a Lorentzian form q we write $[q] =]0, \infty[q$ for the *conformal equivalence class* of q. The space M of all such equivalence classes is

$$M := \mathrm{Gl}(n + 1, \mathbb{R}) / \mathbb{R}^* \, \mathrm{O}(1, n) \cong \mathrm{Sl}(n + 1, \mathbb{R})^+ / \mathrm{SO}(1, n).$$

For every conformal equivalence class $[q]$ the double cone

$$C_{[q]} := \{x \in \mathbb{R}^{n+1} : q(x) \geq 0\}$$

is well defined and the standard double cone $C_{[q_0]}$ is fixed by the group $\mathrm{SO}(1, n)$. If, conversely, an element $g \in \mathrm{Sl}(n+1, \mathbb{R})$ fixes this double cone, then, after multiplying with a suitable element of the group $SO(1,n)$, we may assume that g even preserves the *standard Lorentzian cone*

$$C_0 := \{x \in C_{[q_0]} : x_0 \geq 0\}.$$

Then, according to Theorem 1.8,

$$q_0(g.x) = q_0(x)$$

holds for all $x \in \text{int } C$, and, by analytic continuation, this identity holds everywhere on $\mathbb{R}^{n+1}$, so $g \in SO(1,n)$. It follows that M may be identified with the space of all double Lorentzian cones in $\mathbb{R}^{n+1}$.

The stabilizer $\text{SAut}(C_0)$ of the standard cone C_0 in $\text{Sl}(n+1,\mathbb{R})$ is the group $SO(1,n)_0$, the connected component of the identity in $SO(1,n)$, thus the two-sheeted covering space

$$\widetilde{M} := \text{Sl}(n+1,\mathbb{R})/SO(1,n)_0$$

of M can be viewed as the set of all Lorentzian cones in $\mathbb{R}^{n+1}$. Here the covering mapping $\widetilde{M} \to M$ is simply the assignment of the double cone $C \cup -C$ to the cone C. The space $\widetilde{M}$ comes along with a very natural partial order defined by

$$C \leq C' \qquad \Longleftrightarrow \qquad C' \subseteq C.$$

This relation is invariant under the action of the group $G := \text{Sl}(n+1,\mathbb{R})$, and the semigroup

$$S := \{g \in G : g.C_0 \geq C_0\} = \{g \in G : g.C_0 \subseteq C_0\} = \text{SEnd}(C_0),$$

is the semigroup of of all special endomorphisms of the cone C_0.

We show how the results of Section 1.3 can be applied to show that S is a Lie semigroup. Let $s \in S$. Then $s.C_0 \subseteq C_0$ and $s.\text{int } C_0 \subseteq \text{int } C_0$. Using Proposition 1.13, we see that

$$q_0(s.x) = \psi_{C_0}(s.x) \geq \psi_{C_0}(x) = q_0(x) \qquad \forall x \in C.$$

Set $q_1(x) := q_0(s^{-1}.x)$. Then $q_1(x) \leq q_0(x)$ holds for all $x \in s.C_0$ and for $x \in C_0 \setminus s.C_0$ we have that $q_1(x) \leq 0 \leq q_0(x)$, so that

$$(1.4) \qquad\qquad q_0(x) \geq q_1(x) \qquad \forall x \in C_0.$$

Now, for $\lambda \in [0,1]$, we define

$$q_\lambda := \lambda q_1 + (1-\lambda)q_0.$$

Then (1.4) implies that

$$q_\lambda(x) \leq q_\mu(x) \qquad \forall x \in C_0$$

whenever $\lambda \geq \mu$.

On the other hand, the quadratic forms q_λ are positive on $s.\operatorname{int} C_0$ and negative on $\{0\} \times \mathbb{R}^n$. This shows that they are positive on one-dimensional subspace orthogonal at $\{0\} \times \mathbb{R}^n$, hence the quadratic forms q_λ are Lorentzian for all $\lambda \in [0,1]$. Let

$$C_\lambda := \{c \in C_0 : q_\lambda(c) \geq 0\}.$$

Since $q_\lambda(x) \leq 0$ holds for all $x \in \partial C_0$, it follows that C_λ is one half of the double cone associated with the Lorentzian form q_λ. So

$$\gamma : [0,1] \to \widetilde{M}, \qquad \lambda \mapsto C_\lambda$$

is a smooth monotone curve in the ordered space $\widetilde{M}$. By Proposition 4.14(vi) we find a curve $\widetilde{\gamma} \colon [0,1] \to G$ such that

$$\widetilde{\gamma}(0) = 1 \quad \text{and} \quad \widetilde{\gamma}(\lambda).C_0 = C_\lambda.$$

Whence $\widetilde{\gamma}(1) \in S$ and $\widetilde{\gamma}$ is monotone with respect to the order

$$g \leq_S g' \quad \Leftrightarrow \quad g' \in gS \quad \Leftrightarrow \quad g'.C_0 \subseteq g.C_0.$$

So $\widetilde{\gamma}(1) \in \overline{\langle \exp \mathbf{L}(S) \rangle}$ is a consequence of Lemma 5.14.

We have proved the following theorem:

Theorem 1.15. *Let $C \subseteq \mathbb{R}^{n+1}$ be a Lorentzian cone. Then the semigroup*

$$\operatorname{SEnd}(C) := \{g \in \operatorname{Sl}(n+1, \mathbb{R}) : g.C \subseteq C\}$$

is a Lie semigroup.   ∎

Later we will see some methods to get even more information about the semigroup S. Its tangent wedge can be computed with the Invariance Theorem for Vector Fields (Theorem 5.8) to be

$$\mathbf{L}(S) = \{X \in \mathfrak{sl}(n+1, \mathbb{R}) : (\forall c, q_0(c) = 0)\ B_0(X.c, c) \geq 0\},$$

where B_0 is the bilinear form obtained from q_0 by polarization, and, using Lawson's Theorem (Theorem 7.34), one can even show that $S = H(S) \exp C$, where C is a convex cone of selfadjoint matrices with respect to the Lorentzian form q_0.

Affine compressions of a ball

Let $\|x\| := \sqrt{x_1^2 + \ldots + x_n^2}$ denote the euclidean norm on $\mathbb{R}^n$ and set $B := \{x \colon \|x\| \leq 1\}$. Further let

$$G := \operatorname{Aff}(n, \mathbb{R})^+ \cong \mathbb{R}^n \rtimes \operatorname{Gl}(n, \mathbb{R})^+$$

denote the affine group (with orientation preserving linear part) acting on $\mathbb{R}^n$ via

$$(b, g).x := g.x + b.$$

Let

$$S := \operatorname{compr}(B) \cap \operatorname{Aff}(n, \mathbb{R})^+$$

be the semigroup of affine compressions of B. We claim that S is a Lie semigroup in G. To see this, let

$$\pi : G \to \operatorname{Sl}(n, \mathbb{R})$$

$$(b, g) \mapsto (\det g)^{\frac{1}{n+1}} \begin{pmatrix} 1 & 0 \\ b & g \end{pmatrix}$$

denote the embedding of G into $\operatorname{Sl}(n+1, \mathbb{R})$. Now let

$$C_0 := \{(x_0, x) \in \mathbb{R} \times \mathbb{R}^n : x_0 \geq \|x\|\}$$

denote the standard Lorentzian cone in $\mathbb{R}^{n+1}$. Then

$$S = \pi^{-1}\big(\operatorname{SEnd}(C_0)\big)$$

since $g.B + b \subseteq B$ is equivalent to

$$\pi(b, g).C_0 = \pi(b, g).\big(\mathbb{R}^+(e_0 + B)\big) = \mathbb{R}^+\big((e_0 + b) + g.B\big) \subseteq C_0,$$

where $e_0 = (1, \mathbf{0}) \in \mathbb{R} \times \mathbb{R}^n$. Let us denote the adjoint with respect to the Lorentzian form q_0 of an element $g \in \operatorname{Gl}(n, \mathbb{R})$ by $g^\sharp$. If $s \in \operatorname{SEnd}(C_0)$, then

$$s^\sharp.C_0 = s^\sharp.C_0^* \subseteq C_0^* = C_0,$$

and $s^\sharp.e_0 \in s^\sharp.\operatorname{int} C_0 \subseteq \operatorname{int} C_0$, so that there exists $h \in H := \operatorname{SO}(1, n)_0$ and $\lambda \in \mathbb{R}^+$ such that

$$hs^\sharp.e_0 = \lambda e_0.$$

It follows that

$$sh^{-1}(\mathbb{R}^n) = (hs^\sharp)^\sharp(\mathbb{R}^n) \subseteq \mathbb{R}^n,$$

i.e.,

$$sh^{-1} \in \pi\big(\operatorname{Aff}(n, \mathbb{R})^+\big) \cap \operatorname{SEnd}(C_0) = \pi(S).$$

This shows that

$$\operatorname{SEnd}(C_0) = \pi(S)H.$$

Using Theorem 1.15 and Theorem 5.19, we see that $\pi(S) \cong S$ is a Lie semigroup because its group of units

$$H(S) = \pi^{-1}(H) \cong \operatorname{SO}(n)$$

is connected.

That the Lie semigroup property of the compression semigroup of a ball is somehow related to the roundness of the ball, and therefore to the roundness of the corresponding Lorentzian cone, is illustrated by the following example. Let

$$Q := \{(x, y) \in \mathbb{R}^2 : |x| + |y| \leq 1\}.$$

Let

$$T := \mathrm{compr}(Q) \cap \mathrm{Aff}(2, \mathbb{R})^+$$

denote the semigroup of orientation preserving affine compressions of Q. Then $H(T)$ is the four element group of rotations of the square, hence not connected. So it follows from the Unit Group Theorem (Corollary 5.1) that T cannot be a Lie semigroup.

In this case we can even show more, namely that even

$$H(T)S_{\mathbf{L}(T)} = H(T)\overline{\langle \exp \mathbf{L}(T) \rangle}$$

is strictly smaller than T. To see this, let $A := \frac{1}{2}\begin{pmatrix} 1 & -1 \\ 1 & 1 \end{pmatrix}$. Then $A.Q \subseteq Q$. If $T = H(T)S_{\mathbf{L}(T)}$, then the results from Section 5.5 show that there exists a monotone curve $\gamma: [0,1] \to T$ with $\gamma(0) = \mathrm{id}$ and $\gamma(1) = A$. We show that this is impossible. Since $\gamma(t).Q$ is a convex set with four extreme points lying between Q and $A.Q$, we see that either $\gamma(t).Q$ equals Q, $A.Q$, or $\gamma(t).Q$ possesses at least two different extreme points on one side of the square Q. This is a contradiction.

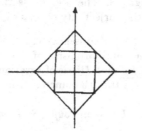

Figure 1.4

1.5. Functorial relations between Lie semigroups and Lie wedges

A morphism $\varphi: (W, \mathfrak{g}) \to (W', \mathfrak{g}')$ of Lie wedges is a morphism $\varphi: \mathfrak{g} \to \mathfrak{g}'$ of Lie algebras satisfying $\varphi(W) \subseteq W'$. We write $\underline{LWed}$ for the so defined category of Lie wedges. A morphism $\varphi: (S, G) \to (S', G')$ of Lie semigroups is a homomorphism $\varphi: G \to G'$ of Lie groups such that $\varphi(S) \subseteq S'$. We denote the category of Lie semigroups with $\underline{LSg}$.

Proposition 1.16. *The prescription*

$$\mathbf{L}: \underline{LSg} \to \underline{LWed}, \quad (S, G) \to (\mathbf{L}(S), \mathbf{L}(G)), \quad \varphi \mapsto \mathbf{L}(\varphi) := d\varphi(\mathbf{1})$$

defines a covariant functor from the category $\underline{LSg}$ to the category $\underline{LWed}$. Moreover, the following assertions hold:

(i) *Suppose that $\alpha: \mathfrak{g} \to \mathfrak{g}_1$ is a surjective morphism of Lie algebras which induces a morphism $\alpha: (W, \mathfrak{g}) \to (V, \mathfrak{g}_1)$ of Lie wedges and that W is Lie generating. Then V is Lie generating.*

(ii) *Let (S,G) be a Lie semigroup and $\varphi: G \to G_1$ a morphism of Lie groups. Then $\overline{(\varphi(S)}, G_1)$ is a Lie semigroup and $\varphi: (S,G) \to (\overline{\varphi(S)}, G_1)$ defines a morphism of Lie semigroups. If (S,G) is a generating Lie semigroup and φ is surjective, then $(\overline{\varphi(S)}, G_1)$ is a generating Lie semigroup.*

Proof.   The functoriality of $\mathbf{L}$ is a trivial consequence of the fact that

$$\mathbf{L}(\varphi_2 \circ \varphi_1) = \mathbf{L}(\varphi_2) \circ \mathbf{L}(\varphi_1),$$

for $\varphi_1: (S_1, G_1) \to (S_2, G_2)$, $\varphi_2: (S_2, G_2) \to (S_3, G_3)$. This is the chain rule for the derivatives of homomorphisms of Lie groups. It is clear that $\varphi(S_1) \subseteq S_2$ implies $\mathbf{L}(\varphi)\mathbf{L}(S_1) \subseteq \mathbf{L}(S_2)$.

(i) Let $\mathfrak{a} \subseteq \mathfrak{g}_1$ be a subalgebra containing V. Then $\varphi^{-1}(\mathfrak{a})$ is a subalgebra of $\mathfrak{g}$ containing W. Thus $\varphi^{-1}(\mathfrak{a}) = \mathfrak{g}$ and therefore $\mathfrak{a} = \varphi(\varphi^{-1}(\mathfrak{a})) = \mathfrak{g}_1$.

(ii) Set $S_1 := \overline{\varphi(S)}$. Then $\mathbf{L}(\varphi)\mathbf{L}(S) \subseteq \mathbf{L}(S_1)$ and therefore

$$\overline{\langle \exp \mathbf{L}(S_1) \rangle} \supseteq \overline{\langle \exp \mathbf{L}(\varphi)\mathbf{L}(S) \rangle} = \varphi(\langle \exp \mathbf{L}(S) \rangle) = S_1.$$

Hence (S_1, G_1) is a Lie semigroup. The second assertion follows from (i) because the surjectivity of φ implies the surjectivity of $\mathbf{L}(\varphi): \mathbf{L}(G) \to \mathbf{L}(G_1)$.   ∎

In the following we say that a subsemigroup S of a group G is *invariant*, if it is invariant under all inner automorphisms of G. The subcategories *ILSg* of invariant Lie semigroups in *LSg* and *IWed* of invariant wedges in *LWed* are full. The following proposition shows that the functor $\mathbf{L}$ maps *ILSg* into *IWed*.

Proposition 1.17.   *For a Lie semigroup (S,G) the following conditions are equivalent*

(1) *S is invariant under all inner automorphisms of G.*

(2) *$\mathbf{L}(S)$ is an invariant wedge.*

Suppose that (1) holds. Then the following are equivalent:

(3) *(S,G) is a generating Lie semigroup.*

(4) *$\mathbf{L}(S) - \mathbf{L}(S) = \mathbf{L}(G)$.*

Proof.   (1) $\Rightarrow$ (2): For $X \in \mathbf{L}(G)$ we have that

$$\mathbf{L}(S) = \mathbf{L}\left(I_{\exp X}(S)\right) = e^{\operatorname{ad} X} \mathbf{L}(S).$$

Now the connectedness of G shows that $\mathbf{L}(S)$ is invariant under all inner automorphisms of $\mathbf{L}(G)$.

(2) $\Rightarrow$ (1): Suppose that $e^{\operatorname{ad} X} \mathbf{L}(S) = \mathbf{L}(S)$ for all $X \in \mathbf{L}(G)$. Then

$$I_{\exp X}(S) = \overline{\langle I_{\exp X}(\exp \mathbf{L}(S)) \rangle} = \overline{\langle \exp(e^{\operatorname{ad} X} \mathbf{L}(S)) \rangle} = \overline{\langle \exp \mathbf{L}(S) \rangle} = S$$

and therefore $I_{\exp X}(S) = S$ which proves the invariance of S since G is connected.

(3) $\Leftrightarrow$ (4): If $\mathbf{L}(S)$ is an invariant wedge, then $\mathbf{L}(S) - \mathbf{L}(S)$ is an ideal, hence a subalgebra. Therefore $\mathbf{L}(S)$ is Lie generating in $\mathbf{L}(G)$ if and only if $\mathbf{L}(S) - \mathbf{L}(S) = \mathbf{L}(G)$.   ∎

In Section 2.7 we will see how invariant Lie semigroups arise very naturally as semigroups of future displacements acting on space-times.

It is a well known fact that for every finite dimensional real Lie algebra $\mathfrak{g}$ there exists a simply connected real Lie group G such that $L(G) = \mathfrak{g}$ (Lie's Third Theorem). This group is unique up to isomorphism. To see that it is possible to choose such a group in a functorial way, we have to take the bypass over germs of local groups.

Let $\mathfrak{g}$ be a finite dimensional real Lie algebra. Then there exists an open symmetric 0-neighborhood U in $\mathfrak{g}$ such that the Baker-Campbell-Hausdorff series

$$X * Y := X + Y + \frac{1}{2}[X, Y] + \dots$$

converges on $U \times U$ and that the partial multiplication

$$m_U : \{(X, Y) \in U \times U : X * Y \in U\} \to U, \quad (X, Y) \mapsto X * Y$$

together with the inversion $\sigma_U(X) = -X$ defines a Lie group germ $(U, 0, \sigma_U, m_U)$ over $\mathbb{R}$ ([Bou71a, Ch. II, §7.2]). For such an open symmetric 0-neighborhood $U \subseteq \mathfrak{g}$ we define the group G_U as the free group on U modulo the relations

$$\{X \circ Y \circ Z^{-1} = 1 : X, Y, Z \in U, X * Y = Z\}.$$

We endow these groups with the topology defined by the basis

$$\{A \subseteq G_U : (\forall a \in A)(A \circ a^{-1}) \cap U \in \mathcal{U}(1)\},$$

where the unit element $1 \in G_U$ is identified with $0 \in U$, and $\mathcal{U}(1)$ denotes the filter of 1-neighborhoods. If $V \subseteq U$ is another symmetric neighborhood of 0 in $\mathfrak{g}$ such that $(V, 0, \sigma_V, m_V)$ is a Lie group germ over $\mathbb{R}$, then the homomorphism

$$\pi_U^V : G_V \to G_U$$

is a covering homomorphism of topological groups (This is a slight generalization of [Ti83, p.62]). Let $\mathbf{G}(\mathfrak{g})$ denote the projective limit of this projective system. Further let G_1 be a simply connected Lie group with $L(G_1) = \mathfrak{g}$ and $U \subseteq \mathfrak{g}$ be a 0-neighborhood as above which is so small that $\exp|_U : U \to \exp(U)$ is a diffeomorphism. Then $\exp|_U : (U, m_U) \to (\exp(U), \cdot)$ is a homomorphic homeomorphism of local groups and therefore it induces for each $V \subseteq U$ a covering $G_V \to G_1$. Since G_1 is simply connected, it follows that $G_1 \cong G_V$ for each $V \subseteq U$. We have in particular that the induced homomorphism

$$\pi_\infty : \mathbf{G}(\mathfrak{g}) \to G_1$$

is a homeomorphism (cf. [Ti83, p.69]). In this way we have described a functor

$$\mathbf{G} : \underline{LAlg} \to \underline{LGrp}, \quad \mathfrak{g} \to \mathbf{G}(\mathfrak{g}).$$

Then it follows immediately from the definitions that

$$\mathbf{L} \circ \mathbf{G} = \mathrm{id}_{\underline{LAlg}}$$

and that the functor $\mathbf{G}$ is a left adjoint of the functor $\mathbf{L}$.

Proposition 1.18.   *The prescription*

$$\mathbf{S} : \underline{LWed} \to \underline{LSg}, \quad (W, \mathfrak{g}) \to (S_W, G(\mathfrak{g})), \quad \alpha \mapsto G(\alpha)$$

defines a functor from the category $\underline{LWed}$ to the category $\underline{LSg}$ which is a left adjoint of the functor

$$\mathbf{L} : \underline{LSg} \to \underline{LWed}, \quad (S, G) \to \big(\mathbf{L}(S), \mathbf{L}(G) \big), \quad \varphi \mapsto \mathbf{L}(\varphi).$$

Proof.   If $\alpha : (W, \mathfrak{g}) \to (W', \mathfrak{g}')$ is a morphism of Lie wedges, then $\alpha(W) \subseteq W'$ implies that

$$\mathbf{S}(\alpha)(S_W) \subseteq S_{\alpha(W)} \subseteq S_{W'}$$

because $\mathbf{L}\big(\mathbf{G}(\alpha) \big) = \alpha$. This shows that

$$\mathbf{S}(\alpha) : \big(S_W, G(\mathfrak{g}) \big) \to \big(S_{W'}, G(\mathfrak{g}') \big)$$

is a morphism of Lie semigroups. The functoriality of $\mathbf{S}$ is a trivial consequence of the fact that

$$\mathbf{G} \circ \mathbf{L} : \underline{LGrp} \to \underline{LGrp}$$

is a functor.

Finally, we have for each Lie wedge $(W, \mathfrak{g})$ a morphism

$$\eta_{(W, \mathfrak{g})} : (W, \mathfrak{g}) \to \big(\mathbf{L}(S_W), \mathfrak{g} \big)$$

which defines a natural transformation

$$\eta : \mathrm{id}_{\underline{LWed}} \to \mathbf{L} \circ \mathbf{S}$$

such that the assignment

$$\beta \mapsto \mathbf{L}(\beta) \circ \eta_{(W, \mathfrak{g})}$$

defines a bijection

$$\mathrm{Hom}_{\underline{LSg}} \big(\mathbf{S}(W, \mathfrak{g}), (S, G) \big) \to \mathrm{Hom}_{\underline{LWed}} \Big((W, \mathfrak{g}), \big(\mathbf{L}(S), \mathbf{L}(G) \big) \Big).$$

∎

We note that, on the level of Lie algebras, the functor $\mathbf{L} \circ \mathbf{G}$ does not "forget" any information. This is different for the functor $\mathbf{L} \circ \mathbf{S}$ because in general the Lie wedge $\mathbf{L}(S_W)$ is different from W.

1.6. Globality of Lie wedges

We have already seen that in general semigroups cannot be recovered from their tangent wedges and therefore introduced the concept of Lie semigroups. But the situation is even worse: In sharp contrast to the group case not every tangent object belongs to a global one. So for a Lie wedge $(W, \mathfrak{g})$ we say that W is *global* if there exists a Lie semigroup (S, G) such that $\mathbf{L}(S) = W$. If $\mathfrak{g}$ is the Lie algebra of the given Lie group G, we say that $(W, \mathfrak{g})$ is *global in* G if there exists a Lie semigroup $S \subseteq G$ such that $\mathbf{L}(S) = W$. There are different reasons why a Lie wedge can fail to be global. The simplest non-trivial Lie group G whose Lie algebra contains global and non-global wedges is the cylinder $G := \mathbb{R} \times \mathbb{R}/\mathbb{Z}$. If $W \subseteq \mathbf{L}(G) \cong \mathbb{R}^2$ is a pointed cone, then W is global in G if and only if $W \cap (\{0\} \times \mathbb{R}) = \{0\}$ (cf. Figure 1.5). Here it is the topology that prevents certain Lie wedges from being global. But there are also contractible Lie groups which have non-global Lie wedges, for instance the Heisenberg group and the universal covering group of $\mathrm{Sl}(2, \mathbb{R})$ (cf. Chapter 2 for details).

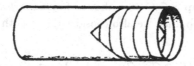

Figure 1.5

Let $(W, \mathfrak{g})$ be a Lie wedge with $\mathfrak{g} = \mathbf{L}(G)$. Using the fact that S_W is the smallest Lie semigroup in G which contains $\exp W$, we see that the globality of W in G is equivalent to each of the following statements

(1.5a) $$\mathbf{L}(S_W) = W$$

(1.5b) $$\mathbf{L} \circ \mathbf{S}(W, \mathfrak{g}) = (W, \mathfrak{g}).$$

Moreover, we note that a Lie subalgebra $\mathfrak{h}$ of $\mathbf{L}(G)$ is global in G if and only if the corresponding analytic subgroup $\langle \exp \mathfrak{h} \rangle$ is closed. This motivates the following definition: A Lie wedge $(W, \mathfrak{g})$ is called *weakly global* if W is global in the Lie algebra $\langle\langle W \rangle\rangle$ generated by W. Then every Lie subalgebra of $\mathfrak{g}$ is weakly global.

It is a difficult task to determine whether a given Lie wedge is global. Useful tools can be developed by using a reformulation in terms of orderings of groups and, dually, function spaces on the group. This is what we study in the next section.

1.7. Monotone functions and semigroups

A *left ordered group* $(G, \leq)$ is a group G endowed with a quasiorder $\leq$ which satisfies

$$x \leq y \implies gx \leq gy \qquad \forall g, x, y \in G.$$

Then $S := \{g \in G : 1 \leq g\}$ is a submonoid of G and if, conversely, S is a submonoid of G, then the assignment

$$g \leq_S g' :\Longleftrightarrow g' \in gS$$

defines a left invariant quasiorder on G such that $S := \{g \in G : 1 \leq_S g\}$. This establishes a one-to-one correspondence between left ordered groups and submonoids of groups. The simplest example for a left ordered group is the group $(\mathbb{R}, +)$ with the natural order.

Dually, for any topological group G with a submonoid $S \subseteq G$ we define the set $\text{Mon}(S)$ of S-*monotone functions* as the set of all continuous monotone mappings

$$(G, \leq_S) \to (\mathbb{R}, \leq).$$

For a subset $M \subseteq C(G)$ we define

$$SG(M) = \{s \in G : f(g) \leq f(gs) \text{ for all } f \in M, g \in G\},$$

the largest subsemigroup $S \subseteq G$ such that all functions $f \in M$ are S-monotone.

If G is a Lie group, then we also consider the smooth and analytic monotone functions

$$\text{Mon}^\infty(S) := \text{Mon}(S) \cap C^\infty(G) \quad \text{and} \quad \text{Mon}^\omega(S) := \text{Mon}(S) \cap C^\omega(G).$$

The mappings $S \mapsto \text{Mon}(S)$ and $M \mapsto SG(M)$ are inclusion reversing and

$$(1.6) \qquad\qquad M \subseteq \text{Mon}(S) \quad \Leftrightarrow \quad S \subseteq SG(M),$$

i.e., Mon and SG define a Galois connection from the subsemigroups of G to the subsets of $C(G)$. Therefore we have

$$(1.7a) \qquad S \subseteq SG\big(\text{Mon}(S)\big) \quad \text{and} \quad \text{Mon}(S) = \text{Mon}\Big(SG\big(\text{Mon}(S)\big)\Big)$$

$$(1.7b) \qquad M \subseteq \text{Mon}\big(SG(M)\big) \quad \text{and} \quad SG(M) = SG\Big(\text{Mon}\big(SG(M)\big)\Big).$$

The following proposition indicates that monotone functions are a powerful tool in the study of closed monoids in topological groups.

Proposition 1.19. *Let G be a topological group.*

(i) *If $M \subseteq C(G)$, then $SG(M)$ is a closed submonoid of G.*

(ii) *The submonoids of G which are fixed points of $SG \circ \text{Mon}$ are precisely the closed submonoids of G.*

Proof. (i) For $f \in M$ and $g \in G$ set $X_{f,g} := \{s \in G : f(gs) \geq f(g)\}$. Then $X_{f,g}$ is closed. Therefore $SG(M) = \bigcap_{f \in M, g \in G} X_{f,g}$ is closed. If $s, s' \in SG(M)$ and $g \in G$, then $f(gss') \geq f(gs) \geq f(g)$ and therefore $ss' \in SG(M)$. Trivially, $1 \in SG(M)$.

(ii) We first show that every closed submonoid S of G is a fixed point of $SG \circ \text{Mon}$. For $s \in G \setminus S$ we have to find a function $f \in \text{Mon}(S)$ with $f(s) < f(1)$. Set $U_0 := G \setminus s^{-1}S$. This is an open neighborhood of 1. Inductively we define a sequence $(U_n)_{n \in \mathbb{N}}$ of symmetric open 1-neighborhoods in G such that $U_{n+1}^3 \subseteq U_n$ for every $n \in \mathbb{N}_0$. For $n \in \mathbb{N}$ we set

$$\mathcal{U}_n := \{(x,y) \in G \times G : x^{-1}y \in U_n\}.$$

Then $(\mathcal{U}_n)_{n \in \mathbb{N}}$ is a filter base of symmetric subsets of $G \times G$ which contain the diagonal and which satisfy the condition that

$$\mathcal{U}_{n+1}^3 := \{(x,x') : (\exists y, z \in G)\, (x,y),(y,z),(z,x') \in \mathcal{U}_{n+1}\} \subseteq \mathcal{U}_n.$$

Hence the filter $\mathcal{U}$ generated by this basis defines a uniform structure on G. Using the construction in [Sch75, p.117] we obtain a quasimetric $d \colon G \times G \to \mathbb{R}^+$ such that the uniform structure induced by d agrees with $\mathcal{U}$. Moreover the condition that $(x,y) \in \mathcal{U}_n$ implies that $(gx,gy) \in \mathcal{U}_n$ for every $n \in \mathbb{N}$ shows

$$d(gx,gy) = d(x,y) \qquad \forall g,x,y \in G.$$

We set

$$f(g) := d(gS,s) := \inf_{s' \in S} d(gs',s) = \inf_{s' \in S} d(s',g^{-1}s).$$

Then f is an S-monotone function on G with

$$f(s) = d(sS,s) = d(S,1) = 0 < d(s^{-1}S,1) = d(S,s) = f(1)$$

because $U_1 \cap s^{-1}S = \emptyset$. Moreover, since

$$|f(x) - f(y)| = |d(S,x^{-1}s) - d(S,y^{-1}s)| \leq d(x^{-1}s,y^{-1}s)$$

and the original topology on G is finer than the topology induced by $\mathcal{U}$, the function f is continuous on G. We conclude that

$$SG \circ \mathrm{Mon}(S) \subseteq S \subseteq SG \circ \mathrm{Mon}(S).$$

Now the claim follows immediately with (i). ∎

The following corollary will be useful on various occasions.

Corollary 1.20.   *Let $S \subseteq G$ be a closed submonoid of the topological group G. If $S/H(S)$ is compact, then $S = H(S)$ is a group.*

Proof.   Let $s \in S$. To show that $s \in H(S)$, in view of Proposition 1.19, it suffices to show that $f = f \circ \rho_s$ holds for all $f \in \mathrm{Mon}(S)$. Replacing f by $\tanh \circ f$ we even may assume that f is bounded. So let $s \in S$ and $f \in \mathrm{Mon}(S)$ be bounded. We consider the sequence $f_n := f \circ \rho_{s^n}$. Then

$$f_n \leq f_{n+1} \leq \sup\{f(g) : g \in G\} < \infty.$$

Thus $f_n(g)$ is a bounded increasing sequence for each $g \in G$ and therefore $h := \lim_{n \to \infty} f_n$ exists pointwise. Since $\{s^n : n \in \mathbb{N}\} \subseteq S$ is relatively compact modulo $H(S)$, there exists a cluster point $s_0 H(S) \in S/H(S)$. Then the monotonicity of the sequence $f_n(g)$ shows that

$$h(g) = \lim_{n \to \infty} f(gs^n) = f(gs_0) \qquad \forall g \in G.$$

Let $s_0 H(S) = \lim_I s^{n_i} H(S)$. Then $h = f \circ \rho_{s_0}$ satisfies

$$h \circ \rho_{s_0} = \lim_I h \circ \rho_{s^{n_i}} = \lim_I \lim_{n \to \infty} f \circ \rho_{s^{n+n_i}} = h.$$

Thus $f \circ \rho_{s_0^2} = f \circ \rho_{s_0}$ and therefore

$$f = f \circ \rho_{s_0^2} \circ \rho_{s_0^{-1}} = f \circ \rho_{s_0} = h.$$

It follows that $f \leq f \circ \rho_s \leq h = f$, so that $f = f \circ \rho_s$. ∎

Corollary 1.21. *Let G be a compact topological group and $S \subseteq G$ an open subsemigroup. Then S is a group and S contains the connected component G_0 of the identity.*

Proof.   In view of Corollary 1.20, the semigroup $T := \overline{S}$ is a group. Let $s \in S^{-1} \subseteq T \subseteq \overline{S}$. Then there exists $s' \in S \cap S^{-1}$ arbitrary near to s. We conclude that $1 = s's'^{-1} \in S$, so that S is an open set containing 1. On the other hand the semigroup G_0 is generated by every open neighborhood of the identity, whence $G_0 \subseteq S$. The semigroup S even contains an open subgroup H of G. Now S/H is a subsemigroup of the finite group G/H, hence a group. Thus $S = SH$ is a group.   ∎

1.8. Smooth and analytic monotone functions on a Lie group

The objective of this section is to show that Proposition 1.19(ii) can be improved for a Lie group insofar as the monotone functions may be replaced by smooth and even analytic monotone functions. We start with a lemma that shows that there is quite a supply of smooth monotone functions.

Lemma 1.22. *Let G be a Lie group, $S \subseteq G$ a closed submonoid, $f \in \mathrm{Mon}(S)$, h a non-negative smooth function with compact support, and m a left Haar measure on G. Then $h * f \in \mathrm{Mon}^\infty(S)$.*

Proof.   First we note that $h * f$ is well defined because $x \mapsto h(x)f(x^{-1}g)$ is a continuous function with compact support. Moreover,

$$h * f(gs) = \int_G h(x)f(x^{-1}gs)\,dm(x) \le \int_G h(x)f(x^{-1}g)\,dm(x) = h * f(g)$$

shows that $h * f \in \mathrm{Mon}(S)$. To see that $h * f \in C^\infty(G)$, we note that

$$h * f(g) = \int_G h(gx)f(x^{-1})\,dm(x)$$

implies that we may differentiate under the integral to obtain

$$d(h * f)(g) = \int_G f(x^{-1})dh(gx) \circ d\rho_x\,dm(x).$$

Now the same argument applies to this function, thus $h * f \in C^\infty(G)$.   ∎

Theorem 1.23. *If G is a Lie group, then the fixed points of $\mathrm{SG} \circ \mathrm{Mon}^\infty$ are exactly the closed submonoids of G.*

Proof.   That $S = \mathrm{SG} \circ \mathrm{Mon}^\infty(S)$ implies that S is a closed submonoid of G, follows immediately from Proposition 1.19. For the converse we assume that $s \in G \setminus S$, where S is a closed submonoid of G. Then, according to Proposition 1.19, we find a continuous S-monotone function h such that $h(s) < h(1)$. Let U_n be a fundamental sequence of 1-neighborhoods in G and φ_n smooth non-negative functions with $\mathrm{supp}(\varphi_n) \subseteq U_n$ and $\int_G \varphi_n(x)\,dm(x) = 1$, where m is a left Haar measure on G. Then $h_n := \varphi_n * h$ is a sequence of functions in $\mathrm{Mon}^\infty(S)$ (Lemma 1.22) with $h_n \to h$. If n is sufficiently large, we find that $h_n(s) < h_n(1)$. This implies that $s \notin \mathrm{SG}\big(\mathrm{Mon}^\infty(S)\big)$ and therefore that $S = \mathrm{SG}\big(\mathrm{Mon}^\infty(S)\big)$.   ∎

Now we can formulate a few principles which show how properties of semi-groups are reflected by properties of monotone functions.

Proposition 1.24.   *Let S be a closed submonoid of a Lie group G. Then the following assertions hold:*

(i) $S = \{g \in G : (\forall f \in \text{Mon}^\infty(S)) f(g) \geq f(1)\}$.

(ii) *Every function in $\text{Mon}^\infty(S)$ factors to a smooth function on the homogeneous space $G/H(S)$.*

(iii) $H(S) = \{1\}$ *if and only if $\text{Mon}^\infty(S)$ separates the points of G.*

(iv) $H(\text{Mon}^\infty(S)) := \text{Mon}^\infty(S) \cap -\text{Mon}^\infty(S) \cong C^\infty(G/\overline{G_S})$.

(v) $\overline{G_S} = G$ *if and only if $H(\text{Mon}^\infty(S)) = \mathbb{R}1$.*

(vi) $S \neq G$ *if and only if $\text{Mon}^\infty(S)$ contains non-constant smooth functions.*

(vii) $S \neq \{1\}$ *if and only if $\text{Mon}^\infty(S) \neq C^\infty(G)$.*

Proof.   (i) This follows from Theorem 1.23.

(ii) For $h \in H(S)$ and $g \in G$ it is immediate from the definition of Mon^∞ that

$$f(gh) \geq f(g) = f(ghh^{-1}) \geq f(gh)$$

for all $f \in \text{Mon}^\infty(S)$.

(iii) From (ii) it is clear that $H(S) = \{1\}$ if $\text{Mon}^\infty(S)$ separates the points of G. Let us assume that $H(S) = \{1\}$ and that $g \neq g' \in G$. Then $g^{-1}g' \notin S$ or $g'^{-1}g \notin S$. Assume the first case. Then, using (i), we find a function $f \in \text{Mon}^\infty(S)$ with $f(g^{-1}g') < f(1)$. Now $\tilde{f} := f \circ \lambda_{g^{-1}} \in \text{Mon}^\infty(S)$ with $\tilde{f}(g) \neq \tilde{f}(g')$.

(iv) Since, for every function $f \in C^\infty(G)$ the relation $\overline{G_S} \subseteq \text{SG}(f)$ is equivalent to $S \cup S^{-1} \subseteq \text{SG}(f)$, according to (1.6), we have the following identity

$$H(\text{Mon}^\infty(S)) = \text{Mon}^\infty(S) \cap -\text{Mon}^\infty(S)$$
$$= \text{Mon}^\infty(S) \cap \text{Mon}^\infty(S^{-1}) = \text{Mon}^\infty(\overline{G_S}).$$

For a closed subgroup $H \subseteq G$ the set $\text{Mon}^\infty(H)$ consists precisely of those functions which are constant on the left cosets gH of H, i.e., the smooth functions on the quotient G/H.

(v) This is a consequence of (iv).

(vi) This is immediate from (i).

(vii) This is immediate from (iv) and the fact that the smooth functions separate the points of G.   ∎

Next we prove the existence of sufficiently many S-monotone analytic functions for a given closed submonoid of G. This is much more difficult than the corresponding proof in the continuous or the smooth case.

The strategy is to approximate a smooth S-monotone function by a one-parameter family $p_t * f$ of analytic functions. Here the p_t are analytic density functions of a one-parameter semigroup of probability measures on G which are also fundamental solutions of a heat equation on G. These functions are provided by Hunt's theory. For details we refer to Heyer's book ([He77]).

We write $C_c^\infty(G)$ (respectively $C_c(G)$) for the set of smooth (respectively continuous) functions with compact support on G. We also write $C_0(G)$ for the Banach space of continuous functions vanishing at infinity. This is the completion of $C_c(G)$ with respect to the norm

$$\|f\|_\infty := \max\{|f(x)| : x \in G\}.$$

For $X \in \mathfrak{g} = \mathbf{L}(G)$ we write $D(X)$ for the set of all functions in $f \in C_0(G)$ for which the limit

$$\tilde{X}f := \lim_{t \to 0} \frac{1}{t}(f \circ \rho_{\exp tX} - f)$$

exists in the norm-topology. We define

$$C_0^2(G) := \{f \in C_0(G) : (\forall X, Y \in \mathfrak{g})f \in D(X), \tilde{X}f \in D(Y)\}.$$

Proposition 1.25. *Let G be a connected Lie group, $\Delta_G \colon G \to \mathbb{R}^+$ its modular function, $\{X_i : i = 1, \ldots, n\}$ a basis of the Lie algebra $\mathfrak{g}$, $\mathcal{X}_i$ the associated left invariant vector fields on G, and N the left invariant differential operator defined by*

$$Nh := \sum_{i=1}^n \left(\mathcal{X}_i^2 + \langle d\Delta_G(1), X_i\rangle \mathcal{X}_i\right)h \quad for \quad h \in C^\infty(G).$$

Then there exists a one parameter family $(p_t)_{t>0}$ of analytic functions with the following properties:

(i) $p_t(g) > 0$ *for all* $g \in G$.

(ii) $\int_G p_t(x)\, dm(x) = 1$ *with respect to a left invariant Haar measure m on G.*

(iii) $p_t * p_s = p_{t+s}$ *for* $t, s > 0$.

(iv)

$$\lim_{t \to 0} \frac{1}{t}(f * p_t - f) = Nf$$

exists in the norm-topology in $C_0(G)$ whenever $f \in C_0^2(G)$.

Proof.   From [He77, p.263, p.296] we get a convolution semigroup $(\mu_t)_{t \geq 0}$ of probability measures on G such that the operators

$$S_t : f \mapsto (y \mapsto \mu_t(f \circ \lambda_y)) \quad \text{on} \quad C_0(G)$$

are a strongly continuous one-parameter semigroup of contractions on the Banach space $(C_0(G), \|\cdot\|_\infty)$ and that

$$\lim_{t \to 0} \frac{1}{t}(S_t(f) - f) = Nf \quad \forall f \in C_0^2(G),$$

where the limit exists in the norm-topology on $C_0(G)$. Moreover, [He77, p.446] implies that there exist strictly positive analytic functions p_t, $t > 0$, on G such that $\mu_t = p_t m$, where m is a left invariant Haar measure on G. Hence

$$S_t(f)(y) = \int_G p_t(x)f(yx)\, dm(x) = \int_G f(x)p_t(y^{-1}x)\, dm(x) = (f * \check{p}_t)(y),$$

where $\check{p}_t(g) = p_t(g^{-1})$. Now $S_t \circ S_{t'} = S_{t+t'}$ implies that

$$p_{t+t'} = (\check{p}_t * \check{p}_{t'})\check{} = (\check{p}_{t'} * \check{p}_t)\check{} = p_t * p_{t'}.$$ ∎

The following lemma illustrates that one cannot expect that S-monotone functions are small at infinity.

Lemma 1.26. *Let S be a closed submonoid of G. Then every continuous S-monotone function f which vanishes at infinity is also G_S-monotone, i.e., $S \subseteq H(SG(f))$.*

Proof.   Let $s \in S$. First we assume that $K := \overline{\{s^n : n \in \mathbb{N}_0\}} \subseteq S$ is compact. This is a compact subsemigroup of the group G and therefore a group (Corollary 1.20). Hence $s \in H(S) \subseteq H(SG(f))$ and therefore $f(gs) = f(g)$ holds for all $g \in G$. Next we assume that K is not compact. Let $a_k := s^{n_k}$ be a subsequence which eventually leaves every compact subset of G. Then

$$0 = \lim_{k \to \infty} f(ga_k^{-1}) \leq f(g) \leq f(gs) \leq \lim_{k \to \infty} f(ga_k) = 0$$

for every $g \in G$ implies that $S \subseteq H(SG(f))$.   ■

The next problem is to see which conditions on f we need to guarantee that the functions $p_t * f$ are analytic. As we have seen in the preceding lemma, one cannot expect that these properties follow from Proposition 1.25 because in general $f \notin C_0(G)$. In the following we apply results of Nelson and Garding ([Ncl70] and [Ga60]) to deduce the analyticity properties of these functions. We first define the convenient growth condition.

Let G be a Lie group, d a left invariant Riemannian metric on G, and V a Banach space. A continuous function $F : G \to V$ is said to be of *moderate growth* if there exist positive real numbers M, λ such that

$$(1.8) \qquad \qquad \|F(g)\| \leq M e^{\lambda d(g,1)} \quad \text{for all} \quad g \in G.$$

Note that $d(g^{-1}, 1) = d(1, g) = d(g, 1)$ implies that F is of moderate growth if and only if $\check{F} : g \mapsto F(g^{-1})$ is of moderate growth.

Proposition 1.27.   *Let G be a connected Lie group and p_t as in Proposition 1.25. Then $\check{p}_t = p_t$ and if $f \in C_c^\infty(G)$, V is a Banach space, and $F : G \to V$ is a function of moderate growth, then for every $t > 0$ the function*

$$u_t : x \mapsto \int_G F(y)(f * p_t)(y^{-1}x) \, dm(y) = (F * f * p_t)(x)$$

is an analytic V-valued function and $\lim_{t \to 0} u_t = F * f$ holds pointwise on G. The same statement holds for the functions $p_t * f * F$.

Proof.   Let N be the left invariant differential operator defined by

$$Nh := \sum_{i=1}^{n} \left(\mathcal{X}_i^2 + \langle d\Delta_G(1), X_i \rangle \mathcal{X}_i \right) h \quad \text{for} \quad h \in C^\infty(G).$$

Set $N_0 := N \big|_{C_c^\infty(G)}$. Then N_0 is a densely defined symmetric operator on the Hilbert space $L^2(G)$ with respect to left invariant Haar measure m on G and $\Lambda := N_0^*$ is a negative self-adjoint operator on $L^2(G)$ ([Ga60, p.81]). Now [Paz83, p.15] guarantees the existence of a strongly continuous one-parameter semigroup $(U_t)_{t \geq 0}$ of contractions on $L^2(G)$ with Λ as infinitesimal generator. We claim that

$$U_t(f) = S_t(f) := f * \check{p}_t \quad \forall f \in L^2(G).$$

We note that it is not a priori clear that $S_t(f) \in L^2(G)$ for all $t > 0$. So we use Garding's method of Banach space valued solutions of the heat equation to get a connection between $C_0(G)$ and $L^2(G)$. Let $f, h \in C_c^\infty(G)$ and $F \in C(G, C_0(G))$ defined by $F(g) = h \circ \rho_g$, where $\rho_g \colon G \to G, x \mapsto xg$ is right multiplication with g. One of the main results in [Ga60, pp.84,85] is that

$$U \colon (t,x) \mapsto \left(F * U_t(f)\right)(x) \in C^\infty(\mathbb{R}^+ \times G, C_0(G)),$$

and that U is a $C_0(G)$-valued solution of the heat equation $\frac{\partial}{\partial t} U = NU$ with

$$U(0,x) = F * f(x) \quad \text{and} \quad (NU)(t,x) = \left(F * \Lambda U_t(f)\right)(x) = \left(F * U_t(Nf)\right)(x).$$

Moreover,

(1.9)
$$\widetilde{X} U(t,\cdot) = F * \widetilde{X} U_t(f) \qquad \forall X \in \mathfrak{g}.$$

We consider the curve $\gamma(t) := U(t,1)$ in $C^\infty(\mathbb{R}^+, C_0(G))$. Then

$$\gamma(0) = (F * f)(1) = \int_G (h \circ \rho_x) f(x^{-1}) \, dm(x) = h * f,$$

$$\gamma(t) = \left(F * U_t(f)\right)(1) = h * U_t(f)$$

and

$$\gamma'(t) = (NU)(t,1) = \left(F * U_t(Nf)\right)(1) = h * U_t(Nf).$$

Furthermore, the left invariance of the operator Λ on $L^2(G)$ implies the left-invariance of U_t. Hence U_t commutes with left convolutions and

$$\gamma'(t) = U_t(h * Nf) = U_t\left(N(h * f)\right) = \Lambda U_t(h * f) = \Lambda\left(h * U_t(f)\right) = \Lambda \gamma(t) = N\gamma(t)$$

because a second application of [Ga60, p.84] shows that $(t,x) \mapsto \left(h * U_t(f)\right)(x) \in C^\infty(\mathbb{R}^+ \times G)$. Next

$$\left(F * U_t(f)\right)(g) = \left((F \circ \lambda_g) * U_t(f)\right)(1) = \left(h * U_t(f) \circ \rho_g\right) = \left(h * U_t(f)\right) \circ \rho_g = \gamma(t) \circ \rho_g$$

and (1.9) implies that the function $g \mapsto \gamma(t) \circ \rho_g$ is in $C^\infty(G, C_0(G))$, so that

$$\gamma(t) \in C_0^2(G) \qquad \forall t > 0.$$

Now $\gamma \colon \mathbb{R}^+ \to C_0(G)$ is a solution of the differential equation

$$\gamma(0) = h * f \quad \text{and} \quad \gamma'(t) = N\gamma(t).$$

Consequently the uniqueness theorem for these equations ([Paz83, p.101]) implies that $\gamma(t) = S_t(h * f)$ because the differential operator N agrees with the infinitesimal generator of the semigroup S_t on the space $C_0^2(G)$ (Proposition 1.25). Whence $\gamma(t) = h * U_t(f) = h * f * \check{p}_t$. Since h was arbitrary, we find that

$$U_t(f) = f * \check{p}_t.$$

To show that u_t is real analytic with $u_t \to F * f$, in view of [Ga60, p.84], it remains to show that $\check{p}_t = p_t$. This follows easily from the fact that the operator $f \mapsto f * \check{p}_t$ on $L^2(G)$ is selfadjoint.

The last statement follows from $(p_t * f * F)\check{} = \check{F} * \check{f} * p_t$ and the fact that $\check{F}$ is of moderate growth whenever this holds for F.   ∎

Corollary 1.28.   *Let G be a connected Lie group and p_t as in* Proposition *1.25. If for $h \in C(G)$ there exists $f \in C_c^\infty(G)$ and a bounded continuous function $f_1 \in C(G)$ such that $h = f * f_1$, then for every $t > 0$ the function $p_t * h$ is analytic and $\lim_{t \to 0} p_t * h = h$ holds pointwise.*

Proof.   One only has to apply Proposition 1.27 with the function $f_1 : G \to \mathbb{R}$ which is trivially of moderate growth because it is bounded.   ∎

Now we obtain the expected characterization of the fixed points of the mapping $\mathrm{Sg} \circ \mathrm{Mon}^\omega$ in the analytic category.

Theorem 1.29.   *If G is a Lie group, then the fixed points of $\mathrm{SG} \circ \mathrm{Mon}^\omega$ are exactly the closed submonoids of G.*

Proof.   With the above information on analytic functions at hand, the proof is nearly the same as that of Theorem 1.23. If S is a closed submonoid of G and $s \notin S$, then it follows from Theorem 1.23 that we find a bounded smooth S-monotone function f such that $f(s) < f(1)$. So we only have to show that every bounded smooth S-monotone function f is the pointwise limit of analytic S-monotone functions. To see this, let U_n be a base of 1-neighborhoods in G and h_n smooth non-negative functions with support in U_n such that $\int_G h_n(x) \, dm(x) = 1$ and $\check{h}_n = h_n$. Now we apply Corollary 1.28 to see that the functions $p_t * h_n * f$ are analytic. The fact that $p_t * h_n \geq 0$ implies that they are also S-monotone (Lemma 1.22). Now the facts that $\lim_{t \to 0} p_t * h_n * f = h_n * f$ and $\lim_{n \to \infty} h_n * f = f$ conclude the proof.   ∎

1.9. W-Positive functions and globality

In this section we will achieve a characterization of globality which is in general not easy to check, but yields important and effective tools for the treatment of the globality question. We take the additional structure into account which is at hand if G is a Lie group: Let $\mathfrak{g} = L(G)$ be the Lie algebra of G. For an element $X \in \mathfrak{g}$ we write $\mathcal{X}$ for the associated left invariant vector field on G. This vector field satisfies $\mathcal{X}(g) = d\lambda_g(1)X$ and it represents a derivation of the algebras $C^\infty(G)$ and $C^\omega(G)$ by

$$\mathcal{X} f(g) := \frac{d}{dt}\bigg|_{t=0} f\big(g \exp(tX)\big).$$

For a subset $\Omega \subseteq \mathfrak{g} := L(G)$ we set

$$\operatorname{Pos}(\Omega) := \{f \in C^\infty(G) : (\forall X \in \Omega)\mathcal{X} f \geq 0\}$$

and $\operatorname{Pos}^\omega(\Omega) := \operatorname{Pos}(\Omega) \cap C^\omega(G)$. For a subset $M \subseteq C^\infty(G)$ we define

$$\operatorname{ev}(M) := \{df(g) \circ d\lambda_g(1) : f \in M, g \in G\} \subseteq \mathfrak{g}^*,$$

and $\operatorname{ev}^*(M) := \operatorname{ev}(M)^* = \{X \in \mathfrak{g} : \langle X, \operatorname{ev}(M)\rangle \subseteq \mathbb{R}^+\} \subseteq \mathfrak{g}$. Then the mappings $\Omega \mapsto \operatorname{Pos}(\Omega)$ and $M \mapsto \operatorname{ev}^*(M)$ are inclusion reversing and

$$(1.10) \qquad\qquad M \subseteq \operatorname{Pos}(\Omega) \quad \text{if and only if} \quad \Omega \subseteq \operatorname{ev}^*(M),$$

i.e., Pos and $\operatorname{Pos}^\omega$ with ev^* define a Galois connection from the subsets of $\mathfrak{g}$ to the subsets of $C^\infty(G)$ and $C^\omega(G)$ respectively.

To see how to describe the fixed points of $\operatorname{Pos} \circ \operatorname{ev}^*$ and $\operatorname{ev}^* \circ \operatorname{Pos}$ we relate these operators to $\operatorname{Mon}^\infty$ and SG. For the following lemma we recall the definition of the semigroups

$$S_{\mathbb{R}^+\Omega} := \overline{\langle \exp \mathbb{R}^+\Omega\rangle}$$

for a subset Ω of the Lie algebra $L(G)$ of G.

Lemma 1.30.

(i) *For $M \subseteq C^\infty(G)$ we have $\operatorname{ev}^*(M) = L\big(\operatorname{SG}(M)\big)$.*

(ii) *For a subset $\Omega \subseteq \mathfrak{g}$ we have*

$$\operatorname{Pos}(\Omega) = \operatorname{Mon}^\infty(S_{\mathbb{R}^+\Omega}) \quad and \quad \operatorname{Pos}^\omega(\Omega) = \operatorname{Mon}^\omega(S_{\mathbb{R}^+\Omega}).$$

Proof. (i) Because $\operatorname{SG}(M)$ is a closed subsemigroup of G and $\mathcal{X} f \geq 0$ is equivalent to $f\big(g \exp(tX)\big) \geq f(g)$ for all $g \in G$ and $t \in \mathbb{R}^+$, we have, according to (1.6) and (1.9), the following equivalences for $X \in L(G)$

$$X \in L\big(\operatorname{SG}(M)\big) \Leftrightarrow \exp(\mathbb{R}^+X) \subseteq \operatorname{SG}(M) \Leftrightarrow M \subseteq \operatorname{Mon}^\infty\big(\exp(\mathbb{R}^+X)\big)$$

$$\Leftrightarrow \operatorname{ev}(M) \subseteq (\mathbb{R}^+X)^* \Leftrightarrow X \in \operatorname{ev}(M)^*.$$

(ii) For $\Omega \subseteq \mathfrak{g}$ we obtain from this chain of equivalences that $f \in \operatorname{Pos}(\Omega)$ is equivalent to $\exp(\mathbb{R}^+\Omega) \subseteq \operatorname{SG}(f)$ for $f \in C^\infty(G)$. This is the same as $\operatorname{Pos}(\Omega) = \operatorname{Mon}^\infty(S_{\mathbb{R}^+\Omega})$. The second equality follows by intersection with $C^\omega(G)$. ∎

Note that Lemma 1.30 shows that the fixed points of $\mathrm{Pos}\circ\mathrm{ev}^{*}$, i.e., the images of Pos, are in particular fixed points of $\mathrm{Mon}^{\omega}\circ\mathrm{SG}$, i.e., in the image of Mon^{ω} ((1.7) and Lemma 1.30). So the images of the fixed points of $\mathrm{Pos}\circ\mathrm{ev}^{*}$ under SG constitute a certain class of closed submonoids of G (Lemma 1.19). In the following proposition we show that this class coincides with the class of Lie semigroups and at the same time we give a first characterization of global Lie wedges in terms of fixed point properties.

Theorem 1.31. *Let G be a Lie group with Lie algebra $\mathfrak{g}$ and $\Omega\subseteq\mathfrak{g}$. Then*

(i) $\mathrm{SG}\big(\mathrm{Pos}^{\omega}(\Omega)\big)=S_{\mathbb{R}^{+}\Omega}$ *and the fixed points of $\mathrm{SG}\circ\mathrm{Pos}^{\omega}$ are precisely the Lie semigroups in G.*

(ii) *For a closed submonoid $S\subseteq G$ we have*

$$\mathrm{ev}^{*}\big(\mathrm{Mon}^{\omega}(S)\big)=\mathbf{L}(S) \quad and \quad \mathrm{ev}^{*}\big(\mathrm{Pos}^{\omega}(\Omega)\big)=\mathbf{L}(S_{\mathbb{R}^{+}\Omega}).$$

Moreover the fixed points of $\mathrm{ev}^{*}\circ\mathrm{Pos}^{\omega}$ are precisely the global Lie wedges in $\mathbf{L}(G)$.

Proof. (i) With Theorem 1.29 and Lemma 1.30 we obtain that

$$\mathrm{SG}\big(\mathrm{Pos}^{\omega}(\Omega)\big)=\mathrm{SG}\big(\mathrm{Mon}^{\omega}(S_{\mathbb{R}^{+}\Omega})\big)=S_{\mathbb{R}^{+}\Omega}.$$

It follows that the fixed points of $\mathrm{SG}\circ\mathrm{Pos}^{\omega}$ are those closed submonoids $S\subseteq G$ for which there exists a subset $\Omega\subseteq\mathbf{L}(G)$ such that $S=\overline{\langle\exp\mathbb{R}^{+}\Omega\rangle}$, or equivalently $S=\overline{\langle\exp\mathbf{L}(S)\rangle}$. This condition defines the Lie semigroups in G.

(ii) In view of Theorem 1.29 and Lemma 1.30,

$$\mathrm{ev}^{*}\circ\mathrm{Mon}^{\omega}(S)=\mathbf{L}\Big(\mathrm{SG}\big(\mathrm{Mon}^{\omega}(S)\big)\Big)=\mathbf{L}(S).$$

This implies by Lemma 1.30 that $\mathrm{ev}^{*}\big(\mathrm{Pos}^{\omega}(\Omega)\big)=\mathbf{L}(S_{\mathbb{R}^{+}\Omega})$. This shows that every fixed point of $\mathrm{ev}^{*}\circ\mathrm{Pos}$ is a global Lie wedge. Let us assume that W is global in G. Then the calculation from above shows that

$$W=\mathbf{L}(S_{W})=\mathrm{ev}^{*}\big(\mathrm{Pos}^{\omega}(W)\big)$$

is a fixed point of $\mathrm{ev}^{*}\circ\mathrm{Pos}^{\omega}$. ∎

It is possible to characterize the globality of Lie wedges by the existence of positive fuctions. But for that it is necessary to have a stronger type of positivity. Let $W\subseteq\mathbf{L}(G)$ be a Lie wedge and $f\in C^{\infty}(G)$. Then f is said to be *strictly W-positive* if

$$f'(g)\in\mathrm{algint}\,W^{*} \qquad \forall g\in G.$$

If $f\in C^{\infty}(G)$ we use the notation $f':G\to\mathfrak{g}^{*}$ for the function $g\mapsto df(g)d\lambda_{g}(1)$.

Lemma 1.32. *If $S\subseteq G$ is a closed submonoid, then*

$$\mathrm{algint}\,\mathbf{L}(S)^{*}\subseteq\mathrm{ev}\big(\mathrm{Mon}^{\omega}(S)\big),$$

i.e., for every $\omega \in \mathrm{algint}\,\mathbf{L}(S)^*$ *there exists an* S*-monotone analytic function* f *on* G *such that* $f'(1) = \omega$.

Proof.   First we note that $\mathrm{Mon}^\omega(S)$, and therefore $\mathrm{ev}\left(\mathrm{Mon}^\omega(S)\right)$, is a convex cone and $\mathrm{ev}^*\left(\mathrm{Mon}^\omega(S)\right) = \mathbf{L}(S)$ (Theorem 1.31(ii)). Hence Proposition 1.1(v) shows that

$$\mathrm{algint}\,\mathbf{L}(S)^* = \mathrm{algint}\left(\mathrm{ev}^*\left(\mathrm{Mon}^\omega(S)\right)^*\right) = \mathrm{algint}\left(\overline{\mathrm{ev}\left(\mathrm{Mon}^\omega(S)\right)}\right)$$

$$= \mathrm{algint}\left(\mathrm{ev}\left(\mathrm{Mon}^\omega(S)\right)\right).$$

∎

If we apply Lemma 1.32 with $S = \{\mathbf{1}\}$, it implies for every linear functional $\omega \in \mathfrak{g}^* = \mathbf{L}(S)^*$ the existence of a global analytic function $f \in C^\omega(G)$ with $f'(1) = \omega$. In order to show the existence of strictly $\mathbf{L}(S)$-positive functions for a given monoid S we need another lemma:

Lemma 1.33. *Let* S *be a closed submonoid of* G, f_1 *a non-negative smooth function with compact support containing* $\mathbf{1}$, *and* f_2 *a bounded smooth* S*-monotone function. Then* $f := f_1 * f_2$ *is an* S*-monotone function such that* f' *is of moderate growth. Moreover, if* f_2 *is strictly* $\mathbf{L}(S)$*-positive in* g, *then the same holds for* f.

Proof.   It is clear that f is bounded smooth and S-monotone (Lemma 1.22). Using [Ga60, p.76], we find for every norm on $\mathfrak{g}^*$ and every Riemannian metric d on G that

$$\|df(g)d\lambda_g(\mathbf{1})\| = \|df(g)d\rho_g(\mathbf{1})\,\mathrm{Ad}(g)\|$$

$$\leq \|\,\mathrm{Ad}(g)^*\|\;\|\int_G df_1(gx)d\rho_x(g)d\rho_g(\mathbf{1})f_2(x^{-1})\,dm(x)\|$$

$$\leq M_1 e^{\lambda d(g,\mathbf{1})}\|\int_G df_1(gx)d\rho_{gx}(\mathbf{1})f_2(x^{-1})\,dm(x)\|$$

$$\leq M_1 e^{\lambda d(g,\mathbf{1})}m\left(\mathrm{supp}\,f_1\right)\max_{y\in G}\|df_1(y)d\rho_y(\mathbf{1})\|\max_{x\in G}|f_2(x)|.$$

This f' is of moderate growth.

Pick $g \in G$, $X \in \mathbf{L}(S) \setminus H(\mathbf{L}(S))$, and suppose that f_2 is strictly $\mathbf{L}(S)$-positive in g. Then

$$\langle df(g), d\lambda_g(\mathbf{1})X\rangle = \int_G f_1(x)\langle df(x^{-1}g), d\lambda_{x^{-1}g}(\mathbf{1})X\rangle\,dm(x)$$

$$= \int_G f_1(x)\langle df(x^{-1}g), d\lambda_{x^{-1}g}(\mathbf{1})X\rangle\,dm(x) > 1$$

because $\langle df(g), d\lambda_g(\mathbf{1})X\rangle > 0$, $f_1 \geq 0$, and $\mathbf{1} \in \mathrm{supp}(f_1)$. ∎

Proposition 1.34.   *For every closed submonoid* $S \subseteq G$ *there exists a strictly* $\mathbf{L}(S)$*-positive* S*-monotone analytic function.*

Proof.   According to Lemma 1.32, there exists a smooth S-monotone function f_1 such that $df_1(\mathbf{1}) \in \mathrm{algint}\,\mathbf{L}(S)^*$. Replacing f_1 by $\tanh \circ f_1$, we may assume that f_1

is bounded and by Lemma 1.33 also that f_1' is of moderate growth. Let $f \in C_c^\infty(G)$ be non-zero and non-negative. Then, with the notations of Proposition 1.25, the function $h := p_t * f * f_1$ is analytic and, because f_1' is of moderate growth, we may differentiate under the integral to see that $h' = p_t * f * f_1'$ ([Ga60, p.84]). The fact that the function p_t is strictly positive implies that $p_t * f$ is strictly positive on G because $f \neq 0$. Let $X \in \mathbf{L}(S)$. Then $f_1'(g)(X) > 0$ for all $g \in G$. Hence $h'(g)(X) > 0$ for all $g \in G$ because $p_t * f$ is strictly positive. This shows that h is strictly $\mathbf{L}(S)$-positive and S-monotone (Lemma 1.22).   ∎

In the proof of the following theorem we use a result which will be proved in Chapter 4.

Theorem 1.35. *Let G be a connected Lie group and $W \subseteq \mathbf{L}(G)$ a Lie wedge. Then the following conditions are equivalent:*

(1) *There exists a strictly W-positive analytic function f on G.*

(2) *There exists a smooth W-positive function f on G such that*

$$f'(1) \in \text{algint } W^*.$$

(3) *The convex cone $W + \mathbf{L}(S_{H(W)})$ is a global Lie wedge.*

Proof. (1) $\Rightarrow$ (2): trivial.

(2) $\Rightarrow$ (3): Let $H := S_{H(W)}$ and f be as above. Then f is constant on H (Proposition 1.24), and therefore $F := \mathbf{L}(H) \subseteq \ker f'(1)$. Hence

$$F \cap W \subseteq \ker f'(1) \cap W \subseteq H(W)$$

because $f'(1) \in \text{algint } W^*$. We conclude that $F \cap W = H(W)$ and that $V := W + F$ is a wedge with $H(V) = H(W) + F = F$ (Proposition 1.4). To see that V is a Lie wedge, we have to show that $\text{Ad}(H)V = V$. This follows from the fact that $\langle \exp H(W) \rangle$ is dense in H, the closedness of V, and

$$e^{\text{ad } H(W)}V \subseteq e^{\text{ad } H(W)}F + e^{\text{ad } H(W)}W \subseteq F + W = V.$$

So V is a Lie wedge, $f'(1) \in \text{algint } V^*$, and the analytic subgroup $\langle \exp H(V) \rangle = H$ is closed in G. Now Corollary 4.22 entails that V is global in G.

(3) $\Rightarrow$ (1): Proposition 1.34.   ∎

Corollary 1.36. (Characterization of the global Lie wedges) *Let G be a connected Lie group and $W \subseteq \mathbf{L}(G)$ a Lie wedge. Then the following conditions are equivalent:*

(1) *W is global in G.*

(2) *$H(W)$ is global in G and there exists a strictly W-positive analytic function f on G.*

(3) *$H(W)$ is global in G and there exists a smooth W-positive function f on G such that*

$$f'(1) \in \text{algint } W^*.$$

Proof. Since the globality of $H(W)$ is equivalent to $\mathbf{L}(S_{H(W)}) = H(W)$, this follows immediately from Theorem 1.35.   ∎

1.10. Globality criteria

Even though Corollary 1.36 characterizes globality via an existence condition which is by no means easy to check, we can derive from it criteria which are quite effective. The first criterion deals with Lie wedges which sit nicely in global ones.

Proposition 1.37.   *Let G be a connected Lie group, $W \subseteq V \subseteq \mathbf{L}(G)$ Lie wedges, and suppose that*

(a) $W \cap H(V) \subseteq H(W)$,

(b) $H(W)$ *is global in G, and*

(c) V *is global in G.*

Then W is global in G.

Proof.   Using Corollary 1.36 we find a V-positive function $f \in C^{\infty}(G)$ such that $df(\mathbf{1}) \in \text{algint } V^{*}$. According to our assumption we have

$$\text{algint } V^{*} \subseteq \text{algint } W^{*}$$

(Proposition 1.1) and therefore f is a W-positive function with

$$df(\mathbf{1}) \in \text{algint } W^{*}.$$

Applying again Corollary 1.36 we conclude that W is global in G.   ∎

We have seen in Corollary 1.20 that compact subgroups intersect closed monoids in a compact group. This of course has consequences for global Lie wedges as well. To be more precise, if K is a compact subgroup of G leaving a Lie wedge W invariant, then one shows that one may replace the wedge W by $W + \mathbf{L}(K)$ in order to check globality. First we need a lemma on the action of compact groups on the set of monotone functions.

Lemma 1.38.   *Let G be a connected Lie group.*

(i) *Let $K \subseteq \text{Aut}(G)$ be a compact subgroup, m_K a normalized Haar measure on K, $S \subseteq G$ a closed submonoid which is invariant under K, and $f \in \text{Mon}^{\infty}(S)$. Then the function $\tilde{f}$ defined by*

$$\tilde{f}(g) = \int_{K} f(k.g) \, dm_K(k)$$

is in $\text{Mon}^{\infty}(S)$ and satisfies $\tilde{f} \circ k = \tilde{f}$ for all $k \in K$. Suppose that $f'(g) \in \text{algint } \mathbf{L}(S)^{*}$. Then

$$\tilde{f}'(g) \in \text{algint } \mathbf{L}(S)^{*}.$$

(ii) *Let $K \subseteq G$ be a compact subgroup, m_K a normalized Haar measure on K, $S \subseteq G$ a closed submonoid, and $f \in \text{Mon}^{\infty}(S)$. Then the function $\tilde{f}$ defined by*

$$\tilde{f}(g) = \int_{K} f(kg) \, dm_K(k)$$

is in $\mathrm{Mon}^\infty(S)$ and satisfies $\widetilde{f} \circ \lambda_k = \widetilde{f}$ for all $k \in K$. Suppose that $f'(g) \in \mathrm{algint}\, \mathbf{L}(S)^*$. Then

$$\widetilde{f}'(g) \in \mathrm{algint}\, \mathbf{L}(S)^*.$$

Proof. (i) That $\widetilde{f} \in \mathrm{Mon}^\infty(S)$ follows from

$$\widetilde{f}(gs) = \int_K f(k.(gs))\, dm_K(k) = \int_K f((k.g)(k.s))\, dm_K(k)$$

$$\geq \int_K f(k.g)\, dm_K(k) = \widetilde{f}(g)$$

for all $g \in G$ and $s \in S$.

By [Bou71a, Ch. III, §2, no. 10, Th. 1] the mapping $K \times G \to G, (k, g) \mapsto k(g) = k.g$ is analytic and because K is compact, we may differentiate under the integral to get

$$\widetilde{f}'(g) = \int_K d(f \circ k)(g) d\lambda_g(1)\, dm_K(k)$$

$$= \int_K df(k.g) dk(g) d\lambda_g(1)\, dm_K(k)$$

$$= \int_K df(k.g) d\lambda_{k.g}(1) dk(1)\, dm_K(k)$$

$$= \int_K f'(k.g) \circ dk(1)\, dm_K(k).$$

Let $f'(g) \in \mathrm{algint}\, \mathbf{L}(S)^*$ and $X \in \mathbf{L}(S) \setminus H(\mathbf{L}(S))$. Then the invariance of $\mathbf{L}(S)$ under the automorphisms $dk(1)$ of $\mathbf{L}(G)$ entails that

$$\langle \widetilde{f}'(g), X \rangle = \int_K \langle f'(k.g) dk(1), X \rangle\, dm_K(k) > 0$$

because the integrand is non-negative and positive at $k = 1$.

(ii) That $\widetilde{f} \in \mathrm{Mon}^\infty(S)$ follows from

$$\widetilde{f}(gs) = \int_K f(kgs)\, dm_K(k) \geq \int_K f(kg)\, dm_K(k) = \widetilde{f}(g)$$

for all $g \in G$ and $s \in S$.

We may differentiate under the integral to get

$$\widetilde{f}'(g) = \int_K d(f \circ \lambda_k)(g) d\lambda_g(1)\, dm_K(k)$$

$$= \int_K f'(kg)\, dm_K(k).$$

Let $f'(g) \in \mathrm{algint}\, \mathbf{L}(S)^*$ and $X \in \mathbf{L}(S) \setminus H(\mathbf{L}(S))$. Then

$$\langle \widetilde{f}'(g), X \rangle = \int_K \langle f'(k.g), X \rangle\, dm_K(k) > 0$$

because the integrand is non-negative and positive at $k = 1$. ■

Proposition 1.39.   *Let G be a connected Lie group, $K \subseteq G$ a compact subgroup, and $W \subseteq \mathfrak{g}$ a Lie wedge.*

(i) *If W is global in G, then*

(1.11) $$\mathbf{L}(K) \cap W \subseteq H(W).$$

(ii) *Suppose, in addition to (1.11), that $\mathrm{Ad}(K)W = W$ and $H(W)$ is global in G. Then the Lie wedge $V := W + \mathbf{L}(K)$ is global in G if and only if W is global in G.*

Proof.   (i) It follows from Corollary 1.20 that $S_W \cap K \subseteq H(S_W)$. Taking tangent wedges, this proves the assertion because $L\big(H(S_W)\big) = H(W)$ (Proposition 1.14).

(ii) Suppose that W is global and invariant under $\mathrm{Ad}(K)$. From (1.11) and Proposition 1.4 we get that V is a wedge. For $X \in H(V) = H(W) + \mathbf{L}(K)$ we clearly have that $e^{\mathrm{ad}\,X}V = V$, so V is a Lie wedge. Using Lemma 1.38 we find a function $f \in \mathrm{Pos}(W) = \mathrm{Mon}(S_W)$ such that $f'(\mathbf{1}) \in \mathrm{algint}\,W^*$, $f \circ I_k = f$ for all $k \in K$, and $f(Kg) = \{f(g)\}$ holds for all $g \in G$. Therefore $f(gK) = \{f(g)\}$ for all $g \in G$. This leads to $f'(g) \in \mathbf{L}(K)^\perp$ for all $g \in G$, and consequently $f \in \mathrm{Pos}(V)$ and $f'(\mathbf{1}) \in \mathrm{algint}\,W^* \cap \mathbf{L}(K)^\perp = \mathrm{algint}\,V^*$. An application of Theorem 1.35 shows that V is global in G because the group

$$\langle \exp H(V)\rangle = \langle \exp \mathbf{L}(K)\rangle\langle \exp H(W)\rangle = K_0 S_{H(W)}$$

is closed as a product of a compact and a closed subgroup of G.

If, conversely, V is global in G and $H(W)$ is global, then the globality of W follows from $(*)$ and Proposition 1.37.   ∎

Next we show which halfspace Lie wedges are global.

Proposition 1.40.   *Let G be a connected Lie group, $H \subseteq G$ be a closed connected subgroup of codimension 1, and $W \subseteq \mathbf{L}(G)$ a halfspace Lie wedge with $H(W) = \mathbf{L}(H)$. Then W is global in G if and only if G/H is not compact, i.e., diffeomorphic to the circle. These conditions are satisfied if G is simply connected.*

Proof.   If G is simply connected and H is connected, then G/H is also simply connected and therefore diffeomorphic to $\mathbb{R}$. Assume this. Let $\pi: G \to G/H$ be the canonical projection and $\tilde{f}: G/H \to \mathbb{R}$ be a diffeomorphism such that $d\tilde{f}(d\pi(\mathbf{1})W) \subseteq \mathbb{R}^+$. For $f := \tilde{f} \circ \pi$ we conclude that $f'(g) \in H(W)^\perp \setminus \{0\}$ which is the union of the two open half lines $\mathrm{int}\,W^*$ and $-\mathrm{int}\,W^*$. Since f has no singular points and $f'(\mathbf{1}) \in W^*$, we get $f'(g) \in \mathrm{algint}\,W^*$ for all $g \in G$ because G is connected. Theorem 1.35 is now applicable and implies that W is global. If G/H is compact, it is impossible to find a function $f \in C^\infty(G/H)$ without singular points, hence W is not global in G (Theorem 1.35, Proposition 1.24).   ∎

In contrast to the previous globality criteria the following two propositions are completely elementary.

Proposition 1.41.   *Let $\alpha: H \to G$ be a morphism of connected Lie groups and $W \subseteq \mathbf{L}(G)$ global in G. Then $V := d\alpha(\mathbf{1})^{-1}(W)$ is global in H.*

Proof.   Let $S \subseteq G$ be a closed subsemigroup of G such that $\mathbf{L}(S) = W$ and $\tilde{S} := \alpha^{-1}(S)$. Then $\tilde{S} \subseteq H$ is a closed subsemigroup and

$$
\begin{aligned}
\mathbf{L}(\tilde{S}) &= \{X \in \mathbf{L}(H) : \exp(\mathbb{R}^+ X) \subseteq \tilde{S}\} \\
&= \{X \in \mathbf{L}(H) : \alpha(\exp(\mathbb{R}^+ X)) \subseteq S\} \\
&= \{X \in \mathbf{L}(H) : \exp(\mathbb{R}^+ d\alpha(\mathbf{1})X) \subseteq S\} \\
&= d\alpha(\mathbf{1})^{-1} \mathbf{L}(S) = V.
\end{aligned}
$$

Hence V is global in H.   ∎

Proposition 1.42.   *Let G be a connected Lie group, $W \subseteq \mathbf{L}(G)$ be a Lie wedge and $H \subseteq H(S_W)$ a closed normal subgroup of G. Then $W/\mathbf{L}(H)$ is global in G/H if and only if W is global in G.*

Proof.   That the globality of $W/\mathbf{L}(H)$ implies the globality of W follows from Proposition 1.41. Let $p: G \to G/H$ be the canonical projection and $T := p(S)$. The closedness of T follows from $SH = S$ and the closedness of $S \subseteq G$. Let $W = \mathbf{L}(S)$ be global in G. Then we find that

$$
\begin{aligned}
\mathbf{L}(T) &= \{X \in \mathbf{L}(G/H) : \exp(\mathbb{R}^+ X) \subseteq T\} \\
&= \{X \in \mathbf{L}(G/H) : p^{-1}(\exp(\mathbb{R}^+ X)) \subseteq S\} \\
&= dp(\mathbf{1})\{Y \in \mathbf{L}(G) : \exp(\mathbb{R}^+ Y) \subseteq S\} = dp(\mathbf{1})\mathbf{L}(S) = W/\mathbf{L}(H).
\end{aligned}
$$

Therefore $W/\mathbf{L}(H)$ is global in G/H.   ∎

From Proposition 1.42 we immediately derive that if G is a connected Lie group, and $W \subseteq \mathbf{L}(G)$ an invariant wedge such that $H(W)$ is global in G, we have that W is global in G if and only if the pointed invariant Lie wedge $V := dp(\mathbf{1})W$ is global in G/H, where $p: G \to G/S_{H(W)}$ denotes the canonical projection.

The last two globality criteria of this section again depend on Theorem 1.35 (and the last two reductions).

Proposition 1.43.   *Let G be a connected Lie group, $W \subseteq \mathbf{L}(G)$ be a Lie wedge and $N \subseteq G$ a closed connected subgroup containing the commutator group $G' = (G, G)$. Then the following conditions are sufficient for W to be global in G:*

(a) $H(W)$ *is global in G,*

(b) *the fundamental group $\pi_1(G/N)$ is finite, and*

(c) $W \cap \mathbf{L}(N) \subseteq H(W)$.

Proof.   From Lemma 1.4 we know that $V := W + \mathbf{L}(N)$ is a wedge. But $H(V)$ contains the commutator algebra $\mathbf{L}(G)' = [\mathbf{L}(G), \mathbf{L}(G)]$ and therefore it is invariant and a Lie wedge because for $Y \in V$ and $X \in \mathbf{L}(G)$ we have

$$
e^{\operatorname{ad} X} Y \subseteq Y + \mathbf{L}(G)' \subseteq V.
$$

The factor group G/N is an abelian Lie group with finite fundamental group, hence a vector group. Consequently the wedge $V/\mathbf{L}(N)$ is global in G/N. According to the observation we made preceding this proposition, the wedge V is global in G. Now Proposition 1.37 implies the globality of W in G because $W \cap H(V) \subseteq H(W)$ and $H(W)$ is global.   ∎

Proposition 1.44.   *Let G be a connected Lie group with compact Lie algebra $\mathfrak{g} = \mathbf{L}(G)$ and $W \subseteq \mathfrak{g}$ a Lie wedge. Suppose that $H(W)$ is global in G. Then W is global in G iff*

$$\mathbf{L}(K) \cap W \subseteq H(W)$$

for every compact subgroup $K \subseteq G$.

Proof.   The necessity follows from Proposition 1.39. To prove sufficiency we notice that the Lie algebra $\mathfrak{g}$ is reductive and therefore $\mathfrak{g} = Z(\mathfrak{g}) \oplus [\mathfrak{g}, \mathfrak{g}]$ where $\mathfrak{g}' = [\mathfrak{g}, \mathfrak{g}]$ is a compact semisimple Lie algebra. According to [Ho65, p.180] we find an $n \in \mathbb{N}$ and a maximal compact subgroup $K \subseteq G$ such that G is diffeomorphic to $K \times \mathbb{R}^n$. Consequently K is connected. The subgroup $G' = \langle \exp \mathfrak{g}' \rangle$ of G is compact, semisimple and normal. This implies that $G' \subseteq K$ because every compact subgroup is conjugate to a subgroup of K under an inner automorphism of G. The manifold G/K is diffeomorphic to n-space and has trivial fundamental group. Now an application of Proposition 1.43 completes the proof.   ∎

Notes

The material of the first two sections is standard and only presented here for easy reference. Our presentation is essentially taken from [HHL89]. The characteristic function of a cone was introduced by Vinberg in [Vin63]. Some of the material presented here can also be found in [FK92]. The concept of a Lie wedge has been introduced independently by various authors in the late seventies and early eighties (cf. the introduction of [HHL89] for more details). Our definition of a Lie semigroup has been chosen to denote closed semigroups in order to avoid the complications arising from the semigroup analogs of dense winds in tori. Globality theorems, S-monotone and W-positive functions can be traced back to the work of Vinberg ([Vi80]) and Ol'shanskiĭ ([Ols82b]). More general results can be found in [HHL89]. The approach using the full duality given by the S-monotone and W-positive functions was developed in [Ne90b], where the existence of strictly positive smooth functions was proved. The results in the analytic setting are taken from [Ne91e].

2. Examples

This chapter is devoted to the study of various examples of Lie semigroups, which are instructive and important in the general theory. We have seen in Chapter 1 that already the abelian case provides us with some examples which indicate the fundamental differences between the local-global-principle in Lie groups and Lie semigroups. But in that case all the obstructions to the existence of a global semigroup with prescribed tangent wedge were of topological nature. We will see in this chapter that there are also algebraic obstructions. We start with the simplest group in which such obstructions are present, the Heisenberg group. In Section 2.2 we study the easiest but nevertheless very important example of a simple Lie group, namely $Sl(2, \mathbb{R})$. In this example it is very easy to compute the compression semigroups of open subsets of the real projective line very explicitly and to show that these are maximal subsemigroups. These ideas will be taken up in Section 8.1 in greater generality.

The one-sheeted 2-dimensional hyperboloid is a typical example of an ordered symmetric space which is dealt with in Section 2.3. It has a two-sheeted ordered covering which is the space of all pointed generating cones in $\mathbb{R}^2$. Hence the corresponding semigroup which defines the order is closely related to the semigroup of linear compressions of a cone in $\mathbb{R}^2$.

The section on the Ol'shanskiĭ semigroup in $Sl(2,\mathbb{C})$ is preparatory for the general theory of Ol'shanskiĭ semigroup in complex groups developed in Sections 8.2 – 8.6.

As we have already seen in other examples, affine compressions of subsets of vector spaces can be interesting examples of Lie semigroups. In Section 2.5 we discuss this concept in a greater generality and show that many compression semigroups have compact order intervals. In the setting of ordered homogeneous spaces this property is called global hyperbolicity and will be taken up in Section 4.4.

Since we will see in Section 8.5 how to compute the semigroup of contractions and expansions for an indefinite non-degenerate symmetric form using the theory of Ol'shanskiĭ semigroups, we describe the case of definite forms which is considerably easier in Section 2.6.

Gödels cosmological model explained in Section 2.7 is a homogeneous Lorentzian manifold which permits closed causal curves and which can be viewed as the universal covering group of $Sl(2, \mathbb{R})$ with a left-invariant Lorentzian metric. A more general framework for these ideas is described in Section 2.8.

The last three sections are devoted to some typical and instructive examples of

Lie semigroups in low dimensional solvable groups: Convex cones in almost abelian groups, the whirlpot and the parking ramp in the universal covering group of the euclidean motion group, and the invariant subsemigroups of the four dimensional oscillator group.

2.1. Semigroups in the Heisenberg group

The Heisenberg group G is usually represented as the group of all real 3×3-matrices of the form

$$(a,b,c) = \begin{pmatrix} 1 & a & c \\ 0 & 1 & b \\ 0 & 0 & 1 \end{pmatrix}, \quad a,b,c \in \mathbb{R},$$

but it will turn out to be useful to identify it with its Lie algebra $L(G)$ using the exponential function. In that representation the multiplication is given by the Campbell-Hausdorff formula, which in this example reads:

$$(2.1) \qquad X * Y = X + Y + \frac{1}{2}[X,Y] \quad \text{for all} \quad X,Y \in L(G).$$

If we use the above form of G then the Lie algebra $L(G)$ consists of the real 3×3-matrices of the form

$$(x,y,z) = \begin{pmatrix} 0 & x & z \\ 0 & 0 & y \\ 0 & 0 & 0 \end{pmatrix}, \quad x,y,z \in \mathbb{R}.$$

Then we have

$$\exp(x,y,z) = (x,y,z + \frac{1}{2}xy)$$

and

$$\log(a,b,c) = (a,b,c - \frac{1}{2}ab).$$

We will show that if the interior of a wedge W in $L(G)$ meets the center of $L(G)$ then the semigroup generated by $\exp W$ is all of G. We prove this by showing the following slightly more general proposition:

Proposition 2.1. *Let S be a subsemigroup of the Heisenberg group G containing central elements in its interior. Then $S = G$.*

Proof. If we set $X = (1,0,0)$, $Y = (0,1,0)$ and $Z = (0,0,1)$, then $\mathbb{R} \cdot Z$ is the center of G. For $W = \xi \cdot X + \eta \cdot Y$ and $W' = \eta \cdot X - \xi \cdot Y$ we calculate for $Z_0 = \lambda \cdot Z$ and $n \in \mathbb{N}$

$$U := (n \cdot (Z_0 + W)) * (n \cdot (Z_0 + W'))$$

$$= n \cdot (\xi + \eta) \cdot X + n(\eta - \xi) \cdot Y + n\big(2\lambda - \frac{1}{2}n \cdot (\xi^2 + \eta^2)\big) \cdot Z.$$

If S is a semigroup containing a whole neighborhood of Z_0, this calculation shows that for a suitable choice of n, ξ, and η, the element U lies both in $\operatorname{int} S$ and in the XY-plane. Rotating W and W' around the Z-axis we also find $-U \in \operatorname{int} S$. But then $1 = U * -U \in \operatorname{int} S$, so that $S = G$.   ∎

Note that taking the inverse image of a wedge in $\mathbb{R}^2 = G/Z(G)$ in G we find plenty of subsemigroups in G. In fact, for each wedge W in $\mathbf{L}(G)$ whose edge contains the center of $\mathbf{L}(G)$ gives us a semigroup S with $\mathbf{L}(S) = W$.

There are also subsemigroups in the Heisenberg group which do not contain a nontrivial group: Consider

$$S = \{(a,b,c) \in G \;:\; 0 \le a, b; \; 0 \le c \le ab\}.$$

Then

$$\log(S) = \left\{(x,y,z) \in \mathbf{L}(G): 0 \le x,y, |z| \le \frac{1}{2}xy\right\}$$

and hence

$$\mathbf{L}(S) = \{(\alpha,\beta,\gamma) \in \mathbf{L}(G) \;:\; \gamma = 0, \; 0 \le \alpha, \beta\}$$

Identifying G with a three dimensional vector space, we may visualize S as the region in the first octant bounded by the surface $z = xy$ and the xy-plane (cf. Figure 2.1).

Figure 2.1

The one-parameter semigroups $\sigma(t) = \exp t \cdot (1,0,0) = (t,0,0)$ and $\tau(t) = \exp t \cdot (0,1,0) = (0,t,0)$ generate S. In fact, if $(a,b,c) \in S$ and $b > 0$, then

$$(a,b,c) = \sigma\left(\frac{c}{b}\right)\tau(b)\sigma\left(a - \frac{c}{b}\right).$$

If $b = 0$, then $c = 0$, hence $(a,b,c) = \sigma(a)$.

2.2. The groups $\mathrm{Sl}(2,\mathbb{R})$ and $\mathrm{PSl}(2,\mathbb{R})$

Recall that $\mathrm{Sl}(2,\mathbb{R})$ is the set of real 2×2-matrices with determinant 1. Its Lie algebra $\mathfrak{sl}(2,\mathbb{R})$ is the set of real 2×2-matrices with trace zero. The exponential function $\exp\!: \mathfrak{sl}(2,\mathbb{R}) \to \mathrm{Sl}(2,\mathbb{R})$ is given by the usual matrix exponential function.

Before we start discussing semigroups in $\mathrm{Sl}(2,\mathbb{R})$, we recall some facts about the group itself. Consider the following subgroups of $\mathrm{Sl}(2,\mathbb{R})$:

$$N = \left\{\begin{pmatrix} 1 & r \\ 0 & 1 \end{pmatrix} : r \in \mathbb{R}\right\}$$

$$A = \left\{ \begin{pmatrix} t & 0 \\ 0 & t^{-1} \end{pmatrix} : t > 0 \right\}$$

$$K = \left\{ \begin{pmatrix} c & s \\ -s & c \end{pmatrix} : c^2 + s^2 = 1 \right\} = \mathrm{SO}(2).$$

These groups are the ingredients of the standard Iwasawa decomposition of $\mathrm{Sl}(2,\mathbb{R})$ which is given by

$$K \times A \times N \to KAN = \mathrm{Sl}(2,\mathbb{R})$$

$$\left(\begin{pmatrix} c & s \\ -s & c \end{pmatrix}, \begin{pmatrix} t & 0 \\ 0 & t^{-1} \end{pmatrix}, \begin{pmatrix} 1 & r \\ 0 & 1 \end{pmatrix} \right) \mapsto \begin{pmatrix} ct & ctr + st^{-1} \\ -st & -str + ct^{-1} \end{pmatrix}.$$

We also mention the Cartan decomposition of $\mathrm{Sl}(2,\mathbb{R})$. To that end we define

$$A^+ = \left\{ \begin{pmatrix} t & 0 \\ 0 & t^{-1} \end{pmatrix} : t > 1 \right\}.$$

Then the Cartan decomposition is

$$\mathrm{Sl}(2,\mathbb{R}) = KA^+K \cup K.$$

Apart from these classical decompositions we will also use the sets HAN and HA^+H, where

$$H = \left\{ \begin{pmatrix} c & s \\ s & c \end{pmatrix} : c^2 - s^2 = 1 \right\} = \mathrm{SO}(1,1).$$

If we set

$$X_0 = \begin{pmatrix} 1 & 0 \\ 0 & -1 \end{pmatrix}, \quad X_+ = \begin{pmatrix} 0 & 1 \\ 0 & 0 \end{pmatrix}, \quad X_- = \begin{pmatrix} 0 & 0 \\ 1 & 0 \end{pmatrix},$$

then $\{X_0, X_+, X_-\}$ is a basis of $\mathfrak{sl}(2,\mathbb{R})$ and it is easy to check that

$$K = \exp \mathbb{R}(X_+ - X_-), \ H_0 = \exp \mathbb{R}(X_+ + X_-), \ A = \exp \mathbb{R}X_0, \ N = \exp \mathbb{R}X_+$$

and

$$A^+ = \exp(]0, \infty[X_0),$$

where H_0 denotes the identity component of H.

The Cartan-Killing form κ of $\mathfrak{sl}(2,\mathbb{R})$ is a multiple of the trace form

$$k(X,Y) = \frac{1}{2} \mathrm{tr}(XY)$$

on $\mathfrak{sl}(2,\mathbb{R})$. With respect to the basis (X_0, X_+, X_-) the form k is given by the matrix

$$\frac{1}{2} \begin{pmatrix} 2 & 0 & 0 \\ 0 & 0 & 1 \\ 0 & 1 & 0 \end{pmatrix}.$$

The zero set of this form is the double cone obtained by rotating the line $\mathbb{R}X_+$ around the axis $\mathbb{R}(X_+ - X_-)$. The interior points of this Lorentzian double cone are precisely the elements $X \in \mathfrak{sl}(2,\mathbb{R})$ with negative k-length $k(X) := k(X,X)$. Note that because of the invariance of the Killing form, all the level sets of the quadratic form k are invariant under the adjoint action, which in our example is simply conjugation of matrices. The orbits under the adjoint action are precisely (cf. Figure 2.2) the connected components of the level sets (with the one exception that the zero set consists of three orbits: the zero and the two connected components of the remainder of the double cone).

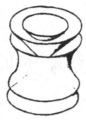

Figure 2.2

The first example of a subsemigroup of $\mathrm{Sl}(2,\mathbb{R})$ is given by

$$\mathrm{Sl}(2,\mathbb{R})^+ := \left\{ \begin{pmatrix} a & b \\ c & d \end{pmatrix} \in \mathrm{Sl}(2,\mathbb{R}): a,b,c,d \geq 0 \right\}.$$

It will turn out that its Lie wedge is given by

$$\mathfrak{sl}(2,\mathbb{R})^+ := \left\{ \begin{pmatrix} x & y \\ z & -x \end{pmatrix} \in \mathfrak{sl}(2,\mathbb{R}): y,z \geq 0 \right\}.$$

In fact, even more is true: The restriction of the exponential map to $\mathfrak{sl}(2,\mathbb{R})^+$ is a homeomorphism onto $\mathrm{Sl}(2,\mathbb{R})^+$. In order to show that, we need to take a closer look at the exponential function: For any X in $\mathfrak{sl}(2)$, we have $X^2 = k(X) \cdot 1$, whence all odd powers of X are scalar multiples of X. We define the power series

$$C(z) = 1 + \frac{z}{2!} + \frac{z^2}{4!} + \ldots \quad \text{and} \quad S(z) = 1 + \frac{z}{3!} + \frac{z^2}{5!} + \ldots$$

Then we have

$$\exp X = C\big(k(X)\big) \cdot 1 + S\big(k(X)\big) \cdot X.$$

Note that $C'(z) = \frac{1}{2} S(z)$,

$$C(x) = \begin{cases} \cosh\sqrt{x} & \text{for } 0 \leq x, \\ \cos\sqrt{-x} & \text{for } 0 > x, \end{cases} \qquad \sqrt{|x|}S(x) = \begin{cases} \sinh\sqrt{x} & \text{for } 0 \leq x, \\ \sin\sqrt{-x} & \text{for } 0 > x, \end{cases}$$

and

$$C(z)^2 - zS(z)^2 = 1$$

Thus $C'(x) > 0$ for $x > -\pi^2$ and hence $C: [-\pi^2, \infty[\to [-1, \infty[$ is invertible. We denote the inverse function by $\tilde{C}$ and use it to invert the exponential function on a suitable domain.

Lemma 2.2. *Consider the sets $\Omega = \{X \in \mathfrak{sl}(2, \mathbb{R}): k(X) > -\pi^2\}$ and $U = \{g \in \mathrm{Sl}(2, \mathbb{R}): \mathrm{tr}\, g > -2\}$ then $\exp: \Omega \to U$ is a diffeomorphism whose inverse is given by*

$$\mathrm{Log}(g) = \frac{1}{S(\widetilde{C}(\frac{\mathrm{tr}(g)}{2}))} \left(g - \frac{\mathrm{tr}(g)}{2} 1 \right) \quad \forall g \in U.$$

Proof. Let $g \in U$. Note first that $k(g - \frac{\mathrm{tr}(g)}{2} 1) = -\det(g) + (\frac{\mathrm{tr}(g)}{2})^2$, whence

$$k(\mathrm{Log}\, g) = S\left(\widetilde{C}(\frac{\mathrm{tr}(g)}{2})\right)^{-2} ((\frac{\mathrm{tr}(g)}{2})^2 - 1).$$

Then $C(z)^2 - zS(z)^2 = 1$ implies

$$-\pi^2 < \widetilde{C}(\frac{\mathrm{tr}(g)}{2}) = S\left(\widetilde{C}(\frac{\mathrm{tr}(g)}{2})\right)^{-2} ((\frac{\mathrm{tr}(g)}{2})^2 - 1) = k(\mathrm{Log}\, g),$$

i.e., $\mathrm{Log}\, g \in \Omega$. Now we have

$$\exp \mathrm{Log}\, g = C\big(k(\mathrm{Log}\, g)\big)1 + S\big(k(\mathrm{Log}\, g)\big) \mathrm{Log}\, g = \frac{\mathrm{tr}(g)}{2}1 + (g - \frac{\mathrm{tr}(g)}{2}1) = g.$$

Conversely,

$$\mathrm{tr}\big(C\big(k(X)\big)1 + S\big(k(X)\big)X\big) = 2C\big(k(X)\big)$$

so that $\exp$ maps Ω to U. Finally we have

$$\mathrm{Log}(\exp X) = \mathrm{Log}(C\big(k(X)\big)1 + S\big(k(X)\big)X)$$

$$= \frac{1}{S(k(X))}(C\big(k(X)\big)1 + S\big(k(X)\big)X - C\big(k(X)\big)1)$$

$$= X.$$

∎

Note that Lemma 2.2 implies that the exponential function for $\mathrm{PSl}(2, \mathbb{R}) = \mathrm{Sl}(2, \mathbb{R})/\{\pm 1\}$ is surjective and open since for $g \in \mathrm{Sl}(2, \mathbb{R})$ either $\mathrm{tr}(g)$ or $-\mathrm{tr}(g)$ is larger then -2.

Proposition 2.3. *The exponential map induces a homeomorphism*

$$\exp: \mathfrak{sl}(2, \mathbb{R})^+ \to \mathrm{Sl}(2, \mathbb{R})^+.$$

Proof. Obviously we have $\mathfrak{sl}(2, \mathbb{R})^+ \subseteq \Omega$ and $\mathrm{Sl}(2, \mathbb{R})^+ \subseteq U$. Thus it suffices to show that $\exp(\mathfrak{sl}(2, \mathbb{R})^+) \subseteq \mathrm{Sl}(2, \mathbb{R})^+$ and $\mathrm{Log}(\mathrm{Sl}(2, \mathbb{R})^+) \subseteq \mathfrak{sl}(2, \mathbb{R})^+$. Let $X = xX_0 + yX_+ + zX_-$ with $y, z \geq 0$. Then $k(X) = x^2 + yz$. If $k(X) = 0$, then X is either a nonnegative multiple of X_+ or of X_-, so that $\exp X \in \mathrm{Sl}(2, \mathbb{R})^+$. If $k(X) > 0$, then $S\big(k(X)\big)$ and $C\big(k(X)\big)$ are positive. In fact, we even have

$$C\big(k(X)\big) = \cosh \sqrt{k(X)} > \sinh \sqrt{k(X)}(\frac{x}{\sqrt{x^2 + yz}}) = S\big(k(X)\big)x$$

which shows that all components of $\exp X$ are positive.

For the second claim we let $g = \begin{pmatrix} a & b \\ c & d \end{pmatrix} \in \mathrm{Sl}(2,\mathbb{R})^+$, and note that

$$\frac{\mathrm{tr}(g)}{2} = \frac{a+d}{2} \geq \sqrt{ad} = \sqrt{1+cb} \geq 1,$$

whence

$$\tilde{C}(\frac{\mathrm{tr}(g)}{2}) = (\mathrm{arcosh}\frac{\mathrm{tr}(g)}{2})^2 \geq 0$$

and

$$S(\tilde{C}(\frac{\mathrm{tr}(g)}{2})) \geq 1.$$

Now the claim follows from the formula for Log in Lemma 2.2. ∎

The semigroup $\mathrm{Sl}(2,\mathbb{R})^+$ plays a special role in $\mathrm{Sl}(2,\mathbb{R})$ which will be clear from our subsequent study of its homomorphic image $\mathrm{PSl}(2,\mathbb{R})^+$ in $\mathrm{PSl}(2,\mathbb{R})$.

We denote the projection of $\mathrm{Sl}(2,\mathbb{R})$ onto $\mathrm{PSl}(2,\mathbb{R})$ by p and note that for any subsemigroup S of $\mathrm{Sl}(2,\mathbb{R})$ with non-empty interior $S_d := p^{-1}(p(S)) = \{\pm 1\}S$ is a subsemigroup which is proper in $\mathrm{Sl}(2,\mathbb{R})$ if and only if S is. Since the exponential function for $\mathrm{PSl}(2,\mathbb{R})$ is surjective and open, we find an open set $\Omega_S \subseteq \mathfrak{sl}(2,\mathbb{R})$ such that $\exp \Omega_S \subseteq S_d$. We may even assume that $\exp \Omega_S \subseteq S$ if we replace Ω_S by $2\Omega_S$. If Ω_S contains an element of negative k-length, then $\mathrm{int}\, S \cap K \neq \emptyset$ and hence $S = \mathrm{Sl}(2,\mathbb{R})$. Thus Ω_S can always be chosen in such a way that it contains an element of positive k-length, i.e., we find a $g \in \mathrm{Sl}(2,\mathbb{R})$ with

$$g(\mathrm{int}\, S)g^{-1} \cap A \neq \emptyset$$

because for every element X in Ω_s the semigroup $\exp \mathbb{R}^+ X$ is compact (cf. Corollary 1.21). Note that $(g(\mathrm{int}\, S)g^{-1}N) \cap A$ is an open subsemigroup in A. In fact, if $a_i = s_i n_i$ for $i = 1,2$, $a_i \in A$, $n_i \in N$ and $s_i \in g(\mathrm{int}\, S)g^{-1}$, then

$$a_1 a_2 = s_1 n_1 a_2 = s_1 a_2 (a_2^{-1} n_1 a_2) = s_1 s_2 (n_2 a_2^{-1} n_1 a_2).$$

Next we consider the action of $\mathrm{Sl}(2,\mathbb{R})$ respectively $\mathrm{PSl}(2,\mathbb{R})$ on the projective space $\mathbb{P}^1(\mathbb{R})$ given by the fractional linear transformations

$$\begin{pmatrix} a & b \\ c & d \end{pmatrix}.(r:s) = (ar+bs : cr+ds).$$

For each semigroup S in $\mathrm{Sl}(2,\mathbb{R})$ or $\mathrm{PSl}(2,\mathbb{R})$ with interior points we set

$$C_S = \bigcap_{m \in \mathbb{P}^1(\mathbb{R})} \overline{S.m}.$$

We choose g as above and set $S_g := gSg^{-1}$. Then we know that

$$\begin{pmatrix} t & 0 \\ 0 & t^{-1} \end{pmatrix} \in \mathrm{int}\, S_g$$

either for large t's or for small t's. Suppose the first possibility - the other can be dealt with analogously. Then we have for $r \neq 0$

$$\lim_{t \to \infty} \begin{pmatrix} t & 0 \\ 0 & t^{-1} \end{pmatrix} . (r:1) = (1:0) = \lim_{t \to \infty} \begin{pmatrix} t & 0 \\ 0 & t^{-1} \end{pmatrix} . (1:0).$$

Since S is open, $S.(0:1)$ contains an element $(r:1)$ with $r \neq 0$. This shows that $(1:0) \in C_{S_g}$, so that $C_S \neq \emptyset$ for any subsemigroup S of $\mathrm{Sl}(2, \mathbb{R})$ with non-empty interior.

We note for later use that the definition implies that

$$S \subseteq \mathrm{compr}(C_S),$$

where

$$\mathrm{compr}(C) = \{g \in \mathrm{Sl}(2, \mathbb{R}) : g.C \subseteq C\}$$

for any subset C of $\mathbb{P}^1(\mathbb{R})$.

Lemma 2.4. *Let S be a proper subsemigroup of $\mathrm{Sl}(2, \mathbb{R})$ with interior points. Then C_S is strictly contained in $\mathbb{P}^1(\mathbb{R})$.*

Proof. $C_S = \mathbb{P}^1(\mathbb{R})$ implies $(\mathrm{int}\, S)^{-1}.m \cap S.m \neq \emptyset$ for all $m \in \mathbb{P}^1(\mathbb{R})$. Therefore there exists a $g_m \in \mathrm{int}\, S$ such that $g_m.m = m$ and hence $\mathrm{int}\, S.m$ is an open neighborhood of m. But then $(\mathrm{int}\, S)(S.m)$ is an open neighborhood of $\overline{S.m} = \mathbb{P}^1(\mathbb{R})$ and thus $(\mathrm{int}\, S).m = \mathbb{P}^1(\mathbb{R})$ for each m. Now we choose a $g \in \mathrm{Sl}(2, \mathbb{R})$ such that $\mathrm{int}\, S_g \cap A \neq \emptyset$. We claim that

$$(2.2) \qquad\qquad A \subseteq ((\mathrm{int}\, S_g)N) \cap A.$$

To see this set $m_0 = (1:0)$ and $m_1 = (0:1)$. Then we can choose $g_0, g_1 \in \mathrm{int}\, S_g$ such that $g_1.m_1 = m_0$ and $g_0.m_0 = m_1$. We have

$$g_0 = \begin{pmatrix} 0 & b_0 \\ c_0 & d_0 \end{pmatrix}, \quad g_1 = \begin{pmatrix} a_1 & b_1 \\ c_1 & 0 \end{pmatrix}.$$

Recall that $h_t := \begin{pmatrix} t & 0 \\ 0 & t^{-1} \end{pmatrix} \in \mathrm{int}\, S_g$ for large (or for small) t. Thus

$$g_1 h_t g_0 = \begin{pmatrix} t^{-1} b_1 c_0 & t a_1 b_0 + t^{-1} d_0 b_1 \\ 0 & t b_0 c_1 \end{pmatrix} \in \mathrm{int}\, S_g,$$

whence

$$\begin{pmatrix} t^{-1} b_1 c_0 & 0 \\ 0 & t(b_1 c_0)^{-1} \end{pmatrix} \in (g_1 h_t g_0 N) \cap A$$

for large (small) t. Taking the square, we find elements

$$\begin{pmatrix} t^{-2} r & 0 \\ 0 & t^2 r^{-1} \end{pmatrix} \in (g_1 h_t g_0 N) \cap A$$

with positive r. But now we have shown that the open subsemigroup

$$((\mathrm{int}\, S_g)N) \cap A$$

contains elements of the form $\begin{pmatrix} a & 0 \\ 0 & a^{-1} \end{pmatrix}$ with $a < 1$ as well as with $a > 1$, and that proves (2.2).

Now we find $s_i \in \mathrm{int}\, S_g$ and $n_i \in N$ for $i = 1, 2$ with

$$s_1 n_1 = (s_2 n_2)^{-1} =: a_2^{-1} \in A$$

so that

$$s_1 s_2 = a_2^{-1} n_1^{-1} a_2 n_2^{-1} \in N \cap \mathrm{int}\, S_g.$$

Therefore $\exp^{-1}(\mathrm{int}\, S_g)$ contains elements of negative k-length which implies $S_g = \mathrm{Sl}(2,\mathbb{R})$ contradicting our hypothesis on S. ∎

It is an elementary calculation to check that

$$C_{\mathrm{Sl}(2,\mathbb{R})^+} = \{(r:1): r \geq 0\} \cup \{(1:0)\}.$$

We denote this set by C.

Theorem 2.5. *Let S be a proper connected subsemigroup of $\mathrm{PSl}(2,\mathbb{R})$ with non-empty interior. Then the following statements are equivalent*

(1) *S is a maximal connected proper subsemigroup of $\mathrm{PSl}(2,\mathbb{R})$.*

(2) *S is conjugate to $\mathrm{PSl}(2,\mathbb{R})^+$.*

Proof. (1) $\Rightarrow$ (2): Note first that for connected S the set C_S must also be connected since it is the closure of (any) S-orbit in C_S. Since S is proper, Lemma 2.4 tells us that C_S is an "interval" in $\mathbb{P}^1(\mathbb{R})$. We have

$$C_{gS g^{-1}} = g.C_S$$

for any $g \in \mathrm{PSl}(2,\mathbb{R})$, so that conjugation with an element from $p(K)$ allows us to assume that m_1 is one endpoint of C_S (cf. Figure 2.3).

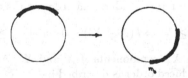

Figure 2.3

Next we apply the transformation

$$\begin{pmatrix} 1 & 0 \\ c & 1 \end{pmatrix}.(1:s) = (1:c+s)$$

which, for suitable c, maps our interval to the set C. Thus we may assume that $C_S = C$. But then $\mathrm{compr}(C)$ is a semigroup containing S and the maximality of S shows that $S = \mathrm{compr}(C)$, provided we can show that $\mathrm{compr}(C)$ is connected. We can do even better and show that $\mathrm{compr}(C) = \mathrm{PSl}(2,\mathbb{R})^+$ which then finishes the first part of the proof: The inclusion $\mathrm{PSl}(2,\mathbb{R})^+ \subseteq \mathrm{compr}(C)$ is clear in view of $C = C_{\mathrm{PSl}(2,\mathbb{R})^+}$. Conversely, if $\begin{pmatrix} a & b \\ c & d \end{pmatrix}$ satisfies $g.C \subseteq C$, then

$$(ar + b)(cr + d) \geq 0 \quad \forall r \geq 0$$

and elementary calculus shows that for $ac > 0$ this parabola takes its minimum value $-(4ac)^{-1}$ at $r = -(ad+bc)(2ac)^{-1}$. Thus, in that case, we have $2ad - 1 = ad + bc > 0$ and hence $ac, ad, bd \geq 0$ which shows that a, b, c and d all have the same sign. In the case $ac = 0$ we have $ad + bc \geq 0$ right away and we have shown that $p(g) \in \mathrm{PSl}(2,\mathbb{R})^+$.

(2) $\Rightarrow$ (1): If $\mathrm{compr}(C) = \mathrm{PSl}(2,\mathbb{R})^+$ is strictly contained in S, then the definition of C_S shows that C is contained in C_S. This containment is even strict since otherwise $S \subseteq \mathrm{compr}(C_S) = \mathrm{compr}(C)$. Thus there exists an $r < 0$ with $(r : 1) \in C_S$, whence $(a^2 r : 1) \in C_S$ for all positive a - let $A \subseteq \mathrm{compr}(C) \subseteq S$ act - which finally implies $C_S = \mathbb{P}^1(\mathbb{R})$ which means $S = G$ by Lemma 2.4.   ∎

Note that in the second part of the proof it was not necessary to assume that S was connected. Thus we have

Corollary 2.6.   *Any maximal connected proper subsemigroup of $\mathrm{PSl}(2,\mathbb{R})$ is a maximal subsemigroup.*   ∎

Corollary 2.7. $\{\pm 1\}\, \mathrm{Sl}(2,\mathbb{R})^+$ *is a maximal subsemigroup of $\mathrm{Sl}(2,\mathbb{R})$.*   ∎

2.3. The hyperboloid and its order structure

The semigroup $\mathrm{Sl}(2,\mathbb{R})^+$ and its conjugates occur in various contexts. For instance it is of importance in the harmonic analysis on the hyperboloid

$$\mathbf{X} = \{X \in \mathfrak{sl}(2,\mathbb{R}) : k(X) = 1\}.$$

Note that the Lorentzian form k induces an order on the space $\mathfrak{sl}(2,\mathbb{R})$ via

$$X \geq X' \Leftrightarrow k(X - X') \leq 0 \quad \text{and} \quad y \geq y',$$

where y and y' are the X_+-components of X and X'. An elementary calculation shows that in $\mathbf{X}$ the induced order is described by

$$X \geq X' \Leftrightarrow k(X, X') \geq 1 \quad \text{and} \quad y \geq y'.$$

Let $\mathrm{Sl}(2,\mathbb{R})$ act on $\mathbf{X}$ via conjugation (i.e., the adjoint action) and fix a base point

$$x_0 = \begin{pmatrix} 0 & 1 \\ 1 & 0 \end{pmatrix} \in \mathbf{X}.$$

Further let (cf. Figure 2.4)

$$X^+ = \{x \in X : x \geq x_0\}.$$

Figure 2.4

Then $H \cong SO(1,1)$ is the stabilizer of x_0 and we have $X \cong Sl(2, \mathbb{R})/H$. This space carries the structure of a pseudo Riemannian symmetric space via the involution

$$\tau : Sl(2, \mathbb{R}) \to Sl(2, \mathbb{R}), \qquad \begin{pmatrix} a & b \\ c & d \end{pmatrix} \mapsto \begin{pmatrix} d & c \\ b & a \end{pmatrix}.$$

In fact, τ is just conjugation by

$$w = \begin{pmatrix} 0 & 1 \\ 1 & 0 \end{pmatrix}$$

and H is the group of fixed points $Sl(2, \mathbb{R})^\tau$.

If $\pi : Sl(2, \mathbb{R}) \to X$ is the quotient map then $S = \pi^{-1}(X^+)$ is a semigroup containing H. In fact, $g, g' \in S$ implies

$$\pi(gg') = g\,\pi(g') \geq ge = \pi(g) \geq x_0.$$

A simple calculation in $Sl(2, \mathbb{R})$ shows that

$$\text{int } S = HA^+H.$$

Moreover, explicit matrix calculation shows that this decomposition is direct. At this point we note that the Lie wedge $L(S)$ contains $\mathbb{R}(X_+ + X_-)$ and $\mathbb{R}^+X_0$. Letting $\exp \mathbb{R}(X_+ + X_-)$ act on X_0 we see that it even contains $\mathbb{R}^+(\pm(X_+ - X_-)+X_0)$ so that Theorem 2.5 implies that $L(S)$ is conjugate to $sl(2, \mathbb{R})^+$ (cf. Figure 2.5). We can even determine the element with which we have to conjugate: it is the rotation by $\frac{\pi}{2}$ around the axis $\mathbb{R}(X_+ - X_-)$ given by the element

$$k_o := \exp\left(\frac{\pi}{4}(X_+ - X_-)\right) = \frac{1}{\sqrt{2}} \begin{pmatrix} 1 & -1 \\ 1 & 1 \end{pmatrix} \in K.$$

Thus we have

$$(2.3) \qquad\qquad S_{k_o} = k_o S k_o^{-1} = \{\pm 1\}\, Sl(2, \mathbb{R})^+.$$

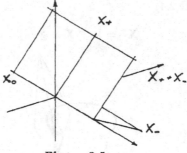

Figure 2.5

We have an analogue of the Iwasawa decomposition with H instead of K:

$$H \times A \times N \to HAN \subseteq \mathrm{Sl}(2, \mathbb{R})$$

$$\left(\begin{pmatrix} c & s \\ s & c \end{pmatrix}, \begin{pmatrix} t & 0 \\ 0 & t^{-1} \end{pmatrix}, \begin{pmatrix} 1 & r \\ 0 & 1 \end{pmatrix} \right) \mapsto \begin{pmatrix} ct & ctr + st^{-1} \\ st & str + ct^{-1} \end{pmatrix}$$

is a diffeomorphism onto an open subset of $\mathrm{Sl}(2, \mathbb{R})$. This decomposition also defines an "Iwasawa decomposition" of the semigroup S since it turns out that $S \subseteq HAN$. In order to show this, we consider

$$X^- = S^{-1}/H$$

and

$$NAH/H \subseteq X.$$

The N-orbits on X, described by

$$\begin{pmatrix} 1 & r \\ 0 & 1 \end{pmatrix} \begin{pmatrix} x & y \\ z & -x \end{pmatrix} \begin{pmatrix} 1 & -r \\ 0 & 1 \end{pmatrix} = \begin{pmatrix} x + rz & -2xr - zr^2 + y \\ z & -zr - x \end{pmatrix}$$

simply are the parabolas which occur as intersections of X with the planes parallel to $\mathfrak{a} + \mathfrak{n}$ (cf. Figure 2.6).

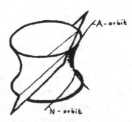

Figure 2.6

The two sheeted covering $\widetilde{X} := \mathrm{Sl}(2, \mathbb{R}) / \mathrm{SO}(1, 1)_0$ is also an interesting symmetric space. Since $\mathrm{SO}(1, 1)_0$ coincides with the group $\mathrm{SAut}(C_0)$ of automorphisms of the cone $C_0 := \{ (x_1, x_2) : x_1 \geq |x_2| \}$ in $\mathbb{R}^2$, the space $\widetilde{X}$ may be identified with the set of all pointed generating cones in $\mathbb{R}^2$. The corresponding order relation simply is the opposite order for set inclusion, i.e.,

$$C \leq C' \quad \Leftrightarrow \quad C' \subseteq C.$$

Note also that

$$k_o^{-1}\,\mathrm{Sl}(2,\mathbb{R})^+ k_o = \mathrm{SEnd}(C) = \{g \in \mathrm{Sl}(2,\mathbb{R}) : g.C_0 \subseteq C_0\}$$

is the semigroup of special endomorphisms of the cone $C_0 \subseteq \mathbb{R}^2$ (cf. Section 1.3).

2.4. The Olshanskiĭ semigroup in $\mathrm{Sl}(2,\mathbb{C})$

Consider the action of $\mathrm{Sl}(2,\mathbb{C})$ on $\mathbb{P}^1(\mathbb{C})$ via fractional linear transformations and the disk $D := \{z \in \mathbb{C} : |z| < 1\}$ which we view as a subset of $\mathbb{P}^1(\mathbb{C})$ via $z \mapsto (z : 1)$. Then the *Olshanskiĭ semigroup* is the semigroup of compressions

$$\mathrm{compr}(D) := \{g \in \mathrm{Sl}(2,\mathbb{C}) : g.D \subseteq D\}.$$

Note that the group

$$\mathrm{SU}(1,1) = \left\{ \begin{pmatrix} a & b \\ \bar{b} & \bar{a} \end{pmatrix} : |a|^2 - |b|^2 = 1 \right\}$$

acts transitively on D with stabilizer

$$K^c := \left\{ \begin{pmatrix} a & 0 \\ 0 & \bar{a} \end{pmatrix} : |a| = 1 \right\} = \{g \in \mathrm{SU}(1,1) : g.(0 : 1) = (0 : 1)\}.$$

The Lie algebra of $\mathrm{SU}(1,1)$ is given by

$$\mathfrak{su}(1,1) = \left\{ \begin{pmatrix} ix & w \\ \bar{w} & -ix \end{pmatrix} : w \in \mathbb{C}, x \in \mathbb{R} \right\} = \mathbb{R}(X_+ + X_-) + i\mathbb{R}(X_+ - X_-) + i\mathbb{R}X_0,$$

where $w \mapsto \bar{w}$ denotes complex conjugation. Conjugation by the element

$$c := \frac{1}{\sqrt{2}} \begin{pmatrix} 1 & i \\ i & 1 \end{pmatrix} \in \mathrm{Sl}(2,\mathbb{C})$$

yields

$$cSU(1,1)c^{-1} = \mathrm{Sl}(2,\mathbb{R})$$

and

$$c.D = \mathcal{H} := \{(z : 1) \in \mathbb{P}^1(\mathbb{C}) : \mathrm{Re}\, z > 0\} \cup \{(1 : 0)\}.$$

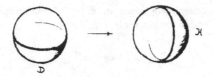

Figure 2.7

Therefore $\mathrm{Sl}(2, \mathbb{R}) \subseteq S_{\mathcal{H}} = c^{-1} \operatorname{compr}(D)c$. Now suppose that $X \in \mathfrak{sl}(2, \mathbb{R})$ with $X \geq 0$ (cf. Section 2.3). Then X is conjugate under an element of $\mathrm{Sl}(2, \mathbb{R})$ either to $r(X_+ - X_-)$ or to rX_+ with $r \geq 0$. But

$$\exp ir(X_+ - X_-) = \begin{pmatrix} \cosh r & i \sinh r \\ -i \sinh r & \cosh r \end{pmatrix} \in S_{\mathcal{H}}$$

as is most easily seen by calculating with D using

$$c \exp(rX_0)c^{-1} = \exp\left(- ir(X_+ - X_-) \right).$$

Moreover, it is easy to see that $\exp(irX_+) \in S_{\mathcal{H}}$. Thus, if we set

$$W := \{X \in \mathfrak{sl}(2, \mathbb{R}) : X \geq 0\},$$

then $\mathrm{Sl}(2, \mathbb{R}) \exp(iW) \subseteq S_{\mathcal{H}}$. In fact we even have equality:

Proposition 2.8. $S_{\mathcal{H}} = \mathrm{Sl}(2, \mathbb{R}) \exp(iW)$.

Proof.   Let $g \in S_{\mathcal{H}}$, then $c^{-1}gc \in \operatorname{compr}(D)$ maps D onto some disk $D_g \subseteq D$ which can be moved via an element $c^{-1}g_1c \in SU(1, 1)$ to be concentric with D

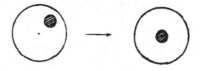

Figure 2.8

But then $D_g = \exp(-rX_0).D$ with $r \geq 0$ so that $\exp(rX_0)c^{-1}g_1gc.D = D$. This shows $\exp\left(- ir(X_+ - X_-) \right)g_1g \in \mathrm{Sl}(2, \mathbb{R})$, and the invariance of W under the action of $\mathrm{Sl}(2, \mathbb{R})$ then implies $g \in \mathrm{Sl}(2, \mathbb{R}) \exp(iW)$. ∎

We will see in Section 8.6 in a more general setting how to show that the Olshanskiĭ semigroup is a maximal semigroup in $\mathrm{Sl}(2,\mathbb{C})$).

There is an interesting connection between the Olshanskiĭ semigroup and the semigroup S constructed from the ordering of the hyperboloid $\mathbf{X}$:

Both $\mathrm{Sl}(2,\mathbb{R})$ and $\mathrm{SU}(1,1)$ have the same complexification $\mathrm{Sl}(2,\mathbb{C})$. Note that we can identify $\mathbb{P}^1(\mathbb{C})$ with the homogeneous space

$$\mathrm{Sl}(2,\mathbb{C})/K_{\mathbb{C}}^c(P_{\mathbb{C}}^c)^-,$$

where

$$K_{\mathbb{C}}^c = \left\{ \begin{pmatrix} a & 0 \\ 0 & a^{-1} \end{pmatrix} : a \in \mathbb{C}^\times \right\}$$

and

$$(P_{\mathbb{C}}^c)^- = \left\{ \begin{pmatrix} 1 & 0 \\ c & 1 \end{pmatrix} : c \in \mathbb{C} \right\}$$

and hence

$$K_{\mathbb{C}}^c(P_{\mathbb{C}}^c)^- = \{g \in \mathrm{Sl}(2,\mathbb{C}): g.(0:1) = (0:1)\}.$$

We extend τ to $\mathrm{Sl}(2,\mathbb{C})$ by $\tau(g) = \overline{wgw^{-1}}$, where $g \mapsto \overline{g}$ again denotes the complex conjugation. Then τ induces a complex conjugation on D and $\mathbb{P}^1(\mathbb{C})$. Restricting the embedding $D \to \mathbb{P}^1(\mathbb{C})$ to the real points yields

$$H/\{\pm 1\} \cong D_\tau := \{r \in \mathbb{R}: |r| < 1\} \to \mathbb{P}^1(\mathbb{R}) \cong \mathrm{Sl}(2,\mathbb{R})/(\pm A\overline{N}),$$

where $\overline{N} = \tau N$.

From (2.3) above we know that $\tau(S) = wSw^{-1} = \{\pm 1\}\,\mathrm{Sl}(2,\mathbb{R})^+_{wk_o^{-1}}$. But

$$wk_o^{-1}.C = D_\tau$$

and hence

$$S = \tau(S)^{-1} = \mathrm{compr}(D_\tau)^{-1}.$$

Now we see that

$$S^{-1} = \mathrm{compr}(D) \cap \mathrm{Sl}(2,\mathbb{R}).$$

Note that the above yields another proof of $S \subseteq HAN$. In fact, we have $S^{-1} \subseteq HA\overline{N}$ and then

$$S \subseteq HAN$$

because of $\tau(S) = S^{-1}$ and $\tau(HA\overline{N}) = HAN$.

2.5. Affine compression semigroups

Let G be a Lie group and $\mathrm{Aff}(G) = G \rtimes \mathrm{End}\,G$, where $\mathrm{End}\,G$ is the semigroup of group endomorphisms of G. Then $\mathrm{Aff}(G)$ is a semigroup with respect to the multiplication

$$(g,\gamma)(g',\gamma') = (g\gamma(g'),\gamma \circ \gamma').$$

Now let Ω be a compact subset of G and set

$$\mathrm{compr}(\Omega) := \{s \in \mathrm{Aff}(G): s.\Omega \subseteq \Omega\}.$$

Here we use

$$(g, \gamma).x := g\gamma(x).$$

The relevance of semigroups of type $\mathrm{compr}(\Omega)$ comes from the fact that under certain circumstances they yield a compactification of the semigroup $\mathrm{compr}(\Omega) \cap (G \rtimes H)$, where H is some subgroup of $\mathrm{Aut}\, G$. In particular in the study of ordered symmetric spaces this is useful as it will turn out that in the notation of the previous sections we have
(2.4)
$$\mathrm{compr}(D_\tau) = H\big(\mathrm{compr}(D_\tau) \cap NA\big), \quad \text{and} \quad \mathrm{compr}(D_\tau) \cap NA = \mathrm{compr}(\Omega) \cap NA$$

for a suitable compact subset Ω of N. Here we view the group A as a subgroup of $\mathrm{Aut}\, N$.

Proposition 2.9.   *Suppose that Ω is a compact neighborhood of 1 in G for which the set $\exp^{-1}(\Omega\Omega^{-1})$ is bounded. Then $\mathrm{compr}(\Omega)$ is compact. Moreover, if $s, s' \in \mathrm{compr}(\Omega) \cap (G \rtimes \mathrm{Aut}(G))$ and $s'.\Omega \subseteq s.\Omega$, then the interval*

$$[s, s'] = \{t \in \mathrm{compr}(\Omega) : s'.\Omega \subseteq t.\Omega \subseteq s.\Omega\}$$

is compact.

Proof.   We have $\gamma(\Omega) \subseteq g^{-1}\Omega \subseteq \Omega^{-1}\Omega$ for any $s = (g, \gamma) \in \mathrm{compr}(\Omega)$ since $s.1 \in \Omega$ for all such s. Now $\gamma \circ \exp = \exp \circ d\gamma(1)$ shows that $d\gamma(1)$ maps the bounded neighborhood $\exp^{-1}(\Omega)$ of $0 \in \mathbf{L}(G)$ into the bounded set $\exp^{-1}(\Omega^{-1}\Omega)$. Hence the set $F := \{d\gamma(1): (g, \gamma) \in \mathrm{compr}(\Omega)\}$ is bounded in $\mathrm{End}\,\mathbf{L}(G)$. Thus

$$\mathrm{compr}(\Omega) \subseteq \Omega \times \{\gamma \in \mathrm{End}(G): d\gamma(1) \in F\}$$

is compact.

The second accertion follows from the fact that $[s, s']$ is a closed subset of the compact semigroup $\mathrm{compr}(\Omega)$ (cf. [HiNe92, I.5]).   ∎

To have a concrete example, let $G = \mathbb{R}^n$ and $H = \mathbb{R}$ with $t(x) = e^{\lambda t}U(t)(x)$, where λ is a fixed negative number and $t \mapsto U(t) \in \mathrm{SO}(n)$ a one parameter group. If now Ω is the closed unit ball in $\mathbb{R}^n$ we have

$$S := \mathrm{compr}(\Omega) \cap (G \rtimes H) = \{(x, t): \|x\| \leq 1 - e^{\lambda t}\}.$$

Using the invariance theorem for vector fields (cf. Theorem 5.8) we find

$$\mathbf{L}(S) = \{(x, t): \|x\| \leq \lambda t\}.$$

If $n = 1$, $\lambda = -1$, and $U(t)$ is trivial, we call the semigroup S the *extended affine triangle*. This is the semigroup alluded to above in the context of ordered symmetric spaces. In fact, we may view D_τ as a domain in N via

$$\begin{pmatrix} 1 & r \\ 0 & 1 \end{pmatrix} \quad \leftrightarrow \quad r$$

whose closure is the compact set Ω we are looking for. It is easy to check that

$$\mathrm{compr}(\Omega) \cap (NA) = \left\{ \begin{pmatrix} a & r \\ 0 & a^{-1} \end{pmatrix} : |r| \leq 1 - a \right\}$$

which is seen to be isomorphic to the extended affine triangle if we write $a = e^{-t}$. Since D_r is the interior of its closure, we may replace D_r by its closure in the definition of $\mathrm{compr}(D_r)$, hence the second equality in (2.4) follows. The first equality follows trivially from the inclusion $\mathrm{compr}(D_r) \subseteq HAN$.

2.6. The euclidean compression and contraction semigroups

Let V be a finite dimensional real vector space endowed with a non-degenerate symmetric bilinear form J. There are four semigroups which are canonically related to such a form. Let

$$V_+ := \{v \in V : J(v,v) \geq 0\} \quad \text{and} \quad V_- := \{v \in V : J(v,v) \leq 0\},$$

and Ω_+ and Ω_- the corresponding subsets of the projective space $\mathbb{P}(V) = \{[v] : v \in V \setminus \{0\}\}$, where $[v] = \mathbb{R}^* v$.

The first semigroup is

$$\mathrm{compr}(\Omega_+) := \{s \in \mathrm{Gl}(V) : s.\Omega_+ \subseteq \Omega_+\},$$

the compression semigroup of Ω_+, the second one is

$$\mathrm{compr}(\Omega_-) := \{s \in \mathrm{Gl}(V) : s.\Omega_- \subseteq \Omega_-\},$$

the compression semigroup of Ω_-, the third one is the *contraction semigroup*

$$\mathcal{C}(J) := \{s \in \mathrm{Gl}(V) : (\forall v \in V) J(s.v, s.v) \leq J(v,v)\},$$

and the fourth one is the *expansion semigroup*

$$\mathcal{E}(J) := \{s \in \mathrm{Gl}(V) : (\forall v \in V) J(s.v, s.v) \geq J(v,v)\}.$$

Note that we have the following inclusions

$$\mathcal{E}(J) \subseteq \mathrm{compr}(\Omega_+) \quad \text{and} \quad \mathcal{C}(J) \subseteq \mathrm{compr}(\Omega_-).$$

The same definitions also apply when V is a complex vector space endowed with a non-degenerate hermitean form.

We have already seen in Section 1.4 that the component of $\mathbf{1}$ in the compression semigroup $\mathrm{compr}(\Omega_+)$ associated with a Lorentzian form J is a Lie semigroup, namely the semigroup of all invertible endomorphisms of a Lorentzian cone. To see that in this case, the expansion semigroup may be strictly smaller than the compression semigroup $\text{compr}(\Omega_+)$, let $V = \mathbb{R}^2$, $J(x,x) = x_1^2 - x_2^2$, and $A := \begin{pmatrix} 3 & 0 \\ 0 & 2 \end{pmatrix}$. Then $x_1^2 - x_2^2 \geq 0$ implies that

$$J(A.x, A.x) = 3x_1^2 - 2x_2^2 \geq 2(x_1^2 - x_2^2) \geq 0,$$

but

$$J(A.(0,1), A.(0,1)) = -2 < -1 = J((0,1),(0,1)).$$

It is also interesting to note that, for Lorentzian forms, the closed convex set $B := \{x \in V : J(x,x) \geq 1\}$ has somehow the meaning of a "unit ball", and that even every element in the **1**-component of $\text{compr}(\Omega_+)$ maps B into itself (Proposition 1.13).

In this section we consider the positive definite case. Then the compression semigroups coincide with the whole group $\text{Gl}(V)$. Let $B := \{v \in V : J(v,v) \leq 1\}$ denote the unit ball in V. Then it is standard linear algebra that

$$\text{compr}_{\text{lin}}(B) := \{g \in \text{Gl}(V) : g.B \subseteq B\} = \mathcal{C}(J).$$

We denote the intersections of these semigroup with $\text{Gl}(V)^+$ by the superscript $^+$. In Section 1.4 we have shown that the semigroup $\text{compr}_{\text{aff}}(B)$ of affine compression of the ball B is a Lie semigroup. Let us consider the homomorphism

$$\alpha : \text{Aff}(V) \cong V \rtimes \text{Gl}(V) \to \text{Gl}(V), \qquad (v,g) \mapsto g.$$

If $(v,g) \in \text{compr}_{\text{aff}}(B)$, then $v + g.B \subseteq B$ yields that the ellipsoid $g.B$ must be contained in B since its diameters are smaller than 2. We conclude that

$$\alpha\big(\text{compr}_{\text{aff}}(B)^+\big) = \text{compr}_{\text{lin}}(B)^+,$$

so that $\mathcal{C}(J)^+$ is a Lie semigroup by Proposition 1.16.

Anyway, in this case there exists a more direct proof for the fact that $\mathcal{C}(J)^+$ is a Lie semigroup. Let $g \in \mathcal{C}(J)^+$ and $g = up$ its polar decomposition, i.e., $u \in \text{SO}(J)$ is orthogonal and $p = p^*$ is symmetric. Then $p \in \mathcal{C}(J)$ has all its eigenvalues less or equal to 1. Whence there exists a one-parameter semigroup e^{tX} in $\mathcal{C}(J)$ with $e^X = p$. Since $\text{SO}(J)$ is connected, we conclude that $\mathcal{C}(J)^+$ is a Lie semigroup, and that

$$\mathcal{C}(J)^+ = \text{SO}(J)\exp(C),$$

where C is the cone of all negative semidefinite symmetric matrices. From this it is clear that the semigroup

$$\mathcal{C}(J) = \text{O}(J)\exp(C),$$

has two components, namely those containing the two components of the group of units $\text{O}(J)$.

2.7. Gödel's cosmological model and the universal covering of $\mathrm{Sl}(2,\mathbb{R})$

Gödel's cosmological model is a low dimensional Lorentzian manifold which is diffeomorphic to $\mathbb{R}^4$ and admits closed timelike curves. This means that there are differentiable closed paths whose derivative has strictly negative square length with respect to the (Lorentzian) metric. There are several possibilities to describe this model as a Lie group with left invariant Lorentz metric. One of these is to use the group $G \times \mathbb{R}$ where G is the universal covering group of $\mathrm{Sl}(2,\mathbb{R})$. All essential features of the model are encoded in G so we will restrict ourselves to a description of G as a three dimensional Lorentzian manifold.

Note that the determinant is a quadratic form on the space $M(2,\mathbb{R})$ of all 2×2-matrices which clearly is left and right invariant under the action (matrix multiplication) of $\mathrm{Sl}(2,\mathbb{R})$. Thus the determinant defines an $\mathrm{Sl}(2,\mathbb{R})$-biinvariant metric β_1 on $M(2,\mathbb{R})$ via

$$\beta_1(X,X) = -\det(X)$$

and polarisation. On the tangent space $\mathfrak{sl}(2,\mathbb{R})$ of $\mathrm{Sl}(2,\mathbb{R})$ in **1** we have $\det(X) = -k(X,X)$ so that it is Lorentzian there. Now the invariance shows that the restriction of β_1 to the respective tangent spaces of $\mathrm{Sl}(2,\mathbb{R})$ yields a biinvariant Lorentz metric on $\mathrm{Sl}(2,\mathbb{R})$. Using the basis $\{X_+ - X_-, X_0, X_+ + X_-\}$, the matrix of β_1 in **1** is

$$\begin{pmatrix} -1 & 0 & 0 \\ 0 & 1 & 0 \\ 0 & 0 & 1 \end{pmatrix}.$$

This is not yet Gödel's metric and we will see below that it does not admit closed timelike curves. We define a family of left invariant metrics β_ρ on $\mathrm{Sl}(2,\mathbb{R})$ with $\rho > 0$ by saying that in **1** the matrix of β_ρ with respect to the above basis is

$$\begin{pmatrix} -\rho & 0 & 0 \\ 0 & 1 & 0 \\ 0 & 0 & 1 \end{pmatrix}.$$

The metric β_2 is the one Gödel used.

Figure 2.9

Proposition 2.10.   *The curve* $s \colon \mathbb{R} \to \mathrm{Sl}(2, \mathbb{R})$ *given by*

$$s(\varphi) = \begin{pmatrix} \cos\varphi & \sin\varphi \\ -\sin\varphi & \cos\varphi \end{pmatrix} \begin{pmatrix} r & 0 \\ 0 & \frac{1}{r} \end{pmatrix} \begin{pmatrix} \cos\varphi & -\sin\varphi \\ \sin\varphi & \cos\varphi \end{pmatrix}$$

is timelike for the metric β_ρ, $\rho > 1$ *and* r *sufficiently large.*

Proof.   In order to prove this, we note that because of the invariance the square length of $\frac{ds}{d\varphi}$ at $s(\varphi)$ is the same as that of $s(\varphi)^{-1} \frac{ds}{d\varphi}$ calculated with respect to the metric on $\mathfrak{sl}(2, \mathbb{R})$. A simple calculation shows

$$\frac{ds}{d\varphi} = (\frac{1}{r} - r) \begin{pmatrix} 2\sin\varphi\cos\varphi & \cos^2\varphi - \sin^2\varphi \\ \cos^2 - \sin^2\varphi & -2\sin\varphi\cos\varphi \end{pmatrix},$$

and hence

$$s(\varphi)^{-1} \frac{ds}{d\varphi}$$
$$= (\frac{1}{r} - r)[(\frac{1}{r} + r)(\sin\varphi\cos\varphi)X_0 + \frac{\cos^2\varphi - \sin^2\varphi}{2}(X_+ + X_-)$$
$$+ \frac{1}{2}(\frac{1}{r} - r)(X_+ - X_-)],$$

so that the square length is

$$\frac{1}{4}(\frac{1}{r} - r)^2 [(\frac{1}{r} + r)^2 - \rho(\frac{1}{r} - r)^2] = \frac{1}{4}(\frac{1}{r} - r)^2 [2(1 + \rho) - (\rho - 1)(r^2 + \frac{1}{r^2})]$$

which is negative for large r.   ∎

   So far we have constructed our metrics on $\mathrm{Sl}(2, \mathbb{R})$ and not on G, but it is clear that we can pull back the metrics and obtain a family of left invariant Lorentzian metrics $\tilde{\beta}_\rho$ on G. The closed timelike curve from Proposition 2.10 also yields such curves in G. To see this, let P denote the set of positive definite symmetric elements of $\mathrm{Sl}(2, \mathbb{R})$ and recall that any element g of $\mathrm{Sl}(2, \mathbb{R})$ may be written in a unique way as $g = ks$ with $k \in K$ and $s \in P$. This decomposition yields a parametrization $\mathbb{R} \times P$ of G which shows that the lifting of the curve γ is again closed. Since it obviously is timelike, we have found closed timelike curves also in G. We remark here that this type of parametrization of G will be considered in a more general context in Section 6.7, where we use it to deal with controllability problems.

   Note that the Lorentzian manifold G is time orientable, i.e., there is a continuous choice of future directed light cones. Such a choice is given by the choice of one half of the double cone coming from the metric in $\mathfrak{sl}(2, \mathbb{R})$. To fix the notation we set

$$W_\rho = \{a(X_+ - X_-) + bX_0 + c(X_+ = X_-) \colon a \geq 0, -\rho a^2 + b^2 + c^2 \leq 0\}.$$

Now we find

Proposition 2.11. *Let* $\text{Exp}: \mathfrak{sl}(2, \mathbb{R}) \to G$ *be the exponential map. If* $\rho > 1$, *then* $\text{Exp}\, W_\rho$ *generates* G *as a semigroup.*

Proof. It follows from the general theory (cf. Prop. 4.36(iii)) that any piecewise differentiable curve $\gamma(t)$ in G whose derivative at $\gamma(t)$ is contained in the interior of $d\lambda_{\gamma(t)}(\mathbf{1})W_\rho$ moves directly into the interior of the semigroup S generated by $\text{Exp}\, W_\rho$. But then the existence of closed timelike curves shows that $\mathbf{1}$ is contained in int S whence $S = G$. ∎

This proposition yields more examples of non-global Lie wedges in a topologically trivial, i.e., contractible Lie group.

The semigroups S_ρ generated by $\text{Exp}\, W_\rho$ for $\rho \leq 1$ satisfy $\mathbf{L}(S_\rho) = W_\rho$. To see this, consider the Iwasawa decomposition KAN of $\text{Sl}(2, \mathbb{R})$ and use it to write G as $\widetilde{K}AN$ with $\widetilde{K} \cong \mathbb{R}$. Then AN can be viewed as a subgroup of G and the homogeneous space G/AN can be identified with $\mathbb{R}$. The set

$$\text{compr}(\mathbb{R}^+) = \{g \in G : g.\mathbb{R}^+ \subseteq \mathbb{R}^+\}$$

obviously is a semigroup. Using $G = \widetilde{K}AN$ we see that $\text{compr}(\mathbb{R}^+) = \mathbb{R}^+ AN$ and $\mathbf{L}\big(\text{compr}(\mathbb{R}^+)\big) = \mathbb{R}^+(X_+ - X_-) + \mathbb{R}X_0 + \mathbb{R}X_+$. Therefore $W_1 \subseteq \mathbf{L}\big(\text{compr}(\mathbb{R}^+)\big)$.

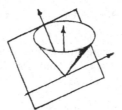

Figure 2.10

The adjoint action of $\text{Exp}\, \mathbb{R}(X_+ - X_-)$ is given by rotation around $\mathbb{R}(X_+ - X_-)$ and hence

$$\bigcap_{t \in \mathbb{R}} e^{\text{ad}\, t(X_+ - X_-)} \mathbf{L}\big(\text{compr}(\mathbb{R}^+)\big) = W_1.$$

In particular

$$\bigcap_{t \in \mathbb{R}} \big(\text{Exp}\, t(X_+ - X_-)\big)\, \text{compr}(\mathbb{R}^+)\big(\text{Exp}\, t(X_- - X_+)\big)$$

is a semigroup containing $\text{Exp}\, W_1$ that has W_1 as tangent wedge. This proves $\mathbf{L}(S_1) = W_1$. Now Proposition 1.39 shows that $\mathbf{L}(S_\rho) = W_\rho$ also for $\rho < 1$.

2.8. The causal action of $SU(n,n)$ on $U(n)$

In the preceding section we have seen how to prove the globality of the invariant cone $W_1 \subseteq \mathfrak{sl}(2,\mathbb{R})$ in the simply connected group $G = \mathrm{Sl}(2,\mathbb{R})\widetilde{}$ by showing that

$$S_{W_1} = \bigcap_{g \in G} g\big(\mathrm{compr}(\mathbb{R}^+)\big)g^{-1},$$

where $\mathbb{R}^+$ is identified with a half line in $\mathbb{R} \cong G/AN$.

In this section we show that this example fits into a general context and also that it is part of an infinite series of examples coming from the action of the pseudo-unitary group $SU(n,n)$ on the unitary group $U(n)$. These examples arise in the cosmological theory of I. E. Segal (cf. [Se76], Section II.3). For a more general setting for these results see [Kan91].

To explain how semigroups arise in such contexts as cosmology we will need some definitions which will be discussed systematically in Chapter 4. Let M be a homogeneous space of the connected Lie group G and suppose that M possesses a G-invariant cone field Θ (cf. Section 4.2). Here a cone field Θ is a function $M \to 2^{T(M)}$ which assigns to each point $x \in M$ a pointed wedge $\Theta(x)$ in the tangent space $T_x(M) \subseteq T(M)$. So the invariance under G means that

$$d\mu_g(x)\Theta(x) = \Theta(g.x) \qquad \forall x \in M,$$

where $\mu_g: M \to M, x \mapsto g.x$ describes the action of G on M. A vector field $\mathcal{X}$ on M is said to be *temporal* if $\mathcal{X}(x) \in \Theta(x)$ holds for all $x \in M$. Let $X \in \mathfrak{g} := \mathbf{L}(G)$. Then we obtain a vector field $\mathcal{X}_X$ on M by

$$\mathcal{X}_X(x) := \frac{d}{dt}\Big|_{t=0} \exp(tX).x.$$

Fix a base point $x_0 \in M$ and let $\pi: G \to M, g \mapsto g.x_0$ denote the orbit mapping. We set

$$W_\Theta := d\pi(\mathbf{1})^{-1}\Theta(x_0) = \{X \in \mathfrak{g} : \mathcal{X}_X(x_0) \in \Theta(x_0)\}$$

and

$$V_\Theta := \{X \in \mathfrak{g} : \mathcal{X}_X \text{ is temporal}\}.$$

Then, since

$$\mathcal{X}_X(g.x) = \frac{d}{dt}\Big|_{t=0} \big(\exp(tX)g\big).x$$
$$= \frac{d}{dt}\Big|_{t=0} g.\big(\exp\big(t\,\mathrm{Ad}(g^{-1})X\big).x\big)$$
$$= d\mu_g(x)\mathcal{X}_{\mathrm{Ad}(g^{-1})X}(x),$$

we have that

$$V_\Theta = \bigcap_{g \in G} \mathrm{Ad}(g^{-1})W_\Theta,$$

so that V_Θ is an invariant wedge in $\mathfrak{g}$.

To connect this infinitesimal data with global objects, we need to endow M with a global order relation corresponding to the cone field Θ, so we assume that M is globally orderable (cf. Section 4.3). This means that there exists a closed partial order $\preceq$ on M such that every differentiable curve $\gamma\colon[a,b] \to M$ such that $\gamma'(t) \in \Theta(\gamma(t))$ holds everywhere, is monotone with respect to the natural order on the real interval $[a,b]$ and the order $\preceq$ on M. We let $\preceq$ denote the smallest order with these properties.

There are two semigroups naturally associated the order $\preceq$. The first one is

$$S_\Theta := \{g \in G : x_0 \preceq g.x_0\}$$

(cf. Proposition 4.16), and the second one is

$$T_\Theta := \{g \in G : (\forall x \in M)\ x \preceq g.x\},$$

the *temporal semigroup*. Now the G-invariance of the order shows that

$$gS_\Theta g^{-1} := \{g' \in G : g.x_0 \preceq g'.(g.x_0)\},$$

so that

$$T_\Theta = \bigcap_{g \in G} gS_\Theta g^{-1}$$

is an invariant subsemigroup of G. The tangent wedges of these semigroups are

$$\mathbf{L}(S_\Theta) = W_\Theta \quad \text{and} \quad L(T_\Theta) = V_\Theta.$$

We want to apply these ideas to the action of the group $G = \mathrm{SU}(n,n)$ on $M = \mathrm{U}(n)$ given by

$$\begin{pmatrix} A & B \\ C & D \end{pmatrix}.U = (AU + B)(CU + D)^{-1}.$$

To see that this is an action and why $(CU + D)$ is always invertible, we explain the wider context where this action comes from.

The action of $\mathrm{SU}(n,n)$ on the euclidean contraction semigroup

We start with a list of those subgroups of $\mathrm{Gl}(2n,\mathbb{C})$ we need in the following.

$$G = \mathrm{SU}(n,n) = \left\{ \begin{pmatrix} A & B \\ C & D \end{pmatrix} : \right.$$

$$\left. A^*A - C^*C = 1, D^*D - B^*B = 1, A^*B - C^*D = 0 \right\},$$

$$G_\mathbb{C} = \mathrm{Sl}(2n,\mathbb{C}),$$

$$K = \mathrm{U}(2n) \cap \mathrm{SU}(n,n) \cong S(\mathrm{U}(n) \times \mathrm{U}(n)),$$

$$P_{\max} = \left\{ \begin{pmatrix} A & 0 \\ C & D \end{pmatrix} \in G_\mathbb{C} \right\},$$

$$N = \left\{ \begin{pmatrix} 1 & Z \\ 0 & 1 \end{pmatrix} \in G_\mathbb{C} \right\},$$

In view of [Hel78, pp.388ff], we know that the mapping

$$N \times P_{\max} \to G_{\mathbb{C}}, \qquad (n,p) \mapsto np$$

is a diffeomorphism onto an open submanifold containing G, and that the projection of G onto N along $P_{\max}$ is the bounded set

$$\Omega := \left\{ \begin{pmatrix} 1 & Z \\ 0 & 1 \end{pmatrix} \in N : Z^*Z < 1 \right\},$$

where $Z^*Z < 1$ means that $1 - Z^*Z$ is a positive definite hermitean matrix ([Hel78, p.527]). Whence

$$\partial \Omega \supseteq \left\{ \begin{pmatrix} 1 & Z \\ 0 & 1 \end{pmatrix} \in N : Z^*Z = 1 \right\} = \left\{ \begin{pmatrix} 1 & U \\ 0 & 1 \end{pmatrix} \in N : U \in \mathrm{U}(n) \right\}.$$

Note that the mapping $Z \mapsto \begin{pmatrix} 1 & Z \\ 0 & 1 \end{pmatrix}$ is a bijection of the euclidean contraction semigroup (Section 2.6) onto $\overline{\Omega}$.

The action of $\mathrm{SU}(n,n)$ on $\overline{\Omega}$ can be described as follows. We identify $\overline{\Omega}$ with a G-invariant subset of the homogeneous space $G_{\mathbb{C}}/P_{\max}$. Then the relation

$$\begin{pmatrix} 1 & Z' \\ 0 & 1 \end{pmatrix} \begin{pmatrix} A' & 0 \\ C' & D' \end{pmatrix} = \begin{pmatrix} A' + Z'C' & Z'D' \\ C' & D' \end{pmatrix}$$

$$= \begin{pmatrix} A & B \\ C & D \end{pmatrix} \begin{pmatrix} 1 & Z \\ 0 & 1 \end{pmatrix}$$

$$= \begin{pmatrix} A & AZ + B \\ C & CZ + D \end{pmatrix}$$

yields that the action of G on $C(J)$ is given by

$$\begin{pmatrix} A & B \\ C & D \end{pmatrix} . Z = (AZ + B)(CZ + D)^{-1}.$$

The stabilizer of $\mathbf{1}$ is

$$G^{\mathbf{1}} = \left\{ \begin{pmatrix} A & B \\ C & D \end{pmatrix} \in \mathrm{SU}(n,n) : A + B = C + D \right\}.$$

To see that G acts transitively on $\mathrm{U}(n)$, let us consider the restriction of the action of G to the compact subgroup K. Then $K \cap G^{\mathbf{1}}$ has dimension $n^2 - 1$, and $\dim K = 2n^2 - 1$, so that the compact orbit of K in $\mathrm{U}(n)$ has dimension n^2, so it is open, and according to the compactness, it must coincide with $\mathrm{U}(n)$. We have shown that $\mathrm{SU}(n,n)$ acts transitively on $\mathrm{U}(n)$.

Next we describe the invariant cone field. Let

$$\Theta(g) := d\lambda_g(\mathbf{1}) iW,$$

where $W = \{X : X \geq 0\}$ is the cone of positive semidefinite hermitean matrices. To see that the action of $SU(n,n)$ preserves this cone field, we have to show that

$$\frac{d}{dt}\Big|_{t=0} (T.U)^{-1}T.(Ue^{tX}) \in iW \qquad \forall X \in iW, U \in U(n), T \in SU(n,n).$$

We first calculate the derivatives involved:

$$\frac{d}{dt}\Big|_{t=0} T.(Ue^{tX})$$
$$= \frac{d}{dt}\Big|_{t=0} (AUe^{tX} + B)(CUe^{tX} + D)^{-1}$$
$$= AUX(CU + D)^{-1} - (AU + B)(CU + D)^{-1}CUX(CU + D)^{-1}.$$

Using

$$(U^*C^* + D^*)((CU + D)(AU + B)^{-1}A - C)$$
$$= (U^*C^* + D^*)(CU + D)(AU + B)^{-1}A - U^*C^*C - D^*C$$
$$= (U^*C^*C^*U + D^*CU + U^*C^*D + D^*D)(AU + B)^{-1}A - U^*C^*C - D^*C$$
$$= \big(U^*(A^*A - 1)U + B^*AU + U^*A^*B + 1 + B^*B\big)(AU + B)^{-1}A$$
$$\quad - U^*C^*C - D^*C$$
$$= (U^*A^* + B^*)(AU + B)(AU + B)^{-1}A - U^*C^*C - D^*C$$
$$= U^*A^*A + B^*A - U^*C^*C - D^*C$$
$$= U^* + B^*A - D^*C = U^*,$$

this leads to

$$\frac{d}{dt}\Big|_{t=0} (T.U)^{-1}T.(Ue^{tX})$$
$$= (CU + D)(AU + B)^{-1}AUX(CU + D)^{-1} - CUX(CU + D)^{-1}$$
$$= ((CU + D)(AU + B)^{-1}A - C)UX(CU + D)^{-1}$$
$$= ((CU + D)^{-1})^*U^*UX(CU + D)^{-1}$$
$$= ((CU + D)^{-1})^*X(CU + D)^{-1}$$

and $-i$ times this matrix is positive semidefinite if and only if $X \in C$. Thus the action of $SU(n,n)$ on $U(n)$ preserves the cone field Θ.

Note that $\gamma(t) = e^{it}\mathbf{1}$ is a closed conal curve in $U(n)$. To obtain a global order, we therefore have to consider the universal covering group

$$\widetilde{U(n)} \cong \{(t,U) \in \mathbb{R} \times U(n) : \det U = e^{it}\}$$

endowed with the pull back $\widehat{\Theta}$ of the cone field Θ on $U(n)$. Now the action of $SU(n,n)$ on $U(n)$ induces an action of the universal covering $SU(n,n)\widetilde{}$ on $\widetilde{U(n)}$ which preserves the cone field.

Let $\chi \colon \widetilde{U(n)} \to \mathbb{R}$ denote the projection on the first component. Then $d\chi(x) \in$ int $\widetilde{\Theta}(x)^*$ for all $x \in \widetilde{U(n)}$, so that Theorem 4.21 shows that $\widetilde{U(n)}$ is globally orderable. Whence

$$S_\Theta = \{g \in \mathrm{SU}(n,n) : \mathbf{1} \preceq g.\mathbf{1}\}$$

is a semigroup in $\mathrm{SU}(n,n)\widetilde{}$ with

$$H(S_\Theta) = G^1,$$

T_Θ is an invariant semigroup with $H(T_\Theta) = \mathbf{1}$, and $\mathbf{L}(T_\Theta)$ is a pointed generating invariant cone in $\mathfrak{su}(n,n)$. Embeddings into these semigroups have been used by Paneitz ([Pa84]) to prove the globality of certain invariant cones in simple simply connected hermitean Lie groups.

There is another interesting feature of the conal homogeneous space $U(n)$, namely that it contains the ordered vector space $\mathbb{H}(n)$ of hermitean symmetric matrices as an open dense conal submanifold. This can be shown as follows. Let

$$\Psi : \mathbb{H}(n) \to U(n), \qquad X \mapsto (X - i\mathbf{1})(X + i\mathbf{1})^{-1}$$

denote the *Cayley transform*. Then Ψ maps $\mathbb{H}(n)$ onto the open subset

$$U_*(n) := \{U \in U(n) : \det(\mathbf{1} - U) \neq 0\}$$

such that

$$d\Psi(X)W = \Theta\big(\Psi(X)\big) \qquad \forall X \in \mathbb{H}(n).$$

To check this, let $Y \in \mathbb{H}(n)$. Then

$$d\Psi(X)(Y) = \frac{d}{dt}\bigg|_{t=0} (X + tY - i\mathbf{1})(X + tY + i\mathbf{1})^{-1}$$
$$= Y(X + i\mathbf{1})^{-1} - (X - i\mathbf{1})(X + i\mathbf{1})^{-1}Y(X + i\mathbf{1})^{-1},$$

so that

$$\Psi(X)^{-1}d\Psi(X)(Y) = (X + i\mathbf{1})(X - i\mathbf{1})^{-1}Y(X + i\mathbf{1})^{-1} - Y(X + i\mathbf{1})^{-1}$$
$$= \big((X + i\mathbf{1})(X - i\mathbf{1})^{-1} - \mathbf{1}\big)Y(X + i\mathbf{1})^{-1}.$$

Furthermore

$$(X + i\mathbf{1})^*\big((X + i\mathbf{1})(X - i\mathbf{1})^{-1} - \mathbf{1}\big)$$
$$=(X - i\mathbf{1})(X + i\mathbf{1})(X - i\mathbf{1})^{-1} - (X - i\mathbf{1})$$
$$=(X + i\mathbf{1}) - (X - i\mathbf{1}) = 2i\mathbf{1},$$

thus

$$\frac{1}{i}\Psi(X)^{-1}d\Psi(X)(Y) = 2\big((X + i\mathbf{1})^{-1}\big)^*Y(X + i\mathbf{1})^{-1}$$

which is in W if $Y \in W$.

There two interesting special cases. For $n = 1$, the group $G = \mathrm{SU}(1,1) \cong \mathrm{Sl}(2,\mathbb{R})$ acts on the closed unit ball in C and $U(1) \cong S^1$ is the unit circle.

For $n = 2$, the group $\mathrm{SU}(2,2)$ acts on the eight dimensional semigroup of contractions in $\mathbb{C}^2$ and the four dimensional group $U(2)$ is an orbit on the boundary of this set. In this case the space $\mathbb{H}(2)$ of hermitean matrices is exactly the four dimensional Minkowski space, so that the Cayley transform provides an embedding of Minkowski space in the the group $U(2)$. This is the reason why $U(2)$ can be viewed as a causal compactification of Minkowski space, so that this group is a more or less reasonable cosmological model for our universe.

2.9. Almost abelian groups

Let G be the group of real $(n+1) \times (n+1)$-matrices of the form

$$\begin{pmatrix} r\mathbf{1}_n & v \\ 0 & 1 \end{pmatrix}$$

where $r > 0$, $v \in \mathbb{R}^n$ and $\mathbf{1}_n : \mathbb{R}^n \to \mathbb{R}^n$ is the identity. Such a group we call an *almost abelian group*. The Lie algebra $L(G)$ is an almost abelian algebra and can be represented by matrices of the form

$$\begin{pmatrix} r\mathbf{1}_n & v \\ 0 & 0 \end{pmatrix}, \qquad r \in \mathbb{R}, \ v \in \mathbb{R}^n.$$

The exponential map of G is given by

$$\exp \begin{pmatrix} r\mathbf{1}_n & v \\ 0 & 0 \end{pmatrix} = \begin{pmatrix} e^r \mathbf{1}_n & \frac{e^r - 1}{r} v \\ 0 & 1 \end{pmatrix}.$$

From this we see that $\exp: L(G) \to G$ is a diffeomorphism. An elementary argument shows that any subspace of $L(G)$ is a subalgebra. In fact, this even characterizes the almost abelian groups (cf. [HHL89], II.2.30). Let now $E \subseteq L(G)$ be a hyperplane and $H = \exp E$ the corresponding subgroup of G. Then $G/H \cong \mathbb{R}$. The decomposition $\mathbb{R} = \{r < 0\} \cup \{0\} \cup \{r > 0\}$ induces a decomposition $G = G^- \cup H \cup G^+$. Since $\{r > 0\}$ and $\{r < 0\}$ are the connected components of $\mathbb{R} \setminus \{0\}$, we know that $H \cup G^\pm =: S^\pm$ is a subsemigroup of G and $G^\pm$ is a semigroup ideal in $S^\pm$. It is clear that $L(S^\pm) = L(G)^\pm$ where $L(G)^\pm$ are the halfspaces bounded by E. Moreover, we see that $\exp: L(S^\pm) \to S^\pm$ is a homeomorphism. Transporting the group multiplication back to the Lie algebra via the exponential function, we have just shown that any halfspace is a subsemigroup. But then any convex cone being the intersection of halfspaces, is a subsemigroup.

2.10. The whirlpot and the parking ramp

Let G be the semidirect product of $\mathbb{C}$ by $\mathbb{R}$ where $\mathbb{R}$ acts on $\mathbb{C}$ by rotation. Then the set $S = \{(c,r) \in G : |c| \le r\}$ is a semigroup with

$$L(S) = \{(\gamma,\rho) \in L(G) : |\gamma| \le \rho, \ \gamma \in \mathbb{C}, \ \rho \in \mathbb{R}\}.$$

The remarkable feature here is that $\exp L(S) \subseteq \{\mathbf{1}\} \cup \operatorname{int} S$.

To prove the claim, note first that we may represent G as the set of 3×3-matrices of the form:

$$(c,r) = \begin{pmatrix} e^{ir} & c & 0 \\ 0 & 1 & 0 \\ 0 & 0 & e^r \end{pmatrix}.$$

Note that the resulting multiplication is given by $(c,r)(c',r') = (c + e^{ir}c', r + r')$. Then the Lie algebra $\mathbf{L}(G)$ is given by

$$(\gamma,\rho) = \begin{pmatrix} i\rho & \gamma & 0 \\ 0 & 0 & 0 \\ 0 & 0 & \rho \end{pmatrix}$$

and the exponential function is the usual matrix exponential function. It is not surjective. Now it is easy to check that S is a closed subsemigroup of G. A one parameter group in G is given by

$$\exp t(\gamma,\rho) = \left(\frac{\gamma(e^{it\rho}-1)}{i\rho}, t\rho \right) \quad \text{for} \quad \rho \neq 0.$$

Hence $\exp\left(\mathbb{R}^+(\gamma,1)\right) \subset S$ if and only if

$$|\gamma(e^{it}-1)| \leq t \quad \text{for all} \quad t \in \mathbb{R}^+.$$

But this is equivalent to

$$2|\gamma|^2(1-\cos t) \leq t^2 \quad \text{for all} \quad t \in \mathbb{R}^+$$

and hence to

$$|\gamma| \leq 1.$$

Since $\exp(\gamma,0) = (\gamma,0)$ is in S if and only if $\gamma = 0$, this shows that $\mathbf{L}(S) = \{(\gamma,\rho) \in \mathbf{L}(G) : |\gamma| \leq \rho\}$ (cf. Figure 2.11).

Figure 2.11

This picture is the reason why we call the semigroup S the *whirlpot*. Now consider the semigroup S_R generated by $\exp \mathbf{L}(S)$. Recall that

$$\exp \mathbf{L}(S) = \left\{ \left(\frac{\gamma(e^{i\rho t}-1)}{i\rho}, \rho t \right) : \rho > 0, |\gamma| \leq \rho, t \geq 0 \right\}.$$

We have $|\gamma|\frac{|\sin t|}{t} < 1$ for $|\gamma| \leq 1$ and $t > 0$. This is equivalent to $|\gamma(e^{it}-1)| < t$. Thus $\exp \mathbf{L}(S) \subset T = \{(c,r) \in G : |c| < r\} \cup \{(0,0)\}$. But it is easy to check that T is a semigroup, so we get $S_R \subset T$.

Conversely, let $(c, r) \in T$. If we set

$$c_k = de^{ia_k}(e^{ir/n} - 1), \quad \text{where} \quad |d| \leq 1, \; a_k \in \mathbb{R},$$

we obtain

$$(c_1, \frac{r}{n}) \cdots (c_n, \frac{r}{n}) = (d(e^{ir/n} - 1) \cdot \sum_{k=1}^{n} e^{i(a_k + \frac{(k-1)}{n})}, r).$$

Thus, if we set $a_k = \frac{-(k-1)}{n}$, we get

$$(c_1, \frac{r}{n}) \cdots (c_n, \frac{r}{n}) = (dn(e^{ir/n} - 1), r).$$

Now we choose n with

$$\frac{|c|}{r} < |e^{ir/n} - 1| \cdot \frac{n}{r} \leq 1.$$

Then we can find $d \in \mathbb{C}$ with $|d| < 1$ such that $c = d(e^{ir/n} - 1)n$. Since $(c_k, \frac{r}{n}) = \exp(\frac{ir}{n} de^{ia_k}, \frac{r}{n})$, this shows that $T = S_R$.

The next subsemigroup of G which we consider, is the *parking ramp*. It is defined as $S := S_1 \cup S_2$ where

$$S_1 := \{(re^{is}, t) : r \in \mathbb{R}^+, 0 \leq s \leq t \leq \pi\}$$

and

$$S_2 := \{(c, t) : c \in \mathbb{C}, t \geq \pi\}$$

(cf. Figure 2.12).

Figure 2.12

It is obvious that $SS_2 \cup S_2 S \subseteq S_2$ so that in order to show that S is a semigroup we have to calculate ab for $a = (re^{is}, t)$ and $b = (r'e^{is'}, t')$ in S_1. Moreover we may assume $t + t' \leq \pi$. But then elementary geometry shows that $ab \in S_1$.

We claim that S is equal to the semigroup generated by the one parameter semigroups $\{(r, 0) : r \geq 0\}$ and $\{(0, t) : t \geq 0\}$. In order to show this, note first that every element $x \in G$ has a unique representation (re^{is}, t) with $r \geq 0$, $t \in \mathbb{R}$ and $0 \leq s < 2\pi$. Suppose that $x \in S$ and $s \leq t$. Then $x = (0, s)(r, 0)(0, t - s)$. If on the other hand $s > t$, then $x \in S_2$ and we may assume $\pi < t < s < 2\pi$. For $t' \in]\pi, t[$ the numbers

$$a = \frac{r \sin s}{\sin t'} \quad \text{and} \quad b = \frac{r \sin(t' - s)}{\sin t'}$$

are positive and a straightforward calculation shows that

$$x = (b,0)(0,t')(a,0)(0,t-t') = (b+e^{it'}a,t)$$

which proves our claim.

Finally we show that the Lie semigroup S_W belonging to

$$W = \{(\gamma,\rho) \in \mathbf{L}(G) : \gamma, \rho \geq 0\} = \mathbf{L}(S)$$

is equal to $S_1 \cup \overline{S}_2$. Let $x = (re^{is},\pi) \in \overline{\langle \exp W \rangle} \setminus S_1 \cup S_2$. Then $s \in]\pi, 2\pi[$. Suppose that $x = x_1 x_2$ with $x_1, x_2 \in S_W$. Then the fact that $S_1 \cup S_2$ is a subsemigroup shows that at least one factor is not contained in S_1. We assume that $x_1 \notin S_1$. Then $x_1 = (r_1 e^{is_1},\pi)$ and $x_2 = (r_2 e^{is_2},0)$. Thus $s_2 = 0$ and $x_2 = (r_2,0)$. Since $x \in \langle \exp W \rangle$ we have a representation $x = \exp w_1 \exp w_2 \cdots \cdot \exp w_n$ with $w_i = (z_i,\varepsilon_i) \in W$. Let i_0 be minimal with $\varepsilon_{i_0} > 0$. Then $x_1 := \exp w_1 \cdots \cdot \exp w_{i_0-1} \exp(\frac{1}{2}w_{i_0}) \in S_W$ and $x_2 = \exp(\frac{1}{2}w_{i_0}) \cdot \ldots \cdot \exp(w_n) \in S_W$ with $x = x_1 x_2$ and $x_1, x_2 \in S_1$, a contradiction. We conclude that

$$\langle \exp W \rangle = S_1 \cup S_2 \neq S_W = S_1 \cup \overline{S}_2.$$

2.11. The oscillator group

Let H be the Heisenberg group, represented as pairs $(v,z) \in \mathbb{R}^2 \times \mathbb{R}$ endowed with the multiplication

$$(v,z) \cdot (v',z') = (v+v', z+z' + \frac{1}{2}\langle Iv \mid v' \rangle),$$

where $\langle \cdot \mid \cdot \rangle : \mathbb{R}^2 \times \mathbb{R}^2 \to \mathbb{R}$ is the scalar product, and $I: \mathbb{R}^2 \to \mathbb{R}^2$ is given by the matrix $\begin{pmatrix} 0 & -1 \\ 1 & 0 \end{pmatrix}$. Note that this is just the Campbell-Hausdorff multiplication on the Heisenberg algebra $\mathbf{L}(H)$ represented as pairs $(\zeta,\xi) \in \mathbb{R}^2 \times \mathbb{R}$ with bracket

$$[(\zeta,\xi),(\zeta',\xi')] = (0, \langle I\zeta \mid \zeta' \rangle).$$

For $t \in \mathbb{R}$ set

$$R(t) = \begin{pmatrix} \cos t & -\sin t \\ \sin t & \cos t \end{pmatrix}$$

and let $\mathbb{R}$ act on H by $\alpha(r)(v,z) = (R(r)v,z)$. The semidirect product $G := H \rtimes_\alpha \mathbb{R}$ with respect to this action is called the *oscillator group*. The product on G is given by

$$(v,z,r)(v',z',r') = (v+R(r)v', z+z' + \frac{1}{2}\langle Iv \mid R(r)v' \rangle, r+r').$$

Then one can show that for any invariant generating cone W in $\mathbf{L}(G)$ the closed subsemigroup S of G, generated by $\exp W$, has tangent cone W and there is a neighborhood U of $\mathbf{1}$ in G such that $S \cap U = (\exp W) \cap U$.

In order to prove this statement, we need to have a good description of the exponential function. We start by calculating the one-parameter subgroups of G. Let $\Phi(t) = (v(t), z(t), r(t))$ be a one parameter subgroup of G. Then $r(t) = tr_0$ and for $t, s \in \mathbb{R}$

$$\left(v(s) + R(sr_0)v(t), z(s) + z(t) + \frac{1}{2}\langle Iv(s) \mid R(sr_0)v(t)\rangle \right) = (v(s+t), z(s+t)).$$

Fixing s and letting t tend to zero we obtain

$$(\dot{v}(s), \dot{z}(s)) = \left(R(sr_0)\dot{v}(0), \dot{z}(0) + \frac{1}{2}\langle Iv(s) \mid R(sr_0)\dot{v}(0)\rangle \right)$$

since $v(0) = 0$ and $z(0) = 0$. But since $R(sr_0) = e^{sr_0 I}$ we conclude $v(s) = (e^{sr_0 I} - 1)v_0$, where $r_0 Iv_0 = \dot{v}(0)$ for $r_0 \neq 0$. Thus we have in this case

$$\begin{aligned}
\dot{z}(s) &= \dot{z}(0) + \frac{1}{2}\langle I(e^{sr_0 I} - 1)v_0 \mid r_0 e^{sr_0 I}I(v_0)\rangle = \\
&= \dot{z}(0) + \frac{r_0}{2}\langle e^{sr_0 I}Iv_0 \mid e^{sr_0 I}Iv_0\rangle - \frac{r_0}{2}\langle Iv_0 \mid e^{sr_0 I}Iv_0\rangle = \\
&= \dot{z}(0) + \frac{r_0}{2}\langle Iv_0 \mid Iv_0\rangle - \frac{r_0}{2}\langle Iv_0 \mid Ie^{sr_0 I}v_0\rangle = \\
&= \dot{z}(0) + \frac{r_0}{2}\|v_0\|^2 - \frac{r_0}{2}\langle v_0 \mid e^{sr_0 I}v_0\rangle
\end{aligned}$$

since I and $e^{sr_0 I}$ are orthogonal. Integration now yields

$$z(s) = s\cdot\left(\dot{z}(0) + \frac{r_0}{2}\cdot\|v_0\|^2\right) - \frac{1}{2}\cdot\langle Iv_0 \mid e^{sr_0 I}\cdot v_0\rangle$$

since I is skew symmetric. Thus the exponential function $\exp: \mathbf{L}(G) \mapsto G$ is given by $\exp(v, z, 0) = (v, z, 0)$, and for $r \neq 0$ by

$$\exp(v, z, r) = \left(\frac{1}{r}(1 - e^{rI})\cdot Iv, z + \frac{1}{2r}\|v\|^2 - \frac{1}{2r^2}\cdot\langle Iv \mid e^{rI}\cdot v\rangle, r \right).$$

In fact, we only need to note that $\frac{d}{dt}\exp t\cdot(v, z, r)\big|_{t=0} = (v, z, r)$ and use the fact that $I^{-1} = -I$ is orthogonal in the above calculations. From this we calculate easily that $\exp\big|_B: B \mapsto B$ is a diffeomorphism, where $B = \mathbb{R}^2 \times \mathbb{R} \times]-2\pi, 2\pi[$. Therefore the set $C = \{((v, z, r), (v', z', r')) \in \mathbf{L}(G) \times \mathbf{L}(G): -2\pi < r + r' < 2\pi\}$ is contained in the set $\{((v, z, r), (v', z', r')) \in \mathbf{L}(G) \times \mathbf{L}(G): \exp(v, z, r)\cdot\exp t\cdot(v', z', r') \in \exp B \text{ for all } t \in [0, 1]\}$. If now W is an invariant cone, one can apply a general theorem (cf. [HHL89], II.2.42) to obtain

$$\exp(v, z, r)\exp(v', z', r') \in \exp W \text{ for all } ((v, z, r), (v', z', r')) \in C \cap (W \times W).$$

Note that $\mathbb{R}^2 \times \mathbb{R} \times [2\pi, \infty[$ is a semigroup ideal in $\mathbb{R}^2 \times \mathbb{R} \times \mathbb{R}^+$. Therefore for any invariant cone W in $\mathbf{L}(G)$ which is contained in $\mathbb{R}^2 \times \mathbb{R} \times \mathbb{R}^+$, the set $S = \exp W \cup (\mathbb{R}^2 \times \mathbb{R} \times [2\pi, \infty[)$ is a subsemigroup of G. But clearly $\mathbf{L}(S) = W$. Note that the form

$$q\big((v, z, r), (v', z', r')\big) = rz' + r'z + \langle v \mid v'\rangle$$

is invariant and Lorentzian so that it defines an invariant cone. One can even show that each invariant cone can be mapped to this one via an algebra automorphism (cf. [HHL89, II.2.15]).

Let $r \in]0, 2\pi[$ and consider $(\exp W) \cap (\mathbb{R}^2 \times \mathbb{R} \times \{r\})$. Note first that for $\exp(v, z, r) = (v', z', r')$ we have $\exp(e^{tI}v, z, r) = (e^{tI}v', z', r')$, i.e., the set $(\exp W) \cap (\mathbb{R}^2 \times \mathbb{R} \times \{r\})$ is invariant under rotations in the v-plane. If now $v = (x, 0)$, then $Iv = (0, x)$ and $e^{rI}v = (x \cos r, x \sin r)$. Therefore $\langle Iv \mid e^{rI}v \rangle = x^2 \sin r$. Moreover,

$$\|(1 - e^{rI})Iv\|^2 = \|(1 - e^{rI})v\|^2 = 2\|v\|^2 - 2\langle v \mid e^{rI}v \rangle = 2\|v\|^2(1 - \cos r)$$

since $\langle v \mid e^{rI}v \rangle = x^2 \cos r$. As $2rz + \|v\|^2 \leq 0$ just means $z + \frac{1}{2r}\|v\|^2 \leq 0$, this shows that $(\exp W) \cap (\mathbb{R}^2, \mathbb{R}, r)$ is the region below the paraboloid given by

$$\left(v, \|v\|^2 \cdot \frac{-\sin r}{4(1 - \cos r)}, r \right), \quad v \in \mathbb{R}^2 \quad \text{(cf. Figure 2.13)}.$$

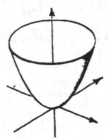

Figure 2.13

Note that $\lim\limits_{r \to 0} \frac{1 - \cos r}{\sin r} = \lim\limits_{r \to 0} \frac{\sin r}{\cos r} = 0$ so that $\frac{\sin r}{2(1 - \cos r)}$ approaches $\pm\infty$ as r approaches $2\pi n$ with $n \in \mathbb{N}$ depending on whether one approaches from the left or from the right (cf. Figure 2.14).

Figure 2.14

It is not possible in the example to replace the oscillator group by another group with the same Lie algebra:

Let G be the oscillator group and $0 \neq Z \in Z(\mathbf{L}(G))$ where $Z(\mathbf{L}(G))$ is the center of the oscillator algebra. Let S be the closed subsemigroup of G generated by $\exp W$ where W is a generating invariant cone in $\mathbf{L}(G)$. Then $N = \exp(\mathbb{Z} \cdot Z)$ is a discrete Lie subgroup of G and the subsemigroup SN/N of G/N has a halfspace bounded by the hyperplane ideal of $\mathbf{L}(G)$ as tangent wedge.

To see this, set $S_0 = \mathbb{R}^2 \times \mathbb{R} \times \mathbb{R}^+$ with $\mathbb{R}^+ = \{r \in \mathbb{R} : 0 \le r\}$. This is a half space semigroup in G which is equal to its tangent wedge. But then we have

$$\mathbf{L}(S_0) = L_{\mathbb{R}Z}\big(\mathbf{L}(S)\big) = \overline{\mathbf{L}(S) + \mathbb{R}Z} \subseteq \mathbf{L}(SN) = \mathbf{L}(S_0),$$

and hence $\overline{SN} = S_0$.

Notes

The Heisenberg example was first calculated by Hofmann and Lawson in [HoLa83] (cf. also [HoLa81]). The calculations for $\mathrm{Sl}(2, \mathbb{R})$ have been carried out in [HiHo85], the presentation of the $\mathrm{PSl}(2, \mathbb{R})$-case and the Ol'shanskiĭ semigroup in $\mathrm{Sl}(2, \mathbb{C})$ draws on San Martin's discussion of invariant control sets in [SM92]. On the other hand the semigroups which we call Ol'shanskiĭ semigroups here appeared first in [Ols82a]. Affine compression groups have been used in [HiNe92] to study Wiener-Hopf operators on ordered symmetric spaces.

Gödel published his cosmological model in [Gö49]. Our presentation of the group theoretical description follows [Hi92]. The causal action of $\mathrm{SU}(n, n)$ on $\mathrm{U}(n)$ has been considered by Segal (cf. [Se76]) and Paneitz (cf. [Pa84]). A thorough discussion of the almost abelian groups was given by Hofmann and Lawson in [HoLa81]. The whirlpot appeared first in [HoLa83], and the parking ramp in [HHL89], whereas the calculations for the oscillator group can be found in [Hi86].

3. Geometry and topology of Lie semigroups

Now we turn to the geometric structure of Lie semigroups (S, G). We want to use the functor L to relate it to the geometric structure of the Lie wedge $(L(S), L(G))$. We recall that this translation process is trivial if, for example, G is a vector group or a compact group because in the latter case the only global Lie wedges are the Lie algebras of closed subgroups (cf. Chapter 1).

For Lie semigroups we have to add to the notions of faces and exposed faces which we know already for wedges from Section 1.1 and the notion of a normal exposed face which in some sense is caused by the non-commutativity of the underlying group G. On the infinitesimal level we discuss the notion of a Lie face of a Lie wedge and show that Lie faces of global Lie wedges are always global. In Section 5.4 we will use the theory of ordered homogeneous spaces to obtain some deeper insight into the structure of faces of Lie semigroups.

Section 3.2 is dedicated to a discussion and proof of the result of Hofmann and Ruppert that whenever the Lie wedge $L(S)$ of a Lie semigroup generates the Lie algebra $L(G)$, then there exists an analytic path from the identity which enters immediately the interior of S. This result plays a crucial role in the construction of the universal covering semigroup in Section 3.4. Another important consequence is that the interior of S is a dense semigroup ideal if $L(S)$ is Lie generating.

To see that there exists a Lie semigroup S whose tangent wedge $L(S)$ is not Lie generating but which has non-empty interior, we construct in Section 3.3 an example of such a semigroup in a 9-dimensional solvable Lie group.

As already mentioned above, the results of Section 3.2 can be used to show that a generating Lie semigroup S has all the properties necessary for the existence of a universal covering semigroup $\tilde{S}$. The theory of coverings of Lie semigroups shows some strong analogy to the theory of Lie group coverings. For example the fundamental group of S can be realized as a discrete central subgroup of the group of units of $\tilde{S}$ and, conversely, for every discrete central subgroup D of $\tilde{S}$ we obtain a semigroup covering $\tilde{S} \to \tilde{S}/D$. Moreover $\tilde{S}$ is cancellative and the left and right multiplication mappings are proper. As we will see in Chapter 9, these semigroup coverings are in particular important for Ol'shanskiĭ semigroups and their representation theory.

Section 3.5 contains a direct proof for the fact that the algebraic and topological free group $G(S)$ on a generating Lie semigroup (S, G) coincide and that $G(S)$ is a covering group of G. As a consequence of this result we show that $G(\tilde{S}) = \tilde{G}$ and therefore we obtain a lot of quotient groups of $\tilde{S}$ which cannot be embedded in any group. The last section describes the relations between covering semigroups and reversability which is also related to the directedness of the corresponding order on the group.

3.1. Faces of Lie semigroups

Let (S, G) be a Lie semigroup and $F \subseteq S$ a subsemigroup. We set

$$L_F(S) := \overline{\langle SF^{-1} \rangle} \quad \text{and} \quad T_F(S) := H(L_F(S)).$$

Suppose, in addition, that F is closed. Then we say that F is

(i) *a face of S* if $S \setminus F$ is an ideal in S.

(ii) *an exposed face of S* if $F = S \cap T_F(S)$.

(iii) *a normal exposed face of S* if $F = S \cap H(S_1)$, where $S_1 \subseteq G$ is a closed subsemigroup with $S \subseteq S_1$ and which has a normal group of units.

We denote the set of faces (exposed faces, normal exposed faces) of S with $\mathcal{F}(S)$, $(\mathcal{F}_e(S), \mathcal{F}_n(S))$.

Proposition 3.1. *Let (S, G) be a Lie semigroup. Then we have the following characterization of exposed and normal exposed faces.*

(i) *A closed subsemigroup $F \subseteq S$ is an exposed face if and only if there exists $f \in \mathrm{Mon}(S) \cap -\mathrm{Mon}(F)$ such that*

$$F = f^{-1}(f(1)) \cap S.$$

(ii) *A closed subsemigroup $F \subseteq S$ is a normal exposed face if and only if there exists a morphism $\varphi : (S, G) \to (S_1, G_1)$ of Lie semigroups such that $H(S_1) = \{1\}$ and*

$$F = \varphi^{-1}(1) \cap S.$$

Proof. (i) Let $f : G \to \mathbb{R}$ be such a function. Then $\mathrm{SG}(f)$ is a closed submonoid of G which contains S and $H(\mathrm{SG}(f)) \subseteq f^{-1}(f(1))$. Moreover

$$F \subseteq H(\mathrm{SG}(f)) \cap S \subseteq f^{-1}(f(1)) \cap S = F.$$

Thus $F \in \mathcal{F}_e(S)$. If, conversely, $F \in \mathcal{F}_e(S)$, then $F = T_F(S) \cap S$ and it suffices to find a continuous $L_F(S)$-monotone function with $f^{-1}(f(1)) \cap L_F(S) = T_F(S)$. We set $f(g) := d(gL_F(S), 1)$, where d is a left invariant metric on G. Then $g \mapsto f(g) = d(L_F(S), g^{-1})$ is continuous, $L_F(S)$-monotone and

$$f^{-1}(f(1)) \cap S = f^{-1}(0) \cap S = \{s \in S : s^{-1} \in L_F(S)\} = T_F(S) \cap S = F.$$

(ii) Suppose that $F \in \mathcal{F}_n(S)$ and $F = S \cap H(S_2)$, where S_2 is a closed subsemigroup of G such that $H(S_2)$ is normal. Then the quotient morphism $\varphi : (S, G) \to (\overline{\varphi(S)}, G/H(S_1))$ has the desired properties because $(\overline{\varphi(S)}, G/H(S_1))$ is a Lie semigroup (Proposition 1.16) and

$$H(\overline{\varphi(S)}) \subseteq H(\varphi(S_2)) = \{1\}.$$

To see the converse, we assume that $\varphi : (S, G) \to (S_1, G_1)$ is a Lie semigroup morphism with $\ker \varphi \cap S = F$ and $H(S_1) = \{1\}$. We set $S_2 := \varphi^{-1}(S_1)$. This is a closed subsemigroup of G with the normal group of units $H(S_2) = \ker \varphi$ and $H(S_2) \cap S = F$. Thus $F \in \mathcal{F}_n(S)$. ∎

Proposition 3.2.   (The Hierarchy of Faces) *For a Lie semigroup* (S, G) *we have the following hierarchy of faces:*

$$normal\ exposed\ face\ \Rightarrow\ exposed\ face\ \Rightarrow\ face\ .$$

Proof.   (1) Let $F \subseteq S$ be a normal exposed face and $\varphi : (S, G) \to (S_1, G_1)$ a morphism of Lie semigroups with $H(S_1) = \{1\}$ and $\ker \varphi \cap S = F$ (Proposition 3.1). Let $S_2 := \varphi^{-1}(S_1)$. Then S_2 is a subsemigroup of G with $F = H(S_2) \cap S$. Therefore $T_F(S) \subseteq H(S_2)$ and it follows that

$$T_F(S) \cap S = H(S_2) \cap S = F.$$

(2) Let $F = T_F(S) \cap S$ be an exposed face. Then $S \setminus F = S \setminus T_F(S)$ and the fact that $L_F(S) \setminus T_F(S)$ is an ideal in $L_F(S)$ shows that $S \setminus F$ is an ideal in S, i.e., F is a face of S.   ■

To see how the concept of a face of a Lie semigroup generalizes the concept of a face of a wedge in a vector space, let us suppose that G is abelian. Then every exposed face is normal because every subgroup of G is normal. Hence

$$\mathcal{F}_n(S) = \mathcal{F}_e(S).$$

Assume, in addition, that G is a vector space, and that $S \subseteq G$ is a Lie semigroup. Then S is a wedge because $\exp_G = \mathrm{id}_G$ and $\exp_G (\mathbf{L}(S))$ is a closed subsemigroup of G (cf. Section 1.4). Let $F \in \mathcal{F}(S)$ and $f \in F$. For each $t \in [0, 1]$ we find that $tf + (1 - t)f \in F$ and $tf, (1 - t)f \in S$. Hence $tf \in F$ and therefore $\mathbb{R}^+ F = F$. Consequently F is a wedge. This shows that the definition of a face in this case is consistent with the definition of a face of a wedge in Section 1.1. A closed subsemigroup $F \subseteq S$ is an exposed face if and only if

$$F = S \cap T_F(S) = S \cap H(\overline{S - F})$$

which is also consistent with the definition from Section 1.1. We notice also that an exposed face F of a wedge W may be defined by the existence of a linear functional $\omega \in W^*$ such that

$$F = \omega^{-1}(0) \cap W$$

(Proposition 1.1). It is interesting to compare this with the characterization in Proposition 3.1 because in the abelian case a functional $\omega \in W^*$ is a W-monotone function $(\mathbb{R}^n, \leq_W) \to (\mathbb{R}, \leq)$ and also a homomorphism of Lie semigroups

$$(W, \mathbb{R}^n) \to (\mathbb{R}^+, \mathbb{R}).$$

So we can identify W^* with the set

$$\mathrm{Hom}\left((W, \mathbb{R}^n), (\mathbb{R}^+, \mathbb{R})\right).$$

This illuminates the splitting of the notion of an exposed face into different ones in the non-abelian case. The set

$$\mathrm{Hom}\left((S, G), (\mathbb{R}^+, \mathbb{R})\right)$$

is in general very small for a Lie semigroup (S, G). Indeed, all such morphisms vanish on the commutator group G' which implies that they are trivial whenever $G = G'$.

Note that the sets $\mathcal{F}(S), \mathcal{F}_e(S)$ and $\mathcal{F}_n(S)$ are complete lattices. $H(S)$ is minimal in $\mathcal{F}(S)$ and $\mathcal{F}_e(S)$ and also in $\mathcal{F}_n(S)$ whenever it is normal. The whole semigroup S is the maximal element in all these lattices.

In Section 1.1 we have seen an example of a wedge W with $\mathcal{F}_e(W) \neq \mathcal{F}(W)$. Now we give an example with $\mathcal{F}_e(S) \neq \mathcal{F}_n(S)$. Let $G = \mathrm{Sl}(2, \mathbb{R})^{\tilde{}}$ and $L(G) = \mathfrak{sl}(2, \mathbb{R})$ (cf. Section 2.2). The cone

$$W := \{hX_0 + t(X_+ + X_-) + x(X_+ - X_-) : x \geq 0, h^2 + t^2 \leq x^2\}$$

is the tangent wedge of the invariant subgroup $S := \overline{\langle \exp W \rangle}$ (Section 2.7). The half-line $F = \exp(\mathbb{R}^+ X_+)$ is an exposed face of S because the subsemigroup $\tilde{S} := \overline{\langle \exp V \rangle}$ with $V = \overline{W - L(F)}$ agrees with $L_F(S)$ and satisfies

$$F = S \cap H(\tilde{S}) = S \cap T_F(S).$$

The simple Lie algebra $L(G)$ contains no non-trivial ideals, hence F is not in $\mathcal{F}_n(S)$.

In the remainder of this section we turn to the structure of faces of a Lie semigroup S and how they are related to faces of $L(S)$.

Let $(W, \mathfrak{g})$ be a Lie wedge. A face $F \in \mathcal{F}(W)$ is called a *Lie face* if F is a Lie wedge. Since $H(W) = H(F)$, this is equivalent to the invariance of F under the group $\langle e^{\mathrm{ad}\, H(W)} \rangle$. Thus we write

$$\mathcal{F}_L(W) := \{F \in \mathcal{F}(W) : e^{\mathrm{ad}\, H(W)} F = F\}$$

for the set of all Lie faces of W.

Proposition 3.3.   *The set $\mathcal{F}_L(W) \subseteq \mathcal{F}(W)$ is stable under arbitrary intersections and therefore a complete lattic in its own right.*

Proof.   trivial   ∎

Corollary 3.4.   *Let (S, G) be a Lie semigroup. Then every Lie face*

$$F \in \mathcal{F}_L(L(S))$$

is global in G.

Proof.   We check the conditions of Proposition 1.37. If F is a Lie face of $L(S)$, then

$$F \subseteq L(S) \quad \text{and} \quad H(F) = H(L(S)).$$

Thus $F \cap H(L(S)) = H(L(S)) \subseteq H(F)$ and $\langle \exp H(F) \rangle = H(S)$ is closed in G.∎

Corollary 3.4 supplies us with Lie semigroups $F \subseteq S$ which are good candidates for faces of S, namely those generated by the Lie faces of $L(S)$. That not every face of S need to come from a Lie face of $L(S)$ will be shown by the following example, where

$$F \in \mathcal{F}(S) \quad \text{with} \quad L(F) \notin \mathcal{F}(L(S)).$$

It is taken from [Su72], where it serves as an example of a control system where not all points are reachable by bang-bang controls. Exactly these points will be the face we are looking for. We define

$$B := \begin{pmatrix} 0 & 0 & 0 & 0 \\ 0 & 0 & 0 & 0 \\ 0 & 0 & 0 & 0 \\ 0 & 0 & 1 & 0 \end{pmatrix} \quad \text{and} \quad C := \begin{pmatrix} 0 & 0 & 0 & 0 \\ 1 & 0 & 0 & 0 \\ 0 & 1 & 0 & 0 \\ 0 & 0 & 0 & 0 \end{pmatrix}.$$

An easy computation shows that the Lie algebra generated by these matrices in $\mathrm{gl}(4, \mathbb{R})$ is isomorphic to $V \rtimes \mathbb{R}C$, where C acts on $V \cong \mathbb{R}^3$ as the matrix

$$\begin{pmatrix} 0 & 0 & 0 \\ 1 & 0 & 0 \\ 0 & 1 & 0 \end{pmatrix},$$

and B corresponds to the first base vector $(1,0,0)^\top$. This Lie algebra belongs to a nilpotent group of unipotent matrices. The Lie semigroup $S = S_W$ in $\mathrm{Gl}(4, \mathbb{R})$ generated by the Lie wedge

$$W := \mathbb{R}^+(B + C) + \mathbb{R}^+(B - C)$$

has the property that every element $s \in S$ is reachable from $\mathbf{1}$ with an absolutely continuous S-monotone curve and every pair of points $s, s' \in S$ with $ss' \in \exp \mathbb{R}^+ B$ satisfies $s, s' \in \exp \mathbb{R}^+ B$. If $u: [0, T] \to [-1, 1]$ is a measurable function, the solution $\gamma_u: [0, T] \to G$ of the initial value problem

$$\gamma_u(0) = \mathbf{1} \quad \text{and} \quad \dot{\gamma}_u(t) = d\lambda_{\gamma_u(t)}(\mathbf{1})\big(B + u(t)C\big)$$

is an S-monotone curve $\gamma_u: ([0, T], \leq) \to (G, \leq_S)$, and

$$\gamma_u(t) = \begin{pmatrix} 1 & 0 & 0 & 0 \\ f(t) & 1 & 0 & 0 \\ \frac{1}{2}f(t)^2 & f(t) & 1 & 0 \\ g(t) & h(t) & t & 1 \end{pmatrix},$$

where

$$f(t) = \int_0^t u(\tau)\, d\tau, \quad h(t) = \int_0^t \tau u(\tau)\, d\tau, \quad \text{and} \quad g(t) = \int_0^t u(\tau)h(\tau)\, d\tau.$$

If $\gamma_{u_n}(T_n)$ is a sequence of points in S which converges to $s \in S$, then the explicit form of γ_u shows that T_n converges to a number $T \geq 0$. Since the set of all measurable functions $u: [0, T] \to [-1, 1]$ is weak-$*$-compact, this topology has a countable base, and convergence of u_n to u in this topology entails that $\gamma_{u_n}(T) \to \gamma_u(T)$ ([CL91]), we conclude that $s = \gamma_u(T)$ for a measurable function $u: [0, T] \to [-1, 1]$. Thus

$$S = \{\gamma_u(T) : T \in \mathbb{R}^+, u : [0, T] \to [-1, 1] \text{ measurable}\}.$$

It is clear that $F := \exp(\mathbb{R}^+ B) \subseteq S$ is closed. If $s = \gamma_u(T)$ and $s' = \gamma_{u'}(T')$ with $ss' \in F$, then $ss' = \gamma_{u''}(T + T')$ with

$$
u''(\tau) := \begin{cases} u(\tau), & \text{for } 0 \leq \tau \leq T \\ u'(\tau - T), & \text{for } T \leq \tau \leq T + T'. \end{cases}
$$

Thus the argument given in [Su72] shows that $u''(\tau) = 0$ almost everywhere on $[0, T + T']$ and therefore $s, s' \in \exp \mathbb{R}^+ B = F$. Consequently $F \in \mathcal{F}(S)$ and $\mathbf{L}(F) = \mathbb{R}^+ B$ is far from being a face of the two dimensional wedge W because it is a ray between the two extremal rays of W.

It is noteworthy that the situation is much better for exposed faces.

Proposition 3.5. *Let S be a Lie semigroup. Then*

$$
\mathbf{L}\left(\mathcal{F}_e(S)\right) \subseteq \mathcal{F}_e\left(\mathbf{L}(S)\right).
$$

Proof. Suppose that $F \in \mathcal{F}_e(S)$ is exposed and set $V := \overline{\mathbf{L}(S) - \mathbf{L}(F)}$. Then

$$
\mathbf{L}(F) \subseteq \mathbf{L}(S) \cap H(V) \subseteq \mathbf{L}(S) \cap \mathbf{L}\left(T_F(S)\right) = \mathbf{L}\left(S \cap T_F(S)\right) = \mathbf{L}(F).
$$

Therefore

$$
\mathbf{L}(F) = \mathbf{L}(S) \cap H(V) = \mathbf{L}(S) \cap H\left(\overline{\mathbf{L}(S) - \mathbf{L}(F)}\right),
$$

so that $\mathbf{L}(F)$ is an exposed face of $\mathbf{L}(S)$. ∎

In view of the above example which shows that in general the tangent wedge of a face is not a face, it is interesting to note that, whenever $\mathbf{L}(F)$ is not a face of $\mathbf{L}(S)$, there exist $X \in \mathbf{L}(F)$ and $Y \in \mathbf{L}(G)$ such that

$$
W := \mathbb{R}^+(X + Y) + \mathbb{R}^+(X - Y)
$$

is global in G and $\mathbf{L}(F) \cap W := \mathbb{R}^+ X$ is the tangent wedge of the face $F \cap S_W$. Thus every example with a face such that $\mathbf{L}(F)$ is not a face provides an example of the $X \pm Y$-type as is Sussman's example above.

In Section 1.1 we have seen that faces of wedges lie on the boundary. The following lemma shows that a similar statement holds for Lie semigroups.

Lemma 3.6. *Let F be a face of a Lie semigroup S and $F \cap \operatorname{int} S \neq \emptyset$. Then $F = S$.*

Proof. Let $f \in F \cap \operatorname{int} S$ and $X \in \mathbf{L}(S)$. Then there exists $t > 0$ such that $\exp(tX) \in fS^{-1}$ because fS^{-1} is a neighborhood of 1 in G. Hence $f \in \exp(tX)S \cap F$ and therefore $\exp(tX) \in F$. The fact that F is a face implies that $\exp(\mathbb{R}^+ X) \subseteq F$. Consequently $S \subseteq F$ since $X \in \mathbf{L}(S)$ was arbitrary. ∎

To obtain further results on the structure of faces of Lie semigroups, we need the theory of ordered homogeneous space which will be developed in Chapter 4. Therefore we shall proceed in Chapter 5.

3.2. The interior of Lie semigroups

In this section we will show that the interior of generating Lie semigroups is a dense semigroup ideal. The method to obtain this result is to construct a family of curves in S, as products of one-parameter semigroups, and then to show that the endpoints of these curves cover an open subset of the semigroups S. We start with a simple lemma on semigroups in topological groups. Note that (ii) is a generalization of Proposition 1.1(v).

Lemma 3.7. *Let S be a subsemigroup of a connected topological group G. Then the following assertions hold:*

(i) $\operatorname{int}(S)$ *is a semigroup ideal.*

(ii) *If $1 \in \overline{\operatorname{int} S}$, then we have that*

$$S \subseteq \overline{\operatorname{int} S} \quad and \quad \operatorname{int} S = \operatorname{int} \overline{S}.$$

(iii) *Suppose that $\operatorname{int} S \neq \emptyset$ and $\overline{S} = G$. Then $S = G$.*

Proof. (i) Let $s \in \operatorname{int} S$, $t \in S$, and U a 1-neighborhood in G such that $UsU \subseteq S$. Then $Ust \subseteq S$. Thus $st \in \operatorname{int} S$ and therefore $\operatorname{int}(S)S \subseteq \operatorname{int} S$. That $S \operatorname{int} S \subseteq \operatorname{int} S$ follows similarly.

(ii) Let $s \in S$ and $s_i \in \operatorname{int} S$ with $s_i \to 1$. Then $ss_i \to s$ and $ss_i \in \operatorname{int} S$ imply that $S \subseteq \overline{\operatorname{int} S}$. To see that $\operatorname{int} \overline{S} \subseteq \operatorname{int} S$ (the other inclusion is trivial), let $x \in \operatorname{int} \overline{S}$ and U a 1-neighborhood with $xU^{-1} \subseteq \overline{S}$. Set $W := U \cap \operatorname{int} S$. Then xW^{-1} is an open subset of $\overline{S}$ and therefore it contains an element of $s \in S$. Then $s = xw^{-1}$ with $w \in W$, so $x = sw \in S \operatorname{int} S \subseteq \operatorname{int} S$.

(iii) Since S is dense in G, there exists an element $x \in (\operatorname{int} S)^{-1} \cap S$. Then, using (i), we find that

$$1 = xx^{-1} \in \overline{S} \operatorname{int} S \subseteq \overline{S \operatorname{int} S} \subseteq \overline{\operatorname{int} S}.$$

Now (i) shows that $\operatorname{int} S = \operatorname{int} \overline{S} = G$ and therefore $S = G$. ∎

The following theorem is one of the main results of this section.

Theorem 3.8. (Hofmann-Ruppert Arc Theorem) *If W is a Lie generating Lie wedge in $\mathbf{L}(G)$, then there exist $X_1, \ldots, X_n \in W$ such that the curve*

$$\gamma : \mathbb{R} \to G, \quad t \mapsto \exp tX_n \cdot \ldots \cdot \exp tX_1$$

satisfies

$$\gamma(0) = 1 \quad and \quad \gamma(t) \in \operatorname{int} S_W \quad \forall t \in]0, 1].$$

Before we come to the proof, we need some lemmas.

Lemma 3.9. *Let* $X_1, \ldots, X_n \in \mathbf{L}(G)$ *and*

$$f : \mathbb{R}^n \to G, \quad (t_1, \ldots, t_n) \to \exp t_n X_n \cdots \exp t_1 X_1.$$

Then $(s_1, \ldots, s_n)$ *is a regular point of* f *if and only if the set*

$$B(X_1, \ldots, X_n) := \{X_n, e^{s_n \, \mathrm{ad} \, X_n} X_{n-1}, \ldots, e^{s_n \, \mathrm{ad} \, X_n} \cdots e^{s_2 \, \mathrm{ad} \, X_2} X_1\}$$

spans $\mathbf{L}(G)$.

Proof. To compute the differential $df(s_1, \ldots, s_n)$ in a convenient way we first compute a different representation of f:

$$
\begin{aligned}
&f(s_1 + h_1, \ldots, s_n + h_n) \\
&= \exp\big((s_n + h_n)X_n\big) \cdots \exp\big((s_1 + h_1)X_1\big) \\
&= \exp(h_n X_n) \cdot \exp\big(e^{\mathrm{ad} \, s_n X_n} h_{n-1} X_{n-1}\big) \cdots \\
&\quad \cdot \exp\big(e^{\mathrm{ad} \, s_n X_n} e^{\mathrm{ad} \, s_{n-1} X_{n-1}} \cdots e^{\mathrm{ad} \, s_2 X_2} h_1 X_1\big) \exp s_n X_n \cdots \exp s_1 X_1 \\
&= \exp(h_n X_n) \cdot \exp\big(e^{\mathrm{ad} \, s_n X_n} h_{n-1} X_{n-1}\big) \cdots \\
&\quad \cdot \exp\big(e^{\mathrm{ad} \, s_n X_n} e^{\mathrm{ad} \, s_{n-1} X_{n-1}} \cdots e^{\mathrm{ad} \, s_2 X_2} h_1 X_1\big) f(s_1, \ldots, s_n).
\end{aligned}
$$

It follows that

$$
\begin{aligned}
df(s_1, \ldots, s_n)(u_1, \ldots, u_n) = d\rho_{f(s_1, \ldots, s_n)}(1)\big(u_n X_n + u_{n-1} e^{s_n \, \mathrm{ad} \, X_n} X_{n-1} + \ldots \\
+ u_1 e^{\mathrm{ad} \, s_n X_n} e^{\mathrm{ad} \, s_{n-1} X_{n-1}} \cdots e^{\mathrm{ad} \, s_2 X_2} X_1\big).
\end{aligned}
$$

■

Lemma 3.10. *Let* $\mathfrak{g}$ *be a finite dimensional real Lie algebra and* $E = \mathbb{R}^+ E$ *be a Lie generating subset. Then there exists* $n \in \mathbb{N}$ *and* $X_1, \ldots, X_n \in E$ *such that the sets*

$$\{X_n, e^{t \, \mathrm{ad} \, X_n} X_{n-1}, \ldots, e^{t \, \mathrm{ad} \, X_n} \cdots e^{t \, \mathrm{ad} \, X_2} X_1\}$$

span $\mathfrak{g}$ *for every* $t \in]0, 1]$.

Proof. For $n \in \mathbb{N}$ and $X_1, \ldots, X_n \in E$ the set

$$B(X_1, \ldots, X_n) := \{X_n, e^{\mathrm{ad} \, X_n} X_{n-1}, \ldots, e^{\mathrm{ad} \, X_n} \cdots e^{\mathrm{ad} \, X_2} X_1\}$$

is called a *basic system*. We say that a subspace $\mathfrak{a} \subseteq \mathfrak{g}$ is *nice* if it is spanned by a basic system and maximal with this property.

We claim that every nice subspace contains E. To see this, suppose that $\mathfrak{a}$ is nice and that $E \not\subseteq \mathfrak{a}$. We choose $X \in E \setminus \mathfrak{a}$. Then $e^{-\mathrm{ad} \, X} X = X \notin \mathfrak{a}$ and thus $X \notin e^{\mathrm{ad} \, X} \mathfrak{a}$. For $\mathfrak{a} = \mathrm{span} \, B(X_1, \ldots, X_n)$ this leads to

$$\dim(\mathbb{R}X + e^{\mathrm{ad} \, X} \mathfrak{a}) = \dim \mathrm{span} \, B(X_1, \ldots, X_n, X) > \dim \mathfrak{a},$$

a contradiction.

Let $\mathfrak{a}$ be a nice subspace. If $X \in E$, then $X \in \mathfrak{a}$ entails that $X \in e^{\mathrm{ad} \, X} \mathfrak{a}$ and therefore

$$B(X_1, \ldots, X_n, X) = \mathbb{R}X + e^{\mathrm{ad} \, X} \mathfrak{a} = e^{\mathrm{ad} \, X} \mathfrak{a}$$

is also a nice subspace. Let $\mathfrak{b}$ denote the intersection of all nice subspaces of $\mathfrak{g}$. Then this discussion shows that $E \subseteq \mathfrak{b}$ and that

$$E \subseteq N_{\mathfrak{g}}(\mathfrak{b}) := \{Y \in \mathfrak{g} : [Y, \mathfrak{b}] \subseteq \mathfrak{b}\} = \{Y \in \mathfrak{g} : e^{\mathbb{R}^+ \operatorname{ad} Y} \mathfrak{b} \subseteq \mathfrak{b}\}.$$

Hence the subspace $N_{\mathfrak{g}}(\mathfrak{b})$ is a Lie subalgebra of $\mathfrak{g}$ containing E. Whence $\mathfrak{g} = N_{\mathfrak{g}}(\mathfrak{b})$ and it follows in particular that $[\mathfrak{b}, \mathfrak{b}] \subseteq \mathfrak{b}$. Since $E \subseteq \mathfrak{b}$ we get that $\mathfrak{b} = \mathfrak{g}$. We conclude that there are $X_1, \ldots, X_n \in E$ with

$$\mathfrak{g} = \operatorname{span} B(X_1, \ldots, X_n).$$

Let $I \subseteq \{1, \ldots, n\}$ be a subset such that

$$B_I := \{e^{\operatorname{ad} X_n} \cdot \ldots \cdot e^{\operatorname{ad} X_{i+1}} X_i : i \in I\}$$

is a basis of $\mathfrak{g}$. Let D_t denote the linear endomorphism of $\mathfrak{g}$ which maps the basis element $e^{\operatorname{ad} X_n} \cdot \ldots \cdot e^{\operatorname{ad} X_{i+1}} X_i$ to $e^{\operatorname{ad} t X_n} \cdot \ldots \cdot e^{\operatorname{ad} t X_{i+1}} X_i$. Then $f : \mathbb{R} \to \mathbb{R}, t \mapsto \det D_t$ is an analytic function with $f(1) = 1$. Therefore the set of zeros of f does not cluster at 0, so there exists $\varepsilon > 0$ such that

$$f(t) \neq 0 \qquad \forall t \in]0, \varepsilon].$$

Now the set $\{\varepsilon X_1, \ldots, \varepsilon X_n\}$ satisfies the requirement of the lemma.   ∎

Proof.   (of Theorem 3.8) We select $X_1, \ldots, X_n \in W$ such that the sets

$$\{X_n, e^{t \operatorname{ad} X_n} X_{n-1}, \ldots, e^{t \operatorname{ad} X_n} \cdot \ldots \cdot e^{t \operatorname{ad} X_2} X_1\}$$

span $\mathfrak{g}$ for every $t \in]0, 1]$ (Lemma 3.10). Then Lemma 3.9 implies that for each $s \in]0, t]$ the element $(s, \ldots, s) \in \mathbb{R}^n$ is a regular point for the mapping

$$f : \mathbb{R}^n \to G, \quad (s_1, \ldots, s_n) \mapsto \exp(s_n X_n) \cdot \ldots \cdot \exp(s_1 X_1).$$

Since $f(\mathbb{R}^{+n}) \subseteq S$, it follows that

$$\gamma(s) \in \operatorname{int} S \qquad \forall s \in]0, 1].$$

∎

Corollary 3.11.   If (S, G) is a generating Lie semigroup, then $\operatorname{int} S$ is a dense semigroup ideal which is contained in $\langle \exp W \rangle$ for every wedge $W \subseteq \mathbf{L}(G)$ with $S_W = S$.

Proof.   This follows from Theorem 3.8 and Lemma 3.7.   ∎

3.3. Non-generating Lie semigroups with interior points

In the preceding section we have seen that the interior of a Lie semigroup S is dense whenever the Lie wedge $\mathbf{L}(S)$ is Lie generating in the Lie algebra $\mathbf{L}(G)$. That this condition is even necessary to have dense interior will be shown in Corollary 5.12. Nevertheless it may happen that the Lie wedge $\mathbf{L}(S)$ is not Lie generating and that S has interior points. This section is devoted to the construction of such an example. We start with a lemma which is the main tool in the construction.

Lemma 3.12. *Let (S, A) be a generating invariant Lie semigroup and $W := \mathbf{L}(S)$ such that $H(S) = \{1\}$ and*

(i) *there exists $X \in Z\big(\mathbf{L}(A)\big)$ with $\mathbb{R}X \cap W = \{0\}$ such that $\exp(\mathbb{R}X)$ is closed,*

(ii) *$V := \mathbb{R}X + W$ is global in A,*

(iii) *and there exists $s \in S$ such that $s \exp \mathbb{R}X \subseteq S$.*

Then there exists a Lie group G and a homomorphism $i : A \to G$ such that

(a) *$i(A) \subseteq G$ is dense,*

(b) *$S_{di(1)W}$ has non-empty interior,*

(c) *$di(1)W$ is global and not Lie generating.*

Proof.   Let $T := \mathbb{R}^2/\mathbb{Z}^2$ be the 2-dimensional torus and pick $Y \in \mathbf{L}(T)$ such that $\exp_T(\mathbb{R}Y)$ is dense in T. We set

$$G := (A \times T)/D \quad \text{with} \quad D = \exp\big(\mathbb{R}(X, -Y)\big).$$

Denote the quotient morphism $A \times T \to G$ by p and define $i : A \to G$ by $i(a) := p(a, 1)$. Note that D is closed because $\exp(\mathbb{R}X)$ is closed. Now (a) is clearly satisfied.

(b) We have that

$$p(T) = p\big(\overline{\exp \mathbb{R}(0, Y)}\big) \subseteq \overline{\exp \mathbb{R}dp(1)(0, Y)}$$
$$= \overline{\exp \mathbb{R}dp(1)(X, 0)} = \overline{i(\exp \mathbb{R}X)} \subseteq i(s)^{-1} S_{di(1)W}$$

Therefore

$$i(s)p(T)i(\operatorname{int}_A S) = i(s)p(\operatorname{int}_A S \times T) \subseteq \operatorname{int} S_1.$$

(c) According to our assumption, the Lie wedge $V := \mathbb{R}X + W$ is global in A, hence $V \oplus \mathbf{L}(T)$ is global in $A \times T$ and Lie generating with

$$\mathbb{R}(X, -Y) \subseteq H\big(V \oplus \mathbf{L}(T)\big) = \mathbb{R}X \oplus \mathbf{L}(T).$$

Thus $\widetilde{V} := dp(1)(V \oplus \mathbf{L}(T))$ is global in G (Proposition 1.42). We have

$$di(1)W \cap H(\widetilde{V}) \subseteq di(1)W \cap dp(1)\big(\mathbb{R}X \oplus \mathbf{L}(T)\big) \subseteq di(1)W \cap \mathbb{R}di(1)(X, 0) = \{0\}.$$

Therefore Proposition 1.37 shows that $di(1)W$ is global in G. Since it is contained in the subalgebra $di(1)\mathbf{L}(A) = \mathbf{L}\big(i(A)\big) \neq \mathbf{L}(G)$ it is not Lie generating. ∎

Now we show how to construct an example of a Lie semigroups S in a connected Lie group G such that $\mathbf{L}(S)$ is not Lie generating in $\mathbf{L}(G)$ and $\operatorname{int} S \neq \emptyset$.

We set $A := G_1 \times G_1$, where G_1 is the 4-dimensional oscillator group (cf. Section 2.11). We represent G_1 as $\mathbb{C} \times \mathbb{R} \times \mathbb{R}$ with the multiplication

$$(v, z, r)(v', z', r') = \left(v + e^{ri}v', z + z' + \frac{1}{2}\langle iv, e^{ri}v'\rangle, r + r'\right).$$

Using $d\exp(0)$, we identify $\mathfrak{g} = \mathbf{L}(G)$ with $\mathbb{C} \times \mathbb{R} \times \mathbb{R}$. The cone

$$W_1 := \{(v, z, r) : r - z \geq 0, 2rz + |v|^2 \leq 0\}$$

is invariant in $\mathfrak{g}$ and global in G (cf. Section 2.11). Let $Z \in \mathbf{L}\left(Z(G_1)\right)$ such that $\mathbb{R}Z \cap W_1 = \mathbb{R}^+Z$. Then we use the explicit description of S_{W_1} given in Section 2.11 to see that there exists $s_1 \in S_1 := S_{W_1}$ such that

$$s_1 Z(G_1) \subseteq S_1.$$

We set
$$W := W_1 \oplus W_1 \subseteq \mathbf{L}(A) \quad \text{and} \quad X := (Z, -Z).$$

Then the following assertions hold:

(i) $\mathbb{R}X \cap W \subseteq W \cap Z(\mathbf{L}(A)) \cap \mathbb{R}X = \left(\mathbb{R}^+(Z,0) \oplus \mathbb{R}^+(0,Z)\right) \cap \mathbb{R}X = \{0\}.$

(ii) $\exp \mathbb{R}X \subseteq A$ is closed and central because $Z(A)_0$ is a 2-dimensional vector group and $X \in \mathbf{L}\left(Z(A)\right)$.

(iii) Let $s := (s_1, s_1)$. Then

$$sZ(A)_0 = s_1 Z(G_1)_0 \times s_1 Z(G_1)_0 \subseteq S_W.$$

(iv) $V := W + \mathbb{R}X$ is global in A because it is a generating invariant wedge in a solvable Lie algebra and A is simply connected (cf. [Ne92b, Ch. VIII]). Now the preceding lemma provides the example we are looking for.

The construction of this example uses the fact that it is possible to obtain points in the interior of a Lie semigroups by closing up a subsemigroup which lies in a dense analytic subgroup. Therefore it is remarkable that there exists an example of a dense analytic subgroup A of a Lie group G whose Lie algebra contains a Lie wedge W which is global in G, Lie generating in $\mathbf{L}(A)$, and S_W is contained in A.

Let G denote the product of the 2-torus and $\mathbb{R}$, i.e., $\mathbb{T}_2 \times \mathbb{R}$, $X_0 \in \mathbf{L}(\mathbb{T}_2)$ a generator of a dense one-parameter subgroup, and set

$$W := \{(\lambda X_0, y) \in \mathbf{L}(\mathbb{T}_2) \times \mathbb{R} : |\lambda| \leq y\}.$$

Then $V := W + \mathbf{L}(\mathbb{T}_2)$ is global in G and

$$W \cap H(V) = W \cap \mathbf{L}(\mathbb{T}_2) = \{0\} \subseteq H(W).$$

Therefore W is global (Proposition 1.37) and $\overline{G_{S_W}} = \overline{\langle \exp W \cup \exp(-W) \rangle} = G$ because

$$(\mathbb{R}X \times \{0\}) + (\{0\} \times \mathbb{R}) \subseteq W - W.$$

3.4. The universal covering semigroup $\widetilde{S}$

If W is a closed convex cone in a vector space V, then W is convex and therefore simply connected. A similar statement for Lie semigroups is false in general. There exist generating Lie semigroups in simply connected Lie groups which are not simply connected (see the end of this section). To find criteria for cases when this is true, one has to consider the homomorphism

$$i_* : \pi_1(S) \to \pi_1(G)$$

induced by the inclusion mapping $i: S \to G$, where S is a generating Lie semigroup in the Lie group G. Our main results concern the description of the image and the kernel of this mapping. We show that the image is the fundamental group of the largest covering group of G, into which S lifts, and that the kernel is the fundamental group of the inverse image of S in the universal covering group $\widetilde{G}$. To get these results, we construct a universal covering semigroup $\widetilde{S}$ of S. If $j: H(S) := S \cap S^{-1} \to S$ is the inclusion mapping of the unit group of S into S, then it turns out that the kernel of the induced mapping

$$j_* : \pi_1(H(S)) \to \pi_1(S)$$

may be identified with the fundamental group of the unit group $H(\widetilde{S})$ of $\widetilde{S}$ and that its image corresponds to the intersection $H(\widetilde{S})_0 \cap \pi_1(S)$, where $\pi_1(S)$ is identified with a central subgroup of $\widetilde{S}$.

Let X be a path connected space and $x_0 \in X$. In the following we write $\Omega(X, x_0)$ for the set of all continuous loops $\gamma: [0, 1] \to X$ with $\gamma(0) = \gamma(1) = x_0$ and $\pi_1(X, x_0)$ for the quotient of $\Omega(X, x_0)$ modulo the homotopy relation with fixed endpoints. This is the *fundamental group of X with respect to x_0*. If $\gamma: [0, 1] \to X$ is a continuous path, which is not necessarily a loop, we write $[\gamma]$ for the homotopy class of γ with fixed endpoints. For paths $\alpha, \beta: [0, 1] \to X$, we set $\widehat{\alpha}(t) := \alpha(1 - t)$, and

$$\alpha \diamond \beta(t) = \begin{cases} \alpha(2t) & \text{for } t \in [0, \frac{1}{2}] \\ \beta(2t - 1) & \text{for } t \in [\frac{1}{2}, 1]. \end{cases}$$

Note that this implies that $[\alpha \diamond \beta] = [\alpha][\beta]$ if α and β are loops. If X is a topological monoid, then we usually use the unit element as base point. Since the isomorphy class of the group $\pi_1(X, x_0)$ is independent of x_0, we also write $\pi_1(X)$ for the fundamental group of X without reference to a base point.

From now on G denotes a connected Lie group and S is a generating Lie semigroup in G.

Proposition 3.13. *For a generating Lie semigroup S in G the following assertions hold*

(i) S *and* $\mathrm{int}(S)$ *are path connected.*

(ii) S *is locally path connected.*

(iii) S *is semi-locally simply connected.*

Proof. (i) It follows from Corollary 3.11 that $\mathrm{int}(S)$ is dense and that it is contained in the path connected semigroup $\langle \exp \mathbf{L}(S) \rangle$. Let $\alpha: [0, 1] \to S$ be a path such that $\alpha(0) = \mathbf{1}$ and $\alpha(]0, 1]) \subseteq \mathrm{int}(S)$ (Theorem 3.8). For $s \in S$ the path

$$\gamma : [0, 1] \to S, \quad t \mapsto s\alpha(t)$$

satisfies $\gamma(0) = s$ and $\gamma(]0, 1]) \subseteq s \, \mathrm{int} \, S \subseteq \mathrm{int} \, S$. Since $\gamma(1)$ is contained in the path component of $\mathbf{1}$, it follows that the same is true for s. Therefore S is path connected.

If $a, b \in \mathrm{int}(S)$, then $U := aS^{-1} \cap bS^{-1}$ is a neighborhood of $\mathbf{1}$ in G. Therefore there exists $s_0 \in \mathrm{int}(S) \cap U$. Hence $a, b \in s_0 S$ and $s_0 S$ is path connected. Therefore a and b are connected by a path lying in $\mathrm{int}(S)$.

(ii) Let $s \in S$ and U be an open subset of G containing s. We have to show that $U \cap S$ contains a path connected neighborhood of s with respect to S. Let $\alpha: [0,1] \to S$ be as in (i) with the additional condition that $s\alpha([0,1]) \subseteq S \cap U$ (reparametrization). Then $s\alpha(1) \in \text{int}(S) \cap U$. Hence there exists a contractible 1-neighborhood W in G such that

$$W s\alpha(1) \subseteq \text{int}(S) \cap U \quad \text{and} \quad (Ws \cap S)\alpha([0,1]) \subseteq S \cap U.$$

Let $x, y \in V := (Ws \cap S)\alpha([0,1])$. Then $x = x'\alpha(t_x)$ and $y = y'\alpha(t_y)$, where $t_x, t_y \in [0,1]$ and $x', y' \in (Ws \cap S)$. To show that V is path-connected, we have to show the existence of a continuous path in V from x to y. First we observe that

$$\alpha_x : [t_x, 1] \to S, \ t \mapsto x'\alpha(t) \quad \text{and} \quad \alpha_y : [t_y, 1] \to S, t \mapsto y'\alpha(t)$$

are paths in V connecting x and y with $x'\alpha(1)$ and $y'\alpha(1)$ respectively. But $x'\alpha(1), y'\alpha(1) \in Ws\alpha(1)$, which is a contractible subset of V. Therefore $x'\alpha(1)$ and $y'\alpha(1)$ may be connected by a path in V. Consequently V is a path-connected neighborhood of s in S which is contained in $S \cap U$.

(iii) We keep the notations from (ii). We show that every loop $\beta: [0,1] \to Ws \cap S$ is in S homotopic to the constant loop. Let

$$F(s,t) := \begin{cases} \beta(0)\alpha(3t) & \text{for } t \in [0, \frac{s}{3}] \\ \beta(\frac{3t-s}{3-2s})\alpha(s) & \text{for } t \in [\frac{s}{3}, 1 - \frac{s}{3}] \\ \beta(0)\alpha(3-3t) & \text{for } t \in [1 - \frac{s}{3}, 1]. \end{cases}$$

Then $F: [0,1] \times [0,1] \to S$ is continuous and satisfies $F(0,t) = \beta(t)$ and $F(s,0) = F(s,1) = \beta(0)$. Moreover, $\gamma: t \mapsto F(1,t)$ is homotopic to $\alpha \diamond \beta\alpha(1) \diamond \widehat{\alpha}$ and $\beta\alpha(1)$ lies in the contractible subset $Ws\alpha(1)$ of S. Consequently

$$[\beta] = [\gamma] = [\alpha \diamond \beta\alpha(1) \diamond \widehat{\alpha}] = [\alpha \diamond \widehat{\alpha}] = 1.$$

∎

Theorem 3.14.   *For every generating Lie semigroup $S \subseteq G$ there exists a locally compact topological monoid $\widetilde{S}$ and a mapping $p: \widetilde{S} \to S$ with the following properties:*

(i) *$\widetilde{S}$ is path connected, locally path connected, and $\pi_1(\widetilde{S}) = \{1\}$.*

(ii) *$p: \widetilde{S} \to S$ is a covering and a semigroup homomorphism.*

(iii) *$\text{int}(\widetilde{S}) := p^{-1}(\text{int}(S))$ is a dense semigroup ideal in $\widetilde{S}$.*

(iv) *If $q: T \to S$ is a covering homomorphism of path connected topological monoids, then there exists a unique covering homomorphism $\widetilde{p}: \widetilde{S} \to T$ such that $\widetilde{p}(1_{\widetilde{S}}) = 1_T$ and $q \circ \widetilde{p} = p$.*

(v) *Let $\alpha: T \to S$ be a continuous homomorphism of locally path connected connected topological monoids and D a subgroup of $\pi_1(S) \subseteq \widetilde{S}$. Then there exists a unique continuous monoid homomorphism $\widetilde{\alpha}: T \to \widetilde{S}/D$ with $p \circ \widetilde{\alpha} = \alpha$ if and only if $\alpha_*(\pi_1(T)) \subseteq D$.*

Proof.   (i) The existence of a universal covering $p: \widetilde{S} \to S$ follows from [Sch75, p.229] because S is path connected, locally path connected, and semi-locally simply connected.

(ii) To define the structure of a monoid on $\widetilde{S}$, we choose $\widetilde{1} \in p^{-1}(1)$. Let $m_s \colon S \times S \to S$ denote the multiplication of S. Then $m_S \circ (p \times p) \colon \widetilde{S} \times \widetilde{S} \to S$ lifts uniquely to a continuous mapping $m_{\widetilde{S}} \colon \widetilde{S} \times \widetilde{S} \to \widetilde{S}$ such that $m_{\widetilde{S}}(\widetilde{1}, \widetilde{1}) = \widetilde{1}$ and $p \circ m_{\widetilde{S}} = m_S \circ (p \times p)$. This follows from [Sch75, p.221] because $\widetilde{S} \times \widetilde{S}$ is path connected, locally path connected, and simply connected ([Sch75, p.203]). We show that $\widetilde{S}$ is a monoid with respect to this multiplication. The mapping $\alpha \colon \widetilde{S} \to \widetilde{S}, s \mapsto \widetilde{1}s$ satisfies $p \circ \alpha = p = p \circ \mathrm{id}_{\widetilde{S}}$ and $\alpha(\widetilde{1}) = \widetilde{1}$. Now the uniqueness of the lift ([Sch75, p.221]) implies that $\alpha = \mathrm{id}_{\widetilde{S}}$. Thus $\widetilde{1}s = s$ holds for all $s \in S$. That $\widetilde{1}$ is a right unit follows similarly. Since

$$p \circ (m_{\widetilde{S}} \times \mathrm{id}_{\widetilde{S}}) \circ m_{\widetilde{S}} = p \circ (\mathrm{id}_{\widetilde{S}} \times m_{\widetilde{S}}) \circ m_{\widetilde{S}}$$

and $(\widetilde{1}\widetilde{1})\widetilde{1} = \widetilde{1} = \widetilde{1}(\widetilde{1}\widetilde{1})$, the fact that $\widetilde{S} \times \widetilde{S} \times \widetilde{S}$ is path connected, locally path connected, and simply connected ([Sch75, p.203]) implies that

$$(m_{\widetilde{S}} \times \mathrm{id}_{\widetilde{S}}) \circ m_{\widetilde{S}} = (\mathrm{id}_{\widetilde{S}} \times m_{\widetilde{S}}) \circ m_{\widetilde{S}},$$

i.e., multiplication on $\widetilde{S}$ is associative ([Sch75, p.221]). That $p \colon \widetilde{S} \to S$ is a homomorphism is a consequence of

$$m_S \circ (p \times p) = p \circ m_{\widetilde{S}}.$$

(iii) As the inverse image of an ideal, the subset $\mathrm{int}(\widetilde{S}) := p^{-1}(\mathrm{int}(S))$ is a semigroup ideal. Since p is a local homeomorphism and $1 \in \overline{\mathrm{int}(S)}$, it follows that $\widetilde{1} \in \overline{\mathrm{int}(\widetilde{S})}$. Therefore $s \in s\,\mathrm{int}(\widetilde{S}) \subseteq \mathrm{int}(\widetilde{S})$ for all $s \in S$.

(iv) That there exists a continuous mapping $\widetilde{p} \colon \widetilde{S} \to T$ such that $q \circ \widetilde{p} = p$ and $\widetilde{p}(\widetilde{1}) = 1_T$ follows again from [Sch75, p.221] and Proposition 3.13. If $m_T \colon T \times T \to T$ denotes multiplication in T, then

$$\widetilde{p} \circ m_{\widetilde{S}}(\widetilde{1}, \widetilde{1}) = m_T \circ (\widetilde{p} \times \widetilde{p})(\widetilde{1})$$

and

$$\begin{aligned} q \circ m_T \circ (\widetilde{p} \times \widetilde{p}) &= m_S \circ (q \times q) \circ (\widetilde{p} \times \widetilde{p}) \\ &= m_S \circ (p \times p) = p \circ m_{\widetilde{S}} \\ &= q \circ \widetilde{p} \circ m_{\widetilde{S}}. \end{aligned}$$

Now the uniqueness assertion of [Sch75, p.221] for the lift of this mapping shows that

$$m_T \circ (\widetilde{p} \times \widetilde{p}) = \widetilde{p} \circ m_{\widetilde{S}},$$

i.e., $\widetilde{p}$ is a morphism of topological monoids.

We claim that $\widetilde{p}$ is surjective. Let $t \in T$ and $\beta \colon [0,1] \to T$ be a path with $\beta(0) = 1_T$ and $\beta(1) = t$. Then there exists a path $\alpha \colon [0,1] \to \widetilde{S}$ such that $\alpha(0) = 1$, $p \circ \alpha(1) = q(t)$, and $p \circ \alpha = q \circ \beta$. Hence $\beta(1) = t = \widetilde{p}(\alpha(1))$ ([Sch75, p.221]) and $\widetilde{p}$ is surjective. Now it follows immediately from the definition that $\widetilde{p}$ is a covering.

(v) In view of [tD91, p.128], we only have to show that whenever a continuous mapping $\widetilde{\alpha}: T \to S' := \widetilde{S}/D$ with $\widetilde{\alpha}(1) = 1$ exists, then it is a monoid homomorphism, i.e., that

$$\widetilde{\alpha} \circ m_T = m_{S'} \circ (\widetilde{\alpha} \times \widetilde{\alpha})$$

holds, where m_T and $m_{S'}$ are the multiplication mappings of T and S' respectively. If $p': S' \to S$ denotes the covering morphism, then this follows immediately from

$$\begin{aligned}
p' \circ m_{S'} \circ (\widetilde{\alpha} \times \widetilde{\alpha}) &= m_S \circ (p' \times p') \circ (\widetilde{\alpha} \times \widetilde{\alpha}) \\
&= m_S \circ \big((p' \circ \widetilde{\alpha}) \times (p' \circ \widetilde{\alpha})\big) \\
&= m_S \circ (\alpha \times \alpha) \\
&= \alpha \circ m_T \\
&= p' \circ \widetilde{\alpha} \circ m_T
\end{aligned}$$

and the fact that

$$\widetilde{\alpha} \circ m_T(\mathbf{1},\mathbf{1}) = 1 = m_{S'} \circ (\widetilde{\alpha} \times \widetilde{\alpha})(\mathbf{1},\mathbf{1}). \qquad \blacksquare$$

In [Ka70] Kahn defines the notion of a covering semigroup $(\overline{S}, \varphi)$ of a topological semigroup S as a pair of a topological semigroup $\overline{S}$ and a covering $\varphi: \overline{S} \to S$ which is a covering. He calls a semigroup S simply connected if for every covering semigroup $(\overline{S}, \varphi)$ of S the mapping φ is a homeomorphism. Now Theorem 3.14 and [Ti83, p.84] show that our $\widetilde{S}$ is simply connected in this sense. Therefore $(\widetilde{S}, p)$ is the simply connected covering semigroup of S in the sense of Kahn ([Ka70, p.430]).$\blacksquare$

Proposition 3.15.   *Let I be a dense path connected semigroup ideal in the path connected topological monoid S. Suppose that there exists a path $\beta: [0,1] \to S$ such that*

$$\beta(0) = 1 \quad and \quad \beta(]0,1]) \subseteq I.$$

Then the inclusion $i: I \to S$ induces an isomorphism

$$i_* : \pi_1(I) \to \pi_1(S).$$

Proof.   Pick a point $x_0 \in I$ which serves as base point for I and S simultaneously.
(1) i_* is injective: Let $\gamma \in \Omega(I, x_0)$ such that $i_*[\gamma] = [i \circ \gamma] = 1$ in $\pi_1(S)$. Then there exists a continuous mapping $F: [0,1] \times [0,1] \to S$ such that

$$F(0,t) = \gamma(t), \quad F(1,t) = x_0, \quad and \quad F(s,0) = F(s,1) = x_0$$

for $s, t \in [0,1]$. We define $G: [0,1] \times [0,1] \to S$ by

$$G(s,t) := F(s,t)\beta\big(s(1-s)t(1-t)\big),$$

where β is a continuous curve $[0,1] \to S$ such that $\beta(0) = 1$ and $\beta(]0,1]) \subseteq I$. Then G is a deformation of γ to the constant path in x_0 and $\operatorname{im} G \subseteq I$ because $F(0,t), F(1,t) \in I$ for all $t \in [0,1]$. Hence $[\gamma] = [x_0]$ and i_* is injective.
(2) i_* is surjective: Let $[\gamma] \in \pi_1(S)$ and $\gamma: [0,1] \to S$ with $\gamma(0) = \gamma(1) = x_0$. Then

$$F: [0,1] \times [0,1] \to S, \quad (s,t) \mapsto \beta\big(t(1-t)s\big)\gamma(t)$$

deforms the path γ into a path which lies entirely in I. Hence $[\gamma] \in \operatorname{im}(i_*)$. $\blacksquare$

Corollary 3.16.

(i) *The inclusions* $i\colon \operatorname{int}(S) \to S$, $\widetilde{i}\colon \operatorname{int}(\widetilde{S}) \to \widetilde{S}$ *induce isomorphisms*

$$i_* : \pi_1\big(\operatorname{int}(S)\big) \to \pi_1(S) \quad \text{and} \quad \widetilde{i}_* : \pi_1\big(\operatorname{int}(\widetilde{S})\big) \to \pi_1(\widetilde{S}).$$

(ii) $\pi_1\big(\operatorname{int}(\widetilde{S})\big) = \{\mathbf{1}\}$.

Proof. The first statement follows from Propositions 3.13 and 3.15. Since $\pi_1(\widetilde{S}) = \{\mathbf{1}\}$, the second statement is a consequence of (i) and Theorem 3.14. ∎

Now we turn to the algebraic structure of the covering semigroups of a given generating Lie semigroup (S, G). The following lemma is a key ingredient.

Proposition 3.17. (Hilton's Lemma for monoids) *Let S be a topological monoid, $\gamma\colon [0,1] \to S$ a continuous path with $\gamma(0) = \mathbf{1}$, and $\gamma' \in \Omega(S, 1)$. Then*

$$[\gamma\gamma'] = [\gamma'\gamma] = [\gamma' \diamond \gamma],$$

where $\gamma\gamma'(t) = \gamma(t)\gamma'(t)$.

Proof. We set

$$\eta'(x) = \begin{cases} \gamma'(2t) & \text{if } t \in [0, \tfrac{1}{2}] \\ \gamma'(1) & \text{if } t \in [\tfrac{1}{2}, 1] \end{cases}$$

and

$$\eta(t) = \begin{cases} 1 & \text{if } t \in [0, \tfrac{1}{2}] \\ \gamma(2t-1) & \text{if } t \in [\tfrac{1}{2}, 1]. \end{cases}$$

Then

$$\eta\eta' = \eta'\eta = \eta' \diamond \eta.$$

Clearly $[\eta] = [\gamma]$ and $[\eta'] = [\gamma']$. Therefore

$$[\gamma\gamma'] = [\eta\eta'] = [\eta'\eta] = [\gamma'\gamma] = [\eta' \diamond \eta] = [\gamma' \diamond \gamma].$$

∎

For the following corollary we recall that an action of a group G on a space X is called *free* if all the orbit mappings $g \mapsto g.x$ are injective, and a continuous action of a topological group G on a topological space X is called *proper* if the mapping

$$\theta : G \times X \to X \times X, \quad (g, x) \mapsto (x, g.x)$$

is *proper*. We recall that, if G and X are locally compact, this means that the inverse images of compact subsets of $X \times X$ under θ are compact.

Corollary 3.18. *Let $p\colon \widetilde{S} \to S$ be as above, then the following assertions hold:*

(i) *Let $\widetilde{\gamma}$ denotes the lift of γ with $\widetilde{\gamma}(0) = \widetilde{1}$. Then the mapping $[\gamma] \mapsto \widetilde{\gamma}(1), \pi_1(S) \to D := p^{-1}(1)$ is an isomorphism of groups.*

(ii) $D \subseteq Z(\widetilde{S}) := \{s \in \widetilde{S} : (\forall t \in \widetilde{S})st = ts\}$.

(iii) $\pi_1(S)$ *is abelian.*

(iv) *The multiplicative action of the discrete group D on $\widetilde{S}$ is free and proper, and it induces an isomorphism of D with the group of deck transformatioms of the covering $\widetilde{S} \to S$.*

(v) *The mapping $\widetilde{S}/D \to S$ is an isomorphism of topological semigroups.*

Proof. (i) Let $\gamma, \gamma' \in \Omega(S, 1)$. Using Hilton's Lemma we find that

$$[\gamma][\gamma'] = [\gamma\gamma'] \mapsto \widetilde{\gamma}(1)\widetilde{\gamma}'(1).$$

Hence $[\gamma] \mapsto \widetilde{\gamma}(1)$ defines a monoid homomorphism. It follows from the construction of $\widetilde{S}$ that it is bijective and therefore $p^{-1}(1)$ is a group isomorphic to $\pi_1(S)$.

(ii) Let $d \in p^{-1}(1)$ and $s \in \widetilde{S}$. Then there exists $\gamma' \in \Omega(S, 1)$ with $\widetilde{\gamma}'(1) = d$ and a path $\gamma: [0, 1] \to S$ such that $\widetilde{\gamma}(1) = s$. Then Hilton's Lemma shows that

$$sd = (\gamma\gamma')\widetilde{}(1) = (\gamma'\gamma)\widetilde{}(1) = ds.$$

Hence d is central in $\widetilde{S}$.

(iii) This is a consequence of (i) and (ii).

(iv) First we note that, according to (ii), we need not distinguish between left and right multiplication with elements of D. Since the group D acts transitively on the fiber $p^{-1}(1)$, and it acts by deck transformations because $p(sd) = p(s)p(d) = p(s)$ for all $s \in \widetilde{S}$, $d \in D$, it follows from [tD91, 6.2] that D acts as the group of deck transformations, and that $\widetilde{S} \to S$ is a D-principal bundle. Hence D acts freely on $\widetilde{S}$ and $\widetilde{S}/D \cong S$. Since the set

$$\{(s, sd) : s \in \widetilde{S}, d \in D\} = \{(u, v) \in \widetilde{S} \times \widetilde{S} : p(u) = p(v)\}$$

is closed, the action of D on $\widetilde{S}$ is proper by [tD91, 6.2].

(v) This follows from the fact that $\widetilde{S} \to S$ is a D-principal bundle. ∎

From now on we identify the fundamental group $\pi_1(S)$ with the discrete central subgroup $p^{-1}(1) \subseteq \widetilde{S}$.

With Corollary 3.18 at hand we can show that the multiplication mapping of $\widetilde{S}$ shares many properties with the multiplication mapping of S.

Proposition 3.19.   *Let $m: \widetilde{S} \times \widetilde{S}$ denote the multiplication mapping of $\widetilde{S}$. Then we have*

(i) *$\widetilde{S}$ is cancellative, i.e., $ab = ac$ or $ba = ca$ implies that $b = c$.*

(ii) *For every compact subset $K \subseteq \widetilde{S}$, the restriction of m to the sets $K \times \widetilde{S}$ and $\widetilde{S} \times K$ is proper.*

(iii) *The left and right multiplication mappings $\lambda_s(x) = sx$ and $\rho_s(x) = xs$ on $\widetilde{S}$ are proper homeomorphisms of S onto sS and Ss respectively.*

(iv) *For every compact subset $K \subseteq \widetilde{S}$ and every neighborhood U of K there exists a $\widetilde{1}$-neighborhood V in $\widetilde{S}$ such that*

$$\lambda_s^{-1}(K) \cup \rho_s^{-1}(K) \subseteq U \qquad \forall s \in V.$$

(v) *For every compact subset $K \subseteq \text{int} \, S$ there exists a $\widetilde{1}$-neighborhood V in $\widetilde{S}$ such that*

$$K \subseteq s(\text{int} \, \widetilde{S}) \cap (\text{int} \, \widetilde{S})s \qquad \forall s \in V.$$

Proof.   (i) Suppose that $ab = ac$ with $a, b, c \in \widetilde{S}$. Then

$$p(a)p(b) = p(ab) = p(ac) = p(a)p(c)$$

implies that $p(b) = p(c)$. Therefore we find $d \in \pi_1(S)$ such that $c = bd$ (Corollary 3.18). Hence $ab = ac = a(bd) = (ab)d$ and Corollary 3.18(iv) shows that $d = \widetilde{1}$, i.e., $b = c$. The other implication follows similarly.

(ii) Let $Q \subseteq \widetilde{S}$ be a compact subset. We have to show that $m^{-1}(Q) \cap (K \times \widetilde{S})$ is compact. Suppose that this is false. Then there exist sequences $s_n \in \widetilde{S}$, $k_n \in K$ such that $q_n := k_n s_n \in Q$ and s_n eventually leaves every compact subset of $\widetilde{S}$. In addition, we may assume that $q_n \to q$ and $k_n \to k$. Then $p(q_n) = p(k_n)p(s_n) \to p(q)$ and since $p(k_n) \to p(k)$, we conclude that $p(s_n) \to s := p(k)^{-1}p(q) \in S$. Now we use the fact that $\widetilde{S}$ is a $\pi_1(S)$-principal bundle (Corollary 3.18(iv)) to find sequences $s_n' \in \widetilde{S}$ and $d_n \in \pi_1(S)$ such that $s_n = s_n' d_n$ and $s_n' \to s'$ with $p(s') = p(s)$. Now $q_n = (k_n s_n')d_n \to q$ and since the action of $\pi_1(S)$ is proper, Corollary 3.18(iv) entails that the sequence d_n lies in a compact subset because $k_n s_n' \to ks'$. This contradicts the assumption that s_n eventually leaves every compact set.

(iii) The properness follows from (ii) with $K = \{s\}$ and that left and right multiplications are homeomorphism onto the image follows from the injectivity (i) and the fact that proper mappings are closed.

(iv) By symmetry, it suffices to show that $\lambda_s^{-1}(K) \subseteq U$ for s sufficiently near to $\widetilde{1}$. Suppose that this is false. Then there exist sequences $s_n \to \widetilde{1}$ and $t_n \in \widetilde{S} \setminus U$ such that $k_n := s_n t_n \in K$. Let V' be a compact $\widetilde{1}$-neighborhood. Then $m^{-1}(K) \cap (V' \times \widetilde{S})$ is compact by (ii). Hence we may assume that $t_n \to t$. Then $k_n = s_n t_n \to \widetilde{1}t = t \in K$, a contradiction.

(v) The set $p(K) \subseteq \text{int} \, S = p(\text{int} \, \widetilde{S})$ is compact. Hence there exists a symmetric 1-neighborhood U in G such that $Up(K)U \subseteq \text{int} \, S$. We set $V := p^{-1}(U)$. Let $k \in K$ and $s \in V$. Then $p(s)^{-1} \in U$ and therefore the elements $s_1 := p(k)p(s)^{-1}$ and $s_2 := p(s)^{-1}p(k)$ are contained in $\text{int} \, S$. Pick $\widetilde{s}_i \in p^{-1}(s_i)$, $i = 1, 2$. Then $k \in p^{-1}(s_1 s) = \pi_1(S)\widetilde{s}_1 s \in (\text{int} \, \widetilde{S})s$ and similarly $k \in s(\text{int} \, \widetilde{S})$.   ∎

Theorem 3.20.   *Let $D \subseteq Z(\widetilde{S})$ be a discrete subgroup. Then the following assertions hold:*

(i) *D acts properly on $\widetilde{S}$.*

(ii) *The quotient mapping*

$$q \colon \widetilde{S} \to S_D := \widetilde{S}/D, \qquad s \mapsto sD$$

is a covering morphism of locally compact semigroups.

(iii) *The assertions of Proposition 3.19 remain true for S_D, where $\text{int} \, S_D := q(\text{int} \, \widetilde{S})$.*

Proof. (i) Let $K \subseteq \widetilde{S}$ be compact and m denote the multiplication mapping of $\widetilde{S}$. Then the set

$$m^{-1}(K) \cap (K \times \widetilde{S}) = \{(k, s) : ks \in K\}$$

is compact by Proposition 3.19(i). Hence the set

$$\{d \in D : (\exists k \in K) dk \in K\}$$

is compact. This shows that D acts properly on $\widetilde{S}$.

(ii) Using [tD91, 5.6], we see that the quotient semigroup S_D endowed with the quotient topology is Hausdorff because D acts properly on $\widetilde{S}$. Since $\widetilde{S}$ is cancellative (Proposition 3.19), D also acts freely on $\widetilde{S}$. Now [tD91, 5.8, 6.1, 6.2] yields that $q \colon \widetilde{S} \to S_D$ is a D-principal bundle, therefore it is a covering mapping. It follows in particular that S_D is a locally compact semigroup.

(iii) (a) S_D is cancellative: Let $s_1, s_2, s_3 \in S_D$ with $s_1 s_2 = s_1 s_3$. Then there exist $\widetilde{s}_i \in \widetilde{S}$ with $q(\widetilde{s}_i) = s_i$, $i = 1, \ldots, 3$. Now $q(\widetilde{s}_1 \widetilde{s}_2) = q(\widetilde{s}_1 \widetilde{s}_3)$ implies the existence of $d \in D$ such that $\widetilde{s}_1 \widetilde{s}_2 = \widetilde{s}_1 \widetilde{s}_3 d$. Now $\widetilde{s}_2 = \widetilde{s}_3 d$ follows from the cancellativity of $\widetilde{S}$. Whence $s_2 = s_3$.

(b) Since q is an open mapping and $\widetilde{S}$ is locally compact, we find for every compact subset $K \subseteq S_D$ a compact subset $\widetilde{K} \subseteq \widetilde{S}$ such that $q(\widetilde{K}) = K$. Let m_D denote the multiplication mapping of S_D. If $Q \subseteq S_D$ is compact, $\widetilde{Q} \subseteq \widetilde{S}$ compact with $q(\widetilde{Q}) = Q$, and $ks \in Q$ with $k \in K$, then there exist $\widetilde{k} \in \widetilde{K}$, and $\widetilde{s} \in q^{-1}(s)$ such that $\widetilde{k}\widetilde{s} \in \widetilde{Q}$. Hence

$$m_D^{-1}(Q) \cap (K \times S_D) \subseteq (q \times q)\big(m^{-1}(\widetilde{Q}) \cap (\widetilde{K} \times \widetilde{S})\big)$$

which is a compact set according to Proposition 3.19(ii). Hence this assertion remains true for S_D.

(c) To check the validity of Proposition 3.19(iv), let $K \subseteq S_D$ be compact, U a neighborhood of K, $\widetilde{K}$ as above, and $\widetilde{U} := q^{-1}(U)$. Then there exists a $\widetilde{1}$-neighborhood $\widetilde{V}$ in $\widetilde{S}$ such that

$$\lambda_s^{-1}(\widetilde{K}) \cup \rho_s^{-1}(\widetilde{K}) \subseteq \widetilde{U} \qquad \forall s \in \widetilde{V}.$$

Set $V := q(\widetilde{V})$ and let $s \in V$ and $\widetilde{s} \in \widetilde{V} \cap q^{-1}(s)$. If $sx \in K$, then there exists $\widetilde{x} \in q^{-1}(x)$ such that $\widetilde{s}\widetilde{x} \in \widetilde{K}$. Hence $\widetilde{x} \in q^{-1}(U)$ and therefore $x \in U$.

(d) Let $K \subseteq \operatorname{int} S_D$ be compact and $\widetilde{K}$ as above. Then $\widetilde{K} \subseteq \operatorname{int} \widetilde{S}$ and the validity of Proposition 3.19(v) for S follows from the validity for $\widetilde{S}$ and from $q(\widetilde{K}) = K$. ∎

Proposition 3.21. *The set* $p^{-1}(H(S))$ *coincides with the unit group* $H(\widetilde{S})$ *of* $\widetilde{S}$. *The mapping* $p|_{H(\widetilde{S})}$ *is a covering of Lie groups.*

Proof. It is clear that $p^{-1}(H(S))$, as the inverse image of a subsemigroup, is a closed subsemigroup of $\widetilde{S}$ and that it contains $H(\widetilde{S})$. Let $x \in p^{-1}(H(S))$. Then there exists $s \in S$ such that $p(x)s = 1$. Let $s = p(y)$. Then $p(xy) = p(x)p(y) = 1$ and therefore $xy \in p^{-1}(1)$. Since $p^{-1}(1)$ is a subgroup of $\widetilde{S}$ (Corollary 3.18), it is contained in $H(\widetilde{S})$. Thus $x \in H(\widetilde{S})$ because $\widetilde{S} \setminus H(\widetilde{S})$ is a semigroup ideal.

Since $p \colon \widetilde{S} \to S$ is a covering, it is obvious that the restriction of p to $H(\widetilde{S})$ is a covering morphism of topological groups. Therefore $H(\widetilde{S})$ is a Lie group and $p|_{H(\widetilde{S})}$ a covering morphism of Lie groups. ∎

We note that the semigroups $\widetilde{S}$ need not be generated, not even topologically, by an arbitrary small neighborhood of 1. To see this, let $G_1 := \mathbb{R}^2$ and $S_1 := \mathbb{R}^+(1, -1) \times \mathbb{R}^+(1, 1)$. Set $G := \mathbb{R} \times \mathbb{R}/\mathbb{Z}$ and write $p : G_1 \to G, (x, y) \mapsto (x, y + \mathbb{Z})$ for the quotient homomorphism. Then the image $S = p(S_1)$ of S_1 in G is a generating Lie semigroup with $\pi_1(S) \cong \mathbb{Z}$ (cf. Theorem 3.37 below). The universal covering $\widetilde{S}$ corresponds to the subsemigroup $\mathbb{Z} + S_1$ of G_1. The subsemigroup S_1 is a closed neighborhood of 1 in $\widetilde{S}$ and $H(\widetilde{S}) \cong \mathbb{Z}$ is not connected.

Proposition 3.22. *For $X \in \mathbf{L}(S)$ we set $\gamma_X : \mathbb{R}^+ \to S, t \mapsto \exp(tX)$. Then $X \mapsto \widetilde{\gamma}_X$ is a bijection $\mathbf{L}(S) \to \mathrm{Hom}(\mathbb{R}^+, \widetilde{S})$. Define $\mathrm{Exp} : \mathbf{L}(S) \to \widetilde{S}, X \mapsto \widetilde{\gamma}_X(1)$. Then the semigroup*

$$\widetilde{S}_L := \overline{\langle \mathrm{Exp}\left(\mathbf{L}(S)\right)\rangle}$$

is a neighborhood of 1 in $\widetilde{S}$. It is the smallest subsemigroup topologically generated by every neighborhood of 1 in $\widetilde{S}$. Moreover

$$\widetilde{S} = \overline{\pi_1(S)\widetilde{S}_L}.$$

Proof. The first statement follows from the fact that $p : \widetilde{S} \to S$ induces a local isomorphism from a neighborhood of $\widetilde{1}$ in $\widetilde{S}$ to a 1-neighborhood in S. Now the second statement follows from the assumption that S is a Lie semigroup, and the last assertion is clear because $\overline{\pi_1(S)\widetilde{S}_L}$ is a $\pi_1(S)$-saturated subset of $\widetilde{S}$ which is mapped surjectively onto S. ∎

Lemma 3.23. *Let $q : \widetilde{G} \to G$ denote the universal covering group of G, identify $\pi_1(G)$ with $\ker q$, and $\pi_1(S)$ with $p^{-1}(1)$. Then there exists a continuous homomorphism $\widetilde{i} : \widetilde{S} \to \widetilde{G}$ such that $q \circ \widetilde{i} = i \circ p$, $\widetilde{i}|_{\pi_1(S)} = i_*$, and the image of $\widetilde{i}$ is the path-component of 1 in $q^{-1}(S)$.*

Proof. The only thing we have to prove is the existence of $\widetilde{i}$. The rest follows from the identification of $\pi_1(S)$ and $\pi_1(G)$ with subgroups of $\widetilde{S}$ and $\widetilde{G}$ respectively. Let S_1 be the path-component of 1 in $q^{-1}(S)$. It follows from Proposition 3.13 that $q^{-1}(S)$ is locally path connected because q is a local homeomorphism. Therefore S_1 is an open closed connected component of $q^{-1}(S)$. Now the universal property of $\widetilde{S}$ (Theorem 3.14(iv)) implies the existence of a surjective semigroup covering $\widetilde{i} : \widetilde{S} \to S_1$ such that $q \circ \widetilde{i} = p$. ∎

Theorem 3.24. *Let $j : H(S) \to S$ be the inclusion mapping and*

$$j_* : \pi_1\left(H(S)\right) \to \pi_1(S)$$

the induced homomorphism of the fundamental groups. Then

$$\ker j_* = \pi_1\left(H(\widetilde{S})\right) \quad and \quad \mathrm{im}\, j_* = H(\widetilde{S})_0 \cap \pi_1(S).$$

Proof. Let $\widetilde{H(S)}$ be the universal covering group of $H(S)$. Then there exists a Lie group homomorphism $q : \widetilde{H(S)} \to H(\widetilde{S})_0$ such that $p \circ q : \widetilde{H(S)} \to H(S)$ is the universal covering morphism of $H(S)$. The homomorphism j_* corresponds to the homomorphism

$$q\big|_{\pi_1(H(S))} : \pi_1\big(H(S)\big) \to \pi_1(S),$$

$\pi_1\big(H(S)\big)$ is identified with the corresponding subgroup of $\widetilde{H(S)}$. Thus $H(\widetilde{S})_0 \cong \widetilde{H(S)}/\ker j_*$ implies that $\ker j_* = \pi_1\big(H(\widetilde{S})_0\big)$.

The image is clearly contained in $D' := H(\widetilde{S})_0 \cap \pi_1(S)$. But $H(S) \cong H(\widetilde{S})_0/D'$ and therefore $D' = \mathrm{im}\, j_*$.   ∎

The situation of Theorem 3.24 is illustrated in the following diagram.

$$
\begin{array}{ccccc}
\pi_1\big(H(S)\big) & \xrightarrow{\;\;j_*\;\;} & \pi_1(S) \cap H(\widetilde{S})_0 & \longrightarrow & 1 \\
\downarrow & & \downarrow & & \downarrow \\
\widetilde{H(S)} & \longrightarrow & H(\widetilde{S})_0 & \xrightarrow{\;\;p\;\;} & H(S) \\
& & \downarrow & & \downarrow \\
& & \widetilde{S} & \xrightarrow{\;\;p\;\;} & S
\end{array}
$$

Corollary 3.25.   *The mapping $j_*\colon \pi_1\big(H(S)\big) \to \pi_1(S)$ is*

(i) *injective if and only if $H(\widetilde{S})$ is simply connected.*

(ii) *surjective if and only if $H(\widetilde{S})$ is connected.*

Proof.   In view of Theorem 3.24 we only have to show that the connectedness of $H(\widetilde{S})$ follows from the surjectivity of j_*. If $\pi_1(S) = \pi_1(S) \cap H(\widetilde{S})_0$, then $\pi_1(S) \subseteq H(\widetilde{S})_0$ and therefore, according to Proposition 3.21,

$$H(\widetilde{S})_0 = p^{-1}\big(H(S)\big) = H(\widetilde{S}).$$

∎

We illustrate the situation in some examples.

First let $G := \mathrm{SU}(2) \times \mathbb{R}$, the universal covering of the unitary group in dimension 2. Then $\mathbf{L}(G) = \mathfrak{su}(2) \oplus \mathbb{R} \cong \mathfrak{u}(2)$ and there exists a pointed invariant wedge $C \subseteq \mathbf{L}(G)$ with non-empty interior (take the matrices with spectrum on the positive imaginary axis in $\mathfrak{u}(2)$). Pick $X \in \mathfrak{su}(2)$. Then $\exp \mathbb{R}X$ is a circle group in G because $\overline{\exp \mathbb{R}X}$ is a torus and the maximal tori in $\mathrm{SU}(2)$ are of dimension 1. We set $W := \mathbb{R}X + C$. Then $S := S_W$ is a generating Lie semigroup in G. That $\mathbf{L}(S) = W$ follows from Proposition 1.44 because $\exp H(W)$ is a closed subgroup of G, $\mathrm{SU}(2)$ is the unique maximal compact subgroup of G, and $\mathfrak{su}(2) \cap W = H(W)$.

According to Theorem 3.43 below, we know that S is simply connected because G is simply connected and $\mathbf{L}(G)$ is a compact Lie algebra. Hence $\widetilde{S}$ agrees with S and we have an example, where $H(\widetilde{S})$ is not simply connected. In view of Corollary 3.25 this is related to the fact that the circle $\exp \mathbb{R}X$ cannot be deformed to a constant loop in $H(S)$, but if one pushes it far enough into the interior of S, for example into a coset of $\mathrm{SU}(2)$, the contraction becomes possible.

That $H(\widetilde{S})$ need not be connected follows from the example given after Proposition 3.21.

Next let $\mathfrak{g}$ be a Lie algebra which contains a pointed generating invariant cone C, $G_{\mathbb{C}}$ the simply connected Lie group with $\mathbf{L}(G_{\mathbb{C}}) = \mathfrak{g}_{\mathbb{C}}$, and $G := \langle \exp_{G_{\mathbb{C}}} \mathfrak{g} \rangle \subseteq G_{\mathbb{C}}$. Then the set $S := G \exp(iC)$ is a generating Lie semigroup in $G_{\mathbb{C}}$ (Lawson's Theorem, Theorem 7.33). We claim that $H(\widetilde{S}) = \widetilde{G}$ and that $\widetilde{S} = \widetilde{G} \operatorname{Exp}(iC)$.

Clearly $\widetilde{G} \times C$ is a simply connected, locally path connected space. Therefore the mapping $\varphi \colon \widetilde{G} \times C \to S, (g,c) \mapsto g \exp(ic)$, which is a covering, lifts to a covering $\widetilde{\varphi} \colon \widetilde{G} \times C \to \widetilde{S}$ with $p \circ \widetilde{\varphi} = \varphi$. Since $\widetilde{S}$ is simply connected and locally path connected (Proposition 3.13), the mapping $\widetilde{\varphi}$ is a homeomorphism. This proves that $H(\widetilde{S}) \cong \widetilde{G}$ and that $\widetilde{S} = \widetilde{G} \operatorname{Exp}(iC)$. Note that the multiplication mapping is holomorphic on the interior of S and that this property lifts to the open subsemigroup $\operatorname{int} \widetilde{S} = \widetilde{G} \operatorname{Exp}(i \operatorname{int} W)$ of S.

We will return to these examples in the end of the following section, where we will see that the semigroups $\widetilde{S}$ are in general not realizable as subsemigroups of groups. In Chapter 9 we will discuss the representation theoretic significance of these semigroups.

3.5. The free group on S

Let us return to the problem from the beginning. Given a generating Lie semigroup $S \subseteq G$, we consider the inclusion mapping $i \colon S \to G$ and the associated homomorphism $i_* \colon \pi_1(S) \to \pi_1(G)$ with respect to the base point 1. The main achievement of the preceding section is the realization of $\pi_1(S)$ as a concrete subgroup of the center of the locally compact semigroup $\widetilde{S}$. In the following we identify $\pi_1(S)$ with this subgroup of $\widetilde{S}$, and similarly $\pi_1(G)$ with the corresponding subgroup of $\widetilde{G}$, the universal covering group of G.

We start with the determination of the image of i_*. To state the first main theorem, we recall the following result from [HHL89, VII.3.28]. In our situation the proof is much more easier than the proof given in [HHL89] because we only consider global semigroups.

Theorem 3.26.   (Free group theorem) *Let $S \subseteq G$ be a generating Lie semigroup. Then there exists a covering group $p \colon G(S) \to G$ and a continuous homomorphism $\gamma_S \colon S \to G(S)$ which has the universal property of the free (topological) group on S, i.e., for every (continuous) homomorphism $\varphi \colon S \to K$, where K is a (topological) group, there exists a (continuous) homomorphism $\widetilde{\varphi} \colon G(S) \to K$ such that $\varphi = \widetilde{\varphi} \circ \gamma_S$. The group $G(S)$ is the largest covering of G in which S lifts.*

Proof.   Let $G(S)$ denote the free group on S (cf. [CP61]) and $\gamma_S \colon S \to G(S)$ the universal morphism of semigroups. Then the universal property of $G(S)$ implies the existence of a group homomorphism $p \colon G(S) \to G$ such that $p(\gamma_S(s)) = s$ holds for all $s \in S$.

Now we construct a group homomorphism from the universal covering group $\widetilde{G}$ onto $G(S)$. Let U be a 1-neighborhood in S. Then, since S is generating, $V := UU^{-1} \cap U^{-1}U$ is a 1-neighborhood in G (Corollary 3.11). Moreover, $1 \in \operatorname{int}(SS^{-1} \cap S^{-1}S)$, so we may choose U so small such that

$$UU^{-1} \cup U^{-1}U \subseteq SS^{-1} \cap S^{-1}S$$

and such that the covering group $\pi_V : G_V \to G$ is the universal covering group of G (cf. Section 1.5). Our strategy is to extend $\gamma_S \mid_U : U \to G(S)$ to the whole 1-neighborhood V such that we have a morphism of local groups. Then we may use the universal property of the group G_V.

Let $g = a_1 a_2^{-1} = b_1 b_2^{-1} \in V$ with $a_1, a_2, b_1, b_2 \in U$. Then $g \in S^{-1}S$ and we find $s_1, s_2 \in S$ with $g = s_1^{-1} s_2$. Now

$$a_1 a_2^{-1} = s_1^{-1} s_2 = b_1 b_2^{-1}$$

implies that

$$s_1 a_1 = s_2 a_2 \quad \text{and} \quad s_1 b_1 = s_2 b_2.$$

Thus

$$\gamma_S(s_1)\gamma_S(a_1) = \gamma_S(s_2)\gamma_S(a_2) \quad \text{and} \quad \gamma_S(s_1)\gamma_S(b_1) = \gamma_S(s_2)\gamma_S(b_2)$$

shows that

$$\gamma_S(a_1)\gamma_S(a_2)^{-1} = \gamma_S(s_1)^{-1}\gamma_S(s_2) = \gamma_S(b_1)\gamma_S(b_2)^{-1}.$$

Consequently

$$\beta : V \to G(S), \quad a_1 a_2^{-1} \mapsto \gamma_S(a_1)\gamma_S(a_2)^{-1}$$

is well defined. If $v = a_1 a_2^{-1} = b_1^{-1} b_2$ with $a_i, b_i \in U$, then $b_1 a_1 = b_2 a_2$ shows that

$$\beta(v) = \gamma_S(a_1)\gamma_S(a_2)^{-1} = \gamma_S(b_1)^{-1}\gamma_S(b_2).$$

To see that β is a homomorphism of local groups, let $v_1, v_2 \in V$ such that $v_1 v_2 \in V$. We have to show that $\beta(v_1 v_2) = \beta(v_1)\beta(v_2)$. We write $v_1 v_2 = a_1^{-1} a_2$, $v_1 = b_1 b_2^{-1}$, and $v_2 = c_1 c_2^{-1}$ with $a_i, b_i, c_i \in U$. Since $b_2^{-1} c_1 \in U^{-1}U \subseteq SS^{-1}$ we also find $d_1, d_2 \in S$ with $b_2^{-1} c_1 = d_1 d_2^{-1}$. Now

$$a_1^{-1} a_2 = v_1 v_2 = b_1 b_2^{-1} c_1 c_2^{-1} = b_1 d_1 d_2^{-1} c_2^{-1}$$

implies that

$$a_2 c_2 d_2 = a_1 b_1 d_1.$$

Thus

$$\gamma_S(a_2)\gamma_S(c_2)\gamma_S(d_2) = \gamma_S(a_1)\gamma_S(b_1)\gamma_S(d_1)$$

implies

$$\beta(v_1 v_2) = \gamma_S(a_1)^{-1}\gamma_S(a_2) = \gamma_S(b_1)\gamma_S(d_1)\gamma_S(d_2)^{-1}\gamma_S(c_2)^{-1}.$$

Moreover $\gamma_S(d_1)\gamma_S(d_2)^{-1} = \gamma_S(b_2)^{-1}\gamma_S(c_1)$ follows from $b_2^{-1} c_1 = d_1 d_2^{-1}$. Putting this in the above formula for $\beta(v_1 v_2)$, we get that $\beta(v_1 v_2) = \beta(v_1)\beta(v_2)$. Now the universal property of G_V implies the existence of a group homomorphism $\beta : G_V \to G(S)$ such that $\beta\mid_U = \gamma_S\mid_U$. For $s \in U \subseteq V$ this yields

$$p(\beta(s)) = p(\gamma_S(s)) = s = \pi_V(s).$$

Thus $\pi_V = p \circ \beta$ because the open set U generates the group G. We claim that β is surjective. Let $u \in \operatorname{int} U$. Then

$$uS \subseteq \operatorname{int}(S) \subseteq \langle \exp \mathbf{L}(S) \rangle \subseteq \langle U \rangle$$

(cf. Corollary 3.11) since for every 1-neighborhood U in S and every $X \in \mathbf{L}(S)$ there exists $\lambda > 0$ such that $\exp([0, \lambda]X) \subseteq U$. With $\gamma_S(U) = \beta(U)$ we conclude that $uS \subseteq \beta(G_V)$. Now $\gamma_S(u)^{-1} \in \beta(G_V)$ shows that $S \subseteq \beta(G_V)$ and therefore $G(S) \subseteq \beta(G_V)$.

The surjectivity of β and the fact that $p \circ \beta = \pi_V$ shows that $D := \ker \beta$ is a discrete central subgroup in $\widetilde{G} \cong G_V$. So $G(S) \cong G_V/D$ carries the structure of a Lie group such that $p: G(S) \to G$ is a covering morphism of Lie groups and $\operatorname{int} \gamma_S(S) \neq \emptyset$ in $G(S)$. If $\alpha: S \to H$ is a continuous homomorphism in a topological group, then the universal property of $G(S)$ as the free group on S implies the existence of a group homomorphism $\alpha': G(S) \to H$ with $\alpha' \circ \gamma_S = \alpha$. Now α' is continuous on the interior of $\gamma_S(S)$ and therefore on $G(S)$ because it is a group homomorphism. Therefore $\gamma_S: S \to G(S)$ also has the universal property of the free topological group on S.

If $q: G_1 \to G$ is a covering group of G such that S lifts into G_1, i.e., there exists a continuous semigroup homomorphism $\alpha: S \to G_1$ with $q(\alpha(s)) = s$ for all $s \in S$, then the universal property of $G(S)$ as the free topological group on S implies the existence of a continuous group homomorphism $\beta: G(S) \to G_1$ with $\beta \circ \gamma_S = \alpha$. Then

$$q \circ \beta \circ \gamma_S = q \circ \alpha = \operatorname{id}_S = p \circ \gamma_S$$

entails that $q \circ \beta = p$. Thus G_1 is a covering group of G lying between $G(S)$ and G. ∎

Our first main result will be the identification of $\pi_1(G(S))$, as a subgroup of $\pi_1(G)$, as the image of i_*.

First we need more detailed information about the situation of this theorem. We start with a general lemma about subsemigroups of metrizable topological groups.

Lemma 3.27. *Let G be a metrizable topological group, $S \subseteq G$ a closed subsemigroup with non-empty interior, and $D \subseteq Z(G)$ a discrete central subgroup. Then the following assertions hold:*

(i) $SS^{-1} = \operatorname{int}(S)\operatorname{int}(S)^{-1}$.

(ii) $D_1 := D \cap SS^{-1} = D \cap S^{-1}S$ *is a subgroup of* D.

(iii) *The semigroup* $S_1 := D_1 S$ *is relatively open and closed in the semigroup* $S_2 := DS$.

(iv) $\overline{S_2} = D\overline{S_1}$.

(v) $dS_1 = d'S_1$ *if and only if* $d \in d'D_1$.

Proof. (i) If $g = s_1 s_2^{-1}$ with $s_1, s_2 \in S$, then $g = s_1 s_0 s_0^{-1} s_2^{-1}$, where $s_0 \in \operatorname{int}(S)$ is arbitrary. Then $s_1 s_0, s_2 s_0 \in \operatorname{int}(S)$ and the assertion follows.

(ii) Set $D_1 := D \cap SS^{-1}$. It is clear that $D_1 = D_1^{-1}$. Let $d = s_1 s_2^{-1} \in D_1$, where $s_1, s_2 \in S$. Then s_1 and s_2 commute with d and therefore with each other.

Hence $d = s_2^{-1} s_1 \in S^{-1}S$. By symmetry we see that $S^{-1}S \cap D$ equals D_1, too. If $d' = s_1' s_2'^{-1}$ with $s_1', s_2' \in S$, then $dd' = s_1 s_2^{-1} d' = s_1 d' s_2^{-1} \in SS^{-1}$. Thus $D_1^2 \subseteq D_1$ and consequently D_1 is a group.

(iii) Let $g = \lim_{n \to \infty} d_n s_n$ with $d_n \in D$ and $s_n \in S$. Suppose first that $d_n \in D_1$ and $g = ds \in DS = S_2$. We choose an element $s_0 \in \text{int}(S)$. Then $gs_0 = dss_0 = \lim_{n \to \infty} d_n s_n s_0$ and $ss_0 \in \text{int}(S)$. Therefore there exists $n_0 \in \mathbb{N}$ such that $d^{-1} d_{n_0} s_{n_0} s_0 \in \text{int}(S)$. Then $d^{-1} d_{n_0} \in SS^{-1} \cap D = D_1$. This shows that $d \in D_1$ and $g \in S_1 = D_1 S$. So we have proved that S_1 is relatively closed in S_2.

To show that S_1 is also relatively open in S_2, we assume that $g = ds \in D_1 S = S_1$. By the same argument as above we find $n_0 \in \mathbb{N}$ such that $d^{-1} d_n \in D_1$ for $n \geq n_0$. But this means that eventually $d_n \in D_1$ and $d_n s_n \in S_1$. Thus S_1 is also relatively open.

(iv) We only have to prove that $\overline{S_1}D$ is closed. So let $g = \lim_{n \to \infty} d_n s_n$ with $d_n \in D$ and $s_n \in \overline{S_1}$. Because G was supposed to be metrizable, we may replace s_n by $d_n' s_n'$, where $d_n' \in D_1$ and $s_n' \in S$. Hence we may assume that $s_n \in S$. Then there exists $m \in \mathbb{N}$ such that $d_m s_m \in gSS^{-1}$ because SS^{-1} is a 1-neighborhood in G. Thus $g \in d_m s_m SS^{-1} \subseteq DSS^{-1} = D\,\text{int}(S)\,\text{int}(S)^{-1}$. Choose $d \in D$ and $a, b \in \text{int}(S)$ such that $g = dab^{-1}$. Then $a = \lim_{n \to \infty} d^{-1} d_n s_n b \in \text{int}(S)$ and there exists $n_0 \in \mathbb{N}$ such that $d^{-1} d_n s_n b \in S$ whenever $n \geq n_0$. In this case $d^{-1} d_n \in S(s_n b)^{-1} \subseteq SS^{-1}$, so $d_n \in dD_1$. Now $g = \lim_{n \to \infty} d_n s_n \in d\overline{D_1 S} = d\overline{S_1}$.

(v) If $dS_1 = dD_1 S = d'D_1 S = d'S_1$, then $d^{-1} d' \in SS^{-1} D_1 \subseteq SS^{-1}$. Therefore $d' \in dD_1$. ∎

Proposition 3.28.   *Let $S \subseteq G$ be a generating Lie semigroup, $p: G_1 \to G$ a covering morphism with $\exp_G = p \circ \exp_{G_1}$, $D := \ker p$, and $S_1 \subseteq G_1$ the Lie semigroup with $\mathbf{L}(S_1) = \mathbf{L}(S)$. Then the following are equivalent:*

(1) *S lifts into G_1.*

(2) *The subsets dS_1, $d \in D$ are the connected components of of the closed semigroup $S_2 := D \cdot S_1$.*

(3) *$S_1 S_1^{-1} \cap D = \{1\}$.*

Proof.   (1) $\Rightarrow$ (2): Suppose that $\gamma: S \to G_1$ is a lift of S into G_1. Then $p \circ \gamma = \text{id}_S$ and $\gamma(S)$ is a locally compact subsemigroup of G_1 with

$$ p(\exp_{G_1} X) = \exp_G X = p \circ \gamma(\exp_G X) \quad \text{for all} \quad X \in \mathbf{L}(S). $$

Thus $\gamma(\exp_G X) = \exp_{G_1} X$ for all $X \in \mathbf{L}(S)$. This proves that $\gamma(S) \subseteq S_1$. On the other hand it is clear that $p(S_1) \subseteq S$. The mapping $\gamma \circ p$ agrees with the identity on the dense subsemigroup $\langle \exp_{G_1} \mathbf{L}(S) \rangle$ of S_1 and therefore $\gamma \circ p|_{S_1} = \text{id}_{S_1}$. In particular $\gamma(S) = S_1$.

This proves that $S_1^{-1} S_1 \cap D = \{1\}$ because $s = s'd$ with $s, s' \in S_1$ implies that $p(s) = p(s')$, so $s = s'$. It is clear that the subsets $dS_1 \subseteq S_2$ are connected. We prove that they are pairwise disjoint. If this is false, we find $d \in \ker p \setminus \{1\}$ such that $dS_1 \cap S_1 \neq \emptyset$. Choose $s, s' \in S_1$ with $ds = s'$. Then $p(s) = p(s')$ which proves that $s = s'$ because $p|_{S_1}$ is injective. It follows from Lemma 3.27 that the sets dS_1 are open closed subsets of the closed semigroup DS_1.

(2) $\Rightarrow$ (3): Assume that the sets dS_1, $d \in \ker p$ are the connected components of the closed semigroup S_2. If $d \in S_1 S_1^{-1} \cap D$, then $dS_1 = S_1$ and therefore $d = \mathbf{1}$.

(3) $\Rightarrow$ (1): Suppose that $D \cap S_1 S_1^{-1} = \{1\}$. Lemma 3.27 implies that the subsets dS_1 of DS_1 are open and closed in the closed semigroup S_2.

Now $p(S_2) = p(S_1) = S$ since this is a closed subsemigroup of G which contains $\exp_G \mathbf{L}(S)$. We claim that $p|_{S_1}$ is injective. To see this, let $s, s' \in S_1$ with $p(s) = p(s')$. Then there exists $d \in \ker p$ with $s' = ds \in S_1 \cap dS_1$, i.e., $d \in S_1 S_1^{-1}$. Thus $d = 1$ and $s = s'$. Therefore the restriction $p|_{S_1}: S_1 \to S$ is a continuous locally homeomorphic bijection, whence an isomorphism of topological semigroups. We conclude that $(p|_{S_1})^{-1}: S \to S_1$ is a lift of S into G_1.   ∎

Note that the proof of Proposition 3.28 even shows that the condition $D \cap S_1 S_1^{-1} = \{1\}$ implies the existence of a closed subsemigroup S of $G = G_1/D$ with $\mathbf{L}(S) = \mathbf{L}(S_1)$ (cf. Section 5.2).

So far these results were not directly related to the fundamental group of S but now the largest part of the work is done and we can put the pieces together.

Proposition 3.29. *Let* $S_1 := \overline{\langle \exp_{\widetilde{G}} \mathbf{L}(S) \rangle}$. *Then*

$$\operatorname{im} i_* = S_1 S_1^{-1} \cap \pi_1(G) \quad \text{and} \quad \ker i_* = \pi_1 \big(\overline{(\operatorname{im} i_*) S_1} \big),$$

where $\overline{(\operatorname{im} i_*) S_1}$ *is the path-component of* 1 *in* $q^{-1}(S)$.

Proof. Let $D_1 := S_1 S_1^{-1} \cap \pi_1(G)$. According to Lemma 3.27, this is a subgroup of $\pi_1(G)$. Let $\widetilde{i}: \widetilde{S} \to \widetilde{G}$ be the homomorphism from Lemma 3.23.

If $d = s_1 s_2^{-1} \in D_1$ with $s_1, s_2 \in \operatorname{int}(S)$, then there exist continuous mappings $\alpha, \beta: [0,1] \to S$ such that $\alpha(0) = 1$, $\alpha(1) = s_1$, $\beta(0) = s_2$, and $\beta(1) = 1$. Thus $p(\alpha(1)) = p(s_1) = p(s_2) = p(\beta(0))$ and therefore $(p \circ \alpha) \diamond (p \circ \beta)$ is a continuous path in S whose homotopy class corresponds to d. Hence $d \in \widetilde{i}(\pi_1(S))$.

If, conversely, $d = \widetilde{i}(x)$, then there exists $\gamma \in \Omega(S, 1)$ such that $d = [\gamma]$. According to Proposition 3.13 and Corollary 3.16 we may assume that $\gamma([0,1]) \subseteq \operatorname{int}(S) \cup \{1\} \subseteq \langle \exp_G \mathbf{L}(S) \rangle$. Therefore $\widetilde{\gamma}([0,1]) \subseteq DS_1$. Using Lemma 3.27(iii), we find that $\widetilde{\gamma}([0,1]) \subseteq D_1 S_1$. Consequently $\widetilde{\gamma}(1) \in D_1$.

With Proposition 3.22 we conclude that

$$S_1 D_1 \subseteq \widetilde{i}(\widetilde{S}) \subseteq \overline{\widetilde{i}(\widetilde{S}_L) \widetilde{i}(\pi_1(S))} \subseteq \overline{S_1 D_1}.$$

According to Lemma 3.23 the semigroup $\widetilde{i}(\widetilde{S})$ coincides with the path-component of 1 in $q^{-1}(S)$ which is open and closed in $q^{-1}(S)$ because $q^{-1}(S)$ is locally path connected (Proposition 3.13). Therefore $\overline{S_1 D_1} = \widetilde{i}(\widetilde{S})$ and $\pi_1(\overline{S_1 D_1}) \cong \ker \widetilde{i}$.   ∎

Theorem 3.30. $\operatorname{im} i_* = \pi_1(G(S))$.

Proof. Let $D' \subseteq D := \pi_1(G) \subseteq \widetilde{G}$ be a subgroup, $G' := \widetilde{G}/D'$, $q': \widetilde{G} \to G'$ the corresponding covering homomorphism, and $S' := \langle \exp_{G'} \mathbf{L}(S) \rangle$, $S_1 := \langle \exp_{\widetilde{G}} \mathbf{L}(S) \rangle$ the Lie subsemigroups of G' and $\widetilde{G}$ generated by $\mathbf{L}(S)$.

Then

$$S' S'^{-1} = \operatorname{int}(S') \operatorname{int}(S')^{-1} \subseteq q'(S_1) q'(S_1)^{-1}.$$

Therefore

$$S' S'^{-1} \cap q'(D) = q'(S_1 S_1^{-1}) \cap q'(D) = q'(S_1 S_1^{-1} \cap D) = q'(\operatorname{im} i_*).$$

Now Proposition 3.28 shows that S lifts to G' if and only if $q'(\operatorname{im} i_*) = \{1\}$, i.e., $\operatorname{im} i_* \subseteq D'$. So the largest covering group of G into which S lifts is $\widetilde{G}/\operatorname{im} i_*$, whence $\operatorname{im} i_* = \pi_1(G(S))$. ∎

Corollary 3.31. *The mapping i_* is surjective if and only if $G(S) = G$, i.e., if S does not lift in a non-trivial covering group of G.* ∎

Corollary 3.32. *Let $q' : G' \to G$ be a covering of Lie groups, $D' := \ker q'$, $S' \subseteq G'$ the Lie semigroup with $\mathbf{L}(S') = \mathbf{L}(S)$, and $q'' : \widetilde{G} \to G'$ the universal covering of G'. Then $q''(\operatorname{im} i_*) = D' \cap S'S'^{-1}$.*

Proof. This follows from the proof of Theorem 3.30. ∎

Corollary 3.33. *If $\pi_1(S) = \{1\}$, then $G(S) = \widetilde{G}$, i.e., every simply connected generating Lie semigroup $S \subseteq G$ lifts in the universal covering group $\widetilde{G}$ of G.* ∎

The following diagram represents graphically most of the situation of the preceding discussion.

$$
\begin{array}{ccccccc}
\pi_1(\overline{D_1 S_1}) & \longrightarrow & \pi_1(S) & \xrightarrow{\;i_*\;} & \pi_1(G(S)) & \longrightarrow & \pi_1(G) \\
 & & \downarrow & & \downarrow & & \downarrow \\
 & & \widetilde{S} & \xrightarrow{\;\widetilde{i}\;} & \widetilde{G} & \xrightarrow{\;\operatorname{id}_{\widetilde{G}}\;} & \widetilde{G} \\
 & & \downarrow{\scriptstyle p} & & \downarrow & & \downarrow{\scriptstyle q} \\
 & & S & \longrightarrow & G(S) & \longrightarrow & G
\end{array}
$$

An interesting example where Corollary 3.33 applies is the following subsemigroup of $G = \mathrm{Sl}(2, \mathbb{R})$. Let

$$
S := \mathrm{Sl}(2, \mathbb{R})^+ := \left\{ \begin{pmatrix} a & b \\ c & d \end{pmatrix} \in G : a, b, c, d \geq 0 \right\}.
$$

Then S is a Lie semigroup and the mapping $\exp : \mathbf{L}(S) \to S$ is a homeomorphism (Section 2.2). Therefore $\pi_1(S) = \{1\}$ and $G(S) = \mathrm{Sl}(2, \mathbb{R})\widetilde{}$ (Corollary 3.33).

Theorem 3.34. *The homomorphism $\widetilde{i} : \widetilde{S} \to \widetilde{G}$ (Lemma 3.23) has the universal property of the free group on $\widetilde{S}$, i.e., for every homomorphism $\alpha : \widetilde{S} \to T$, where T is a group, there exists a unique homomorphism $\alpha_1 : \widetilde{G} \to T$ such that $\alpha_1 \circ \widetilde{i} = \alpha$.*

Proof. Let T be a group and $\alpha : \widetilde{S} \to T$ a homomorphism. Then

$$
\beta : \widetilde{S} \to T \times \widetilde{G}, \quad s \mapsto (\alpha(s), \widetilde{i}(s))
$$

is a homomorphism into a group such that

$$
\ker \beta = \ker \alpha \cap \ker \widetilde{i} \subseteq \ker \widetilde{i}.
$$

Therefore $\ker \beta$ is a discrete subgroup of $\pi_1(S)$. Set $S_1 := \widetilde{S}/\ker \beta$ and $S_1^0 := \operatorname{int}(\widetilde{S})/\ker \beta$. Then S_1 is a covering semigroup of S which can be algebraically embedded in a group. Then $G(S_1^0)$, the free group on S_1^0, admits the structure of a Lie group (Theorem 3.26), and the corresponding universal homomorphism $\alpha_1\colon S_1^0 \to G(S_1^0)$ is an embedding onto an open subsemigroup of $G(S_1^0)$. The homomorphism $\tilde{i}$ is constant on $\ker\beta$. Thus it induces a homomorphism $p_1\colon S_1 \to \widetilde{G}$ which has an extension to a homomorphism $p_1'\colon G(S_1^0) \to \widetilde{G}$ with $p_1' \circ \alpha_1 = p_1\,|_{S_1^0}$. We conclude that p_1' is a surjective covering of Lie groups, because it is continuous on S_1^0 and therefore everywhere, and $p_1(S_1^0)$ is an open subset of $\widetilde{G}$. Since $\widetilde{G}$ is simply connected, p_1' is an isomorphism. Hence $p_1\,|_{S_1^0}$ is injective. This proves that $\ker\beta = \ker\tilde{i}$, i.e., $\ker\tilde{i} \subseteq \ker\alpha$. Whence α factors to a homomorphism $\alpha'\colon \tilde{i}(\widetilde{S}) \to T$ with $\alpha' \circ \tilde{i} = \alpha$.

It remains to prove that α' permits a continuation to a homomorphism $\alpha_1\colon \widetilde{G} \to T$ with $\alpha_1 \circ \tilde{i} = \alpha$. We use Theorem 3.26 and Proposition 3.29 to see that $G(\tilde{i}(\widetilde{S})) = \widetilde{G}$. Then the universal property of $G(\tilde{i}(\widetilde{S}))$ provides a continuation of α' to the whole group $\widetilde{G}$. ∎

Corollary 3.35. *Every quotient $\widetilde{S}/D$ with $\ker i_* \not\subseteq D \subseteq \pi_1(S)$ is not algebraically embeddable in a group.* ∎

We resume the notations from the last example after Corollary 3.25. If $\mathfrak{g}$ is a semisimple Lie algebra containing a pointed generating invariant cone C, then the center of a maximal compactly embedded subalgebra $\mathfrak{k}$ is non-trivial (Section 7.2). Therefore the center of $\widetilde{G}$ is infinite and $Z(G) \subseteq G_{\mathbb{C}}$ is finite. We conclude that $\widetilde{G} \neq G$ and therefore that $\widetilde{S} \neq S$. Now Corollary 3.35 shows that no quotient $\widetilde{S}/D$, i.e., no non-trivial covering semigroup of S is isomorphic to a subsemigroup of a group. The simplest example is the semigroup $S = \mathrm{Sl}(2,\mathbb{R})\exp(iC) \subseteq \mathrm{Sl}(2,\mathbb{C})$, where $\pi_1(S) \cong \mathbb{Z}$ and $\widetilde{S} \cong \widetilde{\mathrm{Sl}(2,\mathbb{R})}\,\mathrm{Exp}(iC)$. Another interesting example is *Howe's oscillator semigroup* (cf. [Hi89] and Section 9.5). Here $S = \mathrm{Sp}(n,\mathbb{R})\exp(iC) \subseteq \mathrm{Sp}(n,\mathbb{C})$, the group $\mathrm{Sp}(n,\mathbb{C})$ is simply connected, and $\pi_1\big(\mathrm{Sp}(n,\mathbb{R})\big) \cong \mathbb{Z}$. Consequently $\pi_1(S) \cong \mathbb{Z}$. The oscillator semigroup is the double cover $\widetilde{S}/\pi_1(S)^2$ of S. Its group of units is the well known metaplectic group $\mathrm{Mp}(n,\mathbb{R})$ which is a double cover of the symplectic group $H(S) = \mathrm{Sp}(n,\mathbb{R})$.

3.6. Groups with directed orders

In the last subsection we have considered the relations between the free group $G(S)$ over S and the homomorphism $i_*\colon \pi_1(S) \to \pi_1(G)$. Now we consider a particular class of generating Lie semigroups, namely those for which $G = S^{-1}S$. We show that this condition implies that i_* is an isomorphism. Note that this is equivalent to $\widetilde{S} \cong q^{-1}(S) \subseteq \widetilde{G}$, where $q\colon \widetilde{G} \to G$ is the universal covering of G (Proposition 3.29).

Let $S \subseteq G$ be a Lie semigroup. We recall the definition of the left invariant quasiorder $\leq_S$ on G by

$$g \leq_S g' \iff g' \in gS.$$

Lemma 3.36. *The quasiorder $\leq_S$ is directed (filtered) if and only if $G = SS^{-1}$*
($G = S^{-1}S$).

Proof.   If $\leq_S$ is directed and $g \in G$, then there exists $s \in G$ such that $1 \leq_S s$ and
$g \leq_S s$. Hence $s \in S$ and $g \in sS^{-1}$. If, conversely, $G = SS^{-1}$ and $g, g' \in G$, then
$g^{-1}g' = s_1 s_2^{-1}$. Thus $g, g^{-1} \leq_S g' s_2 = g s_1$. The proof for the second statement is
similar.   ∎

Note that $\leq_S$ is filtered if and only if $\leq_{S^{-1}}$ is directed.

Lemma 3.37. *If S is generating and the quasiorder $\leq_S$ is filtered, then each*
compact subset $K \subseteq G$ is bounded from below, i.e., there exists $g_0 \in G$ such that
$K \subseteq g_0 S$.

Proof.   It is clear that $K \subseteq \bigcup_{g \in G} g\operatorname{int}(S)$. Let $K \subseteq \bigcup_{i=1}^{n} g_i \operatorname{int}(S)$. Then there
exists $g_0 \in G$ such that $g_0 \leq_S g_1, \ldots, g_n$. Therefore $g_i \operatorname{int}(S) \subseteq g_0 \operatorname{int}(S)$ for
$i = 1, \ldots, n$ and consequently $K \subseteq g_0 \operatorname{int}(S)$.   ∎

Theorem 3.38.   *Let $S \subseteq G$ be a generating Lie semigroup such that $(G, \leq)$*
is filtered (directed). Then the homomorphism $i_*: \pi_1(S) \to \pi_1(G)$ induced by the
inclusion $i: S \to G$ is an isomorphism.

Proof.   Let $[\gamma] \in \pi_1(S)$, where $\gamma: [0,1] \to S$ is a continuous mapping with
$\gamma(0) = \gamma(1) = 1$. Suppose that $i_*([\gamma]) = [i \circ \gamma] = 1$. Then there exists a continuous
mapping $F: [0,1] \times [0,1] \to G$ such that

$$F(0, t) = \gamma(t), \quad F(1, t) = 1, \quad \text{and} \quad F(s, 0) = F(s, 1) = 1.$$

According to Lemma 3.37 there exists $g_0 \in G$ such that $K := F([0,1] \times [0,1]) \subseteq$
$g_0 \operatorname{int}(S)$. In particular we have that $1 \in g_0 \operatorname{int}(S)$, i.e., $g_0^{-1} \in \operatorname{int}(S)$. Hence
there exists a continuous path $\alpha: [0,1] \to S$ such that $\alpha(0) = 1$ and $\alpha(1) = g_0^{-1}$
(Proposition 3.13). Now

$$[\alpha \circ g_0^{-1} \gamma \circ \widehat{\alpha}] = [\alpha \circ g_0^{-1} \circ \widehat{\alpha}] = [\alpha][\alpha]^{-1} = 1$$

holds in $\pi_1(S)$ because $(s, t) \mapsto g_0^{-1} F(s, t)$ is a deformation of $g_0^{-1}\gamma$ to the constant
path g_0^{-1} in S. Now

$$t \mapsto \alpha\big|_{[0,t]} \circ \alpha(t)\gamma \circ \widehat{\alpha}\big|_{[0,t]}$$

defines a continuous deformation of the path $\alpha \circ g_0^{-1}\gamma \circ \widehat{\alpha}$ to γ. Thus $[\gamma] = 1$ in
$\pi_1(S)$. So we have proved that i_* is injective.

Let $[\gamma] \in \pi_1(G)$ and $\gamma: [0,1] \to G$ with $\gamma(0) = \gamma(1) = 1$. Then, by the same
argument as above, there exists a $g_0 \in G$ such that $\gamma([0,1]) \subseteq g_0 \operatorname{int}(S)$. By the
same construction as in the first part of the proof we find a path $\alpha: [0,1] \to S$ such
that

$$[\gamma] = [\alpha \circ g_0^{-1}\gamma \circ \widehat{\alpha}] \in i_*(\pi_1(S)).$$

If $(G, \leq_S)$ is directed, the first part of the proof shows that the inclusion
$S^{-1} \to G$ induces an isomorphism $\pi_1(S^{-1}) \to \pi_1(G)$. Then we apply the inversion
$G \to G, g \mapsto g^{-1}$ to see that the inclusion $S \to G$ also induces an isomorphism
$\pi_1(S) \to \pi_1(G)$.   ∎

Corollary 3.39. *If $S \subseteq G$ is a Lie semigroup such that $G = SS^{-1}$, then $G(S) = G$.*

Proof. This follows from $\operatorname{im} i_* = \pi_1(G(S)) = \pi_1(G)$ (Theorem 3.30, Theorem 3.37). ∎

Let S be an abstract subsemigroup of an abstract group G such that $G = SS^{-1}$. Then $G(S) = G$ holds in this general context. The simple proof can be found in [CP61, p.36].

Note that it was already clear from Proposition 3.28 that $G(S) = S_1 S_1^{-1}$ implies that $G(S) = G$. That $G = SS^{-1}$ even implies that $G(S) = S_1 S_1^{-1}$ follows from the fact that $S_1 S_1^{-1}$ is a subgroup of $G(S)$ (cf. [Ru89, 1.2]).

Corollary 3.40. *If G is a simply connected Lie group and $S \subseteq G$ a generating Lie semigroup such that $\leq_S$ is filtered or directed, then S is simply connected.* ∎

This corollary is a solution of Problem PVII.2 in [HHL89] and also to Problem 3.1 in [Gr83, p.122]: Is a Lie subsemigroup of a simply connected Lie group simply connected? Corollary 3.40 gives a sufficient condition for simply connectedness. That a generating Lie semigroup S in a simply connected Lie group G is not always simply connected can be seen with the example following Corollary 3.40. The Lie semigroup $\mathrm{Sl}(2, \mathbb{R}) \exp(iC)$ in the simply connected group $\mathrm{Sl}(2, \mathbb{C})$ is such an example.

Note that $\leq_S$ is in particular directed if S is invariant. For weaker conditions on S which guarantee that $\leq_S$ is directed, see [Ne91b, 2.12] and Proposition 3.43.

In the remainder of this section we are concerned with the problem to show that the order $\leq_S$ is always directed when the Lie algebra of G is compact or nilpotent. We start with the nilpotent case.

We recall that a semigroup S is called

(1) *left reversible* if $aS \cap bS \neq \emptyset$ for every pair $a, b \in S$.

(2) *right reversible* if $Sa \cap Sb \neq \emptyset$ for every pair $a, b \in S$.

(3) *reversible* if it is both left and right reversible.

(4) A subsemigroup S of a group G is said to be *mid-reversible* if $G = S^{-1} S S^{-1}$. Note that S is mid-reversible if and only if S^{-1} is mid-reversible.

Lemma 3.41. *A subsemigroup S of a group G is left (right) reversible if and only if SS^{-1} $(S^{-1}S)$ is a subgroup of G.*

Proof. Assume that S is left reversible. Let $a, b, a', b' \in S$. Then

$$ab^{-1}a'b'^{-1} \in aSS^{-1}b^{-1} \subseteq SS^{-1}.$$

Since $(SS^{-1})^{-1} = SS^{-1}$, it follows that SS^{-1} is a subgroup of G. The other part of the assertion follows similarly. ∎

Note that the preceding lemma shows in particular that S is left reversible if and only if S^{-1} is right reversible.

Lemma 3.42. *Let S be an open subsemigroup of the topological group G, N a connected normal subgroup of G and $s \in S$ such that $s^{-1}Us \subseteq U$ for all U in a neighborhood basis $\mathcal{U}$ of 1 in N. Then $N \subseteq SS^{-1}$.*

Proof.   Let $n \in N$. Since S is open, there exists a 1-neighborhood $U \in \mathcal{U}$ with $Us \subseteq S$, and since N is connected, there exists a natural number k with $n \in U^k$. Thus

$$ ns^k \in U^k s^k = Us(s^{-1}Us)s(s^{-2}Us^2)s \ldots (s^{-k+1}Us^{k-1})s \subseteq (Us)^k \subseteq S, $$

hence $nS \cap S \neq \varnothing$, i.e., $n \in SS^{-1}$.   ∎

Lemma 3.43. *Let S be a subsemigroup of the group G and suppose that N is a normal subgroup of G such that SN/N is left reversible. Then each of the following conditions implies that S is left reversible.*

(1) *If $aS \cap bS \neq \varnothing$ for all $a,b \in S$ with $a^{-1}b \in N$ (or, equivalently, if $nS \cap S \neq \varnothing$ for all $n \in N \cap S^{-1}S$).*

(2) *There exists a subgroup G_1 of G, containing N such that $S_1 := S \cap G_1$ generates G_1 and S_1 is left reversible.*

Proof.   (1) We assume that G is generated by S and that $aS \cap bS \neq \varnothing$ for all $a,b \in S$ with $a^{-1}b \in S$. Pick elements $u, v \in S$. By assumption, $uSN \cap vSN \neq \varnothing$, so we find elements $s_1, s_2 \in S$ and $n \in N$ such that $us_1 = vs_2n$. Now $(us_1)^{-1}vs_2 \in S^{-1}S \cap N$. Hence, again by assumption,

$$ vs_2 S \cap us_1 S = us_1\big((us_1)^{-1}vs_2 S \cap S\big) \neq \varnothing. $$

Thus $vS \cap uS \neq \varnothing$ and the assertion follows.

(2) In view of (1), we only have to observe that $nS \cap S \supseteq nS_1 \cap S_1$, and that $nS_1 \cap S_1 \neq \varnothing$, for any $n \in N$, if S_1 is left reversible (Lemma 3.41).   ∎

We note that condition (1) is always satisfied if N is central in G. In fact, let $a,b \in S$ with $a^{-1}b \in Z(G)$. Then

$$ ab = a^2(a^{-1}b) = a(a^{-1}b)a = ba \in aS \cap bS. $$

Lemma 3.44. *Let N be a closed connected subgroup of the topological group G and $S \subseteq G$ a subsemigroup. Suppose that SN/N is mid-reversible and that there exists a neighborhood basis $\mathcal{U}$ of 1 in N and $s \in \mathrm{int}\,S$ such that*

$$ s^{-1}Us \subseteq U \qquad \forall U \in \mathcal{U}. $$

Then S is mid-reversible.

Proof.   Since SN/N is a mid-reversible semigroup we have that

$$ G \subseteq S^{-1}SS^{-1}N = S^{-1}SNS^{-1}. $$

In view of Lemma 3.42, we find that

$$ N \subseteq \mathrm{int}(S)\,\mathrm{int}(S)^{-1} \subseteq SS^{-1}. $$

Hence

$$ G \subseteq S^{-1}SNS^{-1} \subseteq S^{-1}SSS^{-1}S^{-1} \subseteq S^{-1}SS^{-1}, $$

i.e., S is mid-reversible.   ∎

Proposition 3.45.   (Ruppert's Theorem)

(i) *A subsemigroup of a nilpotent group is always reversible.*

(ii) *Let G be a connected Lie group with compact Lie algebra and $S \subseteq G$ a subsemigroup with non-empty interior. Then S is reversible.*

(iii) *Let G be a connected solvable Lie group and $S \subseteq G$ a subsemigroup with non-empty interior. Then S is mid-reversible.*

Proof.   (i) We may assume that S is a subsemigroup of a nilpotent group and that S generates G. Let

$$Z_1 \subseteq Z_2 \subseteq \ldots \subseteq Z_n = G$$

be the ascending central series of G. We use induction on n. If $n = 1$, then G is abelian and the assertion is trivial. Suppose that the assertion has been shown for all nilpotent groups with degree of nilpotency $< n$. Then SZ_1/Z_1 is left and right reversible and Z_1 is central. Thus, applying Lemma 3.43(1), we see that S is left reversible. Since S was arbitrary, the remark after Lemma 3.41 finishes the proof.

(ii) Since the Lie algebra of G is compact, there exists a compact subgroup N and a central vector group V such that $G \cong N \times V$. Let $\mathcal{U}$ denote a basis of 1-neighborhoods of N which are invariant under the adjoint action. Then Lemma 3.42 implies that $N \subseteq SS^{-1}$. Moreover, it follows from (i) that the subsemigroup SN/N of the abelian group $G/N \cong V$ is reversible. Now the left reversibility of S follows from Lemma 3.43(1). Since S was arbitrary, the same argument applies to S^{-1}, and the reversibility of S follows.

(iii) We use induction on the dimension of G. The assertion is obvious for $\dim G = 0$. Suppose that it holds for all connected solvable Lie groups of strictly smaller dimension.

If G is abelian, then SS^{-1} is a subgroup of G with non-empty interior. Since G is connected, it follows that $G = SS^{-1} = S^{-1}SS^{-1}$. Now we assume that G is not abelian and write $\mathfrak{g} = \mathbf{L}(G)$ for its Lie algebra. Let $\{0\} \neq \mathfrak{n} \subseteq [\mathfrak{g},\mathfrak{g}]$ be an ideal of minimal dimension. In view of Lie's theorem, $\dim \mathfrak{n} \leq 2$, and, since the center of the nilpotent Lie algebra $[\mathfrak{g},\mathfrak{g}]$ is non-zero, the ideal $\mathfrak{n}$ is contained in the center of $[\mathfrak{g},\mathfrak{g}]$. Let $N := \langle \exp \mathfrak{n} \rangle$ and T the maximal torus of the abelian normal subgroup $\overline{N}$. The subgroup T of $\overline{N}$ is characteristic, hence normal in G. Now the fact that the connected component of the automorphism group of a torus is trivial and the connectedness of G show that $T \subseteq Z(G)$. It follows in particular that there exists a basis $\mathcal{U}$ of 1-neighborhoods in T which is invariant under all inner automorphisms of G. If $T \neq \{1\}$, the induction hypothesis and Lemma 3.44 imply that S is mid-reversible in G. So we may assume that $T = \{1\}$.

Then $\overline{N}$ is a vector group. A path-connected subgroup of a vector group is closed. Hence N is closed in $\overline{N}$ and therefore $N = \exp \mathfrak{n}$ is closed in G. If $\dim \mathfrak{n} = 1$, then for any $g \in G$ the operator $\mathrm{Ad}(g)$ restricted to $\mathfrak{n}$ just amounts to multiplication with a real scalar. If $\dim \mathfrak{n} = 2$, then there exists a complex structure I on $\mathfrak{n}$ such that $\mathrm{Ad}(G)$ acts by complex linear mappings. Replacing S by S^{-1} if necessary, we find $s \in \mathrm{int}(S)$ and a norm $\|\cdot\|$ on $\mathfrak{n}$ such that $\|\mathrm{Ad}(s^{-1})\| \leq 1$. Thus the assumptions of Lemma 3.44 are satisfied by induction hypothesis and the assertion follows. ∎

Theorem 3.46.   *Let S be a generating Lie semigroup in G and suppose that one of the following conditions is satisfied:*

(1) *G is nilpotent.*

(2) *$\mathbf{L}(G)$ is a compact Lie algebra.*

(3) *S is invariant in G.*

Then $G = SS^{-1} = S^{-1}S$ and $i_*: \pi_1(S) \to \pi_1(G)$ is an isomorphism.

(4) *If G is solvable, then i_* is surjective.*

Proof.   For (1) and (2) we refer to Proposition 3.45 and Lemma 3.41. Since the invariance of S implies that SS^{-1} is a group, (3) follows immediately.

(4): Let $p: G(S) \to G$ denote the covering morphism and $q: S \to G(S)$ the lifting of S into $G(S)$. Then $q(S) \cong S$ is mid-reversible in the solvable group $G(S)$ (Proposition 3.45(iii)). We claim that p is an isomorphism. Let $d \in \ker p$. Then there exist elements $s, t, u \in q(S)$ such that $d = s^{-1}tu^{-1}$. It follows that $sdu = t$ and that $p(t) = p(s)p(u)$. Therefore

$$t = q(p(t)) = q(p(s)p(u)) = q(p(s))q(p(u)) = su.$$

Hence $sdu = su$ entails that $d = 1$ because d in central in $G(S)$. Now (4) follows from Theorem 3.30.   ∎

Notes

Faces of Lie semigroups and their relation to the faces of the corresponding Lie wedges were introduced in [Ne92b], where they were extensively applied to study the structure of the almost periodic compactification of an invariant Lie semigroup. The existence of interior points in semigroups with Lie-generating tangent wedge has been known for quite some time in control theory (cf. [SJ72]). The Hofmann-Ruppert Theorem appeared in [HoRu88]. The theory of covering semigroups for Lie semigroups was developed in [Ne91f] which also contains the description of the free group on S in terms of its covering semigroup (Theorem 3.34) which leads to the result that coverings of Ol'shanskiĭ semigroups in simply connected groups are never embeddable in any group. Theorem 3.26 (The free group theorem) is a special case (with especially simple proof) of results that appeared in [HHL89, Ch. VII]. Section 3.6 combines material of Ruppert ([Ru89]) with the results on covering semigroups ([Ne91f]) to draw sharper conclusions for solvable groups or groups with compact Lie algebras.

4. Ordered homogeneous spaces

Ordered homogeneous spaces are closely related to semigroups and the study of the one concept automatically enhances the understanding of the other. In this chapter we consider the ordered homogeneous spaces as a useful tool for the study of Lie semigroups, but it should be understood that the semigroup approach will later on also yield a lot of information about the order structure of homogeneous spaces.

Section 4.1 provides some abstract background to deal with connected chains in a locally compact metric partially ordered space M, where the order is a closed subset of $M \times M$. The crucial idea is that the *chain approximation property* of such a space has strong consequences for the order structure such as local convexity and the connectedness of maximal chains in compact order intervals and the existence of disconnected maximal chains in non-compact order intervals.

In Section 4.2 it is shown how Lie wedges W with global edge $\mathfrak{h}$ are linked with invariant cone fields on homogeneous spaces G/H, where $\mathbf{L}(H) = \mathfrak{h}$. The globality of such a Lie wedge is then equivalent to the existence of a global closed partial order on G/H which is associated to the invariant cone field. The results of Section 4.1 are shown to apply to ordered homogeneous spaces in Section 4.3, where they are used to obtain a characterization of the globally orderable homogeneous spaces by the existence of certain monotone functions. This result completes the proof of the corresponding characterization for Lie wedges in Chapter 1. The main result of Section 4.4 is that maximal chains in ordered homogeneous spaces can be parametrized as conal curves. This links the abstract theory of ordered spaces to the differential geometric setting.

It is easy to show that every homogeneous covering space of an ordered homogeneous space is also globally orderable but the converse is false in general. In Section 4.5 we derive a general characterization of those cases where the converse remains true.

Regular ordered homogeneous spaces are those where the corresponding Lie wedge is Lie generating in $\mathbf{L}(G)$. On the level of ordered spaces this means that the forward sets have dense interior (Section 4.6).

In the last section we generalize the notion of Lorentzian arc-length and Lorentzian distance to general regular ordered homogeneous spaces. We show that there exists a distance maximizing curve from p to q if the order interval $[p,q]$ is compact and that the distance function is continuous for globally hyperbolic spaces.

4.1. Chains in metric pospaces

Throughout this section M denotes a locally compact σ-compact metric space which is a *pospace*, i.e., there exists a closed partial order $\leq$ on M. Later we will have to impose a further condition which will be crucial to obtain the main results. For $x \in M$ we define the *forward set* $\uparrow x := \{y \in M : x \leq y\}$, the *backward set* $\downarrow x := \{y \in M : y \leq x\}$ and the *order interval* $[x, y] := \uparrow x \cap \downarrow y$. For a subset $F \subseteq M$ we set

$$\uparrow F := \bigcup_{f \in F} \uparrow f \quad \text{and} \quad \downarrow F := \bigcup_{f \in F} \downarrow f.$$

Let (K, d) be any compact metric space. We write $\mathcal{CO}(K)$ for the set of compact subsets of K and $\mathcal{CO}_0(K)$ for the set of non-empty compact subsets. For $A \in \mathcal{CO}_0(K)$ and $b \in K$ we set

$$d(A, b) = d(b, A) := \min\{d(a, b) : a \in A\}$$

and for $A, B \in \mathcal{CO}_0(K)$ we define the Hausdorff distance by

$$d(A, B) := \max\left\{ \max\{d(a, B) : a \in A\}, \max\{d(b, A) : b \in B\}\right\}$$

and

$$d(A, \emptyset) = d(\emptyset, A) := \infty.$$

This metric defines a compact topology, called *Vietoris topology*, on $\mathcal{CO}_0(K)$ ([Bou71b, Ch. II, §1, Ex. 15]).

We write $\mathcal{F}(M)$ for the set of closed subsets of M and $\mathcal{CO}(M)$ for the set of compact subsets. To get a compact topology on $\mathcal{F}(M)$, we consider the one-point compactification $M^\omega := M \cup \{\omega\}$ and identify M with the corresponding subset of M^ω. Note that our assumption on M implies that M^ω is metrizable. Let d denote the metric on M and d^ω denote a metric on M^ω which is compatible with the topology. We define the mapping

$$\beta : \mathcal{F}(M) \to \mathcal{CO}_0(M^\omega), F \mapsto F \cup \{\omega\}.$$

Then β is one-to-one and we identify $\mathcal{F}(M)$, via β, with the closed subset $\mathrm{im}\,\beta = \{K \in C_0(M^\omega) : \omega \in K\}$. As a closed subspace of $\mathcal{CO}_0(M^\omega)$, the space $\mathcal{F}(M)$ is a compact metrizable topological space. We collect a few basic facts about the compact space $\mathcal{F}(M)$:

Proposition 4.1.   *The following assertions hold:*

(i) *If $U \subseteq M$ is open, then $\{F \in \mathcal{F}(M) : F \cap U \neq \emptyset\}$ is open and if $A \subseteq M$ is closed, then $\{F \in \mathcal{F}(M) : F \subseteq A\}$ is closed.*

(ii) *Let A_n be a sequence in $\mathcal{F}(M)$ with $A_n \to A$. Then A consists of the set of limit points of convergent sequences a_n with $a_n \in A_n$.*

(iii) *If $A \subseteq X$ is closed, then $\{F \in \mathcal{F}(M) : A \subseteq F\}$ is closed.*

(iv) *If A_n is a sequence of connected sets, $A_n \to A \neq \emptyset$, and for every $n \in \mathbb{N}$ the set $\bigcup_{m \geq n} A_m$ is not relatively compact, then every connected component of A is non-compact.*

Proof.   (i) The first assertion follows from the observation that

$$\{F \in \mathcal{F}(M) : F \cap U \neq \emptyset\} = \mathcal{F}(M) \cap \{F \in C\mathcal{O}(M^{\omega}) : F \cap U \neq \emptyset\}$$

because U is open in M^{ω}. The second assertion follows by applying the first one
with $U := M \setminus A$.

(ii) Let $a \in A$. Then there exist numbers $n_m \in \mathbb{N}$ such that A_n intersects the
$\frac{1}{m}$-ball with respect to d^{ω} around a if $n \geq n_m$. W.l.o.g. we may assume that the
sequence n_m is increasing. For $k = n_m + 1, \ldots, n_{m+1}$ we choose $a_k \in A_k$ with
distance less than $\frac{1}{m}$ from a. Then $a = \lim_{k \to \infty} a_k$.

If, conversely, $a = \lim_{n \to \infty} a_n$ with $a_n \in A$, and $\varepsilon > 0$, then we choose
$n_0 \in \mathbb{N}$ such that $d^{\omega}(A, A_n) < \varepsilon$ for $n > n_0$. Then $d^{\omega}(a_n, A) \leq \varepsilon$ and therefore
$d^{\omega}(a, A) \leq \varepsilon$. Since ε was arbitrary, we conclude that $d^{\omega}(a, A) = 0$, i.e., $a \in A$.

(iii) This is an immediate consequence of (ii).

(iv) Let $a \in A$ be arbitrary and $C(a)$ the connected component of a in A. We
have to show that $C(a)$ is non-compact. Suppose that this is false. Then $C(a)$ is a
compact subset of M and there exists a relatively compact open neighborhood V
of $C(a)$ in M such that $V \cap A$ is closed ([Bou71b, Ch. II, §4.4, Cor.]) and therefore
compact. Let

$$\delta := \min\{d^{\omega}(a, b) : a \in V \cap A, b \in X \setminus V\} > 0$$

and $\varepsilon := \min\{\frac{\delta}{4}, \frac{1}{3}d^{\omega}(\omega, \overline{V})\}$. Since $a \in \lim A_n$, there exists $n_V \in \mathbb{N}$ such that
$d^{\omega}(A_n, a) < \varepsilon$ for all $n \geq n_V$. Moreover, we may assume that $d^{\omega}(A_n, A \cup \{\omega\}) < \varepsilon$.
Let $a_n \in A_n$ with $d^{\omega}(a_n, V \cap A) < \varepsilon$. Then $d^{\omega}(a_n, \omega) > \varepsilon$ because $d^{\omega}(\omega, V \cap A) \geq$
3ε and therefore $d^{\omega}(a_n, A) < \varepsilon$. Let $b \in A$ such that $d^{\omega}(a_n, b) < \varepsilon$. Then
$d^{\omega}(b, V \cap A) < 3\varepsilon < \delta$. Hence $b \in V \cap A$ and this entails that $d^{\omega}(a_n, V \cap A) < \varepsilon$.
Thus

$$\{c \in A_n : d^{\omega}(c, A \cap V) \in\,]\varepsilon, 2\varepsilon[\} = \emptyset.$$

Since A_n is connected, this entails that $A_n \subseteq V$. This contradicts the assumption
that $\bigcup_{m \geq n_0} A_m$ is not relatively compact.   ∎

A subset $C \subseteq M$ is called a *chain* if two elements $x, y \in C$ are *comparable*,
i.e., if either $x \leq y$ or $y \leq x$ holds. We write $\mathcal{C}(M)$ for the set of closed chains in
M. and $\mathcal{C}(x)$ for the set of all chains C in $\mathcal{C}(M)$ for which $x \in C$ and $C \subseteq \uparrow x$ (the
chains starting in x), and $\mathcal{C}(x, y)$ for the set of all chains C in $\mathcal{C}(x)$ which satisfy
$y \in C$ and $C \subseteq \downarrow y$. We use the subscript c for the set of connected chains in these
spaces. The following proposition shows that the Vietoris topology is well suited for
the various sets of chains.

Proposition 4.2.

(i) $\mathcal{C}(M)$ *is a compact subspace of* $\mathcal{F}(M)$.

(ii) *If the subset* $C \subseteq M$ *is a chain, then* $\overline{C}$ *is a chain.*

(iii) *If* $C_n \to C$ *and* $C_n \in \mathcal{C}(x_n)$ *or* $\mathcal{C}(x_n, y_n)$ *with* $x_n \to x$ *and* $y_n \to y$, *then*
$C \in \mathcal{C}(x)$ *or* $\mathcal{C}(x, y)$ *respectively. It follows in particular that* $\mathcal{C}(x)$ *and* $\mathcal{C}(x, y)$
are closed in the compact space $\mathcal{C}(M)$.

(iv) *Suppose that* $C_n \to C$, $C_n \in \mathcal{C}_c(M)$, *and that there exists a compact set*
$K \subseteq M$ *with* $C_n \subseteq K$ *for all* $n \in \mathbb{N}$. *Then* C *is connected.*

(v) *Suppose that $C_n \to C$, $C_n \in \mathcal{C}_c(x_n)$, and that $x_n \to x$. Then $C \in \mathcal{C}(x)$ and either C is compact and connected or the connected component of x in C is non-compact. If C is compact, then there exists a compact subset K of M and $n \in \mathbb{N}$ such that $C_m \subseteq K$ holds for all $m > n$.*

Proof.   (i) Let $C_n \to C$ in $\mathcal{F}(M)$ with $C_n \in \mathcal{C}(M)$ and $a, b \in C$. According to Proposition 4.1(ii) there exist two sequences $a_n, b_n \in C_n$ with $a_n \to a$ and $b_n \to b$. Since either $a_n \leq b_n$ or $b_n \leq a_n$ we may assume that $a_n \leq b_n$ holds for all $n \in \mathbb{N}$. Now the closedness of the order implies that $a \leq b$. Hence C is a chain.

(ii) Let $a = \lim c_n$ and $b = \lim c'_n$. As above we may assume that $c_n \leq c'_n$ holds for all $n \in \mathbb{N}$. Then $a \leq b$.

(iii) From (ii) we get that C is a chain and from Proposition 4.1(ii) that $x \in C$ in the first case and $x, y \in C$ in the second. Let $c \in C$ and $c_n \in C_n$ with $c_n \to c$ (cf. Proposition 4.1(ii)). Then $x_n \leq c_n$, $c_n \leq y_n$, and the closedness of the order show that $x \leq c$, $c \leq y$.

(iv) This is [Bou71b, Ch. II, §4, Ex. 15].

(v) That $C \in \mathcal{C}(x)$ follows from (iii). Suppose that the connected component C_0 of x in C is compact. Then Proposition 4.1(iv) implies the existence of $n_0 \in \mathbb{N}$ such that $\bigcup_{m \geq n_0} C_m$ is relatively compact. Now (iv) applies, so that C is connected. ∎

A central problem for ordered spaces is to determine if the order intervals are compact. So for $x \in M$ we define

$$\mathrm{comp}(x) := \{y \in M : [x, y] \in \mathcal{CO}(M)\}.$$

We say that M is *globally hyperbolic* if all the order intervals are compact, i.e., if $\mathrm{comp}(x) = \uparrow x$ for all $x \in M$.

The essential features of a "well behaved" ordering ought to be captured by the chains. In order to be able to describe how an element $x \in M$ is placed in M in terms of the ordering, we introduce the concept of the *reach* of an element:

$$\mathrm{reach}(x) := \bigcup \{C : C \in \mathcal{C}_c(x)\}.$$

Now we can say what we mean by "well behaved": We want M to satisfy the *chain approximation property*

(CA) $\mathrm{reach}(x)$ is dense in $\uparrow x$.

Note that if we can find for each $x \in M$ a dense subset $S_x \subseteq \uparrow x$ such that for every $y \in S_x$ there exists a continuous monotone mapping $\gamma : [0, T] \to M$ satisfying $\gamma(0) = x$ and $\gamma(T) = y$, then (CA) is satisfied. This follows from the fact that $\gamma([0, T])$ is a connected chain.

From now on we assume that M satisfies the chain approximation property.

Lemma 4.3.

(i) *Let $x, y \in M$ be such that the order interval $[x, y]$ is not compact. Then there exists a non-compact connected chain $C \in \mathcal{C}(x)$ with $C \subseteq [x, y]$.*

(ii) $\text{comp}(x)$ *is open in* $\uparrow x$.

(iii) $\text{comp}(x) \subseteq \text{reach}(x)$.

Proof. (i) Since $[x, y]$ is non-compact, we find for every $n \in \mathbb{N}$ an element $c_n \in [x, y]$ such that $d^\omega(c_n, \omega) < \frac{1}{n}$, and, using (CA), we find $y_n \in M$ with $d^\omega(y_n, \omega) < \frac{1}{n}$ and $C_n \in \mathcal{C}_c(x, y_n)$. Since $\mathcal{C}(x)$ is compact, there exists a convergent subsequence $(C_{n_k})_{k \in \mathbb{N}}$. Let $C := \lim_{k \to \infty} C_{n_k}$. There exists no compact subset $K \subseteq M$ which eventually contains the sets C_{n_k}. Therefore the non-compactness of the connected component C_0 of x in C follows from Proposition 4.2(v). In view of Proposition 4.2(iii), the set C_0 is a non-compact connected chain in $\mathcal{C}(x)$ with $C_0 \subseteq [x, y]$.

(ii) Let $y_n \in \uparrow x \setminus \text{comp}(x)$ with $y_n \to y$. Then there exists a sequence of non-compact connected chains $C_n \in \mathcal{C}(x)$ with $C_n \subseteq [x, y_n]$. We may assume that the sequence C_n converges to $C \in \mathcal{C}(x)$. Then $C \subseteq [x, y]$ is non-compact (Proposition 4.2(v)) and therefore $y \notin \text{comp}(x)$.

(iii) Let $n \in \mathbb{N}$. According to (CA), we find a sequence $c_n \in M$ and $C_n \in \mathcal{C}_c(x, c_n)$ such that $d(c_n, y) < \frac{1}{n}$. Let C_{n_k} be a subsequence which converges to C. Then $C \in \mathcal{C}(x, y)$ (Proposition 4.2(iii)). Thus $C \subseteq [x, y]$ is compact. Now the connectedness of C is a consequence of Proposition 4.2(iv). ∎

Now we can show that *locally* we always have compact order intervals.

Proposition 4.4. *Let $x \in M$. Then there exists a neighborhood U of x such that for all $y, z \in U$ the order interval $[y, z]$ is compact. Moreover, given a neighborhood V of x we may choose U so small such that $[y, z] \subseteq V$ for $y, z \in U$.*

Proof. The first statement follows from the second one because M is locally compact. We also may assume w.l.o.g. that V is compact.

Suppose that the second statement is false. Then we choose for every $n \in \mathbb{N}$ elements $y_n, z_n \in M$ with $d(x, y_n), d(x, z_n) < \frac{1}{n}$ such that $[y_n, z_n] \not\subseteq V$ and $y_n, z_n \in V$. If $[y_n, z_n]$ is compact and $a_n \in [y_n, z_n] \setminus V$, then there exist connected chains $C_n \in \mathcal{C}(y_n, a_n)$ with $a_n \in C_n$ (Lemma 4.3(iii)). Now C_n intersects ∂V in at least one point b_n. If $[y_n, z_n]$ is non-compact, then it contains a non-compact connected chain $C \in \mathcal{C}(y_n)$ (Lemma 4.3(i)). So $C \not\subseteq V$ and therefore we also find $b_n \in \partial V \cap C$. We may assume that the sequence b_n converges to a point $b \in \partial V$. Hence $x = \lim y_n \leq \lim b_n = b \leq \lim z_n = x$ contradicts the partiality of the order. This completes the proof. ∎

A pospace M is said to be *locally convex* if every point $x \in M$ has a base of open neighborhoods U which are *order convex* or simply *convex*, i.e., $y, z \in U$ implies that $[y, z] \subseteq U$. We want to show that M is locally convex. For the sake of completeness we give the proof of the following lemma (cf. [CHK86, p.165]).

Lemma 4.5. *Let X be a compact pospace.*

(i) *Let $D \subseteq X$ be an up directed set. Then $\sup D$ exists and $\lim D = \sup D$.*

(ii) *Let $A = \downarrow A$, $B = \uparrow B$ be closed subsets such that $A \cap B = \emptyset$. Then there exist open sets $U = \downarrow U$ and $V = \uparrow V$ such that $A \subseteq U$, $B \subseteq V$ and $U \cap V = \emptyset$.*

(iii) X *is locally convex.*

Proof.   (i) Let a be a cluster point of D as a directed set, i.e., for every neighborhood U of a and every $d \in D$ there exists $d' \in D \cap U$ such that $d \le d'$. Suppose that $x \not\le a$ for some $x \in D$. Since a is a cluster point of D, there exists $b \in D \setminus {\uparrow}x$. Since D is directed, we find $c \in D$ with $x \le c$ and $b \le c$. Next we use the fact that a is a cluster point to find an element $d \in D \cap {\uparrow}c$ in the open neighborhood $X \setminus {\uparrow}x$ of a. Then the contradiction $x \le c \le d$ proves that $x \le a$ for all $x \in D$, i.e., $D \subseteq {\downarrow}a$.

Now $D \subseteq {\downarrow}b$ implies that $a \le b$ because a is a cluster point of D and ${\downarrow}b$ is a closed set. It follows that $a = \sup D$ exists and, since the cluster point of D is unique, that $\lim D = a$.

(ii) Since X is a normal space, there exist open sets P and Q such that $A \subseteq P$, $B \subseteq Q$, and $P \cap Q = \emptyset$. Let $U := X \setminus {\uparrow}(X \setminus P) \subseteq P$ and $V := X \setminus {\downarrow}(X \setminus Q) \subseteq Q$. Then $U = {\downarrow}U$ and ${\uparrow}V = V$. That these sets are open follows from the fact that ${\uparrow}F$ and ${\downarrow}F$ are closed for a compact set F.

(iii) Let $x \in X$ and P be an open set with $x \in P$. For each $y \in X \setminus P$ either $y \notin {\uparrow}x$ or $y \notin {\downarrow}x$, so that either ${\uparrow}x \cap {\downarrow}y = \emptyset$ or ${\downarrow}x \cap {\uparrow}y = \emptyset$. In either case there are open convex sets U_y and V_y such that $x \in U_y$ and $y \in V_y$ with $U_y \cap V_y = \emptyset$ (cf. (ii)). Now $X \setminus P$ is compact and can be covered by finitely many sets $V_{y_1}, \ldots, V_{y_n}$. Set $U := U_{y_1} \cap \ldots \cap U_{y_n}$. Then $x \in U \subseteq P$ and U is convex.   ∎

Proposition 4.6. M *is locally convex.*

Proof.   Let $x \in M$ and V be a relatively compact neighborhood of x. Using Proposition 4.4, we find a compact neighborhood $U \subseteq V$ of x such that all order intervals $[y, z]$ with $y, z \in U$ are contained in V. By Lemma 4.5(iii), the compact pospace $(\overline{V}, \le)$ is locally convex. Given a neighborhood W of x, we therefore find a neighborhood W' of x in $U \cap W$ such that W' is order convex with respect to the restriction of $\le$ to $\overline{V}$. But the construction of U implies that this order agrees with the original order on W'. Hence W' is order convex and we have proved that M is locally convex.   ∎

The remainder of this section is devoted to a characterization of global hyperbolicity in terms of chains. Recall here that M is globally hyperbolic if and only if *all* order intervals are compact.

Lemma 4.7.   *Let* $C \in \mathcal{C}_c(M)$ *and* $c, d \in C$. *Then the following assertions hold:*

 (i) ${\uparrow}c \cap C$, ${\downarrow}c \cap C$ *and* $[c, d] \cap C$ *are in* $\mathcal{C}_c(M)$.

 (ii) *If* $c \ne d$, *then there exists* $e \in [c, d]$ *such that* $e \ne c, d$.

 (iii) *If* c *is not maximal* (*minimal*) *in* C, *then* $c \in \overline{C \setminus {\downarrow}c}$ $(c \in \overline{C \setminus {\uparrow}c})$.

 (iv) *The topology on* C *is generated by the set*

$$\{C \setminus {\downarrow}c, \ C \setminus {\uparrow}c : c \in C\} \subseteq 2^C.$$

Proof.   Since C is closed in M, it is a locally compact locally convex pospace in its own right and therefore we may assume that $M = C$ is a locally convex totally ordered pospace.

(i) Suppose that $\uparrow c$ is not connected. Then there exist two non-empty disjoint open subsets $O_1, O_2 \subseteq C \cap \uparrow c$ such that

$$\uparrow c \cap C = O_1 \cup O_2.$$

We may assume that $c \in O_1$. Then O_2 is open C and we find that

$$C = ((C \setminus \uparrow c) \cup O_1) \cup O_2$$

is a disjoint union of non-empty open subsets of C, a contradiction. That $\downarrow c$ is connected follows by duality. Therefore

$$[c, d] \cap C = \downarrow d \cap (\uparrow c \cap C)$$

is connected because $\uparrow c \cap C$ is a closed connected chain.

(ii) Suppose that $[c, d] = \{c, d\}$ is a two element set. Then

$$C = \uparrow d \cup \downarrow c$$

is a disjoint union of two non-empty closed subsets, a contradiction.

(iii) If c is not maximal, the set $F := C \setminus \downarrow c$ is non-empty and open in C. Moreover it is contained in the closed set $\uparrow c$. Therefore its closure must contain c because C is connected.

(iv) It is clear that the sets $C \setminus \uparrow c$ and $C \setminus \downarrow c$ are open. We claim that these sets form a basis of the topology. It suffices to show that every convex neighborhood U of a point $c \in C$ contains an open set of the topology generated these sets. To see this, let us first assume that c is neither maximal nor minimal. Then, in view of (iii), there exists $a \in U \setminus \downarrow c$ and $b \in U \setminus \uparrow c$. Now the convexity of U shows that

$$(C \setminus \uparrow a) \cap (C \setminus \downarrow b) \subseteq U.$$

If c is minimal or maximal, then it suffices to take one of these sets. ∎

Lemma 4.8. *Let $C \subseteq M$ be a chain. Then $\lim C$ exists in M^ω in the sense of nets.*

Proof. We assume that $\lim C \neq \omega$. Then there exists a compact subset $K \subseteq M$ such that $K \cap C$ is cofinal in C, i.e., each element in C is majorized by an element of $K \cap C$. Passing to a subnet which converges in K, we find a point $x \in K$ such that every neighborhood U of x intersects C in a cofinal subset. Let U be a relatively compact convex neighborhood of x (Proposition 4.6) and $c \in C \cap U$. If $d \in C \cap \uparrow c$, then there exists $e \in C \cap U$ with $d \leq e$, so $d \in [c, e] \subseteq U$. It follows that $\uparrow c \cap C \subseteq U$. Thus $\lim C = x$. ∎

Lemma 4.9. *Let $C \subseteq M$ be a maximal chain. Then the following assertions hold:*

  (i) *For each pair of elements $x \leq y$ in C the chain $C \cap [x, y]$ is maximal in $[x, y]$.*

  (ii) *C is closed.*

  (iii) *If $[x, y]$ is compact, then $C \cap [x, y]$ is connected.*

  (iv) *The connected components of C are open in C.*

Proof.   (i) Suppose that $z \in [x, y]$ is comparable with each element of $C \cap [x, y]$. Then z is comparable with each element of C. Therefore $C \cap [x, y]$ is maximal in $[x, y]$.

(ii) Proposition 4.2(i).

(iii) Suppose that $C \cap [x, y] = F_1 \cup F_2$, where F_1 and F_2 are closed. Now (ii) implies that the sets F_1 and F_2 are compact because C is closed. Suppose w.l.o.g. that $a \in F_1$. Then $b := \min F_2$ exists by Lemma 4.5. Similarly $c := \max(F_1 \cap \downarrow b)$ exists. It follows that

$$C \cap [x, y] \subseteq \uparrow b \cup \downarrow c$$

and that $b \neq c$ because $F_1 \cap F_2 = \emptyset$. The order interval $[c, b]$ is compact so that there exists $C' \in \mathcal{C}_c(c, b)$ (Lemma 4.3(iii)). Thus $(C \cap [x, y]) \cup C'$ is a chain in $[x, y]$, a contradiction to (i).

(iv) Let $A \subseteq C$ be a connected component, $a \in A$ and $a_n \in C$ with $a_n \to a$. We have to show that eventually $a_n \in A$. Let U be a convex neighborhood of a such that all intervals between elements of U are compact (Lemma 4.5(iii)). We choose $n_0 \in \mathbb{N}$ such that $a_n \in U$ for $n \geq n_0$. If $a_n \leq a$, then, in view of (iii), $[a_n, a] \cap C$ is connected and if $a \leq a_n$, then $[a, a_n] \cap C$ is connected. It follows that $a_n \in A$ whenever $a_n \in U$.   ∎

Theorem 4.10.   *Let M be a metric pospace satisfying the chain approximation property. Then there exist disconnected maximal chains in M if and only if M is not globally hyperbolic.*

Proof.   Suppose that M is globally hyperbolic and that C is a maximal chain in M, then Lemma 4.9(iii) implies that C is connected.

If, conversely, there exist $x \leq y$ in M such that $[x, y]$ is not compact, then we find a connected non-compact chain $C_1 \in \mathcal{C}_c(x)$ with $C_1 \subseteq [x, y]$ (Lemma 4.3(i)). Let C be a maximal chain of $[x, y]$ containing C_1. Then y cannot lie on the same component of C as C_1 because otherwise $\lim C_1$ would exist in M (Lemma 4.8).∎

4.2. Invariant cone fields on homogeneous spaces

Cone fields on manifolds are a generalization of geometric distributions, where, instead of vector subspaces one associates to each point in the manifold a convex cone in the tangent space. More precisely, a *wedge (cone) field* on a manifold N assigns to each point $x \in N$ a (pointed) wedge $\Theta(x)$ in the tangent space $T_x(N)$. A manifold with a cone field is called a *conal manifold* (cf. [La89, p.276]). In this context it is not so easy to define what a smooth cone field should be. But for homogeneous manifolds and cone fields invariant under a transitive group action smoothness is automatically given in any reasonable way. So in this section M always denotes a homogeneous space of the connected Lie group G, $x_0 \in M$ a base point, and $H = G_{x_0} = \{g \in G : g.x_0 = x_0\}$. We define the mappings

$$\mu_g : M \to M, \quad x \mapsto g.x$$

and

$$\pi : G \to M, \quad g \mapsto g.x_0.$$

The Lie algebras of G and H respectively are denoted $\mathfrak{g}$ and $\mathfrak{h}$ respectively.

If M is a conal manifold with cone field Θ, then M is said to be a *conal homogeneous space* if the cone field Θ is invariant under the action of G, i.e., if

$$d\mu_g(x)\Theta(x) = \Theta(g.x) \qquad \forall g \in G, x \in M.$$

In this case we say that the cone field is *regular* if the cone $W_\Theta := d\pi(1)^{-1}\Theta(x_0)$ is Lie generating in $\mathbf{L}(G)$.

Figure 4.1

Example 4.11. (i) Let G be a Lie group and $W \subseteq \mathbf{L}(G)$ a wedge. Then

$$\Xi_W(g) := d\lambda_g(1)W$$

is a left-invariant wedge field on G. If, in addition, W is pointed, then G is a conal homogeneous G-space.

(ii) A particularly interesting case is when $G = V$ is a vector space. Then

$$\Xi_W(v) = d\lambda_v(0)W = W$$

because we identify $T_v(V)$ with $V \cong T_0(V)$ via $d\lambda_v(0)$. These wedges are referred to as the *constant wedge fields* on the vector space V.

(iii) Let $G := \mathbb{R}^{n+1} \rtimes O(n,1)_0$ and write $H := O(n,1)_0$ for the connected component of the *Lorentz group*. Then $M := \mathbb{R}^{n+1} \cong G/H$ is a conal G manifold with respect to the constant cone field given by the *Lorentzian cone*

$$C := \left\{ (x_1, \ldots, x_{n+1}) : x_{n+1} \geq \sqrt{x_1^2 + \ldots + x_n^2} \right\}.$$

This conal homogeneous space is also called the $(n+1)$-*dimensional Minkowski space*.

(iv) The construction in (iii) can be generalized as follows. Let W be a pointed cone in the vector space V, and

$$H := \mathrm{Aut}(W) := \{ g \in \mathrm{Gl}(V) : g.W = W \}.$$

Then the group $G := V \rtimes \mathrm{Aut}(W)$ acts on V by affine transformations and leaves the constant cone field Ξ_W invariant. Therefore V is a homogeneous conal G-space.   ∎

It is clear that an invariant cone field is completely determined by the cone at any given point of the homogeneous space. The following proposition says which cones may occur.

Proposition 4.12.   *The prescription*

$$\Theta \mapsto W_\Theta := \left(d\pi(1) \right)^{-1} (\Theta(x_0))$$

defines a bijection from the set of all G-invariant cone fields on M onto the set of all wedges $W \subseteq \mathfrak{g} = \mathbf{L}(G)$ with the property that

(a) $H(W) = \mathfrak{h}$.

(b) $\mathrm{Ad}(H)W = W$.

Proof.   Let Θ be a G-invariant cone field on M and define $W = W_\Theta$ as above. Then $H(W_\Theta) = \mathfrak{h} = \ker d\pi(1)$ because $\Theta(x_0)$ is pointed. Let $h \in H$. Then

$$d\pi(1)\,\mathrm{Ad}(h)W = d\mu_h(x_0)d\pi(1)W = d\mu_h(x_0)\Theta(x_0) = \Theta(x_0).$$

We conclude that $\mathrm{Ad}(h)W \subseteq W$ and therefore that $\mathrm{Ad}(h)W = W$.

If, conversely, W is a wedge in $\mathfrak{g}$ which satisfies (a) and (b), then

$$
\begin{aligned}
d\pi(gh)d\lambda_{gh}(1)W &= d\pi(gh)d\lambda_g(h)d\lambda_h(1)W \\
&= d\pi(gh)d\lambda_g(h)d\rho_h(1)\,\mathrm{Ad}(h)W \\
&= d\pi(gh)d\rho_h(g)d\lambda_g(1)W \\
&= d\pi(g)d\lambda_g(1)W.
\end{aligned}
$$

Therefore $\Theta_W(\pi(g)) := d\pi(g)d\lambda_g(1)W$ is a well defined cone field on M. The invariance follows from

$$d\mu_{g'}(\pi(g))\Theta_W(\pi(g)) = d\mu_{g'}(\pi(g))d\pi(g)d\lambda_g(1)W$$
$$= d\mu_{g'}(\pi(g))d\mu_g(x_0)d\pi(1)W$$
$$= d\mu_{g'g}(x_0)d\pi(1)W$$
$$= d\pi(gg')d\lambda_{gg'}(1)W$$
$$= \Theta_W(\pi(gg'))$$

The relations $W_{\Theta_W} = W$ and $\Theta_{W_\Theta} = \Theta$ are trivial consequences of the definitions. ∎

From Proposition 4.12 we see that invariant cone fields are closely related to Lie wedges.

Corollary 4.13.

(i) *For every G-invariant cone field Θ on M the cone W_Θ is a Lie wedge with $H(W_\Theta) = \mathfrak{h}$.*

(ii) *If G is a connected Lie group and $W \subseteq \mathbf{L}(G)$ a Lie wedge such that $H(W)$ is global, then $G/S_{H(W)}$ is a conal homogeneous G-space.*

Proof. For a wedge W in $\mathbf{L}(G)$ the Lie wedge condition is equivalent to $\mathrm{Ad}(S_{H(W)}) \subseteq \mathrm{Aut}(W)$. Therefore (i) follows immediately from Proposition 4.12 and (ii) likewise because the homogeneous space G/H with $H = S_{H(W)}$ satisfies $\mathfrak{h} = \mathbf{L}(H) = H(W)$ since $H(W)$ is global. ∎

Let N be a differentiable manifold. A continuous mapping $\gamma : [a, b] \to N$ is said to be *absolutely continuous*, if for any chart $\varphi : U \to \mathbb{R}^n$ the curve

$$\eta := \varphi \circ \gamma : \gamma^{-1}(U) \to \mathbb{R}^n$$

has absolutely continuous coordinate functions and the derivatives of these functions are locally bounded (cf. [Po62, p.241]). For $p, q \in N$ we write Ω_p for the set of all absolutely continuous curves starting in p, and $\Omega_{p,q}$ for the set of all curves in Ω_p with endpoint in q.

A cone field on M gives rise to a *conal order* on M, defined by $x \prec y$, if there exists an absolutely continuous curve $\alpha : [a, b] \to M$ with $\alpha(a) = x$, $\alpha(b) = y$, and

$$\alpha'(t) \in \Theta(\alpha(t))$$

whenever the derivative exists. These curves are said to be *conal*. We define the relation $\preceq$ on M as the closure of the set $\prec$ in $M \times M$, i.e.,

$$x \preceq y :\Longleftrightarrow (x, y) \in \overline{\{(x', y') \in M \times M : x' \prec y'\}}.$$

If G is a Lie group and $W \subseteq \mathbf{L}(G)$ a wedge, then an absolutely continuous curve $\gamma : [a, b] \to G$ is said to be W-*conal* if

$$\gamma'(t) \in \Xi_W(\gamma(t)) := d\lambda_{\gamma(t)}(1)W$$

whenever the derivative exists. We write Ω_p^c and $\Omega_{p,q}^c$ respectively for the set of conal curves in Ω_p and $\Omega_{p,q}$ respectively. We collect a few basic facts about conal curves and their related objects:

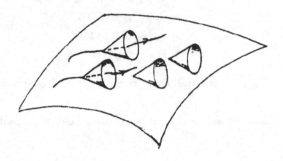

Figure 4.2

Proposition 4.14.   *The following assertions hold:*

(i)   *The sets of conal curves on G and M are invariant under G.*

(ii) *The relations $\prec$ and $\preceq$ are G-invariant.*

(iii) $\uparrow x = \overline{\uparrow_\prec x}$, *where* $\uparrow_\prec x := \{y \in M : x \prec y\}$ *and* $\uparrow x := \{y \in M : x \preceq y\}$.

(iv) *The relations $\prec$ and $\preceq$ are quasi orders on M, i.e., they are reflexive and transitive relations.*

(v)   *A curve $\gamma : [a, b] \to G$ is W_Θ-conal if and only if $\pi \circ \gamma$ is conal.*

(vi) *If $\gamma : [a, b] \to M$ is a conal curve and $g \in G$ with $\pi(g) = \gamma(a)$, then there exists a conal curve $\tilde{\gamma} : [a, b] \to G$ with $\pi \circ \tilde{\gamma} = \gamma$ and $\tilde{\gamma}(a) = g$.*

Proof.   (i) This follows from the G-invariance of Θ.

(ii) This is a consequence of (i).

(iii) Let $x \preceq y$. Then there exists a sequence $(x_n, y_n) \in M \times M$ with $(x_n, y_n) \to (x, y)$ and $x_n \prec y_n$. Let U be a neighborhood of x in M such that there exists a continuous section $\sigma : U \to G$ of π. Then $\sigma(x_n) \to \sigma(x)$ and therefore $g_n := \sigma(x)\sigma(x_n)^{-1} \to \mathbf{1}$. We set $y_n' := g_n . y_n$. Then $y_n' \to y$ and $g_n . x_n = \sigma(x).x_0 = x \prec y_n'$, so $y \in \overline{\uparrow_\prec x}$.

(iv) (cf. [La89, p.299]) The reflexivity of $\prec$ is trivial and the transitivity follows by concatenation of conal curves. Moreover, the reflexivity of $\prec$ implies the reflexivity of $\preceq$. Let $x \preceq y \preceq z$. Then we use (iii) to find $y_n, z_n \in M$ with $x \prec y_n$ and $y \prec z_n$ such that $y_n \to y$ and $z_n \to z$. As above, we find a sequence $g_n \to \mathbf{1}$ in G which satisfies $g_n . y_n = y$. Therefore $g_n . x \prec g_n . y_n = y \prec z_n$ implies that $g_n . x \prec z_n$. Passing to the limit we obtain that $x \preceq z$.

(v) This follows from the fact that

$$\ker d\pi(g) = d\lambda_g(\mathbf{1})\mathfrak{h} \subseteq d\lambda_g(\mathbf{1})W \quad \text{and} \quad d\pi(g)d\lambda_g(\mathbf{1})W_\Theta = \Theta(\pi(g))$$

for all $g \in G$, and from

$$(\pi \circ \gamma)'(t) = d\pi(\gamma(t))\gamma'(t).$$

(vi) Since the quotient mapping $\pi: G \to G/H = M$ permits locally smooth sections, there exists an absolutely continuous curve $\tilde{\gamma}$ with $\tilde{\gamma}(a) = g$ and $\pi \circ \tilde{\gamma} = \gamma$. Now the assertion follows from (v). ∎

We can associate several semigroups to any invariant cone field Θ. On the one hand the transitivity together with the invariance of $\preceq$ shows that

$$S_\Theta := \{g \in G : g.x_0 \succeq x_0\} = \pi^{-1}(\uparrow x_0)$$

is a semigroup. On the other hand we also have the semigroup S_{W_Θ}. We will see shortly how the two semigroups are related.

Lemma 4.15. *Let $W \subseteq \mathbf{L}(G)$ be a wedge and $\gamma: [a, b] \to G$ a W-conal curve, then*

$$\gamma(b) \in \gamma(a) S_W.$$

Proof. We clearly may assume that $\gamma(a) = 1$. Let

$$f \in \mathrm{Mon}^\infty(S_W) = \mathrm{Pos}(W)$$

(Lemma 1.30). Then $f \circ \gamma: [a, b] \to \mathbb{R}$ is monotone because this function is absolutely continuous and

$$(f \circ \gamma)'(t) = df(\gamma(t))\gamma'(t) \in df(\gamma(t)) d\lambda_{\gamma(t)}(1)W \subseteq \mathbb{R}^+.$$

We conclude that $f(\gamma(b)) \geq f(\gamma(a)) = f(1)$. Now Theorem 1.23 shows that $\gamma(b) \in S_W$. ∎

In particular, if C is a wedge in the finite dimensional vector space V and $\gamma: [a, b] \to V$ a C-conal curve, then $\gamma(b) \in \gamma(a) + C$.

Proposition 4.16. *The set*

$$S_\Theta := \{g \in G : g.x_0 \succeq x_0\} = \pi^{-1}(\uparrow x_0)$$

is a closed subsemigroup of G with the following properties:

(i) $S_\Theta = \overline{S_{W_\Theta} H}$.

(ii) $\uparrow x_0 = \pi(S_\Theta) = \overline{\pi(S_{W_\Theta})}$ *and* $\uparrow_{\preceq} x_0 \subseteq \pi(S_{W_\Theta})$.

Proof. The closedness of S_Θ follows from $S_\Theta = \pi^{-1}(\uparrow x_0)$. That S_Θ is a semigroup is a consequence of the invariance of the order $\preceq$. In fact, let $s, s' \in S_\Theta$. Then

$$x_0 \preceq s.x_0 \preceq ss'.x_0$$

because $x_0 \preceq s'.x_0$. The surjectivity of π implies that $\pi(S_\Theta) = \uparrow x_0$. Therefore (ii) is a consequence of (i) and it remains to prove (i).

"$\subseteq$": Let $s \in \pi^{-1}(\uparrow_{\preceq} x_0)$ and $\gamma: [a, b] \to M$ a conal curve with $\gamma(a) = x_0$ and $\gamma(b) = \pi(s)$. Then Proposition 4.14(vi) implies the existence of a conal curve $\tilde{\gamma}: [a, b] \to G$ with $\tilde{\gamma}(a) = 1$. Thus, according to Lemma 4.15, we have $\gamma(b) \in S_W$

for $W = W_\Theta$. We conclude that $s \in \gamma(b)H \subseteq S_W H$. Thus $\uparrow_\prec x_0 \subseteq \pi(S_W)$ and this leads to

$$S_\Theta \subseteq \overline{\pi^{-1}(\uparrow_\prec x_0)} \subseteq \overline{S_W H}.$$

since π permits local sections.

"$\supseteq$": It suffices to show that $\pi(\langle \exp W \rangle) \subseteq \uparrow x_0$. Let $s = \exp w_1 \cdot \ldots \cdot \exp w_n$ with $w_1, \ldots, w_n \in W$. We define

$$\gamma : [0, n] \to G, \quad t \mapsto \exp(w_1) \cdot \ldots \cdot \exp((t - k)w_{k+1}) \qquad \forall t \in [k, k+1].$$

Then γ is a conal curve and therefore $\pi \circ \gamma : [0, n] \to M$ is a conal curve. Thus $\pi(s) \in \uparrow x_0$. ∎

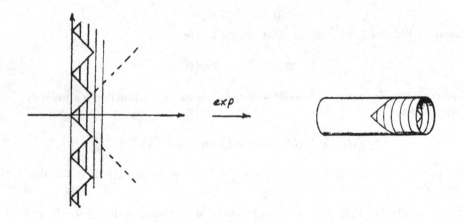

Figure 4.3

4.3. Globality of cone fields

We say that M is *globally orderable* or that the cone field Θ is *global* if $\preceq$ is a partial order.

Proposition 4.17. *Suppose that M is globally orderable and $x \in M$. Then the following assertions hold:*

(i) *M satisfies the chain approximation property.*

(ii) *$(M, \preceq)$ is locally convex.*

(iii) *The set $\text{comp}(x) = \{y \in \uparrow x : [x, y] \text{ is compact}\}$ is open in $\uparrow x$.*

(iv) *$\{(x, y) : y \in \text{comp}(x)\}$ is open in $\{(x, y) : x \preceq y\}$.*

Proof. (i) follows from Proposition 4.14(iii) since $\text{reach}(x) \supseteq \{y \in M : x \prec y\}$. In view of (i) we can now deduce (ii) and (iii) from Proposition 4.6 and Lemma 4.3(ii).

(iv) Let $x_n \preceq y_n$ such that $(x_n, y_n) \to (x, y)$ and $y \in \text{comp}(x)$. We have to show that eventually $y_n \in \text{comp}(x_n)$. Let $g_n \in G$ with $g_n \to 1$ and $x = g_n.x_n$. Then $g_n.y_n \in \uparrow x$ and $g_n.y_n \to y$. Hence there exists $m \in \mathbb{N}$ such that $g_m.y_m \in \text{comp}(x)$. Then $y_m \in \text{comp}(g_m^{-1}.x) = \text{comp}(x_m)$. ∎

We are now heading for the completion of the proof of Theorem 1.35 in which we characterized global Lie wedges in terms of positive functions. We call a function $f \in C^\infty(M)$ Θ-*positive* in x if $df(x) \in \Theta(x)^*$ and *strictly* Θ-*positive in* x if $df(x) \in \text{int}\,\Theta(x)^*$. We simply say that f is (*strictly*) Θ-*positive* if it is everywhere (strictly) Θ-positive.

Lemma 4.18. *Let* $f \in C^\infty(M)$. *Then the set of all points in* M *where* f *is strictly* Θ-*positive is open.*

Proof. Suppose that f is strictly Θ-positive in x. Let $\omega \in \text{int}\,\Theta(x)^*$. Then $K := \omega^{-1}(1) \cap \Theta(x)$ is a compact subset of $\Theta(x)$ such that $\Theta(x) = \mathbb{R}^+K$. For $g \in G$, in view of Proposition 1.1, this shows that $\Theta(g.x) = \mathbb{R}^+ d\mu_g(x)K$ and therefore

$$\text{int}\,\Theta(g.x)^* = \{\nu \in T_{g.x}(M)^* : \langle \nu, d\mu_g(x)K \rangle \subseteq]0, \infty[\}.$$

So we have to show that

$$\langle df(g.x), d\mu_g(x)K \rangle \subseteq]0, \infty[$$

holds for all g in a 1-neighborhood of G. But $df(g.x)d\mu_g(x) = d(f \circ \mu_g)(x)$ and the function

$$G \times K \to \mathbb{R}, \quad (g, v) \mapsto \langle d(f \circ \mu_g)(x), v \rangle$$

is continuous and positive on $\{1\} \times K$. Therefore we find a 1-neighborhood U in G such that $\langle d(f \circ \mu_g)(x), K \rangle \subseteq]0, \infty[$ holds for all $g \in U$. It follows that f is strictly Θ-positive on the neighborhood $U.x$ of x. ∎

Lemma 4.19. *Let* $C \neq \{0\}$ *be a pointed wedge in a finite dimensional vector space* L *and* $K \subseteq \text{int}\,C^*$ *a compact subset. Then we find a pointed wedge* $C' \subseteq L$ *such that*

(i) $K \subseteq \text{int}(C')^*$ *and*

(ii) C' *surrounds* C, *i.e.,* $C \setminus \{0\} \subseteq \text{int}\,C'$.

Proof. We choose a compact convex subset $K' \subseteq \text{int}\,C^*$ which contains K in its interior. This is possible because the convex closure of K is also a compact convex subset of the interior of C^*. But now $0 \notin K' \subseteq \text{int}\,C^*$ because otherwise we would have $C = \{0\}$ which was excluded. Then $\mathbb{R}^+K'$ is a pointed generating wedge in L^*. We set $C' := (\mathbb{R}^+K')^* = K'^*$. Then (i) is immediate from the construction. Suppose that C' doesn't surround C. Then, using the Hahn-Banach Separation Theorem, we find a functional $\omega \in K' \subseteq C'^*$ and a point $x \in C \cap \partial C'$ such that $\langle \omega, x \rangle = 0$. This is a contradiction to $K' \subseteq \text{int}\,C^*$. ∎

Lemma 4.20. *Let* $x \in M$, f *a* Θ-*positive function which is strictly* Θ-*positive in* x *and* U *a neighborhood of* x. *Then there exists* $\varepsilon > 0$ *such that*

$$\uparrow x \subseteq U \cup f^{-1}([f(x) + \varepsilon, \infty[).$$

Proof.   If $\gamma:[a,b] \to M$ is a conal curve, then $f \circ \gamma$ is monotone. Therefore $f:(M, \preceq) \to (\mathbb{R}, \leq)$ is a monotone mapping and $f(\uparrow x) \subseteq [f(x), \infty[$.

We clearly may assume that $f(x) = 0$ and that U is a relatively compact open neighborhood of x such that there exists a chart $\varphi:U \to \mathbb{R}^n$ with $\varphi(x) = 0$. Let $h := f \circ \varphi^{-1}$ and $\Theta'(\varphi(y)) := \Theta(y)$ for all $y \in U$. Then $dh(0) \in \operatorname{int} \Theta'(0)^{*}$. Let K be a compact neighborhood of $dh(0)$ in $\operatorname{int} \Theta'(0)^{*}$. Using Lemma 4.19, we find a pointed cone $C \subseteq \mathbb{R}^n$ which surrounds $\Theta'(0)$ and which satisfies $K \subseteq \operatorname{int} C^{*}$. Let V be a 1-neighborhood in G with $V.x \subseteq U$. Then the mapping

$$F: V \times \mathbb{R}^n \to \mathbb{R}^n, \quad (g, v) \mapsto d\varphi(g.x)d\mu_g(x)d\varphi^{-1}(0)v$$

is continuous and $F(g, \Theta'(0)) = \Theta'(\varphi(g.x))$. Let D be a compact subset of $\Theta'(0) \setminus \{0\}$ such that $\Theta'(0) = \mathbb{R}^+ D$. Then $F(\{1\} \times D) \subseteq \operatorname{int} C$, so there exists a 1-neighborhood $V' \subseteq V$ such that

$$F(V' \times D) \subseteq \operatorname{int} C.$$

We conclude that C surrounds $\Theta'(g.x)$ for all $g \in V'$. Let $U' := V'.x \subseteq U$ such that $\varphi(U')$ is convex, relatively compact in $\varphi(U)$, and $dh(y) \in K$ for all $y \in \varphi(U')$. Then

$$dh(y) \in \operatorname{int} C^{*} \quad \text{and} \quad \Theta'(y) \setminus \{0\} \subseteq \operatorname{int} C$$

for all $y \in \varphi(U')$.

Now we suppose that the assertion of the lemma is false. Then there exists a sequence $x_n \in \uparrow x \setminus U$ with $f(x_n) \to 0$. Since $\uparrow_{\prec} x$ is dense in $\uparrow x$, we even find such a sequence in $\uparrow_{\prec} x$. Let $\gamma_n:[0, t_n] \to M$ be a conal curve with $\gamma_n(0) = x$ and $\gamma_n(t_n) = x_n$. Then there exists $s_n \in [0, t_n]$ such that $z_n := \gamma_n(s_n) \in \partial U'$ and $\gamma_n([0, s_n[) \subseteq U'$. Since $\partial U'$ is compact, we also may assume that $z_n \to z \in \partial U'$. Then

$$0 = f(x) \leq f(z_n) \leq f(x_n) \to 0$$

entails that $f(z) = 0$. On the other hand, applying Lemma 4.15 to C, we have that $\varphi(z_n) \in C$ which implies that $\varphi(z) \in C \setminus \{0\}$. Now

$$\begin{aligned} f(z) &= (f \circ \varphi^{-1})(\varphi(z)) \\ &= \int_0^1 \langle d(f \circ \varphi^{-1})(t\varphi(z)), \varphi(z)\rangle dt \\ &= \int_0^1 \langle dh(t\varphi(z)), \varphi(z)\rangle dt > 0 \end{aligned}$$

because on the one hand $[0, 1]\varphi(z) \in \varphi(U')$ since $\varphi(U')$ is convex and on the other hand $\langle dh(y), \varphi(z)\rangle > 0$ for all $y \in \varphi(U')$. This is a contradiction.   ∎

Theorem 4.21.   (The Globality Theorem) *The following statements are equivalent:*

(1) *M is globally orderable.*

(2) *$H = H(S_\Theta)$.*

(3) *$\mathbf{L}(S_\Theta) = W_\Theta$.*

(4) *There exists a strictly Θ-positive analytic function.*

(5) *There exists a Θ-positive function which is strictly Θ-positive in x_0.*

Proof.   (1) $\Rightarrow$ (2): The inclusion $H \subseteq H(S_\Theta)$ is trivial. If M is globally orderable and $s \in H(S_\Theta)$, then $x_0 \preceq s.x_0$ implies that $s^{-1}.x_0 \preceq x_0 \preceq s^{-1}.x_0$. Therefore $s^{-1}.x_0 = x_0$ and $s \in H$.

(2) $\Rightarrow$ (1): If $H = H(S_\Theta)$, and $x_0 \preceq x \preceq x_0$, we choose $s \in S_\Theta$ with $s.x_0 = x$. Then $x \preceq x_0$ implies that $x_0 \preceq s^{-1}.x_0$ and therefore $s \in H(S_\Theta) = H$. Hence $x = s.x_0 = x_0$ and $\preceq$ is a partial order.

(1) $\Rightarrow$ (3): Let $X \in \mathbf{L}(S_\Theta)$ and set $\gamma(t) := \pi(\exp tX)$. Let U be a neighborhood of x_0 such that $(U, \preceq)$ is convex (Proposition 4.17) and suppose that $X \notin W = W_\Theta$. Then $\gamma'(0) \notin d\pi(1)W = \Theta(x_0)$, so there exists $\omega \in \operatorname{int} \Theta(x_0)^*$ such that $\langle \omega, \gamma'(0) \rangle < 0$. After shrinking U, we may assume that there exists a smooth function $f: U \to \mathbb{R}$ such that $df(x_0) = \omega$. After further shrinking of U, we even may assume that f is strictly Θ-positive on U because f is strictly Θ-positive in x_0 (Lemma 4.18). Since $(f \circ \gamma)'(0) < 0$ there exists $T > 0$ such that $\gamma(T) \in U$ and

$$f(\gamma(T)) < f(x_0).$$

Let $x_n \to \gamma(T)$ with $x_n \in \uparrow_{\prec} x_0$ and $\alpha_n: [0, t_n] \to M$ conal curves with $\alpha_n(0) = x_0$ and $\alpha(t_n) = x_n$. Now the order convexity of U shows that $\alpha_n([0, t_n]) \subseteq U$. Therefore $f \circ \alpha_n$ is well defined and monotone. We conclude that

$$f(\gamma(T)) = \lim_{n \to \infty} f(x_n) = \lim_{n \to \infty} f(\alpha(t_n)) \geq f(\alpha(0)) = f(x_0).$$

This contradicts the construction of T. Consequently $\mathbf{L}(S_\Theta) = W$.

(3) $\Rightarrow$ (4): We apply Proposition 1.34 to find a strictly W-positive, S_Θ-monotone analytic function $\tilde{f}: G \to \mathbb{R}$. Then, according to Proposition 1.24, there exists an analytic funtion $f \in C^\omega(M)$ such that $f \circ \pi = \tilde{f}$. It follows that f is strictly Θ-positive.

(4) $\Rightarrow$ (5): trivial.

(5) $\Rightarrow$ (1): Suppose that $x_0 \preceq x \preceq x_0$ and $x \neq x_0$. Then $f(x_0) \leq f(x) \leq f(x_0)$ entails that $f(x) = f(x_0)$. If U is a neighborhood of x_0 which does not contain x, then Lemma 4.20 implies that $f(x) > f(x_0)$. This is a contradiction which proves that M is globally orderable. ∎

The following corollary to Theorem 4.21 is the result that was used in the proof of Theorem 1.35. It should be noted here that in the proof of Theorem 4.21 we *used* only Proposition 1.34, so that there are no circular conclusions.

Corollary 4.22.   *Let $W \subseteq \mathbf{L}(G)$ be a Lie wedge. Then the following are equivalent:*

(1) *W is global in G.*

(2) *$H(W)$ is global in G and $G/S_{H(W)}$ is globally orderable with respect to Θ_W.*

(3) *$H(W)$ is global and $H(S_W) = S_{H(W)}$.*

(4) *$H(W)$ is global in G and there exists $f \in \operatorname{Pos}(W)$ which is strictly W-positive in 1.*

Proof.   We set $H := S_{H(W)}$.

(1) $\Rightarrow$ (2): If W is global, then $\mathbf{L}(S_W) = W$ and Proposition 1.14(ii) implies that

$$H(W) \subseteq \mathbf{L}(S_{H(W)}) \subseteq \mathbf{L}\big(H(S_W)\big) = H(W),$$

so that $H(W)$ is global in G. Thus H is closed and $S_{\Theta w} = \overline{S_W H} = S_W$ because $H \subseteq S_W$. Whence

$$\mathbf{L}(S_{\Theta w}) = \mathbf{L}(S_W) = W = W_{\Theta w}$$

shows that G/H is globally orderable (Theorem 4.21).

(2) $\Rightarrow$ (3) $\Rightarrow$ (4) $\Rightarrow$ (1): This follows from $S_{\Theta w} = S_W$ and Theorem 4.21.   ∎

4.4. Chains and conal curves

The Globality Theorem provides a very useful tool for the further study of globally ordered homogeneous spaces, namely the strictly Θ-positive functions. We show first how they can be used to pass from connected chains to conal curves. This process makes the results from Section 4.1 available to their full extend.

For the remainder of this section we assume that the conal homogeneous space $M = G/H$ is globally orderable. We start with a simple observation:

Lemma 4.23. $\bigcup_{x \in M} \Theta(x)$ *is closed in* $T(M)$.

Proof.   Let $v_n \in \Theta(x_n)$ with $v_n \to v \in T_x(M)$. Then $x_n \to x$ and there exists a sequence $g_n \to 1$ in G such that $g_n.x_n = x$. Now $d\mu_{g_n}(x_n)v_n \in \Theta(x)$ converges to $d\mu_1(x)v = v$. So $v \in \Theta(x)$ since $\Theta(x)$ is closed.   ∎

Lemma 4.24.

(i) *Let d be a Riemannian metric on M, f strictly Θ-positive, and $K \subseteq M$ compact. Then there exist $C_1, C_2 > 0$ such that*

$$C_1 \langle df(x), v \rangle \leq ||v|| \leq C_2 \langle df(x), v \rangle \qquad \forall x \in K, v \in \Theta(x).$$

If $\alpha : [a, b] \to K$ is conal, it follows that

$$d\big(\alpha(b), \alpha(a)\big) \leq C\Big(f(\alpha(b)) - f(\alpha(a))\Big).$$

If $U \subseteq K$ is open and $\preceq$-convex, then we have

$$d(x, y) \leq C\big(f(y) - f(x)\big) \quad for \quad x, y \in U, x \preceq y.$$

(ii) *If $\gamma : [a, b] \to M$ is a curve which satisfies*

$$d\big(\gamma(t), \gamma(s)\big) \leq L|t - s| \qquad \forall t, s \in [a, b],$$

then γ is absolutely continuous.

Proof.   (i) Let

$$\widetilde{K} := \{w \in T_x(M) : x \in K, w \in \Theta(x), \|w\| = 1\}.$$

In view of Lemma 4.23, this is a closed subset of the unit sphere bundle over the compact space K. Therefore $\widetilde{K}$ is compact. Since $df(x) \in \operatorname{int} \Theta(x)^*$ for all $x \in K$, it follows that

$$\langle df(x), v \rangle > 0 \qquad \forall v \in \widetilde{K}.$$

Since $\widetilde{K}$ is compact, there exist exist positive constants C_1, C_2 with

$$C_2 < \langle df(x), v \rangle < C_1 \qquad \forall v \in \widetilde{K}.$$

Let $\alpha : [a, b] \to K$ be a conal curve. Then

$$d\big(\alpha(b), \alpha(a)\big) \leq \int_a^b \|\alpha'(t)\| \; dt$$

$$\leq C \int_a^b \langle df\big(\alpha(t)\big), \alpha'(t) \rangle \; dt$$

$$= C\big(f(\alpha(b)) - f(\gamma(a))\big)$$

and the convexity of U shows that $\alpha([a, b]) \subseteq U \subseteq K$.

Now let $U \subseteq K$ be open and $\preceq$-convex. If $x, y \in U$ with $x \prec y$, then there exists a conal curve from x to y and the assertion follows from the preceding calculation. If $x \preceq y$, then it follows from the continuity of d and f.

(ii) Using [Hel78, p.54], we find for each $t \in [a, b]$ and $p := \gamma(t)$ a neighborhood $U \subseteq T_p(M)$ of 0 and a local diffeomorphism $\operatorname{Exp}_p : U \to \operatorname{Exp}_p(U)$ with $\operatorname{Exp}_p(0) = p$ such that

$$\|X - Y\| \leq 2d(\operatorname{Exp}_p X, \operatorname{Exp}_p Y) \qquad \text{for all} \quad X, Y \in U.$$

We conclude that

$$\| \operatorname{Exp}_p^{-1} \circ \gamma(s) - \operatorname{Exp}_p^{-1} \circ \gamma(s') \|_p \leq 2d\big(\gamma(s), \gamma(s')\big) \leq 2L|s - s'|$$

for $s, s' \in \gamma^{-1} \operatorname{Exp}_p(U)$. This clearly implies that $\operatorname{Exp}_p^{-1} \circ \gamma$ and therefore γ is absolutely continuous.   ∎

Theorem 4.25.   *Let M be globally orderable and $C \subseteq (M, \preceq)$ a closed connected chain, f a strictly Θ-positive function, and $I := f(C) \subseteq \mathbb{R}$. Then*

$$\gamma : I \to C, \quad f(c) \mapsto c$$

is a conal curve.

Proof.   It follows from Lemma 4.20 that f is injective on C because $x \preceq y$ and $y \neq x$ implies that $f(x) < f(y)$. Therefore γ is well defined. We claim that γ is continuous. In view of Lemma 4.7, this follows from the fact that the sets $\gamma^{-1}(\uparrow c) = I \cap [f(c), \infty[$ and $\gamma^{-1}(\downarrow c) = I \cap] -\infty, f(c)]$ are closed.

To see that γ is absolutely continuous, let $c = \gamma(t) \in C$, U an open $\preceq$-convex relatively compact neighborhood of c (Proposition 4.17), and d be a Riemannian metric on M. Let $s \leq s' \in \gamma^{-1}(U)$. Then Lemma 4.24(i) shows that there exists $C > 0$ such that

$$d\big(\gamma(s), \gamma(s')\big) \leq C\big(f(\gamma(s')) - f(\gamma(s))\big)$$

$$= C(s' - s)$$

and Lemma 4.24(ii) shows that γ is absolutely continuous.   ∎

Corollary 4.26.   *Let M be globally orderable. Then*

(i) *For each $x \in M$ we have $\mathrm{comp}(x) \subseteq \uparrow_{\prec} x$ and $U \cap \uparrow x \subseteq \uparrow_{\prec} x$ for a neighborhood U of x.*

(ii) *Let $I \subseteq \mathbb{R}$ be an interval, and $\gamma: I \to M$ a conal curve. Then*

$$\lim_{t \to \sup I} \gamma(t) \quad and \quad \lim_{t \to \inf I} \gamma(t)$$

exist in the one-point compactification M^ω.

(iii) *Let $x, y \in M$ be such that $[x, y]$ is non-compact. Then there exists a conal curve $\gamma: \mathbb{R}^+ \to M$ in Ω_x such that $\lim_{t \to \infty} \gamma(t) = \omega$ and $\gamma(\mathbb{R}^+) \subseteq [x, y]$.*

Proof.   (i) Let $y \in \mathrm{comp}(x)$. Then there exists a connected chain $C \in \mathcal{C}_c(x, y)$ (Lemma 4.3(iii)) and with Theorem 4.25 we parametrize this chain as a conal curve. The second assertion follows from the fact that $\mathrm{comp}(x)$ is relatively open in $\uparrow x$ and that it contains x (Proposition 4.17).

(ii) Let $C := \overline{\gamma(I)}$. Then C is a chain (Proposition 4.2). Now the assertion follows from Lemma 4.8.

(iii) Using Lemma 4.3(i), we find a non-compact connected chain $C \in \mathcal{C}_c(x)$ in $[x, y]$ and from Theorem 4.25 we get a parametrization $\gamma: I \to C$ as a conal curve. We clearly may assume that $0 = \min I$ and $\gamma(0) = x$. If I is not compact, there exists $T > 0$ with $I = [0, T[$ and with a simple parameter transformation we obtain a parametrization $\gamma: \mathbb{R}^+ \to C$. Now $z := \lim_{t \to \infty} \gamma(t)$ exists in the one-point compactification of M (cf. (ii)). If this limit is different from ω, then C is compact, a contradiction.   ∎

Let $I \subseteq \mathbb{R}$ and $\gamma: I \to M$ a conal curve. Then γ is said to be *infinite* if neither $\lim_{t \to \sup I} \gamma(t)$ nor $\lim_{t \to \inf I} \gamma(t)$ exists.

Theorem 4.27.   (The Chain Theorem) *Let M be globally orderable, $\Theta(x_0) \neq \{0\}$, and $C \subseteq M$ a maximal chain. Then C has at most countably many connected components and each component may be parametrized as an infinite conal curve.*

Proof.   Let f be a strictly Θ-positive function. Then $f \mid_C$ is an injective monotone mapping $C \to \mathbb{R}$. Each connected component A of C is open (Lemma 4.9). We claim that $f(A)$ is an open interval. Suppose that this is false. We may asssume that $f(A)$ has a maximal element $f(a)$. Let $g \in G$ with $a = g.x_0$ and $X \in W_\Theta \setminus \mathbf{L}(H)$. Then $\gamma(t) := g \exp(tX).x_0$ is a conal curve (Proposition 4.14) with $\gamma(0) = a$ and $\gamma'(0) \neq 0$. It follows that $\uparrow a \neq \{a\}$. Therefore there exists $b \in C \cap \uparrow a \setminus \{a\}$. Let

$$m = \inf\{f(c) : a \neq c, c \in C \cap \uparrow a\}$$

and $c_n \in C$ with $f(c_n) \to m$. Since a is maximal in the component A, the order interval $[a, c_n]$ is non-compact (Lemma 4.9(iii)). Therefore we find a sequence $C_n \in \mathcal{C}_c(a)$ such that $C_n \subseteq [a, c_n]$ is non-compact (Lemma 4.3(i)). Let $C_0 \in \mathcal{C}_c(a)$ be a cluster point of this sequence (Proposition 4.1). Then C_0 is non-compact by Proposition 4.2(v).

If $c \in C$ with $f(c) > m$, there exists $c_n \preceq c$ and therefore $C_0 \subseteq [a, c]$. If there exists $c \in C$ with $f(c) = m$, then c is minimal in $\uparrow a \cap C \setminus a$ and in this case we may assume that the sequence $c_n = c$ is constant. In both cases we conclude that $C \cup C_0$ is a chain which contradicts the maximality of C.

So we know that each component is mapped by f onto an open interval in $\mathbb{R}$. Thus there are at most countably many and the assertion follows from Theorem 4.25. ∎

Example 4.28. Let $G = M = \mathrm{Sl}(2, \mathbb{R})$ and $\mathfrak{g} = \mathfrak{sl}(2, \mathbb{R})$. We use the following basis for $\mathfrak{sl}(2, \mathbb{R})$ (cf. Section 2.2):

$$X_0 = \begin{pmatrix} 1 & 0 \\ 0 & -1 \end{pmatrix}, \quad T := X_+ + X_- = \begin{pmatrix} 0 & 1 \\ 1 & 0 \end{pmatrix}$$

and

$$U := X_+ - X_- = \begin{pmatrix} 0 & 1 \\ -1 & 0 \end{pmatrix}.$$

The commutator relations are

$$[U, T] = 2X_0, \quad [U, X_0] = -2T, \quad \text{and} \quad [X_0, T] = 2U.$$

We consider the cone field by

$$\Xi_W(g) = d\lambda_g(1)W,$$

where

$$W := \{hX_0 + tT + xU : x \geq 0, h^2 + t^2 \leq x^2\}.$$

Then G is globally orderable with $\mathrm{comp}(1) \neq \uparrow\mathbf{1}$. We set

$$\alpha_n : \mathbb{R} \to G, \qquad t \mapsto \exp(2n\pi U)\exp\big(t(U + X_0)\big)$$

and

$$\beta_n : \mathbb{R} \to G, \qquad t \mapsto \exp\big((2n + 1)\pi U\big)\exp\big(t(U - X_0)\big).$$

It is clear that all these curves are conal and that the set

$$\mathcal{K} := \bigcup_{n \in \mathbb{Z}} \big(\alpha_n(\mathbb{R}) \cup \beta_n(\mathbb{R})\big)$$

is a chain in G.

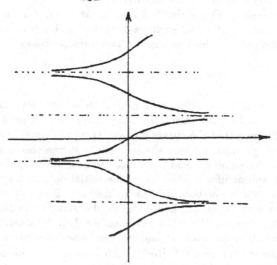

Figure 4.4

We show that this chain is maximal. Let $A := \exp \operatorname{span}\{X_0, T\}$ and $B := \exp \mathbb{R}U$. Then the mapping

$$A \times B \to G, \qquad (a, b) \mapsto ab$$

is a diffeomorphism, the Cartan decomposition of G. We define $f \in C^\infty(G)$ by $f(a \exp(tU)) = t$. To see that f is a Θ-positive, we use the formula for $d\lambda_g(1)$ given in [Ne91a]. We only have to check the positivity of $df(g)$ on $\Theta(g)$ in the plane $\exp(\mathbb{R}X_0)B$. There we get for $g = \exp(hX_0) \exp(xU)$ that

$$df(g)d\lambda_g(1) = \tanh h(-\sin x X_0 - \cos x T) + U,$$

where $\mathbf{L}(G)$ is identified with its dual via the scalar product g_1. For $v = U + \alpha X_0 + \beta T \in W_1$ with $\alpha^2 + \beta^2 \leq 1$ this shows that

$$df(g)d\lambda_g(1)v \geq 1 - \tanh h > 0$$

and consequently f is a Θ-positive. With Corollary I.3 in [Ne91a] we conclude that

$$f(\alpha_n(\mathbb{R})) =](2n - \tfrac{1}{2})\pi, (2n + \tfrac{1}{2})\pi[, \quad \text{and} \quad f(\beta_n(\mathbb{R})) =](2n + \tfrac{1}{2})\pi, (2n + \tfrac{3}{2})\pi[.$$

For every $x \in G \setminus \mathcal{K}$ such that $\mathcal{K} \cup \{x\}$ is a chain this implies that $f(x) \in (\tfrac{1}{2} + \mathbb{Z})\pi$. Suppose that $\widetilde{\mathcal{K}}$ is a maximal chain containing such an x and $\mathcal{K}$. Then the component of x has to lie entirely in a level surface of f which is excluded by the Chain Theorem. This shows that $\mathcal{K}$ is maximal. ∎

The preceding example shows that the maximal chains in a globally orderable space M are not necessarily connected. We recall from Theorem 4.10 that this occurs if and only if G is not globally hyperbolic.

If f is a strictly Θ-positive function on M, then f is in particular regular everywhere on M. So the level surfaces $f^{-1}(t)$ of such a function are codimension one submanifolds of M which satisfy the *antichain condition*

$$x \preceq y, \; f(x) = f(y) \qquad \Rightarrow \qquad x = y.$$

In the simplest space-times of general relativity, for example in four dimensional Minkowski space (Example 4.11), there exists a product decomposition $\varphi: M \to S \times \mathbb{R}$ such that the curves $t \mapsto \varphi^{-1}(t, x)$ are conal and $pr_1 \circ \varphi$ is a time function. These product decompositions serve to define observers and their clocks in space-time. On the other hand space-like codimension one submanifolds are good candidates for submanifolds which carry the initial data of a hyperbolic equation, for example the wave equation in Minkowski-space. To give a precise definition of what we mean with "good candidate" we need a suitable modifictaion of the concept of Cauchy surfaces: We call a hypersurface $T \subseteq M$ a *Cauchy hypersurface* if it satisfies the antichain condition and every infinite conal curve meets T.

The following theorem (cf. [Ne91b, 2.9]) shows how the notion of a Cauchy hypersurface is related to global hyperbolicity.

Theorem 4.29.   *Let M be globally hyperbolic and suppose that W_Θ is Lie generating. Then the following are equivalent:*

(1) *M is globally hyperbolic.*

(2) *There exists a strictly Θ-positive function on M such that $f(\gamma(I)) = \mathbb{R}$ for every infinite conal curve $\gamma: I \to \mathbb{R}$.*

(3) *We find a Cauchy hypersurface $T \subseteq M$ and a diffeomorphism $\varphi: M \to \mathbb{R} \times T$ such that $f := pr_1 \circ \varphi$ is strictly Θ-positive.*

(4) *M possesses a Cauchy hypersurface.*

Proof.   (Sketch) $(1) \Rightarrow (2)$: This time function may be contructed from functions which measure the volume of the forward and backward sets with respect to a Radon probability measure on M whose support coincides with M ([Ne91b, 2.1]).

$(2) \Rightarrow (3)$: It follows from (2) that the level surfaces of f are Cauchy hypersurfaces. To get the product decomposition one has to find a suitable vector field whose flow shifts $T := f^{-1}(0)$ nicely over the manifold M.

$(3) \Rightarrow (4)$: trivial

$(4) \Rightarrow (1)$: Let C be a maximal chain in M. Suppose that C has two different connected components. Then both are infinite conal curves and therefore both intersect T. This is impossible. Hence all maximal chains are connected and therefore G is globally hyperbolic.   ∎

The following proposition is a generalization of Corollary 1.20 to homogeneous spaces.

Proposition 4.30.   *Suppose that M is globally orderable and $K \subseteq G$ is a closed subgroup such that $K.x_0$ is a compact subset of M. Then $K \cap S_\Theta \subseteq H(S_\Theta)$. In particular, M cannot be compact.*

Proof.   The subsemigroup $S := K \cap S_\Theta$ of G is compact modulo $H(S) = K \cap H$ because $S/H(S) \cong S.x_0 = K.x_0 \cap {\uparrow}x_0$. Now Corollary 1.20 implies that S is a group, i.e., $S \subseteq H$.   ∎

4.5. Covering spaces and globality

Let $p: \widetilde{M} \to M$ be a G-equivariant covering of homogeneous G-spaces. To see how we can visualize this on the group level, we assume that $M = G/H$, where $H \subseteq G$ is a closed subgroup. Let $\widetilde{G}$ be the universal covering group of G and $q: \widetilde{G} \to G$ the corresponding covering morphism of Lie groups. Then the mapping

$$\widetilde{G} \times M \to M, \quad (g, x) \mapsto q(g).x$$

defines a transitive $\widetilde{G}$-action on M. Setting $\widetilde{H} := q^{-1}(H)$ we note that the mapping

$$\widetilde{G}/\widetilde{H} \to G/H, \quad g\widetilde{H} \mapsto q(g)H$$

is a $\widetilde{G}$-equivariant diffeomorphism of homogeneous $\widetilde{G}$-spaces. Henceforth we therefore may assume that G is simply connected. It is easy to see with a part of the exact homotopy sequence of the fibre bundle $G \to G/H$, namely

$$\pi_1(G) \cong \{\mathbf{1}\} \to \pi_1(G/H) \to \pi_0(H) \to \pi_0(G) = \{\mathbf{1}\},$$

that $\pi_1(G/H) \cong \pi_0(H)$. We conclude that G/H is simply connected if and only if H is connected and that G/H_0 is the universal covering manifold of M with the G-equivariant covering

$$G/H_0 \to M = G/H, \quad gH_0 \mapsto gH.$$

Since $\widetilde{M} = G/H_1$, we find that $H_0 \subseteq H_1 \subseteq H$ and that the covering p may be written as

$$p : \widetilde{M} = G/H_1 \to M = G/H, \quad gH_1 \mapsto gH.$$

From now on we write $\widetilde{x}_0$ for the base point of $\widetilde{M}$. It is useful to keep this more concrete description of p in mind.

We are interested in the following situation. Let Θ be a G-invariant cone field on M and $\widetilde{\Theta}$ a G-invariant cone field on $\widetilde{M}$ such that $dp(x)\widetilde{\Theta}(x) = \Theta(p(x))$. One may also construct $\widetilde{\Theta}$ from Θ by setting $\widetilde{\Theta}(x) := dp(x)^{-1}\Theta(p(x))$. How is the global orderability of M related to the global orderability of $\widetilde{M}$? Given our results on positive functions, one direction is almost trivial.

Proposition 4.31. *If M is globally orderable, then $\widetilde{M}$ is globally orderable.*

Proof. Since M is globally orderable, we find a strictly Θ-positive function f on M (Theorem 4.21). Then the function $\widetilde{f} := f \circ p$ is a strictly Θ-positive on $\widetilde{M}$ and thus $\widetilde{M}$ is globally orderable by Theorem 4.21. ∎

The preceding proposition splits up the problem to determine whether M is globally orderable or not (The Globality Problem) into two problems:

(1) Determine whether $\widetilde{M}$ is globally orderable or not.

(2) Suppose that $\widetilde{M}$ is globally orderable. When is M globally orderable?

In this section we are mainly interested in the second problem. One should also be aware of the fact that the methods one needs to approach a solution of problem (1) are very different from those one needs for problem (2). If $\widetilde{M}$ is simply connected (this may be assumed in this context) then the Globality Problem for $\widetilde{M}$ is equivalent to the Globality Problem for the Lie wedge $W_\Theta \subseteq \mathbf{L}(G)$ (Corollary 4.22) because the isotropy groups are connected in this case. In the case that W is invariant under the adjoint action of G on $\mathbf{L}(G)$, one can derive rather satisfactory answers to this problem (cf. [Ne92b, Ch. VIII]). The methods to do this are mainly structure theory of Lie algebras which contain such cones and then one has to use many facts on simply connected Lie groups which have no counterpart for general Lie groups.

The main theorem of this section gives a very general answer to the second problem. As we will see below, this theorem can be used in more concrete situations to give more explicit answers to the Globality Problem. We prepare the proof with two lemmas.

Lemma 4.32. *Let $p : \widetilde{M} \to M$ be a covering of conal homogeneous spaces. Then*

$$\uparrow p(x) = \overline{p(\uparrow x)}, \quad and \quad p(\uparrow_\prec x) = \uparrow_\prec p(x) \quad for\ all \quad x \in \widetilde{M},$$

where $\uparrow_\prec y := \{z : y \prec z\}$.

Proof. It follows from $dp(x)\widetilde{\Theta}(x) = \Theta(p(x))$ that composition with p maps conal curves on $\widetilde{M}$ into conal curves on M. Hence $p(\uparrow_\prec x) \subseteq \uparrow_\prec p(x)$ and therefore $\overline{p(\uparrow x)} \subseteq \uparrow p(x)$ for all $x \in \widetilde{M}$ because $\uparrow p(x)$ is closed. Conversely, the set $\uparrow_\prec p(x)$ is dense in $\uparrow p(x)$ and thus it suffices to show that this set is contained in $p(\uparrow_\prec x)$. To see this, let $p(x) \prec y$ and $\gamma \colon [0, T] \to M$ a segment of a conal curve with $\gamma(0) = p(x)$ and $\gamma(T) = y$. Then there exists a lift $\widetilde{\gamma} \colon [0, T] \to \widetilde{M}$ such that $\widetilde{\gamma}(0) = x$. The curve $\widetilde{\gamma}$ is conal with respect to $\widetilde{\Theta}$ and consequently $\widetilde{\gamma}(T) \in \uparrow x$ with

$$p(\widetilde{\gamma}(T)) = \gamma(T) = y.$$

∎

Lemma 4.33. *Suppose that M is not globally orderable. Then there exists a neighborhood U of x_0 and conal curves $\gamma_n \colon [0, T_n] \to M$ such that $\gamma(0) = x_0$, $\gamma_n([0, T_n]) \not\subseteq U$, and $\gamma_n(T_n) \to x_0$.*

Proof. According to our assumption, the order $\preceq$ on M is not partial and therefore we find, by homogeneity, an element $z \in M \setminus \{x_0\}$ such that

$$x_0 \preceq z \preceq x_0.$$

Let $z_n \in \uparrow_\prec x_0$ with $z_n \to z$ and $g_n \in G$ with $g_n \to 1$ and $g_n.z = z_n$. Let U be an open neighborhood of x_0 such that $z_n \notin U$ for all $n \in \mathbb{N}$.

We choose a conal curve α_n from x_0 to z_n, and a conal curve β_n from $z_n = g_n^{-1}.z$ to $g_n^{-1}.x_0$. Then the concatenation of α_n and β_n yields a conal curve $\gamma_n \colon [0, T_n] \to M$ from x_0 over z_n back to $g_n^{-1}.x_0$. It follows that $\gamma_n(T_n) = g_n^{-1}.x_0 \to x_0$ and that $\gamma_n([0, T_n]) \not\subseteq U$. ∎

Theorem 4.34. (The Covering Theorem for Conal Homogeneous Spaces) *Let M be a conal homogeneous space. Suppose that $\widetilde{M}$ is globally orderable and set $D := p^{-1}(x_0)$. Then M is globally orderable if and only if there exists a neighborhood U of 1 in G such that*

$$D \cap U.\uparrow\widetilde{x}_0 = \{\widetilde{x}_0\}.$$

Proof. "⇒": If M is globally orderable, then we find a strictly Θ-positive function f on M such that $f(x_0) = 0$. Set $\widetilde{f} := f \circ p$. Then $\widetilde{f}(D) = \{0\}$ and $\widetilde{f}$ is strictly Θ-positive on $\widetilde{M}$. To prove the assertion, we choose a countable base of neighborhoods U_n of 1 in G and assume that $D \cap U_n.\uparrow\widetilde{x}_0 \neq \{\widetilde{x}_0\}$ for all $n \in \mathbb{N}$. Then we find sequences $g_n \in U_n$ and $y_n \in \uparrow\widetilde{x}_0$ such that $x_n := g_n.y_n \in D \setminus \{\widetilde{x}_0\}$. If the sequence y_n is relatively compact, then we may assume that $y = \lim_{n\to\infty} y_n$ exists in $\widetilde{M}$. So we find that

$$x := \lim_{n\to\infty} x_n = \lim_{n\to\infty} g_n.y_n = y \in \uparrow\widetilde{x}_0.$$

Therefore $\widetilde{f}(y) = \lim_{n\to\infty} \widetilde{f}(x_n) = 0$ and hence $y = \widetilde{x}_0$ (Lemma 4.20). This contradicts the fact that D is discrete and that $x_n \neq \widetilde{x}_0$. Consequently we may assume that the sequence y_n eventually leaves every compact subset of M. Let $C_n \subseteq C_c(\widetilde{x}_0)$ be a chain such that $C_n \in C_c(\widetilde{x}_0, y_n)$ if $[\widetilde{x}_0, y_n]$ is compact (Lemma

4.3(iii)), and $C_n \subseteq [\tilde{x}_0, y_n]$ is non-compact if $[\tilde{x}_0, y_n]$ is non-compact (Lemma 4.3(i)). We may assume that $C_n \to C$ in the compact space $\mathcal{C}(\widetilde{M})$ (Proposition 4.2). Then Proposition 4.1 shows that the sequence $g_n.C_n$ also converges to C. Let C_0 denote the connected component of $\tilde{x}_0$ in C. Then C_0 is non-compact by Proposition 4.2(v). Using $C = \lim_{n \to \infty} g_n.C_n$ and $g_n.C_n \subseteq \downarrow g_n.y_n = \downarrow x_n$, it follows that

$$\tilde{f}(c) \leq \lim_{n \to \infty} \tilde{f}(x_n) = 0.$$

On the other hand we have that $C \subseteq \uparrow \tilde{x}_0$ so that Lemma 4.20 shows that $C = \{\tilde{x}_0\}$, a contradiction.

"$\Leftarrow$": We select a symmetric neighborhood U of 1 in G such that

$$U.\uparrow \tilde{x}_0 \cap D = \{\tilde{x}_0\}.$$

Then, for any V with $V = V^{-1} \subseteq U$ we have that

$$\uparrow \tilde{x}_0 \cap V.D = V.\tilde{x}_0.$$

Suppose that M is not globally orderable. Then Lemma 4.33 implies the existence of a sequence $\gamma_n: [0, T_n] \to M$ of conal curves such that $\gamma_n(0) = x_0$, $\gamma_n(T_n) \to x_0$ and $\gamma([0, T_n]) \not\to \{x_0\}$ in $\mathcal{F}(M)$. Let $\tilde{\gamma}_n$ denote the unique lift to $\widetilde{M}$ with $\tilde{\gamma}_n(0) = \tilde{x}_0$. If $V \subseteq G$ is a symmetric 1-neighborhood with $V \subseteq U$ and $\gamma_n(T_n) \in V.x_0$, then

$$\tilde{\gamma}_n(T_n) \in V.D \cap \uparrow \tilde{x}_0 = V.\tilde{x}_0.$$

We conclude that $\tilde{\gamma}_n(T_n) \to \tilde{x}_0$.

Let f be a strictly $\tilde{\Theta}$-positive function on $\widetilde{M}$ with $f(\tilde{x}_0) = 0$ and $V = V^{-1} \subseteq U$ so small such that $V.\tilde{x}_0$ is a connected component of $p^{-1}(V.x_0)$, no set $\gamma_n([0, T_n])$ is contained in $V.x_0$, and $\varepsilon > 0$ such that

$$\uparrow \tilde{x}_0 \subseteq V.\tilde{x}_0 \cup f^{-1}([\varepsilon, \infty[)$$

(Lemma 4.20).

Fix $n \in \mathbb{N}$. Since $\gamma_n([0, T_n]) \not\subseteq V.x_0$, there exists $s_n \in [0, T_n]$ with $\gamma_n(s_n) \notin V.x_0$. Thus $\tilde{\gamma}_n(s_n) \notin V.\tilde{x}_0$ and consequently $f(\tilde{\gamma}_n(s_n)) \geq \varepsilon$. The monotonicity of the functions $f \circ \tilde{\gamma}_n$ entails that $f(\tilde{\gamma}_n(T_n)) \geq \varepsilon$ for all $n \in \mathbb{N}$. This is a contradiction to $\tilde{\gamma}_n(T_n) \to \tilde{x}_0$.   ∎

4.6. Regular ordered homogeneous spaces

Recall that a conal homogeneous G-space with cone field Θ is said to be *regular* if the Lie wedge $W_\Theta \subseteq \mathbf{L}(G)$ is Lie generating. Note that this does not necessarily imply that the cones $\Theta(x)$ are generating in $T_x(M)$.

Proposition 4.35. *Let M be globally orderable. Then the following are equivalent:*

(1) M *is regular.*

(2) *For every $x \in M$ the set $\operatorname{int} \uparrow x$ is dense in $\uparrow x$.*

Proof. (1) $\Rightarrow$ (2): If M is regular, then the semigroup $S_\Theta = \overline{S_{W_\Theta} H} \subseteq H$ has dense interior (Proposition 4.16, Corollary 3.11). Since $\pi : G \to M$ is a quotient mapping and S_Θ is π-saturated, it follows that $\operatorname{int} \uparrow x_0 = \pi(\operatorname{int} S_\Theta)$ is dense. Now the assertion follows from the G-invariance of the order.

(2) $\Rightarrow$ (1): Let $W := W_\Theta$ and U be a relatively compact open 1-neighborhood in G such that there exists a smooth section $\sigma : \pi(U) \to U$ of π with $\sigma(x_0) = 1$. Let f be a strictly Θ-positive function with $f(x_0) = 0$ and $\varepsilon > 0$ such that

$$\uparrow x_0 \subseteq U \cup f^{-1}([\varepsilon, \infty[)$$

(Lemma 4.20). Take $x \in \operatorname{int}(\uparrow x_0)$ with $f(x) < \varepsilon$. Then

$$[x_0, x] \subseteq f^{-1}([0, f(x)]) \cap \uparrow x_0 \subseteq U$$

is compact and therefore $x_0 \prec x$ (Corollary 4.26(i)). Thus $x \in \pi(S_W)$ (Proposition 4.16). If $\gamma : [a, b] \to M$ is a conal curve with $\gamma(a) = x_0$ and $\gamma(b) = x$, then the unique lift $\widetilde{\gamma}$ of γ with $\widetilde{\gamma}(a) = 1$ is given by $\sigma \circ \gamma$. Therefore $\sigma(x) \in S_W$ and consequently

$$\sigma\big(\operatorname{int} \uparrow x_0 \cap f^{-1}([0, \varepsilon[)\big) \subseteq S_W.$$

The kernel of $d\pi(\sigma(x))$ is given by $d\lambda_{\sigma(x)} \mathbf{L}(H)$. Hence $S_W H_0 = S_W$ is a neighborhood of $\sigma(x)$ in G. Thus $\operatorname{int} S_W$ is dense and we will show below (Corollary 5.12) that this implies that W is Lie generating. $\blacksquare$

The necessity to quote a result from Chapter 5 in the proof of Proposition 4.35 is caused by our somewhat artificial attempt to keep ordered spaces and semigroups apart. The reader should take this as an argument in favour of the claims we made in the introduction to this chapter.

Suppose that M is globally orderable and regular. Then for $x \preceq y \in M$ we say that $x \lll y$ if $y \in \operatorname{int} \uparrow x$. It follows from the preceding proposition that the set $\uparrow_{\lll} x := \{y \in \uparrow x : x \lll y\}$ is dense in $\uparrow x$.

Proposition 4.36. *Let M be globally orderable and regular.*

(i) *If $x \lll y$ and $y \preceq z$, then $x \lll z$.*

(ii) *The relation $\lll$ is transitive and antisymmetric.*

(iii) *Let $\gamma : [a, b] \to M$ be a conal curve and suppose that there exists $t \in]a, b[$ with*

$$\gamma'(t) \in \operatorname{int} \Theta\big(\gamma(t)\big).$$

Then $\gamma(t) \lll \gamma(b)$.

Proof. (i) We recall the semigroup S_Θ from Proposition 4.16 and choose $a, b, c \in G$ with $a.x_0 = x$, $b.x_0 = y$, and $c.x_0 = z$. Then $x_0 \lll a^{-1}.y$ implies that $a^{-1}b \in \operatorname{int}(S_\Theta)$ and $b^{-1}.z \in \uparrow x_0$ that $b^{-1}c \in S_\Theta$. It follows that

$$a^{-1}c = a^{-1}bb^{-1}c \in (\operatorname{int} S_\Theta)S_\Theta \subseteq \operatorname{int} S_\Theta$$

(Lemma 3.7). Thus $x_0 \not\ll a^{-1}.z$ and $x \not\ll z$.

(ii) This is a trivial consequence of (i).

(iii) In view of (i) and the monotonicity of the mapping $\gamma: ([a, b], \leq) \to (M, \preceq)$, it suffices to assume that $a = 0$, and to show that there exists $s > 0$ with $\gamma(0) \not\ll \gamma(s)$.

Let $\varphi: U \to \varphi(U) \subseteq \mathbb{R}^n$ be a chart with $\varphi(\gamma(0)) = 0$ such that $\varphi(U)$ is convex. We set $\Theta'(\varphi(y)) := d\varphi(y)\Theta(y)$ for $y \in U$ and $\alpha := \varphi \circ \gamma|_{\gamma^{-1}(U)}$. We choose a compact convex neighborhood B of $\alpha'(a)$ in the interior of $C := \Theta'(0)$ and set $W := \mathbb{R}^+ B$.

Let V be a 1-neighborhood in G with $V.x \subseteq U$. Then the mapping

$$F : V \times \mathbb{R}^n \to \mathbb{R}^n, \quad (g, v) \mapsto d\varphi(g.x)d\mu_g(x)d\varphi^{-1}(0)v$$

is continuous and $F(\{g\} \times C) = \Theta'(\varphi(g.x))$. Moreover, $B \subseteq \text{int } F(\{0\} \times C) = \text{int } C$. Thus there exists a 1-neighborhood V in G such that $\Theta'(y)$ surrounds W for all $y \in U' := \varphi(V.\gamma(0))$ and U' is convex.

Next we use the fact that

$$\alpha'(0) = \lim_{t \to 0} \frac{1}{t} \alpha(t) \in \text{int}(W).$$

This implies the existence of $t_0 > 0$ such that

$$\alpha(t) \in \text{int}(W) \qquad \forall t \in [0, t_0]$$

and $\alpha([0, t_0]) \subseteq U'$. It follows in particular that $\alpha(t_0) \in \text{int}(W)$. Let B' be an open neighborhood of $\alpha(t_0)$ in $W \cap U$. For each $v \in B'$ the curve

$$\gamma_v : [0, 1] \to V, \quad t \mapsto \varphi^{-1}(tv)$$

is conal because $\gamma_v'(t) \in d\varphi^{-1}(tv)W \subseteq \Theta(\varphi^{-1}(tv))$. We conclude that $\varphi(B') \subseteq \uparrow\gamma(a)$ and therefore that $\gamma(0) \not\ll \gamma(t_0)$. ∎

4.7. Extremal curves

In this section we introduce a distance on ordered homogeneous spaces which generalizes the distance one has for Lorentzian manifolds and show that we always have geodesics (for this distance) between two points $p, q \in M$ with compact order interval $[p, q]$. To this end we recall the characteristic function φ and the length functional ψ from Section 1.3.

In the following $M = G/H$ denotes a homogeneous space, $\pi: g \mapsto gH$ the canonical projection, $x_0 := \pi(1)$ the base point, and $\mathbf{q} := T_{x_0}(M) \cong \mathbf{L}(G)/\mathbf{L}(H)$ the tangent space at x_0. We recall that the tangent bundle $T(G/H)$ may be represented as

$$T(G/H) \cong G \times_H \mathbf{q},$$

where H acts on $G \times \mathbf{q}$ by $h.(g, X) := (gh^{-1}, d\mu_h(x_0)X)$. For an H-invariant cone C in $\mathbf{q}$ we define the cone field Θ on M by

$$\Theta(\pi(g)) = d\mu_g(x_0)C.$$

(cf. Proposition 4.12).

We say that the homogeneous G-space $M = G/H$ is *unimodular* if it carries an invariant measure, i.e., if

$$|\det d\mu_h(x_0)| = 1 \qquad \forall h \in H.$$

For the following proposition we recall the definition of the length functional Ψ_C of a cone C (Section 1.3).

Proposition 4.37. *Let $M = G/H$ be a conal homogeneous unimodular G-space such that $C := \Theta(x_0)$ is generating. Let $g \in G$, $x = \pi(g)$ and $v \in T_x(M)$. Then the prescription*

$$\Psi(v) := \psi_C\big(d\mu_{g^{-1}}(x)v\big)$$

defines a continuous function on $T(M)$ which is smooth on the interior of the set Θ (identified with a subset of $T(M)$).

Proof. We define the function

$$\Psi' : G \times \mathfrak{q} \to \mathbb{R}, \quad (g, x) \mapsto \psi_C(x).$$

Then it follows from Theorem 1.10 that Ψ' is continuous on $G \times \mathfrak{q}$ and smooth on $G \times \operatorname{int} C$. Moreover,

$$\Psi'\big(h.(g, x)\big) = \psi_C\big(d\mu_h(x_0)x\big) = |\det \big(d\mu_h(x_0)\big)|^{\frac{1}{n}} \psi_C(x) = \psi_C(x).$$

Therefore Ψ' factors to a continuous function on $T(M)$ which is smooth on the image $\operatorname{int}\Theta$ of $G \times \operatorname{int} C$. ∎

For a conal curve $\gamma \colon [a, b] \to M$ we define the Ψ-*length* by

$$L_\Psi(\gamma) := \int_a^b \Psi\big(\gamma'(t)\big)\, dt.$$

Lemma 4.38. *Let $\gamma \colon [a, b] \to M$ be a conal curve.*

(i) *If $\alpha \colon [a', b'] \to [a, b]$ is an absolutely continuous monotone mapping, then $L_\Psi(\gamma) = L_\Psi(\gamma \circ \alpha)$, i.e., the Ψ-length of a curve does not depend on the parametrization.*

(ii) *If f is a strictly Θ-positive function on M and $\tilde\gamma \colon f\big(\gamma([a, b])\big) \to M$ is defined by $\tilde\gamma \circ f \circ \gamma = \gamma$, then*

$$L_\Psi(\gamma) = L_\Psi(\tilde\gamma).$$

Proof. (i) In view of Theorem 1.10(iii), we have that

$$L_\Psi(\gamma \circ \alpha) = \int_{a'}^{b'} \Psi\big((\gamma \circ \alpha)'(t)\big)\, dt$$

$$= \int_{a'}^{b'} \Psi\Big(\gamma'(\alpha(t))\alpha'(t)\Big)\, dt$$

$$= \int_{a'}^{b'} \Psi\Big(\gamma'(\alpha(t))\Big)\alpha'(t)\, dt$$

$$= \int_a^b \Psi\big(\gamma'(t)\big)\, dt$$

(cf. [Rud86, p.156]).

(ii) It is clear that the mapping

$$f \circ \gamma : [a,b] \to [f(\gamma(a)), f(\gamma(b))]$$

is monotone and absolutely continuous. Thus (ii) follows from $\gamma = \tilde{\gamma} \circ (f \circ \gamma)$ and (i).   ∎

If $G = \mathbb{R}^{n+1}$, $H = \{0\}$ and

$$C = \left\{ (x_0, x_1, \ldots, x_n) : x_0 \geq \sqrt{x_1^2 + \ldots + x_n^2} \right\},$$

then

$$\psi(x_0, \ldots, x_n) = \sqrt{x_0^2 - x_1^2 - \ldots - x_n^2} \, \psi(1, 0, \ldots, 0).$$

This example shows in particular that, if G/H is a homogeneous Lorentz manifold, then L_Ψ spezializes to the Lorentzian arc-length.

Let $p, q \in M$ and $\Omega_{p,q}^c$ denote the set of all conal curves from p to q. For p, q in M we set

$$d'(p,q) := \begin{cases} 0 & \text{for } p \not\prec q \\ \sup\{L_\Psi(\gamma) : \gamma \in \Omega_{p,q}^c\} & \text{for } p \prec q. \end{cases}$$

Proposition 4.39.   *Let M be globally orderable and f be strictly Θ-positive. Then the distance function d' has the following properties:*

(i) *If $p \prec q \prec r$, then $d'(p,r) \geq d'(p,q) + d'(q,r)$.*

(ii) *If $K \subseteq M$ is compact, then there exists $C_K > 0$ such that*

$$d'(p,q) \leq C_K |f(q) - f(p)|$$

for all $p, q \in K$ with $[p,q] \subseteq K$.

(iii) *If the order interval $[p,q]$ is compact, then $d'(p,q) < \infty$.*

(iv) *$d'(p,q) > 0$ if and only if $q \in \text{int}(\uparrow p)$.*

(v) *$d'(g.p, g.q) = d'(p,q)$ for $g \in G, p, q \in M$.*

Proof.   (i) trivial.

(ii) First we note that the compactness of $[p,q]$ implies that $p \prec q$ (Corollary 4.26(i)) for $p, q \in K$ with $[p,q] \subseteq K$. Let f a strictly Θ-positive function on M (Theorem 4.21) and d a Riemannian metric on M. Using Lemma 4.24 we choose d such that

$$d(x,y) \leq f(y) - f(x) \qquad \forall x, y \in K, x \preceq y$$

and that

$$\|v\| \leq \langle df(x), v \rangle \qquad \forall v \in \Theta(x), x \in K.$$

We set

$$C_K := \max\{\Psi(v) : v \in T_x(M), x \in K, \langle df(x), v \rangle \leq 1\}$$
$$\leq \max\{\Psi(v) : v \in T_x(M), x \in K, \|v\| \leq 1\} < \infty.$$

Then
$$L_\Psi(\gamma) \le C_K L(\gamma) \le C_K\big(f(q) - f(p)\big) < \infty$$

for every conal curve from p to q.

(iii) This is an immediate consequence of (ii).

(iv) (cf. [BE81, p.83]) If $p \prec q$, $q \in \partial{\uparrow}p$, and $\gamma \in \Omega^c_{p,q}$, then $\gamma'(t) \in \partial\Theta\big(\gamma(t)\big)$ whenever this derivative exists (Proposition 4.36). Now suppose that $q \in \text{int}({\uparrow}p)$. Choose a vector $v \in \text{int}\big(\Theta(p)\big)$ and a C^1-conal curve $\gamma\colon [0,T] \to M$ with $\gamma'(0) = v$ and $\gamma(0) = p$. Then there exists $t_0 > 0$ such that $\gamma(t_0) \in {\downarrow}q$. This implies that

$$d'(p,q) \ge d'\big(p,\gamma(t_0)\big) + d'\big(\gamma(t_0),q\big) \ge d'\big(p,\gamma(t_0)\big) > 0.$$

(v) This follows from the fact that the function Ψ satisfies

$$\Psi\big(d\mu_g(p)v\big) = \Psi(v) \qquad \forall g \in G, p \in M, v \in T_p(M).$$

∎

Lemma 4.40. *Let $I \subseteq \mathbb{R}$ be a compact interval, $\Theta\colon t \to \Theta(t) \subseteq \mathbb{R}^n$ a cone field such that*
$$\{(t,v) \in \mathbb{R} \times \mathbb{R}^n : v \in \Theta(t)\}$$

is closed, and $\Psi\colon I \times \mathbb{R}^n \to \mathbb{R}$ a continuous function. For a bounded measurable function $u\colon I \to \mathbb{R}^n$ we set

$$L_\Psi(u) := \int_I \Psi\big(t, u(t)\big)dt.$$

Suppose that the sequence $u_n\colon I \to \mathbb{R}^n$ is uniformly bounded, satisfies $u_n(t) \in \Theta(t)$, converges in the weak-$*$-topology to $u \in L^\infty(I,\mathbb{R}^n)$, and that $\Psi(t,\cdot)$ is concave on $\Theta(t)$ for all $t \in I$. Then

$$L_\Psi(u) \ge \limsup L_\Psi(u_n).$$

Proof. We write X for the Banach space $L^\infty(I,\mathbb{R}^n)$. Then the set

$$X_+ := \{u \in X : u(t) \in \Theta(t) \text{ a.e.}\}$$

is a closed convex cone in X and L_Ψ is a concave continuous functional on X_+ because Ψ is continuous. Therefore it is upper weak-$*$-semicontinuous on X_+ because the sets $\{L_\Psi \ge \lambda\}$ are closed convex subsets of X_+ and therefore are weak-$*$-closed. ∎

Let $p,q \in M$ and $C \in \mathcal{C}_c(p,q)$. We define

$$L_\Psi(C) := L_\Psi(\gamma),$$

where $\gamma \in \Omega^c_{p,q}$ is defined as in Theorem 4.25. We note that, in view of Lemma 4.38, $L_\Psi(\gamma)$ is independent of the choice of γ.

Lemma 4.41. *Let* $p \in M$ *and* $q \in \text{comp}(p)$. *If* $p_n \to p$, $q_n \to q$, *and* $C_n \in C_c(p_n, q_n)$ *with* $C_n \to C$, *then*

$$\limsup_{n \to \infty} L_\Psi(C_n) \leq L_\Psi(C).$$

Proof. Let $g_n \in G$ such that $g_n \to 1$ and $p = g_n.p_n$. Then $g_n.C_n \in C_c(p, g_n.q_n)$, $g_n.C_n \to C$ and $L_\Psi(C_n) = L_\Psi(g_n.C_n)$ (Proposition 4.39). So we may assume that $p_n = p$ for all $n \in \mathbb{N}$.

Now we choose a strictly Θ-positive function f on M. Let $a := f(p)$, $b_n := f(q_n)$ and $\gamma_n: [a, b_n] \to C_n \subseteq M$ be defined by the construction in Theorem 4.25. Similarly we define $b = f(q)$ and $\gamma: [a, b] \to C$. We claim that

$$(\dagger) \qquad\qquad s_n \in [a, b_n], \quad s_n \to s \Longrightarrow \gamma_n(s_n) \to \gamma(s).$$

In view of Proposition 4.2, we know that the chains C_n lie in a compact set Q. Take a subsequence $\gamma_{n_k}(s_{n_k}) \to x$. Then $x \in C$ and therefore $x = \gamma(f(x))$ with

$$f(x) = \lim_{k \to \infty} f\big(\gamma_{n_k}(s_{n_k})\big) = \lim_{k \to \infty} s_{n_k} = s.$$

Thus $x = f(s) = \lim_{n \to \infty} \gamma_n(s_n)$.

Let $\varepsilon > 0$. We have to show that there exists $n_0 \in \mathbb{N}$ such that

$$L_\Psi(\gamma_n) < L_\Psi(\gamma) + \varepsilon$$

holds for all $n \geq n_0$. Take C_Q as in Proposition 4.39(ii) for the compact set Q and $b' < b + \frac{1}{4C_Q}\varepsilon$. Then

$$L_\Psi\big(\gamma_n([b', b_n])\big) < (b_n - b')C_Q < 2(b' - b)C_Q < \frac{\varepsilon}{2}$$

whenever $b_n < b + (b - b')$. Similarly $L_\Psi\big(\gamma([b', b])\big) \leq (b - b')C_Q < \frac{\varepsilon}{4}$. Therefore, after replacing C_n by $\gamma_n([a, b'])$, we may assume that $b_n = b$ holds for all $n \in \mathbb{N}$.

Let $t_0 = a < t_1 < \ldots < t_m = b$ be a subdivision of $[a, b]$, such that there exist neighborhoods U_i of $\gamma(t_i)$, $i = 0, \ldots, m - 1$ and charts $\varphi_i: U_i \to \mathbb{R}$, with $\gamma([t_i, t_{i+1}]) \subseteq U$. Suppose that we know already that

$$L_\Psi\big(\gamma_n([t_i, t_{i+1}])\big) \leq L_\Psi\big(\gamma([t_i, t_{i+1}])\big) + \frac{\varepsilon}{m}.$$

Then

$$L_\Psi(\gamma_n) < L_\Psi(\gamma) + \frac{m\varepsilon}{m} = L_\Psi(\gamma) + \varepsilon.$$

Thus we may assume that $\gamma([a, b]) \subseteq U$. We choose a compact neighborhood $U' \subseteq U$ of $\gamma([a, b])$. Now $(\dagger)$ implies that there exists $n_1 \in \mathbb{N}$ such that $\gamma_n([a, b]) \subseteq U'$ holds for all $n \geq n_1$. From now on we assume that $n \geq n_1$.

Set $\alpha_n := \varphi \circ \gamma_n$, $h := f \circ \varphi^{-1}$, $\Psi(x, v) := \Psi\big(d\varphi^{-1}(x)v\big)$, and $\Theta'(\varphi(y)) = d\varphi(y)\Theta(y)$ for all $y \in U$. First we show that $\alpha_n \to \alpha := \varphi \circ \gamma$ uniformly. If this is false, there exists $\delta > 0$, $n_k \in \mathbb{N}$, and a sequence $s_k \in [a, b]$ such that

$\|\alpha_{n_k}(s_k) - \alpha(s_k)\| \geq \delta$. Again (†) leads to a contradiction. Hence α_n converges uniformly to α.

The set

$$K := \{v \in \mathbb{R}^n : (\exists y \in \varphi(U'))v \in \Theta'(y), \langle dh(y), v \rangle = 1\}$$

is compact. The closedness follows Lemma 4.23, the smoothness of f, the compactness of $\varphi(U')$, and the boundedness from Lemma 4.24.

We conclude that the set of curves α'_n is a uniformly bounded subset of the Banach space $L^\infty([a, b], \mathbb{R}^n)$. Since $L^1([a, b], \mathbb{R}^n)$ is separable ([Alt85, p.140]), the unit ball of $L^\infty([a, b], \mathbb{R}^n)$ is compact separable metric in the weak-$*$-topology.

Consequently we find a weak-$*$-convergent subsequence, which we again denote with α'_n, such that

$$\lim_{n \to \infty} \alpha'_n = \beta \in L^\infty([a, b], \mathbb{R}^n).$$

Weak-$*$-convergence implies that

$$\lim_{n \to \infty} \int_{[a,b]} \langle h(t), \alpha'_n(t) \rangle \, dt = \int_a^b \langle h(t), \beta(t) \rangle \, dt \quad \text{for all} \ \ h \in L^1([a, b], \mathbb{R}^n).$$

Applying this to step functions h, we find in particular that

$$\alpha(t) - \alpha(a) = \lim_{n \to \infty} \alpha_n(t) - \alpha_n(a)$$

$$= \lim_{n \to \infty} \int_a^t \alpha'_n(\tau) \, d\tau = \int_a^t \beta(\tau) \, d\tau.$$

This shows that $\alpha' = \beta$ holds almost everywhere on $[a, b]$.

Now the compactness of the set Q, Lemma 4.40, and the fact that α_n converges uniformly imply that for each subsequence α_{n_k} such that $L_\Psi(\alpha_{n_k})$ converges, we have that

$$\lim_{k \to \infty} L_\Psi(\gamma_{n_k}) = \limsup_{k \to \infty} \int_a^b \Psi(\alpha_{n_k}(t), \alpha'_{n_k}(t)) \, dt$$

$$= \limsup_{k \to \infty} \int_a^b \Psi(\alpha(t), \alpha'_{n_k}(t)) \, dt$$

$$\leq \lim_{k \to \infty} \int_a^b \Psi(\alpha(t), \alpha'(t)) \, dt$$

$$= L_\Psi(\gamma).$$

∎

The following theorem is one of the main results of this section.

Theorem 4.42. (Existence Theorem for maximizing curves) *If the order interval $[p, q]$ is compact, then there exists a d'-distance maximizing conal curve $\gamma \in \Omega^c_{p,q}$.*

Proof. It follows from the definition that there exist conal curves $\gamma_n : [a_n, b_n] \to M$ in $\Omega^c_{p,q}$ such that $L_\Psi(\gamma_n) \to d'(p, q) < \infty$ (Proposition 4.39). Let $C_n := \gamma([a_n, b_n]) \in \mathcal{C}_c(p, q)$ and $C \in \mathcal{C}_c(p, q)$ be a cluster point of the sequence C_n (Proposition 4.2). Then

$$d'(p, q) \geq L_\Psi(C) \geq \limsup_{n \to \infty} L_\Psi(C_n) = \lim_{n \to \infty} L_\Psi(\gamma_n) = d'(p, q).$$

Now the assertion follows from Theorem 4.25. ∎

Lemma 4.43.

(i) *If $\gamma \in \Omega_{p,q}^c$ is a d'-maximizing curve, $\gamma: [a, b] \to M$, and $[c, d] \subseteq [a, b]$, then $\gamma|_{[c,d]}$ is d'-maximizing in $\Omega_{\gamma(c),\gamma(d)}^c$.*

(ii) *If $q \in \mathrm{comp}(p)$, $p_n \to p$, and $q_n \to q$ such that $d'(p_n, q_n)$ converges, then*

$$d'(p, q) = \lim_{n \to \infty} d'(p_n, q_n).$$

Proof.   (i) This is an immediate consequence of Proposition 4.39.

(ii) Let $g_n \in G$, $g_n \to 1$ with $p = g_n.p_n$. Then $d'(p_n, q_n) = d'(p, g_n.q_n)$ and $g_n.q_n \to q$. So we may assume that $p_n = p$ for all $n \in \mathbb{N}$.

If $q_n \notin \uparrow p$ for an infinite set, then $d'(p, q_n) \to 0$ and $q \in \partial\uparrow p$ satisfies $d'(p, q) = 0$ (Proposition 4.39).

Now we may assume that $q_n \in \uparrow p$ for all $n \in \mathbb{N}$. Then it follows from Proposition 4.17 that we even may assume that $q_n \in \mathrm{comp}(p)$. Let $C_n \in \mathcal{C}_c(p, q_n)$ be d'-maximizing chains (Theorem 4.42). Every cluster point C of this sequence is in $\mathcal{C}_c(p, q)$ (Proposition 4.2). Hence

$$d'(p, q) \geq L_\Psi(C) \geq \limsup_{n \to \infty} L_\Psi(C_n) = \lim_{n \to \infty} d'(p, q_n)$$

(Lemma 4.41).

If $d'(p, q) = 0$ the proof is complete. So let us now assume that $d'(p, q) > 0$, i.e., $p \prec\!\!\!\prec q$ (Proposition 4.39). We take $C' \in \mathcal{C}_c(p, q)$ with $d'(p, q) = L_\Psi(C')$ (Theorem 4.42). The set

$$C_0' := \{c \in C' : c \prec\!\!\!\prec q\}$$

is open in C'. Let $c_0 := \sup C_0'$. Then $q \in \partial\uparrow c_0$ and $d'(c_0, q) = 0$ (Proposition 4.39). Let $\varepsilon > 0$. Then, in view of Proposition 4.39(ii), there exists $c_1 \in C_0'$ with $d'(c_1, c_0) < \varepsilon$. Then

$$d'(p, c_1) = d'(p, q) - d'(c_1, q) = d'(p, q) - d'(c_1, c_0) > d'(p, q) - \varepsilon.$$

If n is sufficiently large, then $c_1 \prec\!\!\!\prec q_n$ and therefore

$$d'(p, q_n) \geq d'(p, c_1) + d'(c_1, q_n) > d'(p, q) - \varepsilon.$$

Thus $\liminf_{n \to \infty} d'(p, q_n) \geq d'(p, q)$.   ∎

Theorem 4.44.   *If M is globally hyperbolic, then $d': M \times M \to \mathbb{R}^+$ is continuous.*

Proof.   Let $(p_n, q_n) \to (p, q)$. Take $a, b \in M$ with $a \prec\!\!\!\prec p$ and $q \prec\!\!\!\prec b$. Then eventually $a \preceq p_n$ and $q_n \preceq b$ so that

$$d'(p_n, q_n) \leq d'(a, b) < \infty$$

(Proposition 4.39). Now the assertion follows from Lemma 4.43.   ∎

For further result in this direction, in particular for existence theorems of geodesics etc., we refer to [MN92a].

Notes

The axiomatic approach to conal homogeneous spaces presented in Sections 4.1 and 4.2 is essentially new. It is based on some methods from [Ne91g] where it has been used to show that a large class of ordered homogeneous spaces consists of conal homogeneous spaces. Here it serves mainly to deduce the main facts about the structure of ordered homogeneous spaces which have been proved in [Ne91b] out of very few general principles.

The concept of a monotone curve in the context of ordered homogeneous spaces has been used in [HHL89]. The presentation of the material on cone fields owes a lot to [HHL89] and [La89]. The concept of global hyperbolicity appears - in special cases - throughout the literature on hyperbolic differential equations and Lorentzian geometry (cf. [CB67], [BE81], [Pe72]). The characterization in the general case was given in [Ne91b]. Examples of globally hyperbolic symmetric spaces were discussed in [Fa87], where one also finds a proof for the global hyperbolicity of regular ordered symmetric space. A proof of this result based on monotone functions is given in [Ne91c]. Similarly the results on extremal curves have predecessors in Lorentzian geometry (cf. [BE81]). Their versions for general conal ordered homogeneous space presented in Section 4.7 are new. The idea to replace the Lorentzian arc-length by length functionals is due to Vinberg (Seminar Lecture, November 14, 1991, Technische Hochschule Darmstadt). It has already been applied to obtain some deeper insights into surjectivity properties of the exponential functions of an affine manifold (cf. [MN92a,b]). The Covering Theorem (Theorem 4.34) is taken from [Ne92a].

5. Applications of ordered spaces to Lie semigroups

In this chapter we show how the results of Chapter 4 may be used to derive further information (apart from Theorem 1.35 which depended on Corollary 4.22) about Lie semigroups. We start with some consequences of the Globality Theorem (Theorem 4.21), namely the connectedness of the group of units of a Lie semigroup and the fact that every Lie semigroup is contained in the union of a small uniform tube about its group of units and a closed proper right ideal.

The globality criterion for coverings of ordered homogeneous space is carried over to Lie semigroups in Section 5.2.

In Sections 5.3 and 5.4 we relate the notion of a Lie semigroup to the notion of an infinitesimally generated semigroup as defined in [HHL89]. It is shown in particular that every Lie semigroup S is the closure of a certain completely infinitesimally generated subsemigroup reach(S). Moreover we show that faces of completely infinitesimally generated semigroups are also completely infinitesimally generated.

We conclude this chapter in Section 5.5 with a characterization of the class of Lie semigroups by the existence of certain monotone curves which provides in many interesting cases an effective tool to decide whether a given semigroup is a Lie semigroup or not.

5.1. Consequences of the Globality Theorem

The first two results concern the unit group of a Lie semigroup:

Corollary 5.1. (The Unit Group Theorem) *The unit group $H(S)$ of a Lie semigroup S is connected.*

Proof. The tangent wedge $\mathbf{L}(S)$ is a global Lie wedge (Proposition 1.14). Therefore Corollary 4.22 shows that $H(\mathbf{L}(S))$ is global and that

$$H(S) = H(S_{\mathbf{L}(S)}) = S_{H(\mathbf{L}(S))} = \langle \exp H(\mathbf{L}(S)) \rangle.$$

We conclude that $H(S)$ is connected. ∎

Recall the category LSg of Lie semigroups from Section 1.4. In view of the preceding corollary we have three functors $LSg \rightarrow ConLGrp$, where the latter denotes the category of connected Lie groups. The first one is the obvious forgetful functor

$$(S, G) \mapsto G,$$

the second one is the generation functor

$$(S, G) \mapsto \overline{G_S} = \overline{\langle\langle\langle \mathbf{L}(S) \rangle\rangle\rangle},$$

and the third and last one is the unit group functor

$$(S, G) \mapsto H(S).$$

Corollary 5.2.   (The Units Neighborhood Theorem) *Let (S, G) be a Lie semigroup, U a 1-neighborhood in G, and $f \in \mathrm{Mon}(S)$ with $df(1) \in \mathrm{algint}\, \mathbf{L}(S)^*$ and $f(1) = 0$. Then there exists $\varepsilon > 0$ such that*

$$S \subseteq U H(S) \cup I_\varepsilon,$$

where $I_\varepsilon = f^{-1}([\varepsilon, \infty[)$ is a closed right ideal of S. It follows in particular that

$$H(S) = S \cap f^{-1}(0).$$

Proof.   Let $M := G/H(S)$ be endowed with the G-invariant cone field $\Theta_{\mathbf{L}(S)}$ (Corollary 4.13, Corollary 4.22). According to Proposition 1.24, there exists $h \in C^\infty(M)$ such that $f = h \circ \pi$. Then Proposition 4.12 implies that h is Θ-positive and strictly Θ-positive in x_0. Using Lemma 4.20, we find $\varepsilon > 0$ such that

$$\uparrow x_0 \subseteq U.x_0 \cup h^{-1}([\varepsilon, \infty[).$$

Thus

$$S = \pi^{-1}(\uparrow x_0) \subseteq \pi^{-1}(U.x_0) \cup \pi^{-1}(h^{-1}([\varepsilon, \infty[)) = U H(S) \cup f^{-1}([\varepsilon, \infty[).$$

That I_ε is closed is clear and that it is a right ideal in S follows from the S-monotonicity of f.   ∎

5.2. Consequences of the Covering Theorem

In this section we discuss the question how the globality of a Lie wedge W in a Lie group is related to the globality of W in a covering group $\widetilde{G}$ of G. So we consider a covering $q: \widetilde{G} \to G$ of connected Lie groups. We identify the Lie algebras of $\widetilde{G}$ and G by the isomorphism $dq(1)$. Then we have that $q \circ \exp_{\widetilde{G}} = \exp_G$. Note that Proposition 1.41 shows that a Lie wedge $W \subseteq \mathbf{L}(G)$ is global in $\widetilde{G}$ if it is global in G.

We want to prove a covering theorem for Lie groups by applying the Covering Theorem for homogeneous space. To do this, we first need a condition for the closedness of the images of closed subgroups under coverings.

Lemma 5.3. *Let* $q:\widetilde{G} \to G$ *be a covering of connected Lie groups,* $H \subseteq \widetilde{G}$ *a closed subgroup and* $D := \ker q$. *Then the group* $q(H)$ *is closed in* G *if and only if there exists a neighborhood* U *of* 1 *in* $\widetilde{G}$ *such that*

$$UH \cap D \subseteq H.$$

Proof.   "$\Rightarrow$": Assume that $q(H)$ is closed in G. Then the inverse image $HD = DH$ is a closed subgroup of $\widetilde{G}$. The multiplication mapping $m: D \times H \to DH, (d,h) \mapsto dh$ is a continuous surjective morphism of Lie groups because D is central in $\widetilde{G}$ and DH is closed. Moreover, the group $D \times H$ is σ-compact and thus m is an open mapping ([Ho65] p.7). If $UH \cap D \nsubseteq H$ for all neighborhoods U of 1 in $\widetilde{G}$, then we find sequences $d_n \in D$, $h_n \in H$ and $g_n \in \widetilde{G}$ such that

$$g_n h_n = d_n \notin H \quad \text{and} \quad \lim_{n \to \infty} g_n = 1.$$

Since the mapping m is open, the subgroup $H = m(\{1\} \times H)$ is open in HD, so that there exists $n_0 \in \mathbb{N}$ such that $g_n^{-1} = h_n d_n^{-1} \in H$ for $n \geq n_0$. We conclude that $g_n \in H$ for $n \geq n_0$ which contradicts the fact that $d_n \notin H$.
"$\Leftarrow$": Let U be a symmetric neighborhood of 1 in $\widetilde{G}$ such that $UH \cap D \subseteq H$. If $x \in U \cap DH$, then we find elements $d \in D$ and $h \in H$ such that $x = dh$. Thus $d^{-1} = hx^{-1} \in D \cap HU \subseteq H$. Hence $x \in H$ and we conclude that $U \cap DH = U \cap H$. Consequently the subgroup DH of $\widetilde{G}$ is locally closed which implies that it is closed. This proves that $q(H) = HD/D$ is closed in G.   ∎

Theorem 5.4.   (The Covering Theorem for Groups) *Let* $q : \widetilde{G} \to G$ *be a covering of Lie groups and suppose that* $W \subseteq \mathbf{L}(G)$ *is a Lie wedge which is global in* $\widetilde{G}$. *Set* $\widetilde{S} := \overline{\langle \exp_{\widetilde{G}} W \rangle}$ *and* $D := \ker q$. *Then* W *is global in* G *if and only if there exists a neighborhood* U *of* 1 *in* $\widetilde{G}$ *such that*

$$U\widetilde{S} \cap D \subseteq H(\widetilde{S}).$$

   Suppose, in addition, that W *is Lie generating. Then* W *is controllable in* G, *i.e.,* $S_W = G$, *if and only if*

$$\operatorname{int} \widetilde{S} \cap D \neq \emptyset.$$

Proof.   "$\Rightarrow$": Suppose that $U\widetilde{S} \cap D \subseteq H(\widetilde{S})$. Set

$$\widetilde{M} := \widetilde{G}/H(\widetilde{S}), \quad \widetilde{\pi} : \widetilde{G} \to \widetilde{M}, \ g \mapsto gH(\widetilde{S})$$

and

$$\widetilde{\Theta}(xH(\widetilde{S})) := d\mu_x(\widetilde{x}_0) d\widetilde{\pi}(1)W.$$

Then $\widetilde{M}$ is globally orderable (Corollary 4.22) because $H(\widetilde{S}) = \langle \exp_{\widetilde{G}} H(W) \rangle$ (Corollary 5.1). In particular we have that

$$UH(\widetilde{S}) \cap D \subseteq H(\widetilde{S})$$

which shows that $q(H(\widetilde{S})) = \langle \exp_G H(W) \rangle$ is closed in G (Lemma 5.3), and $DH(\widetilde{S})$ is closed in $\widetilde{G}$. We set

$$M := G/q(H(\widetilde{S})) = \widetilde{G}/DH(\widetilde{S}), \quad \pi : \widetilde{G} \to M, \ g \mapsto gDH(\widetilde{S}),$$

and

$$\Theta(xDH(\widetilde{S})) := d\mu_x(x_0)d\pi(1)W.$$

We write $p \colon \widetilde{M} \to M, gH(\widetilde{S}) \mapsto gDH(\widetilde{S})$ for the corresponding covering of homogeneous spaces, and

$$D_1 := p^{-1}(x_0) = p^{-1}(DH(\widetilde{S})) = DH(\widetilde{S}) = \widetilde{\pi}(D).$$

Now we have that

$$D_1 \cap U.\!\uparrow\!\widetilde{x}_0 = \widetilde{\pi}(D) \cap \widetilde{\pi}(U.\widetilde{S}) = \widetilde{\pi}(D \cap U\widetilde{S}) \subseteq \widetilde{\pi}(H(\widetilde{S})) = \{\widetilde{x}_0\}.$$

Using the General Covering Theorem, we conclude that M is globally orderable and with Corollary 4.22 we see that this implies that W is global in G.

"$\Leftarrow$": Now we assume that W is global in G. Then $\mathbf{L}(S) = W$ for $S := S_W$, $H(S) = \langle \exp_G H(W) \rangle$ is a closed connected subgroup of G (Corollary 5.1), and the homogeneous space $M = G/H(S)$ is globally orderable (Corollary 4.22). With the Covering Theorem for homogeneous spaces we find a neighborhood U of 1 in G such that

$$D_1 \cap U.\!\uparrow\!\widetilde{x}_0 = \widetilde{\pi}(D \cap U\widetilde{S}) \subseteq \{\widetilde{x}_0\}.$$

We conclude that $D \cap U\widetilde{S} \subseteq H(\widetilde{S})$.

To prove the controllability part of the theorem, we first assume that $s \in \operatorname{int} \widetilde{S} \cap D = \operatorname{int}\langle \exp_{\widetilde{G}} W \rangle \cap D$ (cf. Lemma 3.7). Then

$$q(s) = 1 \in q\big(\operatorname{int}\langle \exp_{\widetilde{G}} W \rangle\big) \subseteq \operatorname{int}\langle \exp_G W \rangle,$$

so $\langle \exp_G W \rangle = G$ because G is connected. If, conversely, W is not controllable in G, then

$$\langle \exp_{\widetilde{G}} W \rangle D \subseteq q^{-1}(\langle \exp_G W \rangle) \neq \widetilde{G}.$$

Since D is a normal subgroup, we conclude with Lemmas VI.6 and VI.7 in [Ne92b] that

$$D \cap \operatorname{int}\langle \exp_{\widetilde{G}} W \rangle = D \cap \operatorname{int} \widetilde{S} = \emptyset.$$

∎

Corollary 5.5. *Let $q \colon \widetilde{G} \to G$ be a finite covering of Lie groups and $W \subseteq \mathbf{L}(G)$ a Lie wedge. Then W is global in G if and only if W is global in $\widetilde{G}$.*

Proof. We have seen at the beginning of this section that the globality in G implies the globality in $\widetilde{G}$. To see that the converse also holds, in view of Theorem 5.4, we have to find a neighborhood U of 1 in G such that

$$U\widetilde{S} \cap D \subseteq H(\widetilde{S}),$$

where $\widetilde{S} := \overline{\langle \exp_{\widetilde{G}} W \rangle}$. If such a neighborhood U does not exist, we find sequences $d_n \in D \setminus H(\widetilde{S})$ and $s_n \in \widetilde{S}$ such that $\lim_{n \to \infty} d_n s_n^{-1} = 1$. Since D is finite, we may assume that d_n is a constant sequence with $d_n = d$ for all $n \in \mathbb{N}$. Hence

$$\lim_{n \to \infty} s_n = d \in D \setminus H(\widetilde{S}).$$

We conclude that $d \in \widetilde{S}$. But the semigroup generated by d is a finite subsemigroup of G and therefore a group (Corollary 1.21). Thus $d \in H(\widetilde{S})$ and this contradicts our assumption from above. $\blacksquare$

Corollary 5.6. *Let $q: \widetilde{G} \to G$ be a covering of connected Lie groups, $D := \ker q$, and $W \subseteq \mathbf{L}(G) = \mathbf{L}(\widetilde{G})$ a Lie generating Lie wedge which is global in $\widetilde{G}$. Suppose that $D_1 \subseteq D$ is a subgroup with*

$$\operatorname{rank} D = \operatorname{rank} D_1.$$

Then W is global in G if and only if W is global in $\widetilde{G}/D_1$.

Proof. Let $p_1: \widetilde{G} \to \widetilde{G}/D_1$ and $p_2: \widetilde{G}/D_1 \to G \cong (\widetilde{G}/D_1)/(D/D_1)$ the canonical projections. Then $p = p_2 \circ p_1$. If W is global in G, then W is global in $\widetilde{G}/D_1$. Conversely, we assume that W is global in G/D_1. According to Corollary 5.5, it suffices to show that $\ker p_2 = D/D_1$ is finite. Since the following sequence of finitely generated abelian groups is exact

$$\{0\} \to D_1 \to D \xrightarrow{p_1} D/D_1 \to \{0\},$$

we conclude that $\operatorname{rank}(D/D_1) = \operatorname{rank} D - \operatorname{rank} D_1 = 0$. Hence D/D_1 is finite. $\blacksquare$

Note that the above corollary reduces the problem to determine whether W is global in the group $\widetilde{G}/D$ or not, where $D \subseteq \widetilde{G}$ is a discrete central subgroup, to the case where D is a free abelian group because D is a direct sum of a free abelian group and a finite abelian group.

5.3. Conal curves and reachability in semigroups

So far we have not said much about the question how much of a Lie semigroup S one can recover from $L(S)$. This question leads to various concepts of infinitesimal generation. *Infinitesimally generated semigroups* in the sense of [HHL89, V.1.11] are the subsemigroups S in a Lie group G which are contained in an analytic subgroup A such that, with respect to the Lie group topology on A, there exists a Lie semigroup S' in A with

$$\langle\langle L(S')\rangle\rangle = L(A) \qquad \text{and} \qquad \exp L(S') \subseteq S \subseteq S'.$$

According to Section 3.3, there exist Lie semigroups which are not infinitesimally generated. Nevertheless there exists a class of infinitesimally generated subsemigroups which will turn out to be in one-to-one correspondence with the class of Lie semigroups: A subsemigroup S of a Lie group G is said to be *completely infinitesimally generated* if $\operatorname{reach}(S) = S$, where for a submonoid $S \subseteq G$ we define the set

$$\operatorname{reach}(S) := \{s \in G : \Omega^c_{1,s}(L(\overline{S})) \neq \emptyset\}$$

and $\Omega^c_{x,y}(W)$ denotes the set of all W-conal curves from x to y (cf. Section 4.2). We set

$$\operatorname{comp}(S) := \{s \in S : [1,s]_{\leq_S} = S \cap sS^{-1} \text{ is compact}\}$$

and

$$\widetilde{\operatorname{comp}}(S) := \{s \in S : \pi([1,s]) = [x_0, \pi(s)] \text{ is compact}\},$$

where $\pi: G \to G/H(S)$ denotes the quotient mapping and $x_0 = \pi(1)$ (cf. Section 4.2).

Proposition 5.7. *For a Lie semigroup $S \subseteq G$ the following assertions hold:*
 (i) *The sets $\operatorname{reach}(S)$, $\operatorname{comp}(S)$ and $\widetilde{\operatorname{comp}}(S)$ are right $H(S)$-invariant.*
 (ii) *$\operatorname{reach}(S)$ is a dense subsemigroup of S.*
 (iii) *$\pi(\widetilde{\operatorname{comp}}(S)) = \operatorname{comp}(x_0)$.*
 (iv) *$\operatorname{comp}(S) \subseteq \widetilde{\operatorname{comp}}(S) \subseteq \operatorname{reach}(S)$.*
 (v) *$\operatorname{reach}(S)$ is a 1-neighborhood in S.*

Proof.   (i) For $h \in H(S)$ we have that $sh \leq_S s \leq_S sh$. Thus $[1,s]_{\leq_S} = [1,sh]_{\leq_S}$ and the $H(S)$-invariance of $\operatorname{comp}(S)$ and $\widetilde{\operatorname{comp}}(S)$ follows. The $H(S)$-invariance of $\operatorname{reach}(S)$ follows from the fact that $H(S)$ is connected (Corollary 5.1).
(ii) To see that $\operatorname{reach}(S)$ is a semigroup, one only has to recall that concatenation and left translations of $L(S)$-conal curves yields $L(S)$-conal curves. The density follows from the inclusion $\langle \exp L(S)\rangle \subseteq \operatorname{reach}(S)$ and Corollary 3.11.
(iii) This follows from the surjectivity of π beause

$$\widetilde{\operatorname{comp}}(S) = \pi^{-1}(\operatorname{comp}(x_0)).$$

(iv) The inclusion $\operatorname{comp}(S) \subseteq \widetilde{\operatorname{comp}}(S)$ is trivial. Let $M := G/H(S)$ be endowed with the cone field $\Theta_{L(S)}$ (Proposition 4.12). Let $s \in \widetilde{\operatorname{comp}}(S)$. Then $\pi(s) \in$

$\mathrm{comp}(x_0) \subseteq \uparrow_{\prec} x_0$ (Corollary 4.26(i)) and by Proposition 4.14 and Lemma 4.15, there exists an $\mathbf{L}(S)$-conal curve $\gamma : [a, b] \to G$ with $\gamma(a) = 1$ and $\pi(\gamma(b)) = \pi(s)$. Hence

$$s \in \gamma(b)H(S) \in \mathrm{reach}(S)H(S) = \mathrm{reach}(S).$$

(v) This follows from (iv),

$$\pi(\widetilde{\mathrm{comp}}(S)) = \mathrm{comp}(x_0),$$

and the fact that $\mathrm{comp}(x_0)$ is open in $\uparrow x_0$ (Proposition 4.17).   ∎

It follows from Proposition 5.7 that the set $S \setminus \mathrm{reach}(S)$, the points in S which are not reachable with any $\mathbf{L}(S)$-conal curve is in some sense a pathological subset of S because it is impossible to use local methods to get control or some knowledge about this set. This is completely different for the semigroup $\mathrm{reach}(S)$ which lies between the semigroup $\langle \exp \mathbf{L}(S) \rangle$ and S. One of the main tools to study this set is the following theorem which we quote without proof (cf. [HHL89, I.5.17]).

Theorem 5.8.   (The Invariance Theorem for Vector Fields) *Let W be a wedge in the vector space L, U open in L and $\mathcal{X}$ a smooth vector fields on U such that*

$$\mathcal{X}(X) \in L_X(W) \qquad \forall X \in U \cap W.$$

If $\gamma : [a, b] \to U$ is a trajectory of $\mathcal{X}$ with $\gamma(a) \in W$, then $\gamma(b) \in W$, i.e., W is invariant under the local flow generated by $\mathcal{X}$. Conversely, if the flow leaves $U \cap W$ invariant, then $\mathcal{X}(X) \in L_X(W)$ holds for all $X \in U \cap W$.   ∎

In our context it is often useful to have the following refinement of this theorem.

Theorem 5.9.   *Let U be an open subset in the vector space L, $W \subseteq L$ a wedge, and $\Phi : U \to \mathrm{Gl}(L)$ smooth. Let $C \subseteq L$ be a convex subset and $\gamma : [a, b] \to U$ an absolutely continuous curve such that*

$$\gamma'(t) \in \Phi(\gamma(t))C \qquad a.e. \ on \quad [a, b].$$

Assume that $\Phi(x)C \subseteq L_x(W)$ holds for all $x \in U \cap W$. Then $\gamma(a) \in W \cap U$ implies that $\gamma(b) \in W$.

Proof.   Let $V \subseteq U$ be a compact subset of U, $\| \cdot \|$ be a norm on L, and $\|d\Phi(x)\| \leq C$ on V.

If $u, v \in V$ and $x \in C$ are given, then

$$\|\Phi(u)x - \Phi(v)x\| \leq \|\Phi(u) - \Phi(v)\| \, \|x\| \leq C\|u - v\| \, \|x\|.$$

Now the assertion follows from Theorem I.5.22 in [HHL89].   ∎

Lemma 5.10. *Let* $\mathfrak{a} \subseteq \mathbf{L}(G)$ *be a subalgebra, and* $\gamma: [a, b] \to G$ *an* $\mathfrak{a}$*-conal curve. Then* $\gamma(b) \in \gamma(a)A$, *where* $A := \langle \exp \mathfrak{a} \rangle$ *is the analytic subgroup of* G *generated by* $\exp \mathfrak{a}$.

Proof. We may assume that $a = 0$. Let $I := \{t \in [0, b]: \gamma([0, t]) \subseteq \gamma(0)A\}$. Then $0 \in I$ and therefore I is non-empty. Let $t_0 \in I$. If $b \in I$, then there is nothing to prove, so we may assume that $t_0 < b$. We claim that there exists a $\delta > 0$ such that

$$\gamma([t_0, t_0 + \delta]) \subseteq \gamma(t_0)A$$

and therefore that $t_0 + \delta \in I$. We may assume that $\gamma(t_0) = 1$ (if not we replace γ by $\lambda_{\gamma(t_0)^{-1}} \circ \gamma$). We find an open neighborhood B of 0 in $\mathbf{L}(G)$ such that $\exp|_B$ is a local diffeomorphism and $\delta > 0$ such that $\gamma([t_0, t_0 + \delta]) \subseteq \exp(B)$. We set

$$\alpha : [t_0, t_0 + \delta] \to B, \quad t \mapsto (\exp|_B)^{-1}(\gamma(t)).$$

Then α is absolutely continuous and

$$\alpha'(t) \in d\exp\left(\alpha(t)\right)^{-1} d\lambda_{\exp \alpha(t)}(1)\mathfrak{a} = g\left(\operatorname{ad} \alpha(t)\right)\mathfrak{a}$$

(cf. [HiNe91, 3.2.26]), where the power series $g(X)$ is defined as the inverse of the power series

$$f(X) = \frac{1 - e^{-X}}{X}.$$

Thus, since $g(\operatorname{ad} Y)\mathfrak{a} \subseteq \mathfrak{a}$ for all $Y \in \mathfrak{a}$, Theorem 5.9 implies that $\alpha([t_0, t_0 + \delta]) \subseteq \mathfrak{a}$ and consequently

$$\gamma([t_0, t_0 + \delta]) \subseteq \gamma(t_0)\exp \mathfrak{a} \subseteq \gamma(t_0)A \subseteq \gamma(0)A.$$

To prove that $I = [0, b]$, is suffices to show that $\sup I \in I$ because this implies, in view of the above argument, that $\sup I = b \in I$. Let $t = \sup I$. Then we know already that $t > 0$. The curve defined by $\alpha(s) := \gamma(t - s)$ for $s \in [0, t]$ satisfies the same assumptions as γ and thus we find $s \in]0, t[$ with $\alpha(s) \in \alpha(0)A$. Hence $\gamma(t) = \alpha(0) \in \alpha(s)A = \gamma(t - s)A \subseteq \gamma(0)A$, so that $t \in I$. ∎

Proposition 5.11. *Let* S *be a Lie subsemigroup of* G, $\mathfrak{a} := \langle\langle \mathbf{L}(S) \rangle\rangle$ *the Lie algebra generated by* $\mathbf{L}(S)$, *and* $A := \langle \exp \mathfrak{a} \rangle$. *Then*

　(i) $\operatorname{reach}(S) \subseteq A$.

　(ii) *There exists a neighborhood* U *of* 1 *in* G *such that* $S \cap U \subseteq A$.

　(iii) *If* S *is completely infinitesimally generated, then it is infinitesimally generated.*

　(iv) *If* G *is connected, then the prescription* $S \to \operatorname{reach}(S)$ *defines a bijection from the set of Lie subsemigroups in* G *onto the set of completely infinitesimally generated subsemigroups in* G.

Proof. (i) Lemma 5.10.

(ii) Proposition 5.7(v).

(iii) We have $S = \operatorname{reach}(S_{\mathbf{L}(\overline{S})})$ so that (i) and (ii) imply the claim.

(iv) Since $\operatorname{reach}(S)$ is dense in S (Proposition 5.7), it is clear that the mapping reach is injective. On the other hand, every completely infinitesimally generated subsemigroup T satisfies $T = \operatorname{reach}(S_{\mathbf{L}(\overline{T})})$, where $S_{\mathbf{L}(\overline{T})}$ is a Lie semigroup. ∎

Corollary 5.12.   *For a Lie semigroup* $S \subseteq G$ *the following are equivalent.*

(1) $1 \in \text{int } \overline{S}$.

(2) $\langle\langle \mathbf{L}(S) \rangle\rangle = \mathbf{L}(G)$, *i.e.,* $\mathbf{L}(S)$ *generates the Lie algebra* $\mathbf{L}(G)$.

Proof.   (1) $\Rightarrow$ (2): Using Proposition 5.11 we see that (1) implies that int $A \neq \emptyset$ for $A = \langle \exp(\langle\langle \mathbf{L}(S) \rangle\rangle) \rangle$. This is true for an analytic subsemigroup if and only if it agrees with G and this holds if and only if $\langle\langle \mathbf{L}(S) \rangle\rangle = \mathbf{L}(G)$.

(2) $\Rightarrow$ (1): This follows from Corollary 3.11.   ∎

Corollary 5.13.   *For a Lie semigroup* $S \subseteq G$ *the set*

$$\text{Ei}(S) = \{ X \in \mathbf{L}(S) : \exp(]0, \infty[X) \subseteq \text{int } S \}$$

is empty if $\mathbf{L}(S)$ *does not generate* $\mathbf{L}(G)$.   ∎

5.4. Applications to faces of Lie semigroups

In this section we show that completely infinitesimally generated semigroups share many properties with wedges. For a submonoid S of a group G we recall that a curve $\gamma \colon [a, b] \to G$ is said S-monotone if $\gamma(t') \in \gamma(t)S$ for $t \leq t'$.

Lemma 5.14.   *Let* S *be a submonoid of the Lie group* G *and* $\gamma \colon [a, b] \to S$ *an absolutely continuous* S-*monotone curve. Then* γ *is* $\mathbf{L}(\overline{S})$-*conal,* $\text{reach}(S)$-*monotone, and* $S_{\mathbf{L}(\overline{S})}$-*monotone.*

Proof.   The condition that γ is S-monotone implies that $f \circ \gamma$ is monotone for all $f \in \text{Mon}(S) = \text{Mon}(\overline{S})$. Let $t \in [a, b]$ such that the derivative $\gamma'(t)$ exists and

$$\omega \in \text{algint}\, \Xi_{\mathbf{L}(\overline{S})}(\gamma(t))^* = (d\lambda_{\gamma(t)}(1) \mathbf{L}(\overline{S}))^*.$$

Then, in view of Theorem 1.31, there exists $f \in \text{Mon}^\infty(\overline{S})$ such that $df(\gamma(t)) = \omega$. It follows that

$$\langle \omega, \gamma'(t) \rangle = df(\gamma(t))\gamma'(t) = (f \circ \gamma)'(t) \geq 0.$$

Thus

$$\gamma'(t) \in \Xi_{\mathbf{L}(\overline{S})}(\gamma(t))$$

and we conclude that γ is $\mathbf{L}(\overline{S})$-monotone.

For the remaining assertion we first note that γ is $S_{\mathbf{L}(\overline{S})}$-monotone by Lemma 4.15. For $t < t'$ in $[a, b]$ this implies that

$$\gamma(t)^{-1}\gamma(t') \in \text{reach}(S_{\mathbf{L}(\overline{S})}) = \text{reach}(S).$$

Thus γ is also $\text{reach}(S)$-monotone.   ∎

Theorem 5.15.   (Main Theorem on Infinitesimally Generated Semigroups) *Let G be a connected Lie group and $S \subseteq G$ completely infinitesimally generated. Then the following assertions hold:*

(i) *If G is a vector space, then S is a wedge, and every wedge in G is completely infinitesimally generated.*

(ii) *The unit element $1 \in S$ has a compact neighborhood.*

(iii) $S = \langle S \cap U \rangle$ *holds for every 1-neigborhood in S, i.e., S is strictly locally generated.*

(iv) *If $F \subseteq S$ is a subsemigroup such that $S \setminus F$ is a semigroup ideal, then F is completely infinitesimally generated.*

Proof.   (i) This follows from Proposition 5.11(iv) and the fact that every wedge $W \subseteq G$ satisfies $\mathrm{reach}(W) = \exp(W) = W$.

(ii) We choose a compact 1-neighborhood U in G such that $U \cap \overline{S} \subseteq S$ (Propositions 5.7, 5.11(iv)). Then $U \cap S$ is a compact 1-neighborhood in S.

(iii) Let U be a 1-neighborhood in S and $s \in S$. Then, since $S = \mathrm{reach}(S)$, there exists an $\mathbf{L}(S)$-conal curve $\gamma : [a, b] \to S$ with $\gamma(a) = 1$ and $\gamma(b) = s$. In view of Lemma 5.14, this curve is also S-monotone because $S = \mathrm{reach}(S)$. We choose a subdivision $a = t_0 < t_1 < t_2 < \ldots < t_n = b$ such that $\gamma(t_{i+1}) \in \gamma(t_i)U$ for $i = 0, \ldots, n-1$. Then it follows that

$$s = \gamma(b) = \gamma(t_n)\gamma(t_{n-1})^{-1}\gamma(t_{n-1}) \cdot \ldots \cdot \gamma(t_1)^{-1}\gamma(t_1)\gamma(t_0)^{-1}\gamma(t_0) \in (S \cap U)^n.$$

(iv) Let $f \in F$. Then $f \in \mathrm{reach}(S) = S$ and there exists an S-monotone curve $\gamma : [a, b] \to S$ with $\gamma(a) = 1$ and $\gamma(b) = f$ (Lemma 5.14). For $t \in [a, b]$ this shows that $f \in \gamma(t)S$ and therefore $\gamma(t) \in F$ and γ is F-monotone since $S \setminus F$ is a semigroup ideal. Now Lemma 5.14 implies that $F \subseteq \mathrm{reach}(F)$. Therefore

$$F \subseteq \mathrm{reach}(F) \subseteq S_{\mathbf{L}(\overline{F})} \subseteq \overline{F}$$

means that F is dense in $\mathrm{reach}(F) = \mathrm{reach}(S_{\mathbf{L}(\overline{F})})$. Let $\mathfrak{a} := \langle\langle \mathbf{L}(\overline{F}) \rangle\rangle$ denote the subalgebra of $\mathbf{L}(G)$ generated by $\mathbf{L}(\overline{F})$ and $A := \langle \exp \mathfrak{a} \rangle$. We write $G(A)$ for the group A endowed with its Lie group topology and $i : G(A) \to G$ for the inclusion mapping. Note that $H(S) = H(F)$ is a closed connected subgroup of G. Therefore we have a $G(A)$-equivariant immersion

$$j : G(A)/H(F) \to G/H(F), \quad aH(F) \mapsto i(a)H(F).$$

Let $U \subseteq G(A)/H(F)$ be a compact 1-neighborhood. Then $j : U \to j(U)$ is a homeomorphism. Now we choose a strictly $\Theta_{\mathbf{L}(\overline{F})}$-positive function f on $G/H(S)$ with $f(x_0) = 0$. Then $f \circ j$ is strictly $\Theta_{\mathbf{L}(\overline{F})}$-positive on $G(A)/H(F)$ and there exists $\varepsilon > 0$ such that

$$\uparrow x_0 \subseteq U \cup (f \circ j)^{-1}([\varepsilon, \infty[)$$

(Lemma 4.20). Let $x \in \mathrm{int}_{G(A)/H(F)} \uparrow x_0$ with $f(j(x)) < \varepsilon$. We choose a sequence $x_n \in \pi(F)$ such that $x_n \to j(x)$. Since $f(x_n) \to f(j(x))$, there exists $n_0 \in \mathbb{N}$ such that $f(x_n) < \varepsilon$ holds for all $n > n_0$. It follows that $j^{-1}(x_n) \in U$ for $n > n_0$. Now the assumption that $j \mid_U$ is a homeomorphism implies that $j^{-1}(x_n) \to x$.

This shows $x \in \overline{j^{-1}(\pi(F))}$ so that $\text{int}_{G(S)} \overline{F} \neq \emptyset$, whence $F \cap \text{int}_{G(S)} \overline{F} \neq \emptyset$. We conclude that there exists $s \in F \cap \text{int}_{G(S)} \overline{F}$. Therefore $[1, s]_{\leq_s}$ is a 1-neighborhood in $\overline{F}$ and consequently $\text{reach}(F) = \text{reach}(\overline{F}) \subseteq F$. This proves that F is completely infinitesimally generated.   ∎

The preceding theorem has interesting consequences for faces of Lie semigroups (cf. Section 3.1).

Corollary 5.16.   *Let $S \subseteq G$ be a Lie semigroup and $F \in \mathcal{F}(S)$. Then*
   (i) *$F \cap \text{reach}(S) = \text{reach}(F)$.*
   (ii) *There exists a 1-neighborhood U in G such that*

$$F \cap U \subseteq \text{reach}(F) \subseteq S_{\mathbf{L}(F)}.$$

   (iii) *Let $H(S) \neq F \in \mathcal{F}(S)$. Then $\mathbf{L}(F) \neq H(\mathbf{L}(S))$.*

Proof.   (i) The subsemigroup $F \cap \text{reach}(S)$ satsfies the condition that

$$\text{reach}(S) \setminus F$$

is a semigroup ideal. Therefore $F \cap \text{reach}(S)$ is completely infinitesimally generated (Theorem 5.15). We conclude that

$$F \cap \text{reach}(S) = \text{reach}(F \cap \text{reach}(S)) = \text{reach}(F).$$

(ii) This follows from (i) and Proposition 5.7(v).

(iii) Assume that $F \neq H(S)$ is a face of S and that $\mathbf{L}(F) = H(\mathbf{L}(S))$. Then Theorem 5.15 shows that $H(S)$ is isolated in F, i.e., there exists a neighborhood U of 1 in G such that

$$U \cap F \subseteq H(S).$$

Let $f \in F \setminus H(S)$. From (i) we see that

$$F \cap \text{reach}(S) \subseteq H(S),$$

thus $f \notin \text{reach}(S)$. Whence $f \notin \widetilde{\text{comp}}(S)$, i.e., $\pi(f) \notin \text{comp}(x_0)$ (Proposition 5.7(iv)). Hence there exists $s \in [1, f]$ such that $x_0 \neq \pi(s) \in \uparrow_{\prec} x_0 = \pi(\text{reach}(S))$ (Lemma 4.3(i)). Since $\text{reach}(S) = \pi^{-1}(\uparrow_{\prec} x_0)$ we have $s \in \text{reach}(S) \cap F \setminus H(S)$, a contradiction.   ∎

Proposition 5.17.   *Let $F \in \mathcal{F}_e(S)$. Then $L_F(S)$ is a Lie semigroup and there exists a sequence of wedges*

$$\mathbf{L}(S) = W_0 \subseteq W_1 \subseteq \ldots \subseteq W_n$$

such that
   (a) *$W_{i+1} = \mathbf{L}(\overline{\langle \exp L_{F_i}(W_i) \rangle})$ for an exposed face $F_i \in \mathcal{F}_e(W_i)$ and*
   (b) *$L_F(S) = \overline{\langle \exp W_n \rangle}$.*
If F and S are invariant, then all the wedges W_i are invariant.

Proof. We set $W_0 := \mathbf{L}(S)$. If $F = H(S)$, then $L_F(S) = S = \overline{\langle \exp W_0 \rangle}$ and we are done. If $F \neq H(S)$, then $\mathbf{L}(F) \neq H(W_0)$ (Lemma 5.16(iii)). We set

$$W_1 := \mathbf{L}\left(\overline{\langle \exp L_{\mathbf{L}(F)}(W_0) \rangle}\right) \supseteq L_{\mathbf{L}(F)}(W_0).$$

Then $\dim H(W_1) > \dim H(W_0)$ and $\exp(W_1) \subseteq L_F(S)$. According to Proposition 3.5, we know that $\mathbf{L}(F) \in \mathcal{F}_e(W_0)$. Suppose that $W_0, \ldots, W_i$ are constructed such that $\dim H(W_j) > \dim H(W_{j-1})$ and

$$W_j := \mathbf{L}\left(\overline{\langle \exp L_{F_{j-1}}(W_{j-1}) \rangle}\right) \supseteq L_{F_{j-1}}(W_{j-1})$$

with $F_{j-1} \in \mathcal{F}_e(W_{j-1})$ for $j = 1, \ldots, i$ and $\exp W_i \subseteq L_F(S)$. There are two cases. If $\langle \exp W_i \rangle$ is dense in $L_F(S)$, the proof is complete. If not, we set

$$T := \overline{\langle \exp W_i \rangle} = \overline{\langle \exp L_{F_{j-1}}(W_{j-1}) \rangle}$$

and $F' := T_F(S) \cap T$. Then $S \subseteq T$ and therefore $F \subseteq F'$. This implies that $F' \not\subseteq H(T)$ because otherwise $F \subseteq H(T)$ and $L_F(S) \subseteq T$. Moreover

$$F' \subseteq T_{F'}(T) \cap T \subseteq T_F(S) \cap T = F'$$

is an exposed face of T. Consequently

$$F_i := \mathbf{L}(F') \in \mathcal{F}_e\big(\mathbf{L}(T)\big) = \mathcal{F}_e(W_i)$$

is different from $H(W_i)$ (Corollary 5.16(iii)). We conclude that $\exp W_{i+1} \subseteq L_F(S)$ for

$$W_{i+1} := \mathbf{L}\left(\overline{\langle \exp L_{F_i}(W_i) \rangle}\right) \supseteq L_{F_i}(W_i)$$

and that $\dim H(W_{i+1}) > \dim H(W_i)$. This procedure has to stop after finitely many steps because $\mathbf{L}(G)$ is a finite dimensional vector space and $\dim H(W_i)$ is enlarged in every step. Therefore there exists an $n \in \mathbb{N}$ such that $\overline{\langle \exp W_n \rangle} = L_F(S)$. Moreover it is clear that all wedges F_i and W_i are invariant if this holds for F and S. ∎

It is not true for general semigroups that the unit group is not isolated in a face. It depends on the special structure of Lie semigroups. To find an example we take

$$G = \mathbb{R}^2 \quad \text{and} \quad S = \mathbb{R}^+ \times \mathbb{R}^+ \setminus \{(x, y) : 0 \leq x < y < 1\}.$$

Then

$$F := \{(0,0)\} \cup \{0\} \times [1, \infty[$$

is a face of S, i.e., $S \setminus F$ is an ideal and F is closed. Moreover, $H(S) = \{0\}$ is isolated in F. The largest Lie semigroup contained in S is $S_L = \mathbb{R}^+(1,0) + \mathbb{R}^+(1,1) \neq S$.

5.5. Monotone curves in Lie semigroups

Let G be a Lie group and $S \subseteq G$ a closed subsemigroup. As before, we endow the group G with the left-invariant quasi-order $\leq_S$ defined by $g \leq_S g'$ if $g' \in gS$. A continuous curve $\gamma : I \to G$ is said to be S-*monotone* if it is a monotone mapping with respect to the natural order on the real interval I and the order $\leq_S$ on G.

Theorem 5.18.   *A closed submonoid S of a Lie group G is a Lie semigroup if and only if the set of all elements in S reachable by absolutely continuous S-monotone curves from the identity is dense in S.*

Proof.   If S is a Lie semigroup, then the dense subsemigroup $\langle \exp \mathbf{L}(S) \rangle$ consists of elements reachable by piecewise one-parameter semigroup, hence by S-monotone curves.

If, conversely, the element $s \in S$ is reachable from 1 by an S-monotone absolutely continuous curve. Using Lemma 5.14, we see that s is contained in the Lie semigroup $S_{\mathbf{L}(S)} \subseteq S$. If this semigroup is dense in S, then it must coincide with S, so that S is a Lie semigroup.   ∎

Corollary 5.19.   (The Lie Semigroup Criterion) *Let S be a closed submonoid of the Lie group G such that $H(S)$ is connected, $M := G/H(S)$ the corresponding homogeneous space endowed with the partial order*

$$xH \leq yH \quad \Longleftrightarrow \quad x \leq_S y,$$

and x_0 the base point in M. Then S is a Lie semigroup if and only if the set of all elements in $\uparrow x_0$ reachable from x_0 by absolutely continuous monotone curves is dense.

Proof.   Let $\pi: G \to M$ denote the canonical projection. Then π is a quotient mapping of quasi-ordered spaces since $g \leq_S g'$ is equivalent to $\pi(g) \leq \pi(g')$. Since π locally admits smooth sections, absolutely continuous monotone curves in M lift to absolutely continuous S-monotone curves in G. Using the connectedness of $H(S)$, we conclude that the condition that the set of all points in $\uparrow x_0$ reachable by absolutely continuous monotone curves is dense, is equivalent to the density of the corresponding set in S. So the assertion follows from Theorem 5.18.   ∎

Corollary 5.20.   *Let S be a Lie semigroup in G and $S' \subseteq S$ a closed submonoid such that $S = S'H(S)$ and $S' \cap H(S) = H(S')$ is connected. Then S' is a Lie semigroup.*

Proof.   The condition $S = S'H(S)$ implies that the ordered spaces $G/H(S)$ and $G/H(S')$ are the same, so that the assertion is a consequence of Corollary 5.19.   ∎

As we have already seen in Section 1.4, this result has interesting applications to affine compression semigroups. For the sake of convenience we formulate the above result in this context.

Proposition 5.21.   *Let Ω be a compact neighborhood of 0 in the finite dimensional real vector space V and $S \subseteq \mathrm{Aff}(V) \cong V \rtimes \mathrm{Gl}(V)^+$ the semigroup of affine compressions of Ω. Then S is a Lie semigroup if and only if there exists for every $s \in S$ an absolutely continuous S-monotone curve $\gamma: [0,1] \to S$ such that $\gamma(0) = 1$ and $\gamma(1) = s$.*

Proof.   In view of Proposition 2.9, the semigroup S satisfies $\mathrm{comp}(S) = S$.

Suppose that S is a Lie semigroup. Then $\mathrm{comp}(S) = \mathrm{reach}(S) = S$ follows from Proposition 5.7, so that every element $s \in S$ may be joined with the identity by an S-monotone curve.

If, conversely, this condition is satisfied, then $\mathrm{reach}(S) = S$, so that S is a Lie semigroup by Theorem 5.18.   ∎

To see how the preceding proposition applies in concrete cases, note that the curve γ yields a decreasing continuous one-parameter family

$$\Omega_t := \gamma(t).\Omega$$

of subsets of Ω. In many cases one can prove or disprove the existence of such a family. If such a family exists, then one has to lift this monotone curve to the semigroup S.

Notes

The Unit Group Theorem and the Units Neighborhood Theorem first appeared in [HHL89], whereas the Covering Theorem for Groups is a consequence of the covering results in Chapter 4 (cf. [Ne92a]). An extensive treatment of reachability can again be found in [HHL89]. Our treatment here is meant to be the bridge between our Lie semigroups and the infinitesimally generated semigroups from [HHL89]. The applications to faces appeared in [Ne91g].

6. Maximal semigroups in groups
with cocompact radical

The characterization of globality via positive functions shows that maximal subsemigroups play a fundamental role in the theory. But also in the applications to control theory and causal structures they are of importance. It is easy to visualize that maximal subsemigroups of vector spaces are halfspaces. It will turn out that a similar statement is true for a large class of Lie groups, the groups which are compact modulo the radical, but not compact. This is why we start in Section 6.1 with a survey of hyperplanes in Lie algebras which are subalgebras. Then we prepare in Sections 6.2 – 6.5 the proof of Lawson's theorem on maximal semigroups in groups with cocompact radical which will be given in Section 6.6.

The $Sl(2, \mathbb{R})$-examples from Chapter 2 show that maximal subsemigroups in non-compact semisimple groups may look more complicated. Nevertheless one can utilize the aforementioned results also for reductive groups via a proper use of the Iwasawa decomposition (cf. Section 6.7). Another tool that is useful in the context of simple groups with finite center are invariant control sets in flag manifolds, a concept borrowed from control theory (cf. Section 8.1).

Even though the theory of maximal subsemigroups is far from being complete, it already yields nice applications to controllability and causality questions which we describe in the later sections of this chapter. For other results concerning maximal subsemigroups in complex simple Lie groups, we refer to Section 8.6.

6.1. Hyperplane subalgebras of Lie algebras

The material of this section is purely Lie algebraic. Therefore we only give the results and refer to [Ho90a] for proofs. Our aim here is to describe an effective method to determine the set $\mathcal{H}(\mathfrak{g})$ of hyperplane subalgebras of a given Lie algebra $\mathfrak{g}$. It is based on several types of reductions. First one factors out the ideal

$$\Delta(\mathfrak{g}) = \bigcap \mathcal{H}(\mathfrak{g})$$

which we call the Δ-*radical* of $\mathfrak{g}$. Note that in the absense of hyperplane algebras this radical is all of $\mathfrak{g}$. The quotient then is Δ-*reduced*, i.e., it has trivial Δ-radical, and the quotient map $\mathfrak{g} \to \mathfrak{g}/\Delta(\mathfrak{g})$ induces a bijection between $\mathcal{H}(\mathfrak{g})$ and $\mathcal{H}(\mathfrak{g}/\Delta(\mathfrak{g}))$.

The set $\mathcal{H}(\mathfrak{g})$ consists of two essentially different types of subalgebras: For any $\mathfrak{h} \in \mathcal{H}(\mathfrak{g})$ the quotient $\mathfrak{g}/ \subset$ with $\subset$ the largest ideal of $\mathfrak{g}$ contained in $\mathfrak{h}$ is either isomorphic to $\mathfrak{sl}(2, \mathbb{R})$, to $\mathbb{R}$, or to the two dimensional non-abelian algebra. Accordingly we call $\mathfrak{h}$ of *simple, abelian* and *solvable type*. We denote the set of hyperplane subalgebras which are of simple type by $\mathcal{H}_s(\mathfrak{g})$ and set $\mathcal{H}_a(\mathfrak{g}) = \mathcal{H}(\mathfrak{g}) \backslash \mathcal{H}_s(\mathfrak{g})$. This gives rise to two more radicals

$$\Delta_s(\mathfrak{g}) := \bigcap \mathcal{H}_s(\mathfrak{g}), \quad \text{and} \quad \Delta_a(\mathfrak{g}) := \bigcap \mathcal{H}_a(\mathfrak{g}).$$

Upon factorization by $\Delta(\mathfrak{g})$, these radicals get mapped to their counterparts for $\mathfrak{g}/\Delta(\mathfrak{g})$. For Δ-reduced algebras, one can show that

$$\mathfrak{g} = \Delta_s(\mathfrak{g}) \oplus \Delta_a(\mathfrak{g})$$

as a Lie algebra and

$$\Delta_a(\mathfrak{g}) \cong \mathfrak{sl}(2, \mathbb{R})^m$$

for some m. In particular $\Delta_s(\mathfrak{g})$ contains the radical of $\mathfrak{g}$. In fact, one can even show that $\Delta_s(\mathfrak{g})$ is equal to the radical of $\mathfrak{g}$ because every simple subalgebra $\mathfrak{a}$ of $\mathfrak{g}$ not isomorphic to $\mathfrak{sl}(2, \mathbb{R})$ is contained in every hyperplane subalgebra since $\mathfrak{a}$ does not permit hyperplane subalgebras. Thus the structure of Δ-reduced algebras, is fairly simple. Before we show how to classify the hyperplane subalgebras in Δ-reduced algebras, we say something about the question how to find the Δ-radical.

It is clear that

$$\Delta(\mathfrak{g}) = \Delta_a(\mathfrak{g}) \cap \Delta_s(\mathfrak{g}),$$

so that it suffices to find the two larger radicals. The simpler part is to determine $\Delta_s(\mathfrak{g})$:

Proposition 6.1. *Let $\mathfrak{r}$ be the radical of $\mathfrak{g}$ and $\pi: \mathfrak{g} \to \mathfrak{g}/\mathfrak{r}$ the quotient map. Then*

$$\Delta_s(\mathfrak{g}) = \pi^{-1}(\mathfrak{j}),$$

where $\mathfrak{j}$ is the ideal in $\mathfrak{g}/\mathfrak{r}$ which is the sum of all simple ideals in $\mathfrak{g}/\mathfrak{r}$ not isomorphic to $\mathfrak{sl}(2, \mathbb{R})$. ∎

Determining $\Delta_a(\mathfrak{g})$ is harder:

Proposition 6.2.

  (i) *Consider $\mathfrak{m} := \mathfrak{g}'/\mathfrak{g}''$ and its dual space $\mathfrak{m}^*$ as $\mathfrak{g}/\mathfrak{g}'$-modules and denote by $\mathfrak{m}_1^*$ the submodule of $\mathfrak{m}^*$ spanned by all joint eigenvectors. Then the annihilator $\mathfrak{m}_0$ of $\mathfrak{m}_1^*$ in $\mathfrak{m}$ is of the form $\mathfrak{m}_0 = \mathfrak{j}/\mathfrak{g}''$ with an ideal $\mathfrak{j}$ in $\mathfrak{g}$.*

  (ii) *Let $\mathfrak{c}$ be a Cartan subalgebra of $\mathfrak{g}/\mathfrak{j}$ and $\pi: \mathfrak{g} \to \mathfrak{g}/\mathfrak{j}$. Then*

$$\Delta_a(\mathfrak{g}) = \pi^{-1}(\mathfrak{c} \cap (\mathfrak{g}/\mathfrak{j})').$$

Proof. [Ho90a, p.216]. ∎

Note that $\mathfrak{g}'' \subseteq \Delta_a(\mathfrak{g})$. For any linear form $\alpha \in \hat{\mathfrak{g}}$ on $\mathfrak{g}$ we consider

$$M_\alpha(\mathfrak{g}) := \{Y \in \mathfrak{g} : (\forall X \in \mathfrak{g}) \, [X, Y] = \alpha(X)Y\}$$

and say that α is a *base root* if $M_\alpha(\mathfrak{g}) \neq \{0\}$. The sum of all base root spaces is an ideal in $\mathfrak{g}$ which we call the *base ideal* and denote by $M(\mathfrak{g})$. Now we can formulate the classification of hyperplane subalgebras for Δ-reduced Lie algebras.

Theorem 6.3.   (Hofmann's Theorem on hyperplane subalgebras) *Let* $\mathfrak{g}$ *be a* Δ-*reduced Lie algebra,* $\mathfrak{r} := \Delta_s(\mathfrak{g})$ *its radical and* $\mathfrak{s} := \Delta_a(\mathfrak{g})$ *the unique Levi complement.*

(i) $\mathfrak{r}$ *is the vectorspace direct sum of* $M(\mathfrak{g})$ *with any Cartan subalgebra* $\mathfrak{c}$ *of* $\mathfrak{r}$.

(ii) *The hyperplane subalgebras of abelian type are precisely the hyperplanes containing* $M(\mathfrak{g}) \oplus \mathfrak{s}$.

(iii) *Any hyperplane of the form*

$$\mathfrak{h} = \mathfrak{c} + \sum_{\beta \neq \alpha} M_\beta(\mathfrak{g}) + H_\alpha,$$

where $\mathfrak{c}$ *is a Cartan subalgebra of* $\mathfrak{r}$, α *is a fixed base root, and* H_α *is a hyperplane in* $M_\alpha(\mathfrak{g})$, *is a hyperplane subalgebra of solvable type and any hyperplane subalgebra of solvable type is of this form. The hyperplane subalgebras of solvable type with* $M_\alpha(\mathfrak{g}) \not\subseteq \mathfrak{h}$ *are all conjugate under* $\mathrm{Aut}(\mathfrak{g})$.

(iv) *Any hyperplane of the form*

$$\mathfrak{h} = \mathfrak{r} + \{Y \in \mathfrak{g} \colon \chi(Y, X) = 0\},$$

where χ *is the Cartan-Killing form of* $\mathfrak{g}$ *and* $X \neq 0$ *is a fixed element of one of the simple factors of* $\mathfrak{s}$, *is a hyperplane subalgebras of simple type and all hyperplane subalgebras of simple type are of this form. The hyperplane subalgebras of simple type corresponding to a fixed simple ideal are all conjugate under* $\mathrm{Aut}(\mathfrak{g})$.

Proof.   [Ho90a, p.220].   ∎

6.2. Elementary facts about maximal subsemigroups

A subsemigroup S in a group G is usually called *maximal* if S is not properly contained in any proper subsemigroup of G. We call S maximal if in addition S is not a group. Since we are really interested in subsemigroups of Lie groups with non-empty interior this is not a serious restriction. Technically our definition has the advantage that $S^\sharp := S \backslash (S \cap S^{-1})$ is non-empty so that any maximal subsemigroup has a largest semigroup ideal.

It is clear that a maximal subsemigroup M contains $\mathbf{1}$. This shows that

$$(6.1) \qquad\qquad\qquad MT = TM = G$$

for any submonoid T of G with $T \not\subseteq M$ and $MT \subseteq TM$. From this one derives that any subsemigroup either intersects M in a peripheral way or otherwise is contained in the inverse of M.

Proposition 6.4.   (The Swallowing Lemma) *Let* M *be a maximal subsemigroup of* G *and* S *a subsemigroup satisfying* $MS^{-1} \subseteq S^{-1}M$ *(which is the case if* S *or* M *is invariant). Then either*

(i) $S \cap I \neq \emptyset$ for every left ideal I of M, or

(ii) $S^{-1} \subseteq M$.

Proof.   Suppose $S^{-1} \not\subseteq M$. Then for $T = S^{-1} \cup \{1\}$ we have $MT \subseteq TM$ so that $TM = G_0$ Let I be a left ideal of M. Pick $x \in M^\sharp$ and $y \in I$. Then $z = xy \in M^\sharp \cap I$. Since $TM = G$, there exist $s \in T$, $m \in M$ such that $z^{-1} = sm$. Hence $s^{-1} = mz \in MI \subseteq I$. Also $s \neq 1$ since $z \in M^\sharp$ implies $z^{-1} \notin M$. Thus $s^{-1} \in S$, and $S \cap I \neq \emptyset$.   ∎

Now we can show that maximal subsemigroups in abelian groups are *total*, i.e., satisfy

(6.2) $$G = M \cup M^{-1}.$$

In fact a little more is true:

Proposition 6.5.   *Let M be a maximal subsemigroup of G and let $x \in G$ satisfy $xM \subseteq Mx$, i.e., $xMx^{-1} \subseteq M$. Then $x \in M \cup M^{-1}$.*

Proof.   Let $T = \{x^n : n \geq 1\}$. Since $Q = \{y : yM \subseteq My\}$ is a subsemigroup, $T \subseteq Q$. Hence $TM \subseteq MT$. Similarly $Mx^{-1} \subseteq x^{-1}M$ implies $MT^{-1} \subseteq T^{-1}M$.

If $T \subseteq M$ or $T^{-1} \subseteq M$, then the proof is complete. If neither of these were to hold, then by Proposition 6.4 and its dual, $T \cap M^\sharp \neq \emptyset$, and $T^{-1} \cap M^\sharp \neq \emptyset$. Let $S = \{m \in \mathbb{Z} : x^m \in M^\sharp\}$. It is immediate that S is a semigroup and S contains both positive and negative numbers. Adding the elements of minimal distance to 0, one finds $0 \in S$. Hence $1 = x^0 \in M^\sharp$, a contradiction. Thus this final case cannot occur.   ∎

Note that not every total subsemigroup of a group is maximal (even in the abelian case). Just consider

$$S = \{(x,y) \in \mathbb{R}^2 : y > 0\} \cup \{(x,0) \in \mathbb{R}^2 : x \geq 0\}.$$

But if a total subsemigroup in a connected group is closed, then it is also maximal:

Proposition 6.6.   *Let G be a connected topological group and M a closed total subsemigroup of G. Then M is maximal.*

Proof.   Suppose that $x \notin M$. Let $U = G \setminus M$. Then U is open and $U \subseteq M^{-1}$ since $G = M \cup M^{-1}$. Thus $U^{-1} \subseteq M$. The subsemigroup T generated by $\{x\} \cup M$ contains xU^{-1}, an open set containing 1. Thus $T = G$ by connectedness of G.   ∎

So far we have not said anything general about the existence of a maximal subsemigroup containing a given semigroup S. If S has non-empty interior, then this is easy to achieve.

Proposition 6.7.   *Let S be a proper subsemigroup of a connected group G with $\text{int}(S) \neq \emptyset$. Then S is contained in a maximal subsemigroup M which is closed and satisfies $M \cap (\text{int } S)^{-1} = \emptyset$.*

Proof.   Let $\mathcal{M}$ be a maximal tower of proper (open) subsemigroups of G containing S and let M be their union. If $U = \text{int}(S)$ and $T \in \mathcal{M}$, then $T \cap U^{-1} = \emptyset$. In fact, suppose $s \in T \cap U^{-1}$. Then $s^{-1} \in U \subseteq T$, and thus $1 = ss^{-1} \in sU \subseteq sT \subseteq T$. Thus T contains the open neighborhood sU of 1, and hence $T = G$, a contradiction. Now we have $M \cap U^{-1} = \emptyset$, so $\overline{M}$ is proper and therefore a member of $\mathcal{M}$ itself. Clearly M does what we want.   ∎

Note that the same proof shows that any proper *open* subsemigroup is contained in a maximal open subsemigroup.

Finally we note that taking the preimage under a surjective group homomorphism preserves maximality and totality.

6.3. Abelian and almost abelian groups

Proposition 6.6 shows that closed halfspaces in $\mathbb{R}^n$, but also in the almost abelian Lie algebra equipped with the group multiplication coming from the multiplication given in Section 2.9 (note that almost abelian Lie algebras are exponential) are maximal subsemigroups. In fact, these are all the maximal subsemigroups. In order to prove this last statement, we start with a technical lemma.

Lemma 6.8. *Let M be a maximal subsemigroup of G, let H be a normal abelian subgroup, and let $y \in H$. If $y^n \in M$ for some $n > 1$, then $y \in M$.*

Proof. Suppose $y \notin M$. Then the subsemigroup generated by $M \cup \{y\}$ is all of G (by maximality of M). Note that since $1 \in M$, any member of G is either in M or has a representation $m_1 y m_2 y \cdots m_{k-1} y m_k$ (since such products together with M form a subsemigroup containing M and y). There exists $g \in M^\sharp \cap H$ by Proposition 6.4. Then for some $m_1, \ldots, m_k \in M$,

$$g^{-1} = m_1 y \cdots y m_k = (m_1 \cdots m_k) y^{m_2 \cdots m_k} \cdots y^{m_k} = mb,$$

where $z^w = w^{-1} z w$, $m = m_1 \cdots m_k$, and b is the product of the remaining factors. Since $y \in H$ and H is normal, $b \in H$. Thus $m = g^{-1} b^{-1} \in H$. So m commutes with b. Hence

$$g^{-n} = m^n b^n = m^{n-1} m (y^n)^{m_2 \cdots m_k} \cdots (y^n)^{m_k} = m^{n-1} m_1 y^n m_2 y^n \cdots y^n m_k \in M$$

since $y^n \in M$. But $g \in M^\sharp$ implies $g^n \in M^\sharp$. Therefore $1 = g^n g^{-n} \in M^\sharp M \subseteq M^\sharp$, a contradiction. ∎

Now let $M \subseteq \mathbb{R}^n$ be a closed maximal subsemigroup and $y \in M$. Then $\frac{1}{n} y \in M$ according to Lemma 6.8, so that $\mathbb{R}^+ M \subseteq M$ since M is closed. Thus the semigroup property of M shows that M is a wedge. But then (6.2) implies that M is a halfspace.

The argument for the almost abelian groups is a little more complicated. Again we do a technical lemma first:

Lemma 6.9. *Let G be the group of positive reals under multiplication. Given $s, t \in G$ with $0 < s < 1 < t$, a positive integer N, and $\varepsilon > 0$, there exist positive integers j, k with $j \geq N$ such that $|s^j t^k - 1| < \varepsilon$.*

Proof. Consider first the additive group $\mathbb{R}$ and positive real numbers x and y. The set $\{nx : n \geq 1\}$ is a cyclic semigroup in the compact group $\mathbb{R}/\mathbb{Z} \cdot y$. Therefore its closure is a compact semigroup and hence, being contained in a group, even a compact group (Corollary 1.20). So the elements nx cluster to the identity of $\mathbb{R}/\mathbb{Z} \cdot y$. Hence there exists $j \geq N$ such that $|jx - ky| < \varepsilon$ for some $k > 0$, that is, $|j(-x) + ky| < \varepsilon$. The lemma now follows from this derivation by applying the exponential function from the additive reals to the multiplicative positive reals. ∎

We write the almost abelian group G as $\{(x,y): x > 0,\ y \in \mathbb{R}^n\}$ with multiplication $(u,v)(x,y) = (ux, uy + v)$. The hyperplane $H = \{(1,y): y \in \mathbb{R}^n\}$ is the only non-trivial connected normal subgroup. In fact,

$$(x, y,)(1, z)(\frac{1}{x}, -\frac{y}{x}) = (1, xz)$$

in addition shows that any wedge W in H with vertex $(1,0)$ is an invariant subsemigroup. Now let M be a closed maximal subsemigroup in G.

By Proposition 6.4 either $H \subseteq M$ or $M^\sharp \cap H \neq \emptyset$. Suppose that $H \subseteq M$. Then G/H is isomorphic to the group of positive multiplicative reals, which in turn is isomorphic to $(\mathbb{R}, +)$, and M/H is a closed maximal subsemigroup. Thus M/H corresponds either to $(0,1]$ or $[1,\infty)$. Then $M = \{(x,y): 0 < x \leq 1\}$ or $M = \{(x,y): 1 \leq x\}$. In either case $\exp^{-1} M$ is a halfspace bounded by the unique codimension one ideal.

In the other case Lemma 6.8 shows that $M \cap H$ is a wedge H^+ with vertex $(1,0)$. Since H^+ is normal, $(H^+)^{-1}M = G$ by (6.1). Multiplying any element of G on the left by $(1,y)$ is a shift by y in the second component, so that $(\{x\} \times \mathbb{R}^n) \cap M \neq \emptyset$ for all $x > 0$.

Let $(s, ms - m)$, $(t, \mu t - \mu) \in M$ with $0 < s < 1 < t$. The powers of these elements are given by $(s^j, ms^j - m)$ and $(t^k, \mu t^k - \mu)$. The product is again in M and is given by

$$(s^j t^k, \mu s^j t^k - \mu s^j + ms^j - m) = (s^j t^k, \mu(s^j t^k - 1) + (m - \mu)(s^j - 1)).$$

If j, k are chosen as in Lemma 6.9, we see that this product can be made arbitrarily close to $(1, \mu - m)$, which must thus be a member of M. Hence $\mu - m \in H^+$. Choose $\alpha \in (H^+)^*$, then from the preceding argument it follows that

$$a := \sup\{\alpha(m): (s, ms - m) \in M \quad \text{for some} \quad s < 1\}$$
$$\leq \inf\{\alpha(\mu): (t, \mu t - \mu) \in M \quad \text{for some} \quad t > 1\} =: b.$$

If d is chosen so that $a \leq d \leq b$, then if follows that the region

$$M_{\alpha,d} := \{(x,y) \in G: \alpha(y) \geq d(x - 1)\}$$

contains M. But $M_{\alpha,d}$ is a subsemigroup, so that $M = M_{\alpha,d}$.

Now we have shown that in simply connected abelian and almost abelian groups closed maximal subsemigroups are halfspaces and in particular total.

6.4. Nilpotent groups

If G is a simply connected nilpotent Lie group, then the commutator subgroup $G' = [G,G]$ is strictly smaller than G and for any hyperplane E in $\mathfrak{g} = \mathbf{L}(G)$ containing $\mathfrak{g}' = [\mathfrak{g}, \mathfrak{g}] = \mathbf{L}(G')$ one finds a unique Lie subgroup H such that $G/H = \mathbb{R}$. The inverse images of $\mathbb{R}^{\pm}$ are total hence maximal closed subsemigroups. In this section we show that these are all the maximal subsemigroups.

Similarly to the study of hyperplane subalgebras it turns out to be useful to first factor the largest normal (w.r.t. G) subgroup of a semigroup S. This group we call the *core of S* and denote it $C(S)$. If M is a maximal subsemigroup of G, then clearly $M/C(M)$ is a maximal subsemigroup of $G/C(M)$. The semigroup $M/C(M)$ is *reduced*, i.e., it does not contain a non-trivial normal subgroup.

For the following two lemmas we don't need to assume that G is nilpotent. We denote by $[g,h]$ the commutator $g^{-1}h^{-1}gh$ in a group G.

Lemma 6.10. *Suppose that G is a group, M is a maximal subsemigroup of G, and $g,h \in G$ are such that $[g,h]$ is in the center of G. Then $[g,h],[g,h]^{-1} \in M$.*

Proof.   Let $w = [g,h]$. If $w = 1$, then the lemma is trivial since $1 \in M$. By Proposition 6.5, $w \in M$ or $w^{-1} \in M$. We suppose without loss of generality that $w \in M$ and show also $w^{-1} \in M$.

Suppose on the contrary that $w^{-1} \notin M$. Let $S = \{w^{-n} : n \geq 0\}$. Then SM is a semigroup containing w^{-1} and M; by maximality of M, $SM = G$. Thus $g = w^{-r}u$, $h = w^{-m}v$, $g^{-1} = w^{-k}x$, $h^{-1} = w^{-p}y$ for some $r,m,k,p \geq 0$ and $u,v,x,y \in M$. Thus $w^r g = u \in M$ and similarly $w^m h, w^k g^{-1}, w^p h^{-1} \in M$. Hence $w^q g, w^q h, w^q g^{-1}, w^q h^{-1} \in M$, where $q = r+m+k+p$ since $w \in M$ by assumption.

Now $gh = hg[g,h]$. Since $[g,h]$ is central, an easy induction yields that $g^n h^n = h^n g^n [g,h]^{n^2}$, i.e., $g^n h^n [g,h]^{-n^2} = h^n g^n$. Then for $z = w^q$, we have
$$(zg^{-1})^n (zh)^n (zg)^n (zh^{-1})^n \in M$$
for all $n \geq 1$ and
$$(zg^{-1})^n (zh)^n (zg)^n (zh^{-1})^n = z^{4n} g^{-n}(h^n g^n)h^{-n} = z^{4n} g^{-n}(g^n h^n [g,h]^{-n^2})h^{-n}$$
$$= z^{4n} w^{-n^2} = w^{4nq} w^{-n^2} = w^{4nq-n^2}.$$

Since for large n, $4nq - n^2 < 0$ and since $w \in M$ is central from Lemma 6.8, we conclude that $w^{-1} \in M$, a contradiction. This completes the proof.   ∎

Lemma 6.11. *Let M be a maximal subsemigroup of G which is reduced in G. Then $G/Z(G)$ has trivial center.*

Proof.   Suppose not. Let $\varphi : G \to G/Z(G)$. Then there exists $g \in G$ such that $\varphi(g)$ is in the center of $G/Z(G)$, but $\varphi(g)$ is not the identity. Since $g \notin Z(G)$, there exists $h \in G$ such that $gh \neq hg$. Then $[g,h] = g^{-1}h^{-1}gh \neq 1$. But
$$\varphi([g,h]) = \varphi(g^{-1})\varphi(h^{-1})\varphi(g)\varphi(h) = \varphi(g^{-1})\varphi(g)\varphi(h^{-1})\varphi(h) = \varphi(1).$$

Thus $[g,h] \in Z(G)$. By Lemma 6.10, $[g,h],[g,h]^{-1} \in M$. Since $[g,h] \in Z(G)$, the subgroup it generates is normal. But this contradicts the assumption that M is reduced in G.   ∎

Theorem 6.12.   *Let M be a maximal subsemigroup of a nilpotent group G. Then M is total and invariant in G and $[G,G] \subseteq H(M)$. Hence $C(M) = H(M)$ is abelian and totally ordered.*

Proof.   We apply Lemma 6.11 to $M/C(M)$ and find that $G/C(M)$ must be abelian since every non-trivial nilpotent group has non-trivial center. Therefore $[G,G] \subseteq C(M) \subseteq H(M)$ and $M/C(M)$ is total. Thus also M is total. For $g \in G$, $m \in M$ we have $(g^{-1}mgm^{-1})m = g^{-1}mg \in M$, so that M and hence $H(M)$ are invariant. But then $C(M) = H(M)$.   ∎

Note that Theorem 6.12 proves the claim we made at the beginning of this section since $M/C(M)$ is a halfspace in $G/C(M)$ if G is simply connected and M is closed. Hence $C(M) = H(M)$ has codimension one.

6.5. Reduction lemmas

In this section we prove a few lemmas which will help us to classify the maximal subsemigroups also for more complicated Lie groups than abelian, almost abelian or nilpotent ones.

Proposition 6.13. *Let H be a connected normal nilpotent subgroup of a connected topological group G and let M be a maximal subsemigroup with $\operatorname{int}(M) \neq \emptyset$. Then $[H, H] \subseteq C(M)$.*

Proof. We first show $[H, H] \cap \operatorname{int}(M) = \emptyset$. Suppose not. Since $1 \notin \operatorname{int}(M)$, $\operatorname{int}(M) \cap H$ is a proper subsemigroup of H with interior in H. By Proposition 6.7 there exists a maximal subsemigroup S of H with $\operatorname{int}(M) \cap H \subseteq S$ and $S \cap (\operatorname{int} M)^{-1} = \emptyset$. By Theorem 6.12, $[H, H] \subseteq S$. If $g \in [H, H] \cap \operatorname{int}(M)$, then $g^{-1} \in [H, H] \cap (\operatorname{int} M)^{-1} \subseteq S \cap (\operatorname{int} M)^{-1}$, a contradiction.

Since $\operatorname{int}(M)$ is an ideal in M, Proposition 6.4 implies $[H, H] \subseteq M$ and hence $[H, H] \subseteq C(M)$. ∎

Lemma 6.14. *Let G be a connected Lie group with Lie algebra $\mathfrak{g}$, let $\exp: \mathfrak{g} \to G$ be the exponential mapping, and let $\mathfrak{a}$ be an abelian ideal of $\mathfrak{g}$. If M is a maximal subsemigroup of G with $\operatorname{int}(M) \neq \emptyset$, then $W = \{X \in \mathfrak{a}: \exp(X) \in M\}$ is a closed wedge in $\mathfrak{g}$ which is invariant under the adjoint action of G and satisfies $W - W = \mathfrak{a}$.*

Proof. Since $\mathfrak{a}$ is an abelian ideal, $\exp$ restricted to $\mathfrak{a}$ is a homomorphism and $\exp(\mathfrak{a})$ is a normal subgroup. If $\exp(\mathfrak{a}) \subseteq M$, then we are finished since $\mathfrak{a}$ is an ideal, and hence invariant under the adjoint action.

If $\exp(\mathfrak{a}) \not\subseteq M$, then $\exp(\mathfrak{a}) \cap \operatorname{int}(M) \neq \emptyset$ by Proposition 6.4. Pick $h = \exp(Y)$, $h \in \operatorname{int}(M)$, $Y \in \mathfrak{a}$.

Since M is closed (cf. Proposition 6.7) and $\exp$ is continuous,

$$W = \exp^{-1}(M) \cap \mathfrak{a}$$

is closed. Since M is a subsemigroup and $\exp$ restricted to $\mathfrak{a}$ is a homomorphism, W is closed under addition. By Lemma 6.8 applied to $H = \exp(\mathfrak{a})$, we conclude that W is closed under scalar multiplication by positive rationals, and then by all positive reals by continuity. It follows that W is a wedge. Since $\exp(Y) \in \operatorname{int}(M)$, it follows that W has Y as an interior point, and hence $\mathfrak{a} = W - W$.

Let X be any other interior point of W. Then $n \cdot X = Y + Z_n$ for some $Z_n \in W$ for all sufficiently large n. To see this, let U be open in I, $X \in U \subseteq W$. Then $\frac{1}{n} \cdot Y \in X - U$ for large n. Thus $\frac{1}{n} \cdot Y + u_n = X$, i.e., $Y + Z_n = n \cdot X$, where $Z_n = n \cdot u_n \in W$.

Since $h \in \operatorname{int}(M)$, there exists an open set $N = N^{-1}$, $1 \in N$ with $Nh \subseteq \operatorname{int}(M)$. Let $g \in N$ and let X be an interior point of W. For large n, pick $Z_n \in W$ such that $n \cdot X = Y + Z_n$. Let $a = \exp(X)$, $b_n = \exp(Z_n)$. We have

$$
\begin{aligned}
\exp\big((\operatorname{Ad} g)(n \cdot X) + Y\big) &= \exp\big((\operatorname{Ad} g)(n \cdot X)\big)\exp(Y) \\
&= g\big(\exp(n \cdot X)\big)g^{-1}h \\
&= g\big(\exp(Y + Z_n)\big)g^{-1}h \\
&= ghb_n g^{-1}h \in Nh \cdot M \cdot Nh \subseteq M.
\end{aligned}
$$

Thus $(\operatorname{Ad} g)(n \cdot X) + Y \in W$ for large n. Since W is a wedge, $(\operatorname{Ad} g)(X) + \frac{1}{n} \cdot Y \in W$ for large n. Since W is closed, $(\operatorname{Ad} g)(X) \in W$. Thus $\operatorname{Ad} g$ carries the interior of W into W and hence preserves W. This is true for all $g \in N$ and by connectivity for all of G (since $\operatorname{Ad}(g_1 \cdots g_n) = \operatorname{Ad} g_1 \circ \cdots \circ \operatorname{Ad} g_n$). ∎

Proposition 6.15. *Let M be a maximal subsemigroup of G with $\operatorname{int}(M) \neq \emptyset$. If H is a compact subgroup, then $H \cap \operatorname{int}(M) = \emptyset$. If H is compact and normal, then $H \subseteq C(M)$.*

Proof.   Suppose that $H \cap \operatorname{int}(M) \neq \emptyset$. Then $S = H \cap \operatorname{int}(M)$ is an open subsemigroup of the compact group H and its closure $\overline{S}$ is a compact subsemigroup hence a group by Corollary 1.20. Therefore $S^{-1} \subseteq \overline{S}$ which in particular implies $1 \in S \subseteq \operatorname{int}(M)$, so $M = G$, a contradiction. Then, by Proposition 6.4, $H \subseteq M$ if H is normal. It follows that $H \subseteq C(M)$ in this case, since $C(M)$ is the largest normal subgroup of M. ∎

A Lie group G is called a *Frobenius-Perron group* if for every continuous linear action of G on a finite dimensional real vector space V that leaves a pointed cone C invariant, there exists $v \in C$ with $G.v \subseteq \mathbb{R}^+ v$. Using duality, this property can also be reformulated as follows: every generating G-invariant wedge in V is contained in a G-invariant halfspace.

Lemma 6.16. *Let G be a finite dimensional connected Lie group which is a Frobenius-Perron group, and let M be a maximal subsemigroup which is reduced in G and satisfies $\operatorname{int} M \neq \emptyset$. Then one of the following holds:*

(i) $\operatorname{Rad}(G) = \{1\}$, *i.e., G is semisimple;*

(ii) $(\operatorname{Rad}(G), \operatorname{Rad}(G) \cap M)$ *is topologically isomorphic to* $(\mathbb{R}, \mathbb{R}^+)$;

(iii) $\operatorname{Rad}(G)$ *is topologically isomorphic to* $\operatorname{Aff}(\mathbb{R})$, *the two dimensional almost abelian group.*

Proof.   If the radical $R = \operatorname{Rad}(G)$ is trivial, then G is semisimple. We consider the case $R \neq \{1\}$, and show that either (ii) or (iii) obtains. The proof consists of a series of reductions.

1. *The nilradical N is abelian.*
The nilradical is the largest connected normal nilpotent subgroup. The assertation then follows immediately from Proposition 6.13.

2. *The radical R is metabelian (i.e., $[R, R]$ is abelian).*
Again it is standard that $[R, R] \subseteq N$, which is abelian.

3. $\dim [\mathbf{L}(R), \mathbf{L}(R)] = 1$ or $[R, R] = \{1\}$.

Let I be the Lie algebra for $[R,R]$. Since $[R,R]$ is normal and abelian, I is an abelian ideal in $\mathbf{L}(G)$. Let $W = \{X \in I : \exp(X) \in M\}$. By Lemma 6.14, W is an invariant wedge which generates I. If $W = I$, then $\exp(I) = [R,R]$ (since I is abelian) is a normal subgroup contained in M. Since M is reduced, $[R,R] = \{1\}$ in this case.

If $W \neq I$, then since G is a Frobenius-Perron group and since W is generating, there exists an invariant half-space Q with $W \subseteq Q$. Then $Q \cap -Q$ is an invariant hyperplane in I, so $F = \exp(Q \cap -Q)$ is a normal subgroup. Now

$$\exp^{-1}(\operatorname{int} M) \cap I \subseteq \operatorname{int}(W) \subseteq \operatorname{int}(Q) \subseteq Q \setminus (Q \cap -Q).$$

Thus $\operatorname{int}(M) \cap F = \varnothing$. By Proposition 6.4, $F \subseteq M$. Since F is normal, $F \subseteq C(M) = \{1\}$. Thus $Q \cap -Q = \{0\}$, so I is one dimensional, as is $\exp(I) = [R,R]$.

4. *If $[R,R] = \{1\}$, then case (ii) obtains.*

In this case, R is abelian. Let $\mathbf{L}(R)$ be the Lie algebra for R. We again obtain an invariant wedge $W = \{X \in \mathbf{L}(R) : \exp(X) \in M\}$. One repeats the arguments of step 3 to conclude that W is an invariant generating wedge, that $W \neq \mathbf{L}(R)$ since M is reduced and we are not in case (i), and finally that $\mathbf{L}(R)$ is one dimensional. By Proposition 6.15, we have that R cannot be the circle group. It follows that the exponential mapping must carry $\mathbf{L}(R)$ onto a copy of $\mathbb{R}$ and W onto a ray, which we take to be $\mathbb{R}^+$.

5. *If $I = [\mathbf{L}(R), \mathbf{L}(R)]$ is one dimensional, then the centralizer of I in $\mathbf{L}(R)$ is I.*

Since I is an abelian ideal, its centralizer is easily verified to be an ideal which contains I. Hence A, the intersection of the centralizer with $\mathbf{L}(R)$, is an ideal containing I. Now $[A,A] \subseteq [\mathbf{L}(R),\mathbf{L}(R)] = I$; thus $[A,[A,A]] \subseteq [A,I] = \{0\}$. Therefore A is a nilpotent ideal, so $\exp(A)$ is a normal nilpotent group. It follows from Proposition 6.13 and the fact that M is reduced that $\exp(A)$ is abelian. Thus A is abelian. One now applies an argument to A completely analogous to that given in Step 3 to conclude that A is one dimensional. Since $I \subseteq A$, we have $I = A$.

6. *If $I = [\mathbf{L}(R), \mathbf{L}(R)]$ is one dimensional, then case (iii) obtains.*

Let $Z \in I$, $Z \neq 0$. Consider $\operatorname{ad} Z : \mathbf{L}(R) \to I = \mathbb{R} \cdot z$. By Reduction 5, the kernel of this mapping is I, which is one dimensional. Thus $\mathbf{L}(R)$ is two dimensional. Since $[\mathbf{L}(R), \mathbf{L}(R)] = I \neq 0$, it follows that $\mathbf{L}(R)$ must be a non-abelian two dimensional Lie algebra, and hence isomorphic to $\mathrm{Aff}(\mathbb{R})$. ∎

6.6. Characterization of maximal subsemigroups

Let G be a connected Lie group with $G/\operatorname{Rad}(G)$ compact. Such a group we call a *group with cocompact radical.* For simply connected groups with cocompact radical one also has a canonical class of maximal subsemigroups: Suppose that G is simply connected and E is a half-space bounded by a hyperplane subalgebra $\mathfrak{h}$ of $\mathfrak{g} = \mathbf{L}(G)$. Then there exists a Lie algebra homomorphism $\mathbf{L}(\varphi)$ either onto $\mathbb{R}$ with kernel $\mathfrak{h}$ or onto the non-abelian two dimensional Lie algebra or onto $\mathfrak{sl}(2,\mathbb{R})$. But the last case is impossible since then $\mathbf{L}(\operatorname{Rad} G)$ must necessarily map to $\{0\}$, and then $\mathfrak{sl}(2,\mathbb{R})$ would be the image of a compact Lie algebra, hence itself compact, a contradiction. Also $\mathbf{L}(\varphi)(E)$ will be a half-space in $\mathbb{R}$ or $\text{Aff}(\mathbb{R})$. Since G is simply connected, there is a corresponding $\varphi: G \to \mathbb{R}$ or $\varphi: G \to \text{Aff}(\mathbb{R})$. Pulling back the subsemigroups of $\mathbb{R}$ or $\text{Aff}(\mathbb{R})$ corresponding to the half-space $\mathbf{L}(\varphi)(E)$, one obtains a maximal subsemigroup M of G containing $\exp(E)$. Since $E \subseteq \mathbf{L}(M) \neq \mathfrak{g}$ and E is a half-space, $E = \mathbf{L}(M)$.

Again we will show that these are all the maximal subsemigroups. The key point in the proof is to note that groups with cocompact radical are Frobenius-Perron groups.

Theorem 6.17.   (Lawson's Theorem on Maximal Subsemigroups) *Suppose that G is a connected Lie group with $G/\text{Rad}(G)$ compact. If M is a maximal subsemigroup of G with $\text{int}(M) \neq \emptyset$, then M is total, $C(M)$ is a normal, connected subgroup containing every semisimple analytic subgroup, and one of the following holds:*

(i) $\left(G/C(M), M/C(M)\right)$ *is topologically isomorphic to* $(\mathbb{R}, \mathbb{R}^+)$,

(ii) $\left(G/C(M), M/C(M)\right)$ *is topologically isomorphic to*

$$(\text{Aff}(\mathbb{R}), \text{Aff}(\mathbb{R})^+)$$

with

$$\text{Aff}(\mathbb{R})^+ = \{(x, y) \in \text{Aff}(\mathbb{R}): y \geq 0\}.$$

Proof.   To simplify our notation, we set $G_R := G/C(M)$ and $M_R := M/C(M)$. If $\varphi: G \to G_R$, then $\varphi(\text{Rad}\, G) = \text{Rad}\, G_R$ ([Bou71a, Ch. III, §7, no. 9, Prop. 24]). Therefore $G_R/\text{Rad}\, G_R$ is a continuous image of $G/\text{Rad}\, G$, so that G_R has cocompact radical. Thus, if K is an analytic semisimple subgroup corresponding to some Levi factor of $\mathbf{L}(G_R)$, then K is compact and normalizes $\text{Rad}\, G_R$. We apply Lemma 6.16 to G_R.

Case (i) of Lemma 6.16 is impossible, for then $G_R = K$ would be compact, an impossibility (see Proposition 6.15). Suppose that case (ii) obtains, i.e., $\text{Rad}\, G_R$ is topologically isomorphic to $\mathbb{R}$. Then K acts on $\mathbb{R}$ by inner automorphisms, and hence must act trivially (since K is compact and connected). Thus elements of $\mathbb{R}$ and K commute, so K is normal (since $G_R = \mathbb{R}K$). Then Proposition 6.15 implies $K = \{1\}$ (since M_R is reduced). So $G_R = \text{Rad}\, G_R$. Since the image of any analytic semisimple subgroup of G must be a semisimple subgroup of $\mathbb{R}$, the image must be trivial. Lemma 6.16 implies the rest.

Finally consider the case of Lemma 6.16 that $\text{Rad}\, G_R$ is topologically isomorphic to $\text{Aff}(\mathbb{R})$. Again K must act on $\text{Aff}(\mathbb{R})$ by inner automorphisms. But the identity component of the automorphism group of $\text{Aff}(\mathbb{R})$ is again topologically isomorphic to $\text{Aff}(\mathbb{R})$ (see [Jac57], p.10), and thus the automorphism group contains no non-trivial compact connected subgroups. Thus again K acts trivially, is thus normal, and hence $K = \{1\}$. So $G_R = \text{Rad}\, G_R$. Since M_R is closed and maximal in G_R, it must be topologically isomorphic to $\text{Aff}(\mathbb{R})^+$ since, according to our results on almost abelian groups and algebras, all maximal subsemigroups whose group of units is not normal are conjugate under $\text{Aut}(G_R)$ (cf. Theorem 6.3(iii)).

Finally we note that $C(M)$, the kernel of φ, is connected since $G/C(M)$ is simply connected.   ∎

Theorem 6.18.   *The maximal subsemigroups M with non-empty interior of a simply connected Lie group with $G/\text{Rad}\, G$ compact are in one-to-one correspondence* *with their tangent wedges $\mathbf{L}(M)$ and the latter are precisely the closed half-spaces with boundary a subalgebra, i.e., the half-space Lie wedges. Furthermore, M is the semigroup generated by $\exp(\mathbf{L}(M))$.*

Proof.   Let M be maximal with $\operatorname{int}(M) \neq \emptyset$. By Theorem 6.17, G_R is either $\mathbb{R}$ or $\mathrm{Aff}(\mathbb{R})$ hence simply connected, so that $C(M)$ is connected. Hence $C(M)$ is generated by the exponential image of its tangent subalgebra. One verifies directly in each case that M_R is (and hence is generated by) the exponential image of its tangent set. Let $\varphi\colon G \to G_R$. It follows from chasing the diagram

$$
\begin{array}{ccccc}
\mathbf{L}(C(M)) & \xrightarrow{\ \mathbf{L}(\mathrm{incl})\ } & \mathbf{L}(G) & \xrightarrow{\ \mathbf{L}(\varphi)\ } & \mathbf{L}(G_R) \\
{\scriptstyle\exp}\big\downarrow & & {\scriptstyle\exp}\big\downarrow & & {\scriptstyle\exp}\big\downarrow \\
C(M) & \xrightarrow[\ \mathrm{incl}\ ]{} & G & \xrightarrow[\ \varphi\ ]{} & G_R
\end{array}
$$

that $\mathbf{L}(M) = \mathbf{L}(\varphi)^{-1}\big(\mathbf{L}(M_R)\big)$ and that M is generated by the exponential image of $\mathbf{L}(M)$. Also since M_R has tangent set a half-space of $\mathbf{L}(G_R)$, this property pulls back so that $\mathbf{L}(M)$ is a half-space of $\mathbf{L}(G)$. Thus associated with each maximal subsemigroup with non-empty interior is a half-space whose exponential image generates it. This guarantees that the assignment is one-to-one.

The converse has been noted already at the beginning of this section.   ■

Let G be a connected nilpotent Lie group and $\mathfrak{g} = \mathbf{L}(G)$ its Lie algebra. Then no quotient of $\mathfrak{g}$ is isomorphic to $\mathfrak{sl}(2, \mathbb{R})$ or the solvable two-dimensional Lie algebra. Hence every hyperplane subalgebra is of the abelian type, i.e., the hyperplane subalgebras of $\mathfrak{g}$ are the hyperplanes containing the commutator algebra $[\mathfrak{g}, \mathfrak{g}]$. Now Theorem 6.18 shows that every maximal subsemigroup of G with interior points contains the commutator group (G, G). This implies in particular the observation from Proposition 2.1.

6.7. Applications to reachability questions

Let G be a Lie group and $\mathfrak{g}$ its Lie algebra. A *control system on* G consists of a family $\widetilde{\mathcal{X}}$ of vector fields on G. The set of endpoints of continuous paths in G, starting in the identity, which are piecewise integral curves for elements of $\widetilde{\mathcal{X}}$ is called the *reachable set* of $\{1\}$. The system is called *controllable* if the reachable set is all of G. We call the control system left *invariant* if it consists of left invariant vector fields. Thus such a control system simply is a subset $\mathcal{X}$ of $\mathfrak{g}$. It is well known (cf. [JS72], [JK81a], [JK81b]) that in order to study controllability properties of such systems it is enough to consider the closed convex cone generated by $\mathcal{X}$.

So let C be a closed convex cone in $\mathfrak{g}$ which we will assume to have non-empty interior. Then the associated left invariant control-system, i.e., the system associated with the set of all left invariant vector fields with values in C, is controllable if and only if the closed semigroup $S(C)$ generated by $\exp C$ is all of G (cf. Lemma 3.7 and Corollary 3.11). In this case we call C *controllable* in G. If G is simply connected, then we omit the reference to G. In this section we will give a simple geometric characterisation of controllability in the case that G is reductive, C is pointed, i.e., satisfies $C \cap -C = \{0\}$, and is invariant under the adjoint action of K, where NAK is an Iwasawa decomposition of G.

The key idea is to describe a reductive group as a homogeneous space of a group with co-compact radical and then use the theory of maximal subsemigroups in such groups which can be reduced to simple linear algebra. Implicitly this way of viewing a reductive group is (for $G = \mathrm{Sl}(2, \mathbb{R})$) contained in [Gö49], where the first simply connected space-time model violating causality was developed.

Now let G be a reductive group and NAK an Iwasawa decomposition of G. If C is an $\mathrm{Ad}\,K$-invariant pointed cone with non-empty interior, averaging over $\mathrm{Ad}\,K$ (which is compact) yields elements in $\mathrm{int}\,C \cap Z(\mathfrak{k})$, where $Z(\mathfrak{k})$ denotes the center of $\mathfrak{k}$. Thus at least for semisimple groups with finite center, C is always controllable in G because in this case the group $\exp Z(\mathfrak{k})$ is compact (Corollary 1.20). We will restrict ourselves mostly to the case of simply connected groups.

The basic observation is that the group $G_a = NA \times K$ acts transitively on G via $(na, k).g = nagk^{-1}$. This action allows us to identify G and G_a as manifolds and hence $\mathfrak{g}$ and $\mathfrak{g}_a = (\mathfrak{n} + \mathfrak{a}) \times \mathfrak{k}$ as vector spaces.

Lemma 6.19. *Let G be a connected Lie group such that there exist closed subgroups K and B with the property that the multiplication mapping*

$$B \times K \to G, \qquad (b, k) \mapsto bk$$

is a diffeomorphism. Let G_a be the group $B \times K$. If C is an $\mathrm{Ad}(K)$-invariant cone in $\mathfrak{g}$, then C is global in G if and only if C is global in G_a.

Proof. G_a operates freely and transitively on G via $(b, k).g = bgk^{-1}$. The left invariant cone field Ξ_C on G generated by C (cf. Section 4.2) is by hypothesis also right invariant with respect to translations from K, i.e., invariant under the action of G_a. Identifying G and G_a as manifolds, we obtain a left invariant cone field $\Xi_C^{(a)}(b, k) = \Xi_C(bk^{-1})$ on G_a which is generated by C. Now the claim follows from Theorem 4.21. ∎

Remark 6.20. The proof of Lemma 6.19 even shows that the closed semigroups generated by $\exp_G C$ and $\exp_{G_a} C$ coincide as sets when we identify G and G_a as manifolds, since the definition of S_Θ for a given cone field on the group only depends on the manifold structure. ∎

Lemma 6.19 shows that in order to check the globality of an $\mathrm{Ad}(K)$-invariant cone in the reductive Lie group G, it suffices to check it in G_a. But G_a is compact modulo its radical and thus, for simply connected G_a, the maximal subsemigroups are known once one has determined all the hyperplanes in its Lie algebra $(\mathfrak{n} + \mathfrak{a}) \oplus \mathfrak{k}$ which are subalgebras (cf. Theorem 6.18).

Lemma 6.21. *Let $\mathfrak{g}$ be a reductive Lie algebra with Iwasawa decomposition $\mathfrak{n} + \mathfrak{a} + \mathfrak{k} = \mathfrak{g}$ and $\mathfrak{g}_a = (\mathfrak{n} + \mathfrak{a}) \oplus \mathfrak{k}$. Then the intersection $\Delta(\mathfrak{g}_a)$ of all hyperplanes in $\mathfrak{g}_a$ which are subalgebras is equal to $\mathfrak{n}' \oplus \mathfrak{k}'$ where $'$ denotes the commutator algebra.*

Proof. We use the notation of Section 6.1 and note first that the s-radical $\Delta_s(\mathfrak{g}_a)$ of $\mathfrak{g}_a$ is all of $\mathfrak{g}_a$ since $\mathfrak{g}_a$ does not contain an isomorphic copy of $\mathfrak{sl}(2, \mathbb{R})$. We note further that $\mathfrak{g}_a' = \mathfrak{n} \oplus \mathfrak{k}'$ and consequently $\mathfrak{g}_a'' = \mathfrak{n}' \oplus \mathfrak{k}'$. Note that $\mathfrak{g}_a/\mathfrak{g}_a' \cong \mathfrak{a} \oplus Z(\mathfrak{k})$. The subalgebra $\mathfrak{n}$ is the sum of one dimensional $\mathfrak{a}$-modules, hence $\mathfrak{g}_a/\mathfrak{g}_a'$-modules. Therefore, in the notation of Proposition 6.2, we have $\mathfrak{m}_1^* = \mathfrak{m}^*$.

and hence $\mathfrak{m}_0 = \{0\}$. Thus $\mathfrak{j} = \mathfrak{g}_a''$. Finally we remark that the image of $\mathfrak{a} \oplus Z(\mathfrak{k})$ in $\mathfrak{g}_a/\mathfrak{g}_a''$ is a Cartan algebra of $\mathfrak{g}_a/\mathfrak{g}_a''$ which has trivial intersection with $(\mathfrak{g}_a/\mathfrak{g}_a'')'$. Now the claim follows from Proposition 6.2. ∎

According to Theorem 6.3, there are two families of hyperplane subalgebras in $\mathfrak{g}_a$. The first family consists of all hyperplanes containing the preimage $\mathfrak{n} + \mathfrak{k}'$ of the base ideal $M(\mathfrak{g}_a/\mathfrak{g}_a'')$ of $\mathfrak{g}_a/\mathfrak{g}_a''$. The second consists of hyperplanes which contain the preimage of a Cartan algebra of $\mathfrak{g}_a/\mathfrak{g}_a''$. Note that $Z(\mathfrak{k})$ is contained in any such preimage. If now C is an $\mathrm{Ad}(K)$-invariant cone in $\mathfrak{g}$, then averaging over K shows that $Z(\mathfrak{k}) \cap \mathrm{int}\, C \neq \emptyset$ provided that $\mathrm{int}\, C$ is non-empty which we always assumed. Thus the only hyperplane subalgebras which can possibly miss $\mathrm{int}\, C$ are the ones containing $\mathfrak{n}$. Now Proposition 1.37 together with Remark 6.20 yields:

Theorem 6.22. *Let G be a simply connected reductive Lie group with Iwasawa decomposition NAK and C an $\mathrm{Ad}(K)$-invariant pointed generating wedge in $\mathfrak{g}$. Let S_C be the closed semigroup generated by $\exp C$. Then the following assertions hold.*

(i) *If $(\mathrm{int}\, C) \cap (\mathfrak{n} + \mathfrak{k}') \neq \emptyset$, then $S_C = G$.*

(ii) *If $C \cap (\mathfrak{n} + \mathfrak{k}') = \{0\}$, then S_C satisfies*

$$C = \{X \in \mathfrak{g} : \exp \mathbb{R}^+ X \subset S_C\},$$

i.e., C is global in G.

(iii) *If $\emptyset \neq (C \cap (\mathfrak{n} + \mathfrak{k}')) \setminus \{0\} \subset \partial C$, then S_C is strictly contained in G.* ∎

It is easy to write down examples of cones which are neither global nor controllable. Take for instance a product of a global and a controllable cone. On the other hand their are situtations in which $S_C \neq G$ automatically implies the globality of C. This is for instance the case if $\mathfrak{g}$ is simple as we will now explain.

Lemma 6.23. *Let G be a simply connected simple hermitean Lie group, $\mathfrak{g} = \mathfrak{k} + \mathfrak{p}$ a Cartan decomposition, and C an $e^{\mathrm{ad}\, \mathfrak{k}}$-invariant generating Lie wedge in $\mathfrak{g}$. If $H(C) = \mathfrak{k}'$, then C is controllable in G if and only if C is global.*

Proof. One implication is trivial. Suppose that C is not controllable. For $S_C = \langle \exp C \rangle$ we have $\mathbf{L}(S_C) \neq \mathfrak{g}$. If $\mathbf{L}(S_C) \neq C$, then $\mathfrak{k}' = H(C)$ is strictly contained in $F := H(\mathbf{L}(S_C))$ because of Proposition 1.37. Thus

$$F = (F \cap \mathfrak{p}) + \mathfrak{k}' + (F \cap Z(\mathfrak{k})).$$

But $\mathfrak{p}$ is an irreducible $\mathfrak{k}$-module, whence $F \cap \mathfrak{p} \neq \{0\}$ would imply $\mathfrak{p} \subseteq F$ and

$$\mathfrak{g} = \mathfrak{p} + [\mathfrak{p}, \mathfrak{p}] = F + [F, F] \subseteq F.$$

Thus $F \cap \mathfrak{p} = \{0\}$, and hence $\mathfrak{k} = F$. Now $\mathbf{L}(S_C) \cap \mathfrak{p}$ is a pointed $e^{\mathrm{ad}\, \mathfrak{k}}$-invariant cone which, in view of Theorem 1.6, is not possible. ∎

Note that in the situation of Theorem 6.22 it follows from $S_C \neq G$ that $(\mathrm{int}(C + \mathfrak{k}')) \cap (\mathfrak{n} + \mathfrak{k}') = \emptyset$ so that Remark 6.20 applied to $C + \mathfrak{k}'$ shows that $S_{C+\mathfrak{k}'} \neq G$. If now $\mathfrak{g}$ is simple, Lemma 6.23 shows that $C + \mathfrak{k}'$ is global. This in turn is equivalent to C being global as is shown by Proposition 1.39. Thus Theorem 6.22 also yields a characterisation of the global invariant cones in simple Lie algebras.

Corollary 6.24.   *Let C be a pointed generating invariant cone in a simple Lie algebra $\mathfrak{g}$. Then C is global if and only if $\operatorname{int} C \cap (\mathfrak{n} + \mathfrak{k}') = \{0\}$, where $\mathfrak{n} + \mathfrak{a} + \mathfrak{k} = \mathfrak{g}$ is an Iwasawa decomposition of $\mathfrak{g}$.*   ∎

Notes

The classification of hyperplane subalgebras in Lie algebras goes back to Lie. A systematic treatment using the various radicals described in Section 6.1 was given by Hofmann in [Ho90a]. The material of the following sections up to Theorem 6.18 is due to Lawson (cf. [La87]). The applications to controllability questions were developed out of the group theoretical description of the Gödel model (cf. [Hi92]). Related results can be found in [Ne90a]. Corollary 6.24 was proved in [Ols82b].

7. Invariant Cones and Ol'shanskiĭ semigroups

Invariant cones in Lie algebras play a role in various fields of mathematics such as representation theory (cf. Chapter 9), symplectic geometry (cf. Chapter 8), and the theory of ordered manifolds (cf. Section 2.8). In Chapter 1 we have already seen how they arise as the tangent cones of invariant Lie semigroups.

In this chapter we study invariant cones and their applications systematically. In the first section we collect some facts about Lie algebras with compactly embedded Cartan algebras and their root decompositions. These results yield applications in Section 7.2 because every Lie algebra containing a pointed generating invariant cone has compactly embedded Cartan algebras. The root decomposition with respect to such a Cartan algebra is a tool which permits us to formulate and characterize the properties of those Lie algebras which contain pointed generating invariant cones. For further results on coadjoint orbits, their convexity properties and complex and Kähler structures on them we refer to [Ne93b] and [Ne93f].

In Section 7.3 we turn from infinitesimal objects to global objects. We prove Lawson's Theorem on Ol'shanskiĭ semigroups, a result which entails for instance for a pointed generating invariant cone W in the Lie algebra $\mathfrak{g}$ the existence of a Lie semigroup S in the simply connected complex Lie group $G_{\mathbb{C}}$ with $L(G_{\mathbb{C}}) = \mathfrak{g}_{\mathbb{C}}$ such that $L(S) = \mathfrak{g} + iW$ and $S = G' \exp(iW)$ is topologically a direct product decomposition, where $G' = \langle \exp_{G_{\mathbb{C}}} \mathfrak{g} \rangle$. Knowing from Section 3.4 that such a semigroup has a universal covering semigroup it follows easily that $\widetilde{S} = \widetilde{G}' \operatorname{Exp}(iW)$, where Exp has to be interpreted in an appropriate way. Moreover we obtain for every discrete central subgroup D of $\widetilde{G}$ a quotient semigroup $\widetilde{S}/D = (\widetilde{G}'/D) \operatorname{Exp}(iW)$. This means that every connected group G with $L(G) = \mathfrak{g}$ arises as the group of units of such a complex Ol'shanskiĭ semigroup. For applications of these semigroups we refer to Chapters 8 and 9. An intrinsic theory of holomorphic representations of Ol'shanskiĭ semigroups which are constructed from generating invariant wedges which are not necessarily pointed has been developed in [Ne93c,d,e].

7.1. Compactly embedded Cartan algebras

In this section $\mathfrak{g}$ always denotes a finite dimensional real Lie algebra. For a subalgebra $\mathfrak{a} \subseteq \mathfrak{g}$ we define

$$\operatorname{Inn}_{\mathfrak{g}}(\mathfrak{a}) := \langle e^{\operatorname{ad} \mathfrak{a}} \rangle \quad \text{and} \quad \operatorname{INN}_{\mathfrak{g}}(\mathfrak{a}) := \overline{\operatorname{Inn}_{\mathfrak{g}}(\mathfrak{a})}.$$

We usually omit the subscript if no confusion is possible. We call an element $X \in \mathfrak{g}$ *compactly embedded* if $\mathrm{INN}_\mathfrak{g}(\mathbb{R}X)$ is compact, and write

$$\mathrm{comp}(\mathfrak{g}) := \{X \in \mathfrak{g} : \mathrm{INN}_\mathfrak{g}(\mathbb{R}X) \text{ is compact}\}$$

for the set of compactly embedded elements of $\mathfrak{g}$. A subalgebra $\mathfrak{a} \subseteq \mathfrak{g}$ is said to be *compactly embedded* if $\mathfrak{a} \subseteq \mathrm{comp}(\mathfrak{g})$. We note that this is equivalent to the compactness of the group $\mathrm{INN}_\mathfrak{g}\,\mathfrak{a}$ ([HiHo89, 2.6]). As an easy argument, using the Jordan normal form shows, an element $X \in \mathfrak{g}$ is compactly embedded if and only if $\mathrm{ad}\,X$ is semisimple and $\mathrm{Spec}(\mathrm{ad}\,X) \subseteq i\mathbb{R}$.

Lemma 7.1.

  (i) *Let* $\mathfrak{a}, \mathfrak{b} \subseteq \mathfrak{g}$ *be compactly embedded subalgebras with* $[\mathfrak{a}, \mathfrak{b}] = \{0\}$. *Then* $\mathfrak{a} + \mathfrak{b}$ *is compactly embedded.*

  (ii) *Let* $\mathfrak{a} \subseteq \mathfrak{g}$ *be an abelian subalgebra with* $\mathfrak{a} \cap \mathrm{int}\,\mathrm{comp}(\mathfrak{g}) \neq \emptyset$. *Then* $\mathfrak{a}$ *is compactly embedded.*

Proof.   (i) Let $X \in \mathfrak{a}$ and $Y \in \mathfrak{b}$. Then $[X, Y] = 0$ and therefore $\mathrm{ad}(X + Y)$ is semisimple with $\mathrm{Spec}\big(\mathrm{ad}(X + Y)\big) \subseteq i\mathbb{R}$. It follows that $X + Y \in \mathrm{comp}(\mathfrak{g})$. Whence $\mathfrak{a} + \mathfrak{b} \subseteq \mathrm{comp}(\mathfrak{g})$ is compactly embedded.

(ii) Let $K := \mathrm{comp}(\mathfrak{g}) \cap \mathfrak{a}$. Then $K = -K$. If $X, Y \in K$, then $[X, Y] = 0$ shows that $X + Y \in K$. Moreover $\mathbb{R}^+ K = K$. Therefore K is a vector subspace of $\mathfrak{a}$. Thus $K = \mathfrak{a}$ because K has interior points in $\mathfrak{a}$.   ∎

From now on $\mathfrak{g}$ denotes a Lie algebra which contains a compactly embedded Cartan algebra and $\mathfrak{t}$ denotes a fixed compactly embedded Cartan algebra of $\mathfrak{g}$.

Lemma 7.2.   *Let* $\mathfrak{b}$ *be a compactly embedded subalgebra of* $\mathfrak{g}$. *Then every* $\mathfrak{b}$-*invariant subspace* $\mathfrak{a} \subseteq \mathfrak{g}$ *decomposes into*

$$\mathfrak{a} = Z_\mathfrak{a}(\mathfrak{b}) \oplus [\mathfrak{b}, \mathfrak{a}].$$

If, in particular, $\mathfrak{t} = \mathfrak{b}$ *is a Cartan algebra, then* $Z_\mathfrak{a}(\mathfrak{t}) = \mathfrak{a} \cap \mathfrak{t}$.

Proof.   Let $K := \mathrm{INN}_\mathfrak{g}\,\mathfrak{b}$. This is a compact group acting on $\mathfrak{g}$ and a subspace is invariant under $\mathfrak{b}$ if and only if it is invariant under K. Therefore Proposition 1.6 provides a direct decomposition $\mathfrak{a} = \mathfrak{a}_{\mathrm{fix}} + \mathfrak{a}_{\mathrm{eff}}$. For $X \in \mathfrak{a}_{\mathrm{fix}}$ and $E \in \mathfrak{b}$ we have that

$$[E, X] = \frac{d}{dt}\bigg|_{t=0} e^{t\,\mathrm{ad}\,E}X = \frac{d}{dt}\bigg|_{t=0} X = 0.$$

Therefore $X \in Z_\mathfrak{a}(\mathfrak{b})$. On the other hand the subspace $\mathfrak{a}' := [\mathfrak{b}, \mathfrak{a}]$ is invariant under $\mathfrak{b}$ (Jacobi identity) and

$$[E, X] = \lim_{t \to 0} \frac{1}{t}(e^{t\,\mathrm{ad}\,E}X - X) \in \mathfrak{a}_{\mathrm{eff}}$$

shows that $[\mathfrak{b}, \mathfrak{a}] \subseteq \mathfrak{a}_{\mathrm{eff}}$. To see that $\mathfrak{a}_{\mathrm{eff}} \subseteq [\mathfrak{b}, \mathfrak{a}]$, it suffices to show that $e^{\mathrm{ad}\,E}X - X \in [\mathfrak{b}, \mathfrak{a}]$ for all $E \in \mathfrak{b}, X \in \mathfrak{a}$. This follows immediately by expansion of the power series defining $e^{\mathrm{ad}\,E}$.

If $\mathfrak{b} = \mathfrak{t}$ is a Cartan algebra, then $\mathfrak{t} = Z_\mathfrak{g}(\mathfrak{t})$ shows that $Z_\mathfrak{a}(\mathfrak{t}) = \mathfrak{t} \cap \mathfrak{a}$.   ∎

For a t-invariant subspace $\mathfrak{a} \subseteq \mathfrak{g}$ we set

$$\mathfrak{a}_{\text{fix}} := \mathfrak{a} \cap t \quad \text{and} \quad \mathfrak{a}_{\text{eff}} := [t, \mathfrak{a}].$$

Proposition 7.3.   *The following assertions hold:*

(i) *Every compactly embedded Cartan algebra of $\mathfrak{g}$ is contained in a unique maximal compactly embedded subalgebra.*

(ii) *Compactly embedded Cartan algebras and maximal compactly embedded subalgebras are conjugate under $\text{Inn}\,\mathfrak{g}$.*

(iii) *For every maximal compactly embedded subalgebra $\mathfrak{k} \subseteq \mathfrak{g}$ there exists a Levi decomposition $\mathfrak{g} = \mathfrak{r} \rtimes \mathfrak{s}$ such that*

   (a) $[\mathfrak{k}, \mathfrak{s}] \subseteq \mathfrak{s}$.

   (b) $[\mathfrak{k} \cap \mathfrak{r}, \mathfrak{s}] = \{0\}$.

   (c) $\mathfrak{k} = \mathfrak{k} \cap \mathfrak{r} + \mathfrak{k} \cap \mathfrak{s}$.

   (d) $\mathfrak{k}' \subseteq \mathfrak{s}$.

   (e) $\mathfrak{k} \cap \mathfrak{s}$ *is maximal compactly embedded in $\mathfrak{s}$.*

(iv) *Let $\mathfrak{n}$ denote the nilradical of $\mathfrak{g}$. Then*

   (a) $\mathfrak{n}_{\text{eff}} = \mathfrak{r}_{\text{eff}}$.

   (b) $\mathfrak{n} \cap t = Z(\mathfrak{g})$, *in particular $\mathfrak{n} = \mathfrak{r}_{\text{eff}} + Z(\mathfrak{g})$.*

   (c) *If t is a Cartan algebra of $\mathfrak{g}$, then $[\mathfrak{r}, \mathfrak{s}] \subseteq \mathfrak{r}_{\text{eff}}$.*

(v) *If $\mathfrak{g}$ contains a compactly embedded Cartan algebra t, then $\text{tr}\,\text{ad}\,X = 0$ for all $X \in \mathfrak{g}$. Every connected Lie group G with $\mathbf{L}(G) = \mathfrak{g}$ is unimodular.*

Proof.   (i) [HiHo89, 3.13].

(ii) [HiHo89, 2.6, 3.5].

(iii) [HiNe91, III.7.15].

(iv) First we note that $\mathfrak{r}_{\text{eff}} = [t, \mathfrak{r}] \subseteq \mathfrak{n}$ shows that $\mathfrak{n}_{\text{eff}} = \mathfrak{r}_{\text{eff}}$. On the other hand $\mathfrak{n} \cap t$ is compactly embedded and nilpotent, therefore central. Thus $Z(\mathfrak{g}) = \mathfrak{n} \cap t$. For the last claim we note that $t + \mathfrak{s}$ is a reductive algebra and $\mathfrak{r}$ a semisimple $(t + \mathfrak{s})$-module. Then $\mathfrak{r}_{\text{eff}}$ obviously is contained in the effective part of $\mathfrak{r}$ as a $(t + \mathfrak{s})$-module. On the other hand $\mathfrak{r}_{\text{fix}} = t \cap \mathfrak{r} \subseteq Z_{\mathfrak{g}}(t + \mathfrak{s})$ since t is a Cartan algebra. Thus the decomposition of $\mathfrak{r}$ in effective and fixed part agrees for the two module structures. In particular, we have $[t + \mathfrak{s}, \mathfrak{r}] \subseteq \mathfrak{r}_{\text{eff}}$.

(v) First we note that $\text{tr}\,\text{ad}\,X = 0$ holds for every compactly embedded element. Since $e^{\text{ad}\,\mathfrak{g}}t$ contains an open subset of $\mathfrak{g}$, the analytic function $X \mapsto \text{tr}\,\text{ad}\,X$ vanishes on $\mathfrak{g}$ because it vanishes on an open subset. Hence $\det(e^{\text{ad}\,X}) = 1$ holds for every $X \in \mathfrak{g}$ so that the modular function $g \mapsto \det\,\text{Ad}(g)$ is constant on G, i.e., G is unimodular.   ∎

Theorem 7.4.   *Let $t \subseteq \mathfrak{g}$ be a compactly embedded Cartan subalgebra, $\mathfrak{g}_{\mathbb{C}}$ the complexification of $\mathfrak{g}$,*

$$\sigma : Z = X + iY \mapsto \overline{Z} := X - iY$$

the corresponding conjugation, and $t_{\mathbb{C}}$ the corresponding Cartan subalgebra of $\mathfrak{g}_{\mathbb{C}}$. For a linear functional $\lambda \in t_{\mathbb{C}}^*$ we set

$$\mathfrak{g}_{\mathbb{C}}^{\lambda} := \{X \in \mathfrak{g}_{\mathbb{C}} : (\forall Y \in t_{\mathbb{C}})[Y, X] = \lambda(Y)X\},$$

$$\Lambda := \Lambda(\mathfrak{g}_{\mathbb{C}}, \mathfrak{t}_{\mathbb{C}}) := \{\lambda \in \mathfrak{t}_{\mathbb{C}}^* \setminus \{0\} : \mathfrak{g}_{\mathbb{C}}^\lambda \neq \{0\}\},$$

and

$$\mathfrak{g}^{[\lambda]} := (\mathfrak{g}_{\mathbb{C}}^\lambda \oplus \mathfrak{g}_{\mathbb{C}}^{-\lambda}) \cap \mathfrak{g},$$

where $[\lambda] := \{\lambda, -\lambda\}$. Then the following assertions hold:

(i) $\mathfrak{g}_{\mathbb{C}} = \mathfrak{t}_{\mathbb{C}} \oplus \bigoplus_{\lambda \in \Lambda} \mathfrak{g}_{\mathbb{C}}^\lambda$.

(ii) $\lambda(\mathfrak{t}) \subseteq i\mathbb{R}$ for all $\lambda \in \Lambda$.

(iii) $\sigma(\mathfrak{g}_{\mathbb{C}}^\lambda) = \mathfrak{g}_{\mathbb{C}}^{-\lambda}$ and $(\mathfrak{g}^{[\lambda]})_{\mathbb{C}} = \mathfrak{g}_{\mathbb{C}}^\lambda + \mathfrak{g}_{\mathbb{C}}^{-\lambda}$.

(iv) Let $X = \operatorname{Re} Z \in \mathfrak{g}^{[\lambda]}$, $\lambda \neq 0$, $Z \in \mathfrak{g}_{\mathbb{C}}^\lambda$, and write $\langle X \rangle$ for the $\mathfrak{t}$-invariant sub-algebra of $\mathfrak{g}$ generated by X. Then $[Z, \overline{Z}] \in i\mathfrak{t}$ and there are four possibilities:

   (a) $\lambda([Z, \overline{Z}]) > 0$. Then $\langle X \rangle \cong \mathfrak{sl}(2, \mathbb{R})$.

   (b) $\lambda([Z, \overline{Z}]) < 0$. Then $\langle X \rangle \cong \mathfrak{so}(3, \mathbb{R}) \cong \mathfrak{su}(2)$.

   (c) $\lambda([Z, \overline{Z}]) = 0$ and $[Z, \overline{Z}] \neq 0$. Then $\langle X \rangle$ is isomorphic to the three-dimensional Heisenberg algebra, and for every $E \in \mathfrak{t}$ with $\lambda(E) \neq 0$ the algebra $\mathbb{R}E \oplus \langle X \rangle$ is isomorphic to the oscillator algebra.

   (d) $[Z, \overline{Z}] = 0$. Then $\langle X \rangle \cong \mathbb{R}^2$ and for every $E \in \mathfrak{t}$ with $\lambda(E) \neq \{0\}$ the algebra $\mathbb{R}E + \langle X \rangle$ is isomorphic to $\operatorname{mot}(2) \cong \mathbb{C} \rtimes \mathbb{R}$.

Proof.   (i) The algebra $\operatorname{ad} \mathfrak{t}_{\mathbb{C}} \subseteq \mathfrak{gl}(\mathfrak{g}_{\mathbb{C}})$ is abelian and semisimple. Therefore this set permits a simultaneous diagonalization, i.e., a decomposition $\mathfrak{g}_{\mathbb{C}} = V_1 \oplus \ldots \oplus V_N$ into one-dimensional $\mathfrak{t}_{\mathbb{C}}$-invariant subspaces. Now each subspace $\mathfrak{g}_{\mathbb{C}}^\lambda$ is a sum of a certain subset of the spaces V_i and the assertion follows.

(ii) This follows from the fact that $\mathfrak{t}$ is compactly embedded.

(iii) For $E \in \mathfrak{t}_{\mathbb{C}}$ and $Z \in \mathfrak{g}_{\mathbb{C}}^\lambda$ we have that

$$[E, \sigma(Z)] = \sigma([\sigma(E), Z]) = \sigma\Big(\lambda(\sigma(E))Z\Big) = \overline{\lambda(\sigma(E))}\sigma(Z).$$

For $E = E_1 + iE_2$ we have that $\sigma(E) = E_1 - iE_2$, so

$$\overline{\lambda(\sigma(E))} = \overline{\lambda(E_1)} + i\overline{\lambda(E_2)} = -\lambda(E_1) - i\lambda(E_2) = -\lambda(E)$$

implies that $\sigma(Z) \in \mathfrak{g}_{\mathbb{C}}^{-\lambda}$.

This shows in particular that $\mathfrak{g}_{\mathbb{C}}^\lambda + \mathfrak{g}_{\mathbb{C}}^{-\lambda}$ is a σ-invariant complex subspace of $\mathfrak{g}_{\mathbb{C}}$. Whence

$$\mathfrak{g}_{\mathbb{C}}^\lambda + \mathfrak{g}_{\mathbb{C}}^{-\lambda} = (\mathfrak{g}_{\mathbb{C}}^\lambda + \mathfrak{g}_{\mathbb{C}}^{-\lambda}) \cap \mathfrak{g} + i((\mathfrak{g}_{\mathbb{C}}^\lambda + \mathfrak{g}_{\mathbb{C}}^{-\lambda}) \cap \mathfrak{g}) = (\mathfrak{g}^{[\lambda]})_{\mathbb{C}}.$$

(iv) Let $Z' := [Z, \sigma(Z)]$. Then, in view of (ii),

$$Z' \in [\mathfrak{g}_{\mathbb{C}}^\lambda, \mathfrak{g}_{\mathbb{C}}^{-\lambda}] \subseteq \mathfrak{g}_{\mathbb{C}}^0 = \mathfrak{t}_{\mathbb{C}}$$

and

$$\sigma(Z') = \sigma([Z, \sigma(Z)]) = [\sigma(Z), Z] = -Z'.$$

It follows that $Z' \in i\mathfrak{t}$.

Let $U := \frac{i}{2}[Z, \overline{Z}]$. Rescaling Z if necessary, we may assume that $i\lambda(U) \in \{-1, 0, 1\}$. Then

$$U = \frac{i}{2}[X + iY, X - iY] = [X, Y]$$

and

$$[U, X + iY] = [U, Z] = \lambda(U)Z = i\lambda(U)Y + \lambda(U)X.$$

Thus

$$[U, X] = i\lambda(U)Y \quad \text{and} \quad [U, Y] = -i\lambda(U)X.$$

Therefore span$\{X, Y, U\}$ is a t-invariant subalgebra and hence

$$\langle X \rangle = \text{span}\{X, Y, U\}.$$

(a) If $\lambda([Z, \overline{Z}]) > 0$, then

$$[U, X] = -Y, \quad [U, Y] = X, \quad \text{and} \quad [X, Y] = U$$

implies that $\langle X \rangle \cong \mathfrak{sl}(2, \mathbb{R})$.
(b) If $\lambda([Z, \overline{Z}]) < 0$, then

$$[U, X] = Y, \quad [Y, U] = X, \quad \text{and} \quad [X, Y] = U$$

implies that $\langle X \rangle \cong \mathfrak{so}(3, \mathbb{R})$.
(c) If $\lambda([Z, \overline{Z}]) = 0$ and $[Z, \overline{Z}] \neq 0$, then

$$[U, X] = 0, \quad [Y, U] = 0, \quad \text{and} \quad [X, Y] = U$$

implies that $\langle X \rangle \cong \mathfrak{h}_1$, the three dimensional Heisenberg algebra. For $E \in \mathfrak{t}$ with $\lambda(E) = i$ we obtain

$$[E, X] = -Y, \quad [E, Y] = X, \quad \text{and} \quad [E, U] = 0.$$

Whence $\langle X \rangle + \mathbb{R}E$ is isomorphic to the Oscillator algebra.
(d) If $[Z, \overline{Z}] = 0$, then $\langle X \rangle \cong \mathbb{R}^2$, and for $E \in \mathfrak{t}$ with $\lambda(E) = i$ we obtain

$$[E, X] = -Y, \quad \text{and} \ [E, Y] = X.$$

Whence $\langle X \rangle + \mathbb{R}E$ is isomorphic to $\text{mot}(2) \cong \mathbb{C} \rtimes \mathbb{R}$, the motion algebra of the two dimensional space. $\blacksquare$

Corollary 7.5. *Let $\mathfrak{t}$ be a maximal compactly embedded subalgebra, $\lambda \in \Lambda$, and $X \in \mathfrak{g}^{[\lambda]}$. Then $\langle X \rangle \cong \mathfrak{so}(3)$ if and only if $X \in \mathfrak{t}$.*

Proof. Let $X \in \mathfrak{t}$. Then $\langle X \rangle + \mathbb{R}E$ is a compact Lie subalgebra of $\mathfrak{t}$ for every $E \in \mathfrak{t}$. Therefore the cases (a), (c) and (d) of Theorem 7.4 are ruled out.

If, conversely, $\langle X \rangle \cong \mathfrak{so}(3)$, then $\mathfrak{h} := \mathfrak{t} + \langle X \rangle$ is a compactly embedded subalgebra of $\mathfrak{g}$ because $\mathfrak{h} \cong \langle X \rangle \oplus \ker \lambda \mid_{\mathfrak{t}}$ (Lemma 7.1). Since $\mathfrak{t}$ is the unique maximal compactly embedded subalgebra containing $\mathfrak{t}$, we conclude that $\langle X \rangle \subseteq \mathfrak{h} \subseteq \mathfrak{t}$. $\blacksquare$

The set Λ is said to be the set of *roots* of $\mathfrak{g}_{\mathbb{C}}$ with respect to $\mathfrak{t}_{\mathbb{C}}$. We say that $\lambda \in \Lambda$ is a *compact root* if $\mathfrak{g}^{[\lambda]} \subseteq \mathfrak{k}$ and that λ is *non-compact* otherwise. We write Λ_k resp. Λ_p for the set of compact and non-compact roots respectively.

A subset $\Lambda^+ \subseteq \Lambda$ is called a *positive system* if there exists $X_0 \in i\mathfrak{t}$ such that

$$\Lambda^+ = \{\lambda \in \Lambda : \lambda(X_0) > 0\}.$$

A positive system is said to be $\mathfrak{k}$-*adapted* if

$$\lambda(X_0) > \mu(X_0) \qquad \forall \mu \in \Lambda_k, \lambda \in \Lambda_p^+.$$

Lemma 7.6. *Let* $\mathfrak{p} := \sum_{\lambda \in \Lambda_p} \mathfrak{g}^{[\lambda]}$. *Then*

$$[\mathfrak{k}, \mathfrak{p}] \subseteq \mathfrak{p} \quad and \quad \mathfrak{k} = \mathfrak{t} \oplus \bigoplus_{\lambda \in \Lambda_k} \mathfrak{g}^{[\lambda]}.$$

Proof. The subspace $\mathfrak{k} \subseteq \mathfrak{g}$ is invariant under $\mathfrak{t}$. Therefore

$$(7.1) \qquad\qquad \mathfrak{k} \subseteq \mathfrak{t} \oplus \bigoplus_{\lambda \in \Lambda_k} \mathfrak{g}^{[\lambda]}.$$

Let $K := \mathrm{INN}_{\mathfrak{g}}\, \mathfrak{k}$ and $\mathfrak{p} \subseteq \mathfrak{g}$ a K-invariant vector space complement to $\mathfrak{k}$ (Proposition 1.6). Then $[\mathfrak{k}, \mathfrak{p}] \subseteq \mathfrak{p}$ follows by differentiation. We have in particular that $[\mathfrak{t}, \mathfrak{p}] \subseteq \mathfrak{p}$ and therefore that $[\mathfrak{t}_{\mathbb{C}}, \mathfrak{p}_{\mathbb{C}}] \subseteq \mathfrak{p}_{\mathbb{C}}$. This shows that

$$\mathfrak{p}_{\mathbb{C}} = \sum_{\lambda \in \Lambda} \mathfrak{g}_{\mathbb{C}}^{\lambda} \cap \mathfrak{p}_{\mathbb{C}}.$$

For $\lambda \in \Lambda_k$ the inclusion $\mathfrak{g}_{\mathbb{C}}^{\lambda} \subseteq \mathfrak{k}_{\mathbb{C}}$ implies that $\mathfrak{g}_{\mathbb{C}}^{\lambda} \cap \mathfrak{p}_{\mathbb{C}} = \{0\}$. Thus the equality

$$\mathfrak{p}_{\mathbb{C}} = \sum_{\lambda \in \Lambda_p} \mathfrak{g}_{\mathbb{C}}^{\lambda}$$

follows from (7.1) and $\dim \mathfrak{p} = \dim \mathfrak{g} - \dim \mathfrak{k}$. ∎

Lemma 7.7. *Let* $\Lambda^+ \subseteq \Lambda$ *be a* $\mathfrak{k}$-*adapted positive system,*

$$\mathfrak{b} := \sum_{\lambda \in \Lambda^+} \mathfrak{g}_{\mathbb{C}}^{\lambda} \quad and \quad \mathfrak{p}_{\mathbb{C}}^{\pm} := \sum_{\lambda \in \Lambda_p^{\pm}} \mathfrak{g}_{\mathbb{C}}^{\lambda}.$$

Then the following assertions hold:

(i) *$\mathfrak{b}$ and $\mathfrak{p}_{\mathbb{C}}^{\pm}$ are subalgebras of $\mathfrak{g}_{\mathbb{C}}$ which are normalized by $\mathfrak{t}_{\mathbb{C}}$ and $\mathfrak{p}_{\mathbb{C}} = \mathfrak{p}_{\mathbb{C}}^+ \oplus \mathfrak{p}_{\mathbb{C}}^-$.* *If $\mathfrak{g}$ is semisimple, then $\mathfrak{p}_{\mathbb{C}}^{\pm}$ is abelian.*

(ii) *The mapping*

$$\mathrm{Re} : \overline{\mathfrak{b}} \to \mathfrak{g}_{\mathrm{eff}}, \quad Z \mapsto \frac{1}{2}(Z + \overline{Z})$$

is bijective and the prescription

$$I \operatorname{Re}(Z) := \operatorname{Re}(iZ) = \frac{1}{2}(iZ - i\overline{Z})$$

defines a t-invariant complex structure on $\mathfrak{g}_{\mathrm{eff}}$.

(iii) *For* $Z = X + iY \in \mathfrak{g}_{\mathbb{C}}^{\lambda}$, $\lambda \in \Lambda^+$ *we have that* $Y = IX$ *and* $i[Z, \overline{Z}] = 2[X, IX]$.

(iv) *Let* $\lambda \in \Lambda^+$. *Then*

$$\mathfrak{g}^{[\lambda]} = \{X \in \mathfrak{g} : (\forall E \in \mathfrak{t}) \ [E, X] = i\lambda(E)IX\}.$$

Proof. (i) This follows from $[\mathfrak{g}_{\mathbb{C}}^{\alpha}, \mathfrak{g}_{\mathbb{C}}^{\beta}] \subseteq \mathfrak{g}_{\mathbb{C}}^{\alpha+\beta}$, $(\Lambda^+ + \Lambda^+) \cap \Lambda \subseteq \Lambda^+$ and Lemma 7.6. If, in addition, $\mathfrak{g}$ is semisimple, then $\mathfrak{g} = \mathfrak{k} + \mathfrak{p}$ is a Cartan decomposition, so that $[\mathfrak{p}, \mathfrak{p}] \subseteq \mathfrak{k}$. Hence

$$[\mathfrak{p}_{\mathbb{C}}^+, \mathfrak{p}_{\mathbb{C}}^+] \subseteq \mathfrak{k}_{\mathbb{C}} \cap \mathfrak{p}_{\mathbb{C}}^+ = \{0\}$$

shows that $\mathfrak{p}_{\mathbb{C}}^+$ is abelian.

(ii) This is a consequence of

$$(\mathfrak{g}_{\mathbb{C}})_{\mathrm{eff}} = \mathfrak{g}_{\mathrm{eff}} + i\mathfrak{g}_{\mathrm{eff}} = \mathfrak{b} + \overline{\mathfrak{b}}$$

(Theorem 7.4), $\mathfrak{b} \cap \overline{\mathfrak{b}} = \{0\}$, and the fact that $\mathfrak{t}$ acts on $\overline{\mathfrak{b}}$ by complex linear mappings.

(iii) First we note that $\overline{Z} = X - iY \in \overline{\mathfrak{b}}$ and

$$\frac{1}{2}(\overline{Z} + \overline{\overline{Z}}) = \frac{1}{2}(Z + \overline{Z}) = X.$$

Thus

$$IX = \frac{1}{2}(i\overline{Z} - iZ) = \frac{1}{2}(iX + Y - iX + Y) = Y$$

and

$$i[Z, \overline{Z}] = 2[X, Y] = 2[X, IX].$$

(iv) Let $X \in \mathfrak{g}^{[\lambda]}$ and $\lambda \in \Lambda^+$. Then, in view of (iii), $Z = X + iIX \in \mathfrak{g}_{\mathbb{C}}^{\lambda}$ and therefore

$$[E, X] = \operatorname{Re}[E, Z] = \operatorname{Re}\left(\lambda(E)Z\right) = i\lambda(E)\operatorname{Im} Z = i\lambda(E)IX.$$

If, conversely, $[E, X] = i\lambda(E)IX$ holds for all $E \in \mathfrak{t}$, then a direct calculation shows that $X + iIX \in \mathfrak{g}_{\mathbb{C}}^{\lambda}$. ∎

Lemma 7.8. *Let* $\lambda \in \Lambda_p^+$, $X \in \mathfrak{g}^{[\lambda]}$ *and* $\alpha := i\lambda([IX, X]) \geq 0$. *Then we have the following formulas:*

(i)

$$(\operatorname{ad} X)^m(E) = \begin{cases} E & \text{for } m = 0 \\ -i\lambda(E)\alpha^n IX & \text{for } m = 2n+1 \\ i\lambda(E)\alpha^n[IX, X] & \text{for } m = 2(n+1). \end{cases}$$

(ii) *If* $p: \mathfrak{g} \to \mathfrak{t} = \mathfrak{g}_{\mathrm{fix}}$ *is the* $\mathfrak{t}$-*equivariant projection, then*

$$p(e^{\operatorname{ad} X} E) = \cosh(\operatorname{ad} X)E = E + i\lambda(E)\frac{\cosh(\sqrt{\alpha}) - 1}{\alpha}[IX, X]$$

and

$$p(e^{\mathbb{R}\operatorname{ad} X} E) = p(e^{\mathbb{R}^+ \operatorname{ad} X} E) = E + i\lambda(E)\mathbb{R}^+[IX, X].$$

Proof.   (i) We prove the assertion by induction with respect to m. For $m = 0$ there is nothing to prove. Suppose that the assertion is true for $m = 2n$. If $m = 0$, then

$$(\operatorname{ad} X)^{m+1} E = [X, E] = -i\lambda(E)IX$$

(Lemma 7.7). If $m = 2n$, then

$$\begin{aligned}
(\operatorname{ad} X)^{m+1} &= \operatorname{ad} X\big(i\lambda(E)\alpha^{n-1}[IX, X]\big) \\
&= i\lambda(E)\alpha^{n-1} \operatorname{ad} X([IX, X]) \\
&= -i\lambda(E)\alpha^n IX,
\end{aligned}$$

and if $m = 2n + 1$, then

$$\begin{aligned}
(\operatorname{ad} X)^{m+1} &= -\operatorname{ad} X\big(i\lambda(E)\alpha^n IX\big) \\
&= -i\lambda(E)\alpha^n[X, IX] \\
&= i\lambda(E)\alpha^n[IX, X].
\end{aligned}$$

(ii) Using (i), we see that $(\operatorname{ad} X)^{2n+1} E \in \mathfrak{g}_{\mathrm{eff}}$. Thus

$$p(e^{\operatorname{ad} X} E) = \cosh(\operatorname{ad} X)E.$$

Now the formula for $\cosh(\operatorname{ad} X)E$ follows from (i). For the remainder of the assertions we distinguish two cases.

If $\alpha = 0$, then $p(e^{t \operatorname{ad} X} E) = E + i\lambda(E)\frac{t^2}{2}[IX, X]$ and if $\alpha \neq 0$, then

$$p(e^{t \operatorname{ad} X} E) = E + i\lambda(E)\frac{\cosh(t\sqrt{\alpha}) - 1}{\alpha}[IX, X].$$

In the first case we use the surjectivity of the square function $\mathbb{R}^+ \to \mathbb{R}^+$, and in the second case we use the surjectivity of the function

$$\mathbb{R}^+ \to \mathbb{R}^+, \quad t \mapsto \cosh(t\sqrt{\alpha}) - 1.$$

∎

7.2. Invariant cones in Lie algebras

In the preceding section we have considered root decompositions of Lie algebras containing compactly embedded Cartan algebras. Since every Lie algebra which admits pointed generating invariant cones is of this type (Proposition 7.10), the root decomposition is a convenient tool to study these algebras. A typical example of a Lie algebra which possesses a compactly embedded Cartan algebra but no pointed generating invariant cone is the Lie algebra of the motion group of the euclidean plane. This example leads immediately to the notion of cone potential, a property shared by all Lie algebras with invariant cones, but this property is not sufficient to guarantee the existence of invariant cones. A further condition, called strong cone potential, on the system of roots is needed. The main result of this section is the characterization of those finite dimensional real Lie algebras which contain pointed generating invariant cones.

Lemma 7.9.   *Let L be a finite dimensional real vector space and $W \subseteq L$ a pointed generating cone. Then the following assertions hold:*

  (i) *For every $w \in \operatorname{int} W$ the group $K := \{g \in \operatorname{Aut}(W): g.w = w\}$ is compact.*

  (ii) *If $K \subseteq \operatorname{Aut}(W)$ is a compact subgroup, then there exists $w \in \operatorname{int} W$ such that $g.w = w$ for all $k \in K$.*

Proof.   (i) This is a reformulation of Lemma 1.11(iii).

(ii) This follows from Proposition 1.6.   ∎

Proposition 7.10.   *Let $W \subseteq \mathfrak{g}$ be a pointed generating invariant wedge. Then the following assertions hold:*

  (i) $\operatorname{int} W \subseteq \operatorname{comp}(\mathfrak{g})$.

  (ii) $\mathfrak{g}$ *contains a compactly embedded Cartan algebra $\mathfrak{t}$.*

  (iii) *If $\mathfrak{k}$ is a maximal compactly embedded subalgebra, then*

$$\operatorname{int} W \cap Z(\mathfrak{k}) \neq \emptyset.$$

Proof.   (i) Let $X \in \operatorname{int} W$. Then $e^{\mathbb{R} \operatorname{ad} X} \subseteq \operatorname{Aut}(W)^X$ is bounded (Lemma 7.9) and therefore $X \in \operatorname{comp} \mathfrak{g}$.

(ii) The open subset W of $\mathfrak{g}$ contains a regular element X. Then, in view of (i), the endomorphism $\operatorname{ad} X$ is semisimple. Therefore $\mathfrak{t} := \ker \operatorname{ad} X$ is a Cartan algebra of $\mathfrak{g}$. Let $Y \in \mathfrak{t}$. Then $\mathfrak{a} := \mathbb{R} X + \mathbb{R} Y$ is commutative and $\mathfrak{a} \cap \operatorname{int} W$ is an open subset of $\mathfrak{a}$ which consists of compactly embedded elements. Thus $\mathfrak{a}$ is compactly embedded by Lemma 7.1.

(iii) Let $\mathfrak{k} \subseteq \mathfrak{g}$ be maximal compactly embedded. Then the group $\operatorname{INN}_{\mathfrak{g}} \mathfrak{k}$ is compact, so there exists $X \in \operatorname{int} W$ with $\gamma(X) = X$ for all $\gamma \in \operatorname{INN}_{\mathfrak{g}} \mathfrak{k}$ (Lemma 7.9). It follows in particular that $[X, \mathfrak{k}] = \{0\}$. Therefore $\mathbb{R} X + \mathfrak{k}$ is a compactly embedded subalgebra (Lemma 7.1). Since $\mathfrak{k}$ is maximal with this property, we conclude that $X \in Z(\mathfrak{k})$.   ∎

Corollary 7.11.   *Let $\mathfrak{g}$ be a Lie algebra which contains a pointed generating invariant cone and $\mathfrak{k}$ a maximal compactly embedded subalgebra. Then*

$$\operatorname{rank} \mathfrak{g} = \operatorname{rank} \mathfrak{k}, \quad Z(\mathfrak{k}) \neq \{0\}, \quad and \quad Z_{\mathfrak{g}}\big(Z(\mathfrak{k})\big) = \mathfrak{k}.$$

Proof.   The first condition means that $\mathfrak{g}$ contains a compactly embedded Cartan algebra. Thus the first two properties follow from Proposition 7.10. Let $W \subseteq \mathfrak{g}$ be a pointed generating invariant cone and $X \in \operatorname{int} W \cap Z(\mathfrak{k})$ (Proposition 7.10), then $\ker \operatorname{ad} X$ is compactly embedded ([HiHo89, 2.9]) and therefore $\mathfrak{k} = \ker \operatorname{ad} X$. This implies in particular that

$$\mathfrak{k} \subseteq Z_\mathfrak{g}\big(Z(\mathfrak{k})\big) \subseteq \ker \operatorname{ad} X = \mathfrak{k}.$$

∎

Let $\mathfrak{g} := \operatorname{mot}(2) := \mathbb{C} \rtimes \mathbb{R}$ with $t.z := itz$ for $t \in \mathbb{R}$, $z \in \mathbb{C}$. This is the motion algebra of the euclidean plane. Then

$$\mathfrak{t} = \mathfrak{k} = Z(\mathfrak{k}) = \{0\} \oplus \mathbb{R}$$

is a compactly embedded Cartan algebra which is maximal compactly embedded and which satisfies $Z_\mathfrak{g}\big(Z(\mathfrak{k})\big) = \mathfrak{k}$. So $\mathfrak{g}$ satisfies the conditions from Corollary 7.11. Suppose that $\mathfrak{g}$ contains a pointed invariant cone W. Then $W^* \subseteq \mathfrak{g}^*$ is invariant under the coadjoint action and $\operatorname{int} W^* \cap \mathbb{C}^\perp \neq \emptyset$ (Proposition 1.6). The orbits of the coadjoint action are cylinders around the axis $\mathbb{C}^\perp$ in $\mathfrak{g}^*$. Thus $W^* = \mathfrak{g}^*$, a contradiction. We conclude that the Lie algebra does not contain any pointed invariant cone.

In view of the preceding example we have to look for an additional condition on a Lie algebra $\mathfrak{g}$ which guarantees that $\mathfrak{g}$ contains a pointed generating invariant cone. To find this condition, we have to use the root decomposition of the Lie algebras $\mathfrak{g}$ and $\mathfrak{g}_\mathbb{C}$ described in Section 7.1. If the Lie algebra $\mathfrak{g}$ contains a pointed generating invariant cone, this has an immediate consequence for the possible types of the subalgebras $\langle X \rangle$.

Proposition 7.12. *If $\mathfrak{g}$ contains a pointed generating invariant cone and $0 \neq X \in \mathfrak{g}^{[\lambda]}$, then $\langle X \rangle$ cannot be abelian.*

Proof. Suppose that $\langle X \rangle \cong \mathbb{R}^2$ and choose $E \in \operatorname{int} W \cap \mathfrak{t}$ such that $\lambda(E) \neq 0$. Then $\mathfrak{a} := \langle X \rangle + \mathbb{R}E$ is isomorphic to $\operatorname{mot}(2)$ and $W \cap \mathfrak{a}$ is a pointed generating invariant cone because $\operatorname{int} W \cap \mathfrak{a} \neq \emptyset$. This contradicts the observation in the example above. ∎

We say that a Lie algebra $\mathfrak{g}$ which contains a compactly embedded Cartan algebra $\mathfrak{t}$ has *cone potential* if the conclusion of Proposition 7.12 holds, i.e., if $\langle X \rangle$ is never abelian. This terminology is justified by the fact that cone potential is a necessary property for Lie algebras containing pointed generating invariant cones.

If $\mathfrak{g} = \mathfrak{r} \rtimes \mathfrak{s}$ is a $\mathfrak{t}$-invariant Levi decomposition (Proposition 7.3), and $X = X_r + X_s \in \mathfrak{g}^{[\lambda]}$ with $X_s \neq 0$, then $[IX, X] \in \mathfrak{r} + [IX_s, X_s]$, $X_s \in \mathfrak{g}^{[\lambda]}$, and $[IX_s, X_s] \neq 0$ follows from $\langle X_s \rangle_\mathbb{C} \cong \mathfrak{sl}(2,\mathbb{C})$. Therefore the condition $[IX, X] \neq 0$ is only essential for elements in the radical.

Lemma 7.13. *Suppose that $\mathfrak{n}$ is a nilpotent Lie algebra of class $m \geq 2$ and*

$$\mathfrak{n}^{[0]} = \mathfrak{n} \supseteq \mathfrak{n}^{[1]} = [\mathfrak{n}, \mathfrak{n}] \supseteq \mathfrak{n}^{[2]} = [\mathfrak{n}, \mathfrak{n}^{[1]}] \supseteq \ldots \supseteq \mathfrak{n}^{[m]} \supseteq \{0\}$$

is the descending central series of $\mathfrak{n}$. Then $\mathfrak{n}^{[m-1]}$ is abelian.

Proof.   Since $m \geq 2$, we have

$$
\begin{aligned}
[n^{[m-1]}, n^{[m-1]}] &\subseteq [n^{[1]}, n^{[m-1]}] \\
&= [[n, n], n^{[m-1]}] \\
&\subseteq [[n, n^{[m-1]}], n] + [n, [n, n^{[m-1]}]] \subseteq n^{[m+1]} = \{0\}.
\end{aligned}
$$

∎

Lemma 7.14. *Let* $\mathfrak{g}$ *be a Lie algebra with cone potential and* $\mathfrak{a} \subseteq \mathfrak{g}$ *an ideal. Then the following assertions hold:*

(i) *If* $\mathfrak{a}$ *is abelian, then* $\mathfrak{a} \subseteq Z(\mathfrak{g})$.

(ii) *If* $\mathfrak{a} \neq \{0\}$, *then* $\mathfrak{a} \cap \mathfrak{t} \neq \emptyset$.

Proof.   As an ideal, $\mathfrak{a}$ is $\mathfrak{t}$-invariant, so $\mathfrak{a} = \mathfrak{a}_{\text{eff}} + \mathfrak{a} \cap \mathfrak{t}$. Suppose that $\mathfrak{a}_{\text{eff}} \neq \{0\}$. Then there exist $0 \neq \lambda \in \Lambda$ and $0 \neq X \in \mathfrak{g}^{[\lambda]} \cap \mathfrak{a}$. Note that $\langle X \rangle \subseteq \mathfrak{a}$ is a non-abelian subalgebra which intersects $\mathfrak{t}$ non-trivially ($\mathfrak{g}$ has cone potential).

(i) If $\mathfrak{a}$ is abelian, we conclude that $\mathfrak{a} \subseteq \mathfrak{t}$. Since $\mathfrak{a}$ is an ideal, this implies that $\mathfrak{a} \subseteq Z(\mathfrak{g})$.

(ii) If $\mathfrak{a} \neq \{0\}$, then $\mathfrak{a} \cap \mathfrak{t} = \{0\}$ would imply that $\mathfrak{a}_{\text{eff}} \neq \{0\}$ which leads to a contradiction since $\langle X \rangle \subseteq \mathfrak{a}$. ∎

Theorem 7.15.   (The Structure Theorem for Lie algebras with cone potential) *Let* $\mathfrak{g}$ *be a Lie algebra with cone potential,* $\mathfrak{t}$ *a compactly embedded Cartan algebra,* $\mathfrak{g} = \mathfrak{r} \rtimes \mathfrak{s}$ *a* $\mathfrak{t}$-*invariant Levi decomposition,* $\mathfrak{n}$ *the nilradical of* $\mathfrak{g}$, *and* $\mathfrak{a} \subseteq \mathfrak{t} \cap \mathfrak{r}$ *a vector space complement of* $Z(\mathfrak{g})$. *Then the following assertions hold:*

(i) $[\mathfrak{n}, \mathfrak{n}] \subseteq Z(\mathfrak{g})$.

(ii) $\mathfrak{g} \cong \mathfrak{n} \rtimes (\mathfrak{a} \oplus \mathfrak{s})$.

Proof.   (i) We consider the descending central series of $\mathfrak{n}$:

$$
n^{[0]} = n \supseteq n^{[1]} = [n, n] \supseteq \ldots \supseteq n^{[m]} \supseteq \{0\}.
$$

We claim that $m < 2$. Assume $m \geq 2$. Then $n^{[m-1]}$ is an abelian ideal of $\mathfrak{g}$ (Lemma 7.13) and therefore central (Lemma 7.14). This contradicts the fact that $n^{[m]} \neq \{0\}$. Thus $m < 2$, so $\mathfrak{n}' = [n, n]$ is abelian. Again we use Lemma 7.14 to see that $\mathfrak{n}'$ is central.

(ii) Since $[\mathfrak{a}, \mathfrak{s}] = \{0\}$, the subspace $\mathfrak{a} + \mathfrak{s}$ is a reductive subalgebra of $\mathfrak{g}$. Since $\mathfrak{n}$ is an ideal, $\mathfrak{n} \cap (\mathfrak{a} + \mathfrak{s}) = \{0\}$, and

$$
\mathfrak{n} + \mathfrak{a} + \mathfrak{s} = \mathfrak{r}_{\text{eff}} + Z(\mathfrak{g}) + \mathfrak{a} + \mathfrak{s} = \mathfrak{g}_{\text{eff}} + \mathfrak{t} = \mathfrak{g},
$$

we have that $\mathfrak{g} \cong \mathfrak{n} \rtimes (\mathfrak{a} + \mathfrak{s})$ (cf. Proposition 7.3). ∎

Proposition 7.16.   *Let* $\omega \in \mathfrak{g}^*$, $\mathcal{O}_\omega = \omega \circ \operatorname{Inn} \mathfrak{g}$ *the corresponding coadjoint orbit, and*

$$
\mathfrak{g}^\omega := \{X \in \mathfrak{g} : \omega \circ \operatorname{ad} X = 0\}
$$

the Lie algebra of the stabilizer of ω. *Then the following assertions hold:*

(i) $\omega \in \mathfrak{g}_{\mathrm{eff}}^{\perp} \cong \mathfrak{t}^*$ *if and only if* $\mathfrak{t} \subseteq \mathfrak{g}^{\omega}$.

(ii) *If* $\mathfrak{g}$ *has cone potential and* $\omega \in \mathfrak{t}^* \cap \mathfrak{s}^{\perp}$, *then* $\mathfrak{t} + \mathfrak{s} \subseteq \mathfrak{g}^{\omega}$ *and* $\mathcal{O}_{\omega} = \omega \circ e^{\mathrm{ad}\, \mathfrak{n}}$.

Proof.   (i) The condition $\mathfrak{t} \subseteq \mathfrak{g}^{\omega}$ is equivalent to $[\mathfrak{t}, \mathfrak{g}] = \mathfrak{g}_{\mathrm{eff}} \subseteq \ker \omega$.

(ii) Suppose that $\omega \in \mathfrak{s}^{\perp}$. Then $\mathrm{ad}^* X(\omega) = 0$ for all $X \in \mathfrak{s}$ so that $\mathfrak{s} \subseteq \mathfrak{g}^{\omega}$. If ω in addition is contained in $\mathfrak{t}^*$ then (i) implies $\mathfrak{t} \subseteq \mathfrak{g}^{\omega}$ so that $\mathfrak{t} + \mathfrak{s} \subseteq \mathfrak{g}^{\omega}$. Since $\mathfrak{n} = \mathfrak{r}_{\mathrm{eff}} + Z(\mathfrak{g})$, we conclude that $\mathrm{Inn}\, \mathfrak{g} = e^{\mathrm{ad}\, \mathfrak{n}} \langle \exp(\mathfrak{a} + \mathfrak{s}) \rangle$. Consequently

$$\mathcal{O}_{\omega} = \omega \circ e^{\mathrm{ad}\, \mathfrak{n}}.$$

■

The idea to obtain invariant cones in a Lie algebra $\mathfrak{g}$ which has cone potential is to use the dual cones to coadjoint orbits of the type $\mathcal{O}_{\omega}$, where $\omega \in \mathfrak{t}^* \cap \mathfrak{s}^{\perp}$. To guarantee that these cones are generating, i.e., that the coadjoint orbit $\mathcal{O}_{\omega}$ is contained in a pointed cone, we need some additional conditions on the Lie algebra $\mathfrak{g}$.

Let $\Lambda^+ \subseteq \Lambda$ be a positive system of roots. For a subset M of a vector space we write wedge(M) for the smallest wedge containing M. We define the *maximal cone* and the *minimal cone*

$$(7.2) \qquad\qquad C_{\max} := C_{\max}(\Lambda^+) := (i\Lambda_p^+)^* \subseteq \mathfrak{t},$$

$$C_{\min} := C_{\min}(\Lambda^+) := \mathrm{wedge}\{[IX, X] : X \in \mathfrak{g}^{[\lambda]}, \lambda \in \Lambda_p^+\}$$
$$= \mathrm{wedge}\{i[\overline{Z}, Z] : Z \in \mathfrak{g}_{\mathbb{C}}^{\lambda}, \lambda \in \Lambda_p^+\} \subseteq \mathfrak{t}.$$

Moreover, we set $\Lambda_{\mathfrak{n}} := \{\lambda \in \Lambda : \mathfrak{g}^{[\lambda]} \cap \mathfrak{n} \neq \{0\}\}$,

$$C_{\max,\mathfrak{n}} := (i\Lambda_{\mathfrak{n}}^+)^* \subseteq \mathfrak{t},$$

and

$$C_{\min,\mathfrak{n}} := \mathrm{wedge}\{[IX, X] : X \in \mathfrak{g}^{[\lambda]} \cap \mathfrak{n}, \lambda \in \Lambda_{\mathfrak{n}}^+\} \subseteq Z(\mathfrak{g}).$$

Lemma 7.17. *Let* $\mathfrak{t} \subseteq \mathfrak{g}$ *be a compactly embedded Cartan algebra,* $p : \mathfrak{g} \to \mathfrak{t} = \mathfrak{g}_{\mathrm{fix}}$ *the projection along* $\mathfrak{g}_{\mathrm{eff}}$, $W \subseteq \mathfrak{g}$ *an invariant wedge, and* $C := W \cap \mathfrak{t}$. *Then the following assertions hold:*

(i) $C = p(W)$.

(ii) $p(\mathrm{int}\, W) = \mathrm{algint}(W \cap \mathfrak{t})$.

(iii) $p(H(W)) = H(W \cap \mathfrak{t}) = H(W) \cap \mathfrak{t}$.

Proof.   This follows from Proposition 1.6 because the group $T := \mathrm{INN}_{\mathfrak{g}} \mathfrak{t}$ is compact and p is the T-equivariant projection onto the submodule $\mathfrak{g}_{\mathrm{fix}} = \mathfrak{t}$ of fixed points. ■

Proposition 7.18. *Suppose that the Lie algebra* $\mathfrak{g}$ *contains a pointed generating invariant cone* W. *Then there exists a* $\mathfrak{t}$-*adapted positive system* Λ^+ *such that*

$$C_{\min} \subseteq W \cap \mathfrak{t} \subseteq C_{\max}.$$

Proof.   We set $C := W \cap \mathfrak{t}$. Let $\lambda \in \Lambda_p$. If $X \in \mathfrak{t} \cap \operatorname{int} W$, then $X \in \operatorname{int} \operatorname{comp} \mathfrak{g}$ and therefore $\ker \operatorname{ad} X$ is compactly embedded (cf. [HiHo89, 2.9]). But $\mathfrak{t} \subseteq \ker \operatorname{ad} X$ so that Proposition 7.3(i) implies that $\ker \operatorname{ad} X \subseteq \mathfrak{k}$. In view of Lemma 7.6, this shows that $\lambda(X) \neq 0$. We conclude that

$$i\Lambda_p \subseteq C^* \cup -C^*.$$

Now we choose $X_1 \in \operatorname{algint} C \cap Z(\mathfrak{k})$, and then $X_0 \in \operatorname{int} C$ sufficiently near to X_1 such that $i\mu(X_0) \geq i\lambda(X_0) \neq 0$ holds for all compact roots λ and all non-compact roots μ which are non-negative on C.

We set

$$\Lambda^+ := \{\lambda \in \Lambda : i\lambda(X_0) > 0\}.$$

Then Λ^+ is a $\mathfrak{k}$-adapted positive system and $i\Lambda_p^+ \subseteq C^*$, i.e., $C \subseteq C_{\max}$.

It remains to prove that $C_{\min} \subseteq C$. To see this, let $0 \neq X \in \mathfrak{g}^{[\lambda]}$ and $E \in \operatorname{algint} C$. Then Lemmas 7.7 and 7.8 imply that

$$p(e^{\mathbb{R} \operatorname{ad} X}) = E + i\lambda(E)\mathbb{R}^+[IX,X] = E + \mathbb{R}^+[IX,X] \subseteq C.$$

Now the closedness of C entails that

$$[IX,X] = \lim_{t \to \infty} \frac{1}{t}(E + t[IX,X]) \in C.$$

Thus $C_{\min} \subseteq C$.   ∎

Corollary 7.19.   *Let $\mathfrak{g}$ be a Lie algebra with a compactly embedded Cartan algebra $\mathfrak{t}$. Suppose that $\mathfrak{g}$ contains a pointed generating invariant cone. Then there exists a choice of a positive system Λ^+ and $\omega \in (\mathfrak{t} \cap \mathfrak{r})^* \cong (\mathfrak{t} \cap \mathfrak{s})^\perp$ such that*

$$\langle \omega, [IX,X] \rangle > 0 \qquad \forall X \in \mathfrak{n}_{\text{eff}}.$$

Proof.   Let $\nu \in \operatorname{int} W^*$ and set $\omega := \nu|_{\mathfrak{t} \cap \mathfrak{r}}$. Using Proposition 7.18 we choose Λ^+ such that $C_{\min} \subseteq W \cap \mathfrak{t}$. Let $\lambda \in \Lambda_p^+$ and $0 \neq X \in \mathfrak{g}^{[\lambda]} \cap \mathfrak{n}$. Then, since $\mathfrak{g}$ has cone potential, $0 \neq [IX,X] \in W \cap \mathfrak{t}$ and therefore

$$\langle \omega, [IX,X] \rangle = \nu([IX,X]) > 0.$$

If $X \in \mathfrak{n}_{\text{eff}}$, then $X = \sum_{\lambda \in \Lambda_p^+} X_\lambda$ with $X_\lambda \in \mathfrak{g}^{[\lambda]} \cap \mathfrak{n}$. Now $[X_\lambda, X_{\lambda'}] = 0$ (cf. Theorem 7.15) for $\lambda \neq \lambda'$ implies that

$$[IX,X] = \sum_\lambda [IX_\lambda, X_\lambda],$$

and the assertion follows.   ∎

We say that the Lie algebra $\mathfrak{g}$ with the compactly embedded Cartan algebra $\mathfrak{t}$ has *strong cone potential* if there exists $\omega \in (\mathfrak{t} \cap \mathfrak{r})^*$ such that

$$\langle \omega, [IX,X] \rangle > 0 \qquad \forall 0 \neq X \in \mathfrak{n}_{\text{eff}}.$$

In view of the preceding corollary this condition is necessary for $\mathfrak{g}$ to contain pointed generating invariant cones.

Lemma 7.20. *The Lie algebra* $\mathfrak{g}$ *with the compactly embedded Cartan algebra* $\mathfrak{t}$ *has strong cone potential if and only if it has cone potential, and there exists a positive system* Λ^+ *such that the cone* $C_{\min,n}$ *is pointed.*

Proof.   If $C_{\min,n} \subseteq Z(\mathfrak{g})$ is pointed and $\mathfrak{g}$ has cone potential, we choose $\nu \in \operatorname{int} C^*_{\min,n}$ and set $\omega := \nu \mid_{\mathfrak{t} \cap \mathfrak{r}}$. For $0 \neq X \in \mathfrak{n}_{\mathrm{eff}}$ it follows from the cone potential of $\mathfrak{g}$ that $0 \neq [IX, X] \in C_{\min,n}$. Whence $\omega([IX, X]) > 0$.

If, conversely, $\mathfrak{g}$ has strong cone potential and $\omega \in (\mathfrak{t} \cap \mathfrak{r})^*$ is chosen such that $\omega([IX, X]) > 0$ holds for all $0 \neq X \in \mathfrak{n}_{\mathrm{eff}}$, then it follows in particular that $[IX, X] \neq 0$ for all $X \in \mathfrak{g}^{[\lambda]} \cap \mathfrak{n}$. Thus $\mathfrak{g}$ has cone potential. Let $S \subseteq \mathfrak{n}_{\mathrm{eff}}$ be a sphere in this vector space. Then $K := \{[IX, X] : X \in S\}$ is a compact subset in the interior of the half space ω^*. Therefore $C_{\min,n} = \mathbb{R}^+ K$ is pointed.   ∎

Proposition 7.21.   *Let* $\mathfrak{g}$ *be a Lie algebra with cone potential,* $\Lambda^+ \subseteq \Lambda$ *a positive system, and* $\omega \in (\mathfrak{t} \cap \mathfrak{r})^* \subseteq \mathfrak{g}^*$. *Then* $W_\omega := \mathcal{O}^*_\omega$ *is an invariant wedge in* $\mathfrak{g}$ *with*
$$W_\omega \cap \mathfrak{t} = \omega^* \cap \{E \in \mathfrak{t} : (\forall \lambda \in \Lambda^+_n)\ i\lambda(E)\omega(C_{\min,n}) \subseteq \mathbb{R}^+\}.$$

Proof.   An element $E \in \mathfrak{t}$ is contained in $\mathcal{O}^*_\omega$ if and only if $\langle E, \mathcal{O}_\omega \rangle \subseteq \mathbb{R}^+$. Using Proposition 7.16 we see that
$$\mathcal{O}_\omega = \omega \circ e^{\operatorname{ad} \mathfrak{n}}.$$
In view Proposition 7.15, the group $e^{\operatorname{ad} \mathfrak{n}}$ is abelian and, since $\mathfrak{n}' \subseteq Z(\mathfrak{g})$ and $\mathfrak{n} = \mathfrak{n}_{\mathrm{eff}} + Z(\mathfrak{g})$, it coincides with $e^{\operatorname{ad} \mathfrak{n}_{\mathrm{eff}}}$. Let $X \in \mathfrak{g}^{[\lambda]} \cap \mathfrak{n}$. Then, for every $E \in \mathfrak{t}$, we use Lemma 7.8 to see that
$$\langle \omega, e^{\operatorname{ad} \mathbb{R} X} E \rangle = \langle \omega, E \rangle + i\lambda(E)\mathbb{R}^+ \langle \omega, [IX, X] \rangle.$$
Whence $E \in W_\omega$ if and only if
$$\omega(E) \geq 0 \quad \text{and} \quad i\lambda(E)\omega(C_{\min,n}) \subseteq \mathbb{R}^+.$$
                                                                                    ∎

Lemma 7.22. *Let* $\mathfrak{g}$ *be a Lie algebra with cone potential and* $\mathfrak{g} = \mathfrak{r} \rtimes \mathfrak{s}$ *a* $\mathfrak{t}$-*invariant Levi decomposition. Then*
(i) $Z_\mathfrak{g}(\mathfrak{r}) = Z(\mathfrak{g}) + Z_\mathfrak{s}(\mathfrak{r}) \cap \mathfrak{t}$, *and*
(ii) $H(C_{\max,n}) = Z(\mathfrak{g}) + Z_{\mathfrak{t} \cap \mathfrak{s}}(\mathfrak{r})$.

Proof.   Let $\mathfrak{g} = (\mathfrak{r} \rtimes \mathfrak{s}_0) \oplus \mathfrak{s}_1$, where $\mathfrak{s} \cong \mathfrak{s}_0 \oplus Z_\mathfrak{s}(\mathfrak{r})$ and $\mathfrak{g}_0 := \mathfrak{r} \rtimes \mathfrak{s}_0$.
(i) "$\subseteq$": Then $Z_\mathfrak{g}(\mathfrak{r}) = Z_{\mathfrak{g}_0}(\mathfrak{r}) \oplus \mathfrak{s}_1$ and the Lie algebra $\mathfrak{g}_0$ has cone potential. We therefore may assume that $\mathfrak{s}_1 = \{0\}$, i.e., that $Z_\mathfrak{s}(\mathfrak{r}) = \{0\}$. Now $Z_{\mathfrak{g}_0}(\mathfrak{r})$ is an ideal and therefore $\mathfrak{t}$-invariant. Suppose that $Z_{\mathfrak{g}_0}(\mathfrak{r}) \not\subseteq \mathfrak{t}$. Then there exists $\lambda \in \Lambda^+$ and $0 \neq X \in \mathfrak{g}^{[\lambda]} \cap Z_{\mathfrak{g}_0}(\mathfrak{r})$. Let $X = X_r + X_s$, where $X_r \in \mathfrak{r}$ and $X_s \in \mathfrak{s}$. For every $E \in \mathfrak{t}$ with $\lambda(E) = -i$ we have $IX = [E, X] \in Z_{\mathfrak{g}_0}(\mathfrak{r})$, $IX_r \in \mathfrak{r}$, $IX_s \in \mathfrak{s}$, and
$$[X, IX] = [X, IX_r + IX_s] = [X, IX_s] = [X_r + X_s, IX_s]$$
$$= [X_r, IX_s] + [X_s, IX_s] \in \mathfrak{t} \cap (\mathfrak{r}_{\mathrm{eff}} + \mathfrak{s}) = \mathfrak{t} \cap \mathfrak{s}.$$
Hence $0 \neq [X, IX] \in Z_\mathfrak{s}(\mathfrak{r}) = \{0\}$, a contradiction. We conclude that $Z_\mathfrak{g}(\mathfrak{r}) \subseteq \mathfrak{t}$ is an abelian ideal. Now $Z_\mathfrak{g}(\mathfrak{r}) \subseteq Z(\mathfrak{g})$ is a consequence of Lemma 7.14.
(i) "$\supseteq$": This inclusion is trivial.
(ii) Since $H(C_{\max,n}) = Z_\mathfrak{t}(\mathfrak{n}_{\mathrm{eff}}) = Z_\mathfrak{t}(\mathfrak{r})$, (i) shows that
$$H(C_{\max,n}) = Z(\mathfrak{g}) + Z_{\mathfrak{t} \cap \mathfrak{s}}(\mathfrak{r}).$$
                                                                                    ∎

Lemma 7.23.   *Let* $\mathfrak{g} = \mathfrak{g}_0 \oplus \mathfrak{k}$, *where* $\mathfrak{k}$ *is a compact Lie algebra and* $W \subseteq \mathfrak{g}_0$ *is a pointed generating invariant cone. Then there exists a pointed generating invariant cone* $W' \subseteq \mathfrak{g}$ *such that* $W' \cap \mathfrak{g}_0 = W$.

Proof.   Let φ_W denote the characteristic function of the cone W (cf. Section 1.3). We set $C := \varphi_W^{-1}(]0, 1])$.

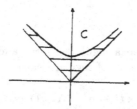

Figure 7.1

This is a closed convex subset of W which is invariant under the action of the group $\mathrm{Inn}\,\mathfrak{g}_0$ because $\det(e^{\mathrm{ad}\,X}) = 1$ holds for all $X \in \mathfrak{g}$ (Theorem 1.8(ii)). Further we pick a compact convex $\mathrm{Inn}\,\mathfrak{k}$-invariant 0-neighborhood B in $\mathfrak{k}$. We define

$$W' := \overline{\mathbb{R}^+(C + B)}.$$

Since the set $C + B$ is convex and invariant under $\mathrm{Inn}\,\mathfrak{g}$, it follows that W' is a generating invariant wedge in $\mathfrak{g}$. Let $p \colon \mathfrak{g} \to \mathfrak{g}_0$ denote the projection along $\mathfrak{k}$. Then

$$W \subseteq p(W') \subseteq \overline{\mathbb{R}^+ p(C + B)} = \overline{\mathbb{R}^+ C} = W$$

implies in particular that $W' \cap \mathfrak{g}_0 = W$ and $H(W') \subseteq \mathfrak{k} = \ker p$. It remains to show that $H(W') = 0$. To see this, let $\omega \in \mathrm{int}\,W^*$ with $\omega(C) = [1, \infty[$ and $\lambda_n(c_n + b_n) \to w' \in H(W') \subseteq \mathfrak{k}$. Then $\lambda_n c_n \to 0$ implies in particular that $\lambda_n \omega(c_n) \to 0$. Hence $\lambda_n \to 0$ and therefore $w' = \lim \lambda_n b_n = 0$.   ∎

Proposition 7.24.   *Let* $\mathfrak{g}$ *be a Lie algebra with strong cone potential such that* $Z_{\mathfrak{s}}(\mathfrak{r})$ *is compact, where* $\mathfrak{g} = \mathfrak{r} \rtimes \mathfrak{s}$ *is a* $\mathfrak{r}$*-invariant Levi decomposition. Then there exists a pointed generating invariant cone in* $\mathfrak{g}$.

Proof.   First we note that $\mathfrak{g} \cong \mathfrak{g}_0 \oplus \mathfrak{k}_0$, where $\mathfrak{k}_0 = Z_{\mathfrak{s}}(\mathfrak{r})$, $\mathfrak{s} = \mathfrak{s}_0 \oplus \mathfrak{k}_0$, and $\mathfrak{g}_0 = \mathfrak{r} \rtimes \mathfrak{s}_0$. In view of Lemma 7.23, it suffices to find a pointed generating invariant cone in $\mathfrak{g}_0$. So we may assume that $\mathfrak{k}_0 = \{0\}$.

The assumption that $\mathfrak{g}$ has strong cone potential means that $C_{\min,n} \subseteq Z(\mathfrak{g})$ is pointed, i.e., that $C_{\min,n}^*$ has non-empty interior. Let $\mathfrak{t} \cap \mathfrak{r} = Z(\mathfrak{g}) \oplus \mathfrak{a}$, and choose a basis $\omega_1, \ldots, \omega_n \in Z(\mathfrak{g})^* \cong (\mathfrak{t} \cap \mathfrak{r})^* \cap \mathfrak{a}^\perp$ which is contained in $\mathrm{int}\,C_{\min,n}^*$. We set

$$W := W_{\omega_1} \cap \ldots \cap W_{\omega_n}.$$

First we show that W is pointed. Suppose that this is false. Then $H(W)$ is a non-zero ideal of $\mathfrak{g}$ and therefore there exists $0 \neq E \in H(W) \cap \mathfrak{t}$ (Lemma 7.14). On the other hand $\omega_i(E) = 0$ holds for $i = 1, \ldots, n$. Hence $E = 0$ and W is pointed.

To see that W is generating, we first use Proposition 7.21 to get

$$W \cap \mathfrak{t} = (\mathbb{R}^+ \omega_1 + \ldots + \mathbb{R}^+ \omega_n)^* \cap C_{\max,n}.$$

That $\mathrm{int}_\mathfrak{t}(W \cap \mathfrak{t}) \neq \emptyset$ follows immediately from the fact that

$$\mathfrak{a} + (\mathfrak{t} \cap \mathfrak{s}_0) \subseteq (\mathbb{R}^+ \omega_1 + \ldots + \mathbb{R}^+ \omega_n)^* \quad \text{and} \quad Z(\mathfrak{g}) = H(C_{\max,n})$$

(Lemma 7.22).

Let $E_0 \in \mathrm{int}_\mathfrak{t} W \cap \mathfrak{t}$ be a regular element and U a neighborhood of E_0 contained in $W \cap \mathfrak{t}$ which consists of regular elements. Then the mapping

$$\Phi : \mathfrak{g} \times U \to \mathfrak{g}, \quad (X, E) \mapsto e^{\mathrm{ad}\, X} E$$

satisfies $d\Phi(0, E_0)(X, E) = E - \mathrm{ad}\, E_0(X)$. Thus $\Phi(\mathfrak{g} \times U)$ is a neighborhood of E_0 which is contained in W, so that W is pointed and generating. ∎

Theorem 7.25. *A simple Lie algebra $\mathfrak{g}$ contains a pointed generating invariant cone if and only if $\mathfrak{g}$ is a Hermitean simple Lie algebra, i.e., if $Z(\mathfrak{k}) \neq \{0\}$ holds for a maximal compactly embedded subalgebra $\mathfrak{k}$.*

In this case there exist, up to sign unique, minimal and maximal invariant cones $W_{\min} \subseteq W_{\max}$ such that for every pointed generating invariant cone either W or $-W$ lies between $W_{\min}$ and $W_{\max}$.

Proof. The necessity of the condition follows from Corollary 7.11. Suppose that $\mathfrak{g}$ is Hermitean simple, and that $\mathfrak{g} = \mathfrak{k} + \mathfrak{p}$ is a Cartan decomposition. Then $Z(\mathfrak{k}) \neq \{0\}$ and there exists a scalar product $\langle \cdot, \cdot \rangle$ on $\mathfrak{g}$ such that the operators $\mathrm{ad}\, X, X \in \mathfrak{k}$ are skew symmetric, and the operators $\mathrm{ad}\, Y, Y \in \mathfrak{p}$ are symmetric. Moreover, $\mathrm{Inn}\, \mathfrak{g} = e^{\mathrm{ad}\,\mathfrak{p}} K$, where $K = \mathrm{Inn}_\mathfrak{g}\, \mathfrak{k} = \mathrm{INN}_\mathfrak{g}\, \mathfrak{k}$. For $g = e^{\mathrm{ad}\, Y} k \in \mathrm{Inn}\, \mathfrak{g}$, $Y \in \mathfrak{p}$ and $Z \in Z(\mathfrak{k})$, we have

$$\langle g.Z, Z \rangle = \langle e^{\mathrm{ad}\, Y}.Z, Z \rangle > 0$$

because $e^{\mathrm{ad}\, Y}$ is positive definite. Thus $W := \mathrm{wedge}(\mathrm{Inn}\, \mathfrak{g}.Z)$ is an invariant wedge in $\mathfrak{g}$ and $\langle W, Z \rangle \subseteq \mathbb{R}^+$. Hence $W \neq \mathfrak{g}$. Since $H(W)$ is an ideal of $\mathfrak{g}$ and $\mathfrak{g}$ is simple, W is pointed. The subspace $W - W$ is a non-zero ideal of $\mathfrak{g}$, so that $W - W = \mathfrak{g}$ also follows from the simplicity of $\mathfrak{g}$.

Suppose that $\mathfrak{g}$ is a simple Hermitean Lie algebra. Fix a non-zero element $Z_0 \in Z(\mathfrak{k})$ and write $W_{\min}$ for the smallest invariant cone in $\mathfrak{g}$ containing Z_0. Using the Cartan Killing form B on $\mathfrak{g}$ which is negative definite on $\mathfrak{k}$, we identify $\mathfrak{g}$ with its dual $\mathfrak{g}^*$. We set $W_{\max} := -W_{\min}^*$. Then $W_{\max}$ is invariant because the Cartan Killing form is invariant under the adjoint action. Thus $W_{\max}$ intersects the one-dimensional subspace $Z(\mathfrak{k})$ non-trivially so that $Z_0 \in W_{\max}$ follows from the fact that B is negative definite on $Z(\mathfrak{k})$. Whence $W_{\min} \subseteq W_{\max}$.

If now W is a pointed generating invariant cone in $\mathfrak{g}$, then, since W intersects $Z(\mathfrak{k})$, it contains either Z_0 or $-Z_0$. Assume that $Z_0 \subseteq W$. Then $W_{\min} \subseteq W$ and the same holds for $-W^*$. We conclude that also $W \subseteq W_{\max}$. ∎

Theorem 7.26.   (Characterization of the Lie algebras with invariant cones) *A finite dimensional real Lie algebra* $\mathfrak{g}$ *contains a pointed generating invariant cone if and only if* $\mathfrak{g}$ *satisfies the following conditions:*

(1) $\mathfrak{g}$ *has strong cone potential.*

(2) *If* $\mathfrak{k}$ *is a maximal compactly embedded subalgebra, then*

   (a) $Z(\mathfrak{k}) \neq \{0\}$,

   (b) $Z_{\mathfrak{g}}\big(Z(\mathfrak{k})\big) = \mathfrak{k}$,

Proof.   The necessity of these conditions follows from Corollary 7.11 and Corollary 7.19.

To see that these conditions are also sufficient, we proceed as follows. Let $\mathfrak{g} = \mathfrak{k} \rtimes \mathfrak{s}$ be a $\mathfrak{k}$-invariant Levi decomposition. Then $\mathfrak{g}$ decomposes as a direct product

$$\mathfrak{g} = \mathfrak{g}_0 \oplus Z_{\mathfrak{s}}(\mathfrak{r}),$$

where $\mathfrak{g}_0 = \mathfrak{r} \rtimes \mathfrak{s}_0$ and $\mathfrak{s}_0$ is an ideal of $\mathfrak{s}$ which acts effectively on $\mathfrak{r}$. Let $\mathfrak{k}_0 \subseteq Z_{\mathfrak{s}}(\mathfrak{r})$ denote the sum of all compact ideals. We apply Proposition 7.24 to find a pointed generating invariant cone C in $\mathfrak{g}_0 \oplus \mathfrak{k}_0$, so we only have to find such a cone in $\mathfrak{s}_1$, where $Z_{\mathfrak{s}}(\mathfrak{r}) = \mathfrak{k}_0 \oplus \mathfrak{s}_1$.

Let $\mathfrak{s}_1 = \mathfrak{a}_1 \oplus \ldots \oplus \mathfrak{a}_n$ denote a decomposition into simple ideals. First we note that the condition $Z_{\mathfrak{g}}\big(Z(\mathfrak{k})\big) = \mathfrak{k}$ implies in particular that $Z(\mathfrak{k}) = \{X \in \mathfrak{g} : [X, \mathfrak{k}] = \{0\}\}$. In view of Lemma 7.2 we therefore have that

$$Z(\mathfrak{k}) = Z(\mathfrak{k}) \cap \mathfrak{g}_0 \oplus Z(\mathfrak{k}) \cap \mathfrak{a}_1 \oplus \ldots \oplus Z(\mathfrak{k}) \cap \mathfrak{a}_n$$

and

$$\mathfrak{k} = \mathfrak{k} \cap \mathfrak{g}_0 \oplus \mathfrak{k}_0 \oplus \mathfrak{k} \cap \mathfrak{a}_1 \oplus \ldots \oplus \mathfrak{k} \cap \mathfrak{a}_n.$$

We conclude that every simple ideal $\mathfrak{a}_i$ satisfies

$$Z_{\mathfrak{a}_i}\big(Z(\mathfrak{k} \cap \mathfrak{a}_i)\big) = \mathfrak{k} \cap \mathfrak{a}_i.$$

It follows in particular that $\mathfrak{a}_i$ is simple Hermitean because $\mathfrak{a}_i$ is not compact.

In the Hermitean simple ideals $\mathfrak{a}_i$ we find pointed generating invariant cones W_i via Theorem 7.25. Now the sum of all these invariant cones is a pointed generating invariant cone in $\mathfrak{g}$.   ∎

Since the preceding theorem describes the class of those Lie algebras containing pointed generating invariant cones, one also would like to have a description of all possibilities for pointed generating invariant cones in these Lie algebras. Such a description is provided by the following result which we state without proof.

Theorem 7.27.   (*The Classification Theorem for invariant cones*) *Let* $\mathfrak{g}$ *be a Lie algebra containing pointed generating invariant cone and* $\mathfrak{t} \subseteq \mathfrak{g}$ *a compactly embedded Cartan algebra. Let further*

$$\mathcal{W} := N_{\text{Inn}(\mathfrak{g})}(\mathfrak{t}) / Z_{\text{Inn}(\mathfrak{g})}(\mathfrak{t})$$

denote the Weyl group of $\mathfrak{t}$. *Then the following assertions hold:*

(i) *Every pointed generating invariant cone $W \subseteq \mathfrak{g}$ is uniquely determined by its intersection $C := W \cap \mathfrak{t}$ with $\mathfrak{t}$. More precisely,*

$$\operatorname{int} W = \operatorname{Inn}(\mathfrak{g}).\operatorname{algint} C.$$

(ii) *A pointed generating cone $C \subseteq \mathfrak{t}$ arises as the trace $W \cap \mathfrak{t}$ of a pointed generating invariant cone $W \subseteq \mathfrak{g}$ if and only if $\mathcal{W}(C) \subseteq C$ and ther exists a $\mathfrak{t}$-adapted positive system $\Lambda^+ \subseteq \Lambda(\mathfrak{g}_\mathbb{C}, \mathfrak{t}_\mathbb{C})$ such that*

$$C_{\min}(\Lambda^+) \subseteq C \subseteq C_{\max}(\Lambda^+).$$

Proof.   [HHL89, III.9.18]   ∎

7.3. Lawson's Theorem on Ol'shanskiĭ semigroups

Symmetric Lie algebras

A *symmetric Lie algebra* is a pair $(\mathfrak{g}, \tau)$ of a Lie algebra $\mathfrak{g}$ together with an involutive automorphism τ. We set

$$\mathfrak{h} := \{X \in \mathfrak{g} : \tau(X) = X\} \quad \text{and} \quad \mathfrak{q} := \{X \in \mathfrak{g} : \tau(X) = -X\}.$$

Note that this implies that

$$[\mathfrak{h}, \mathfrak{h}] \subseteq \mathfrak{h}, \quad [\mathfrak{h}, \mathfrak{q}] \subseteq \mathfrak{q} \quad \text{and} \quad [\mathfrak{q}, \mathfrak{q}] \subseteq \mathfrak{h}.$$

If we use another symbol for the subalgebra of τ-fixed points, we write $(\mathfrak{g}, \mathfrak{h}, \tau)$ for the symmetric Lie algebra $(\mathfrak{g}, \tau)$ to fix the notation. A morphism $\pi: (\mathfrak{g}, \tau) \to (\mathfrak{g}', \tau')$ of symmetric Lie algebras is a homomorphism of Lie algebras which respects the involution, i.e.,

$$\pi \circ \tau = \tau' \circ \pi.$$

This defines the category *SLa* of symmetric Lie algebras. A *symmetric Lie group* is a pair (G, τ) of a Lie group G together with an analytic involutive automorphism τ. For $g \in G$ we set $g^* := \tau(g)^{-1}$. The category *SLg* of symmetric Lie groups is defined similarly to *SLa*.

Let $(\mathfrak{g}, \tau)$ be a symmetric Lie algebra. Then it is clear that the complexification is a symmetric Lie algebra with respect to the involutions:

(1) $\sigma: X + iY \mapsto X - iY$ (complex conjugation).

(2) $\tau_\mathbb{C} : X + iY \mapsto \tau(X) + i\tau(Y)$ (complex linear extension of τ).

(3) $\sigma\tau_\mathbb{C} : X + iY \mapsto \tau(X) - i\tau(Y)$ (complex antilinear extension of τ).

Note that this implies in particular that $\mathfrak{g}^c := \mathfrak{h} + i\mathfrak{q}$ is a symmetric Lie algebra with respect to $\tau_\mathbb{C}|_{\mathfrak{g}^c} = \sigma|_{\mathfrak{g}^c}$ and it is the fixed point algebra of $\sigma\tau_\mathbb{C}$. This symmetric Lie algebra is called the *dual symmetric Lie algebra* .

Lemma 7.28.

(i) *If* $\pi : (\mathfrak{g}, \tau) \to (\mathfrak{g}', \tau')$ *is a morphism of symmetric Lie algebras, then the complex extension* $\pi_{\mathbb{C}}$ *is a morphism* $(\mathfrak{g}_{\mathbb{C}}, \tau_{\mathbb{C}}) \to (\mathfrak{g}'_{\mathbb{C}}, \tau'_{\mathbb{C}})$ *of symmetric Lie algebras.*

(ii) *Let* $\pi : (\mathfrak{g}, \mathfrak{h}, \tau) \to (\mathfrak{g}', \mathfrak{h}', \tau')$ *be a morphism of symmetric Lie algebras. Then* $\ker \pi$ *is a* τ*-invariant ideal and*

$$\pi^{-1}(\mathfrak{h}') = \ker \pi + \mathfrak{h}.$$

Proof. (i) This is a trivial consequence of the definitions.

(ii) If $X \in \ker \pi$, then $\pi(\tau X) = \tau \pi(X) = 0$ proves the first assertion. For the second assertion we have to show that $X \in \mathfrak{q}$ and $\tau(\pi(X)) = \pi(X)$ implies that $\pi(X) = 0$. This is a direct consequence of the symmetry of the morphism as the relation

$$\tau(\pi(X)) = \pi(\tau X) = -\pi(X)$$

shows.   ∎

Ol'shanskiĭ wedges

Let $(\mathfrak{g}, \tau)$ be a symmetric Lie algebra. A Lie wedge $W \subseteq \mathfrak{g}$ is called a *symmetric Lie wedge* if $\mathfrak{h} \subseteq H(W)$. We say that a symmetric Lie wedge W is an *Ol'shanskiĭ wedge* if $W \cap \mathfrak{q}$ is pointed and generating.

To check that a given wedge is a symmetric Lie wedge, the following proposition is a useful tool.

Proposition 7.29. *Let* $(\mathfrak{g}, \tau)$ *be a symmetric Lie algebra,* $W \subseteq \mathfrak{g}$ *a wedge containing* $\mathfrak{h}$, *and* $C := W \cap \mathfrak{q}$. *Then the following are equivalent:*

(1) W *is a symmetric Lie wedge.*

(2) $\mathrm{Inn}_{\mathfrak{g}} \mathfrak{h}(C) = C$.

(3) $[\mathfrak{h}, Y] \subseteq T_Y(C)$ *for all* $Y \in C$.

Proof. (1) $\Rightarrow$ (2): Let $H := \mathrm{Inn}_{\mathfrak{g}} \mathfrak{h}$. Then $\mathfrak{q}$ is invariant under H because $[\mathfrak{h}, \mathfrak{q}] \subseteq \mathfrak{q}$ and W is invariant under H since W is a Lie wedge. Now (2) follows from the fact that the intersection of H-invariant wedges is H-invariant.

(2) $\Rightarrow$ (3): Let $X \in \mathfrak{h}$. Then, in view of (2), the Invariance Theorem for vector fields (Theorem 5.8) implies that

$$[X, Y] \subseteq L_Y(C) \quad \text{and} \quad [-X, Y] \subseteq L_Y(C).$$

It follows that $[X, Y] \in H(L_Y(C)) = T_Y(C)$.

(3) $\Rightarrow$ (1): Let $X = X_1 + X_2 \in H(W)$, where $X_1 \in \mathfrak{h}$, and $X_2 \in H(C)$, and $Y = Y_1 + Y_2 \in W$ with $Y_1 \in \mathfrak{h}$ and $Y_2 \in C$. Since (3) implies that $[\mathfrak{h}, H(C)] \subseteq H(C)$, we conclude that

$$[X, Y] = [X_1 + X_2, Y_1 + Y_2] \in \mathfrak{h} + H(C) + [X_1, Y_2]$$
$$\in H(W) + T_{Y_2}(C) = \mathfrak{h} + T_{Y_2}(C) = T_Y(W).$$

Now the Invariance Theorem for vector fields (Theorem 5.8) applies and shows that $e^{\mathrm{ad} X} Y \in W$, hence W is a Lie wedge.   ∎

A wedge $C \subseteq \mathfrak{q}$ is said to be *regular* if $\mathrm{Spec}(\mathrm{ad}\,Y) \subseteq \mathbb{R}$ holds for all $Y \in C$.

If $(\mathfrak{g}_{\mathbb{C}}, \sigma)$ is the complexification of the Lie algebra $\mathfrak{g}$, and $C \subseteq \mathfrak{g}$ is an invariant wedge, then the preceding proposition shows that

$$W_C := \mathfrak{g} + iC$$

is a symmetric Lie wedge in $\mathfrak{g}_{\mathbb{C}}$. This wedge is said to be the *symmetric Lie wedge* associated with C.

Suppose that W is pointed. Then $C = iW$ is a regular wedge. To see this, we first note that, for $X \in \mathrm{int}\,iW$, the relation

$$\mathrm{Spec}(\mathrm{ad}\,X) = i\,\mathrm{Spec}(-\,\mathrm{ad}\,iX) \subseteq \mathbb{R}$$

follows from Proposition 7.10. Now the assertion follows from the continuous dependence of the set $\mathrm{Spec}(\mathrm{ad}\,X)$ of X and the fact that $\mathrm{int}\,W$ is dense in W.

Lemma 7.30. *Let $C \subseteq \mathfrak{q}$ be a regular wedge such that $W := \mathfrak{h} + C$ is a symmetric Lie wedge. Then $f(\mathrm{ad}\,X) = \frac{1 - e^{-\,\mathrm{ad}\,X}}{\mathrm{ad}\,X}$ is invertible for every $X \in W \cap \mathfrak{q} =: C$ and*

$$g(\mathrm{ad}\,X)W \subseteq L_X(W) \qquad \forall X \in C$$

holds for $g(\mathrm{ad}\,X) = f(\mathrm{ad}\,X)^{-1}$.

Proof.   Let $X \in C$. Since $\mathrm{ad}\,X$ has real spectrum, the spectrum of $f(\mathrm{ad}\,X)$ does not contain 0. Hence $g(\mathrm{ad}\,X) := f(\mathrm{ad}\,X)^{-1}$ exists. Let $Y = Y_1 + Y_2 \in W$ with $Y_1 \in \mathfrak{h}$ and $Y_2 \in C$. Then

$$[X,Y] = [X,Y_1] + [X,Y_2] \in T_X(C) + \mathfrak{h} = T_X(W).$$

Hence $\mathrm{ad}\,X(W - W) \subseteq T_X(W)$ and

$$f(\mathrm{ad}\,X)T_X(W) \subseteq T_X(W).$$

Since $f(\mathrm{ad}\,X)$ is invertible, we conclude that $f(\mathrm{ad}\,X)T_X(W) = T_X(W)$.

Let $Y \in W$. Then

$$Z := f(\mathrm{ad}\,X)Y - Y = -\frac{1}{2}[X,Y] + \frac{1}{3!}(\mathrm{ad}\,X)^2 Y + \ldots \in T_X(W).$$

Pick $Z' \in T_X(W)$ with $Z = f(\mathrm{ad}\,X)Z'$. Then

$$g(\mathrm{ad}\,X)Y = Y - g(\mathrm{ad}\,X)Z = Y - Z' \in W + T_X(W) \subseteq L_X(W).$$

∎

Lemma 7.31. *Suppose that two elements X, Y in the Lie algebra $\mathbf{L}(G)$ of the Lie group G satisfy $\exp X = \exp Y$, and that $\exp$ is non-singular at X. Then $[X,Y] = 0$ and $\exp(X - Y) = 1$.*

Proof.   (cf. [HHL89, V.6.7]) All elements $\exp tY$ commute with $\exp X = \exp Y$. Thus

$$\exp X = \exp(tY)\exp X \exp(-tY) = \exp(e^{t\,\mathrm{ad}\,Y}X) \qquad \forall t \in \mathbb{R},$$

and therefore

$$0 = \frac{d}{dt}\bigg|_{t=0} \exp(e^{t\,\mathrm{ad}\,Y}X) = d\exp(X)[Y,X].$$

Since $\exp$ is non-singular in X, we obtain $[X,Y] = 0$. Then

$$\exp(X - Y) = \exp(X)\exp(-Y) = 1$$

follows.   ∎

In the following (G, τ) denotes a symmetric Lie group, $(\mathfrak{g}, d\tau(1))$ the associated symmetric Lie algebra, and $W \subseteq \mathfrak{q}$ a regular wedge.

Lemma 7.32. *Let $C \subseteq \mathfrak{q}$ be a regular wedge. Then the following are equivalent:*

(1) *The mapping* exp *restricted to C is injective.*

(2) *If $Z \in (C - C) \cap Z(\mathfrak{g})$ and $\exp Z = 1$, then $Z = 0$.*

Proof. $(1) \Rightarrow (2)$: Suppose that $Z \in (C - C) \cap Z(\mathfrak{g})$ and $\exp Z = 1$. Pick $X, Y \in C$ with $Z = X - Y$. Then $X = Y + Z$, and, since $[Y, Z] = 0$, we have

$$\exp(X) = \exp(Y + Z) = \exp(Y) \exp(Z) = \exp(Y).$$

Since exp is injective on C, it follows that $X = Y$, and hence $Z = 0$.

$(2) \Rightarrow (1)$: Suppose that $\exp X = \exp Y$ holds for $X, Y \in C$. Since $\operatorname{ad} X$ has real spectrum, exp is regular in X, and it follows that $[X, Y] = 0$ and $\exp(Y - X) = 1$ (Lemma 7.31). Then $\operatorname{ad}(Y - X) = \operatorname{ad} Y - \operatorname{ad} X$ also has real spectrum and $e^{\operatorname{ad}(X - Y)} = \operatorname{Ad}\big(\exp(X - Y)\big) = \operatorname{id}_{\mathfrak{g}}$. This implies that $\operatorname{ad}(X - Y) = 0$, hence $X - Y \in Z(\mathfrak{g})$. By hypothesis, $Y = X$. ∎

Lemma 7.33. *Let W be a regular wedge in $\mathfrak{q}$ such that* exp *is injective on W. Then the following are equivalent:*

(1) *The mapping* exp *from W to $\exp(W)$ is a homeomorphism, and $\exp(W)$ is closed.*

(2) *For each non-zero $X \in W$ the closure of $\exp(\mathbb{R}X)$ is not compact.*

(3) *For each non-zero $X \in W \cap Z(\mathfrak{g})$ the closure of $\exp(\mathbb{R}X)$ is not compact.*

Proof. $(1) \Rightarrow (2)$: Suppose that for some $X \in W$ the closure $T := \overline{\exp \mathbb{R}X}$ is compact. Then T is a torus, and thus there exists a sequence $t_n \to \infty$ such that $\exp(t_n X) \to 1$. Hence exp is not a homeomorphism on W.

$(2) \Rightarrow (3)$: trivial

$(3) \Rightarrow (1)$: Since exp restricted to W is injective, exp is a homeomorphism on any compact subset $K \subseteq W$ onto a compact subset of G. So if $\exp|_W$ is not a homeomorphism of W onto $\exp(W)$, there exists a sequence $X_n \in W$ converging to infinity such that $\exp(X_n)$ converges to some $g \in G$. Then $\operatorname{Ad}(\exp X_n) = e^{\operatorname{ad} X_n} \to \operatorname{Ad}(g)$, and, since $\operatorname{ad} X_n$ has real spectrum, $e^{\operatorname{ad} X_n}$ has positive real spectrum. Hence the same holds for $\operatorname{Ad}(g)$.

For a linear operator T with positive spectrum we define

$$\log T := \log(\operatorname{tr} T) I - \sum_{n=1}^{\infty} \frac{1}{n} \big(I - \operatorname{tr}(T)^{-1} T\big)^{n},$$

where $\operatorname{tr}(T)$ is the trace of T and I denotes the identity (cf. [Ho65, p.172]). Then $\log e^{\operatorname{ad} X_n} = \operatorname{ad} X_n \to \log \operatorname{Ad}(g)$.

Now we pick a vector space complement $\mathfrak{p}$ to $Z(\mathfrak{g}) \cap (W - W)$ in $W - W$. Then ad restricted to $\mathfrak{p}$ is a vector space isomorphism onto $\operatorname{ad}(W - W)$. Pick Y_n and $Y \in \mathfrak{p}$ such that $\operatorname{ad} Y_n = \operatorname{ad} X_n$ and $\operatorname{ad} Y = \log \operatorname{Ad}(g)$. Then $Z_n := X_n - Y_n \in (W - W) \cap Z(\mathfrak{g})$ and

$$\exp(X_n) = \exp(Z_n + Y_n) = \exp(Z_n) \exp(Y_n).$$

Since $\exp(X_n) \to g$ and $\exp Y_n \to \exp Y$, it follows that $\exp Z_n \to g\exp(-Y) \in Z(G)_0$.

By the structure theorem for connected abelian Lie groups, there exists a vector group $V \subseteq Z(G)_0$ and a torus T such that $Z(G)_0 \cong V \times T$. Let $\mathfrak{t} := \mathbf{L}(T)$.

Fix some norm $\|\cdot\|$ on $\mathfrak{g}$. By passing to a subsequence if necessary, we may assume that $X = \lim_{n\to\infty} \frac{1}{\|X_n\|}X_n$ exists. Note that $\|X\| = 1$ and $X \in W$. Since Y_n is convergent and $\|X_n\| \to \infty$ it follows that $\frac{1}{\|X_n\|}Y_n \to 0$. Thus

$$X = \lim_{n\to\infty} \frac{1}{\|X_n\|}Z_n,$$

so $X \in Z(\mathfrak{g}) \cap W$. By hypothesis $X \notin \mathfrak{t}$, and there exists an analytic homomorphism $\chi: Z(G)_0 \to \mathbb{R}$ such that $d\chi(1)X \neq 0$. Then $\chi(\exp Z_n) = d\chi(1)Z_n$ is a bounded sequence in $\mathbb{R}$ and

$$d\chi(1)(X) = \lim_{n\to\infty} \frac{1}{\|X_n\|}d\chi(1)Z_n$$

entails that $d\chi(1)X = 0$, a contradiction.   ∎

Theorem 7.34.   (Lawson's Theorem on Ol'shanskiĭ semigroups) *Let (G, τ) be a symmetric Lie group, $(\mathfrak{g}, d\tau(1))$ the associated symmetric Lie algebra, and $W \subseteq \mathfrak{q}$ a regular wedge. Then the following are equivalent:*

(1) *The mapping $W \times G^\tau, (X, h) \mapsto \exp(X)h$ is a homeomorphism onto a closed subset of G.*

(2) *The mapping $\mathrm{Exp} := \pi \circ \exp|_{\mathfrak{q}} : \mathfrak{q} \to G/G^\tau$ defines a homeomorphism from C onto a closed subset of G/G^τ.*

(3) *The map $\exp|_W : W \to \exp(W)$ is a homeomorphism onto a closed subset of G.*

(4) *For each non-zero $X \in C \cap Z(\mathfrak{g})$ the closure of $\exp(\mathbb{R}X)$ is not compact and*

$$\exp^{-1}(1) \cap (W - W) \cap Z(\mathfrak{g}) = \{0\}.$$

Proof.   (3) ⇔ (4): This follows from Lemmas 7.32 and 7.33.

(1) ⇔ (3): That (1) implies (3) is is trivial because $W \times \{1\}$ is a closed subset of $W \times G^\tau$. Assume (3). Let $\exp(X_n)h_n \to g \in G$, where $X_n \in W$ and $h_n \in G^\tau$. Then

$$\exp(2X_n) = \exp(X_n)h_n\tau\big(\exp(X_n)h_n\big)^{-1} \to g\tau(g)^{-1} = \exp(Y) \in \exp(W).$$

By hypothesis, $X_n \to \frac{1}{2}Y$, and therefore $h_n \to \exp(-\frac{1}{2}Y)g \in G^\tau$. Thus $g \in \exp(W)G^\tau$, hence $\exp(W)G^\tau$ is closed.

If $\exp(X_n)h_n \to \exp(Z)h$, then the above argument also shows that $X_n \to Z$ and that $h_n \to h$. Hence the mapping

$$C \times G^\tau, (X, h) \mapsto \exp(X)h$$

is a homeomorphism onto a closed subset of G.

$(1) \Rightarrow (2)$: If (1) holds, then the subset $\text{Exp}(W) = \pi(\exp(W)G^\tau)$ is closed because $\exp(W)G^\tau$ is closed and saturated. We consider the mapping

$$\sigma : G/G^\tau \to G, \quad gG^\tau \mapsto g\tau(g)^{-1}.$$

Then $\text{Exp}(X_n) \to \text{Exp}(X)$ implies that

$$\sigma(\text{Exp}(X_n)) = \exp(2X_n) \to \exp(2X) = \sigma(\text{Exp}(Y)).$$

Now $X_n \to X$ follows from (1). This proves that Exp is a homeomorphism from W onto $\text{Exp}(W)$.

$(2) \Rightarrow (3)$: Let $X_n \in W$ be a sequence with $\exp(X_n) \to g \in G$. Then (2) implies that

$$\text{Exp}(X_n) = \pi(\exp X_n) \to \pi(g) = \text{Exp}(X) \in \text{Exp}(W) \quad \text{and} \quad X_n \to X.$$

It follows that $\exp(X_n) \to \exp(X) \in \exp(W)$. We conclude that $\exp(W)$ is closed and that exp restricted to W is a homeomorphism. $\blacksquare$

Corollary 7.35.   *Let (G, τ) be a simply connected symmetric Lie group, $(\mathfrak{g}, d\tau(1))$ the associated symmetric Lie algebra, and $C \subseteq \mathfrak{q}$ a G_0^τ-invariant regular wedge. Then the set*

$$\Gamma(C) := \exp(C)G_0^\tau$$

is a Lie semigroup in G with $\mathbf{L}(S) = \mathfrak{h} + C$.

Proof.   First we note that the assumption that G is simply connected implies that the normal subgroup $Z(G)_0$ is also simply connected and closed ([Ho65, p.135]). Hence $\exp^{-1}(1) \cap \mathfrak{z}(\mathfrak{g}) = \{0\}$ and $\exp \mathbb{R}X \cong \mathbb{R}$ is closed for all $X \in \mathfrak{z}(\mathfrak{g})$. So (4) in Theorem 7.34 is satisfied and this theorem applies.

Let $W := \mathfrak{h} + C$. Then W is a symmetric Lie wedge. As in Chapter 1 we denote the associated Lie semigroup with S_W and set $H := G_0^\tau = \langle \exp \mathfrak{h} \rangle$. Then the inclusion $\exp(C)H \subseteq S_W$ is trivial. To see that the converse holds, in view of the closedness of $\exp(C)H$ and the density of $\langle \exp W \rangle$ in S_W, it suffices to show that $\langle \exp W \rangle \subseteq \exp(C)H$.

Let $\gamma : [a, b] \to G$ be a W-monotone curve with $\gamma(0) = 1$. Then $\beta := \pi \circ \gamma : [a, b] \to G/H$ is a Θ_W-monotone curve (Proposition 4.14). We claim that $\beta([a, b]) \subseteq \text{Exp}(C)$. This fact implies that $\gamma(b) \in \exp(C)H$ and the assertion follows.

Let $T := \sup\{t \in [a, b] : \beta([a, t]) \subseteq \text{Exp}(C)\}$. We have to show that $T = b$. Suppose that $T < b$. Then the closedness of $\text{Exp}(C)$ implies that $\beta(T) = \text{Exp}(X) \in \text{Exp}(C)$. Since $\text{ad}\, X$ has only real eigenvalues, the mapping Exp is regular in X. Therefore there exists a neighborhood U of X such that Exp restricted to U is a diffeomorphism onto $\text{Exp}(U)$. Let $T' \in]T, b]$ such that $\beta([T, T']) \subseteq \text{Exp}(U)$. We define

$$\alpha := (\text{Exp}|_U)^{-1} \circ \beta|_{[T, T']} : [T, T'] \to \mathfrak{q}.$$

Then $\pi \circ (\exp \circ \alpha) = \beta|_{[T, T']}$ shows that $\exp \circ \alpha$ is a W-conal curve in G (Proposition 4.14). Therefore

$$\alpha'(t) \in (d\exp(\alpha(t)))^{-1} d\lambda_{\exp(\alpha(t))}(1)W = g(\text{ad}\,\alpha(t))W.$$

Since

$$g(\operatorname{ad}X)(W)\cap\mathfrak{q} = d\operatorname{Exp}(X)^{-1}\Theta_W(\operatorname{Exp}X)$$
$$= d\operatorname{Exp}(X)^{-1}d\mu_{\exp X}(x_0)C$$
$$\subseteq L_X(W)\cap\mathfrak{q} = L_X(C)$$

for all $X \in C$ (cf. the proof of Lemma 5.10 and Lemma 7.30), the Invariance Theorem for Vector Fields (Theorem 5.8) implies that $\alpha(T') \in C$ and hence $\alpha(b) \in C$. ∎

The semigroups $\Gamma(C)$, where $C \subseteq \mathfrak{q}$ is a regular wedge and

$$\mathbf{L}\left(\Gamma(C)\right) = \mathfrak{h} + \mathfrak{q}$$

are Ol'shanskiĭ wedges, are called *Ol'shanskiĭ semigroups*. We will see in Chapter 9.3 how an extension of this class of semigroups plays an essential role in representation theory (cf. the end of Section 3.4). For further extensions of the concept of a complex Ol'shanskiĭ semigroup to the case where the corresponding invariant cone is not pointed we refer to [Ne93c]. The following corollary is basic for this extension.

Corollary 7.36. *Let $W \subseteq \mathfrak{g}$ be an invariant wedge and $G_{\mathbb{C}}$ a simply connected complex group with $\mathbf{L}(G_{\mathbb{C}}) = \mathfrak{g}_{\mathbb{C}}$. Then the Lie wedge $V := \mathfrak{g} + iW$ is global in $G_{\mathbb{C}}$ and $S_V = H_1\widetilde{G}\exp(W)$, where $H_1 := \langle\exp H(W)_{\mathbb{C}}\rangle$ is a closed connected normal subgroup of $G_{\mathbb{C}}$ and $\widetilde{G} := \langle\exp\mathfrak{g}\rangle \subseteq G_{\mathbb{C}}$.*

Proof. Let $H_1 := \langle\exp H(W)_{\mathbb{C}}\rangle$. Since $G_{\mathbb{C}}$ is simply connected, this is a closed simply connected normal subgroup of $G_{\mathbb{C}}$ such that $G'_{\mathbb{C}} := G_{\mathbb{C}}/H_1$ is simply connected ([Ho65, p.135]). Write $\pi: G_{\mathbb{C}} \to G'_{\mathbb{C}}$ for the quotient homomorphism and set $W' := d\pi(1)W$, $\mathfrak{g}' := d\pi(1)\mathfrak{g}$ etc. Suppose that the theorem holds for $G'_{\mathbb{C}}$ and that $S' \subseteq G'_{\mathbb{C}}$ is a Lie semigroup with

$$\mathbf{L}(S') = \mathfrak{g}' + iW' \quad \text{and} \quad S' = G'\exp W'.$$

Then $S := \pi^{-1}(S')$ is a Lie semigroup because S' is closed and H_1 is connected. Moreover,

$$\mathbf{L}(S) = d\pi(1)^{-1}\left(\mathbf{L}(S')\right) = d\pi(1)^{-1}(\mathfrak{g}' + iW') = \mathfrak{g} + iW$$

and

$$S \subseteq \widetilde{G}\exp(W)H_1 = H_1\widetilde{G}\exp(W).$$

Therefore we may assume that $H_1 = \{1\}$, i.e., that W is pointed. Let $\mathfrak{g}_1 := (W - W)_{\mathbb{C}}$. This is an ideal of $\mathfrak{g}_{\mathbb{C}}$ and therefore $X \in W$ implies that $\operatorname{ad}(X)\mathfrak{g}_{\mathbb{C}} \subseteq \mathfrak{g}_1$. So $\operatorname{Spec}(\operatorname{ad}X) \subseteq \operatorname{Spec}(\operatorname{ad}X \mid_{\mathfrak{g}_1}) \cup \{0\} \subseteq \mathbb{R}$ because iX is contained in the boundary of $\operatorname{comp}(\mathfrak{g}_1)$ and therefore $\operatorname{Spec}(i\operatorname{ad}X) \subseteq i\mathbb{R}$. Now Lawson's Theorem (Theorem 7.34) and its corollary imply that $S_1 := G\exp(W)$ is a closed subsemigroup of $G_{\mathbb{C}}$. Since G is connected, it follows that S is a Lie semigroup. ∎

Notes

The study of invariant cones was initiated by Kostant, Segal and Vinberg (cf. [Se76], [Vi80]). A classification for the case of simple Lie algebras was obtained independently bc Ol'shanskiĭ, Pancitz, and Kumaresan-Ranjan (cf. [Ols82b], [Pa84], [KR82]). The general case was settled in [HHL89]. The concept of a compactly embedded Cartan algebra has of course been around implicitely. A systematic treatment can be found in [HiHo89] and [HiNe91]. The characterization of Lie algebras containing invariant cones appeared in [Ne92b]. Lawson's Theorem on Ol'shanskiĭ semigroups in the special case of simple linear groups is due to Ol'shanskiĭ ([Ol82a]).

8. Compression semigroups

In this chapter we study semigroups of transformations leaving a fixed given set invariant. The principal question here is to decide when such a semigroup is big, which for example can mean, has interior points, or even that it is maximal. We will mainly consider actions on real and complex flag manifolds.

In the first section we deal with the real case and introduce invariant control sets and other concepts from geometric control theory which are useful in this context., We prove San Martin's result that every subsemigroup S of a linear reductive Lie group with non-empty interior has a unique invariant control set C_S on every real flag manifold. For a maximal flag manifolds M one even knows that $C_S = M$ is equivalent to $S = G$. This is the basic tool to prove that a given semigroup is maximal.

In Section 8.2 we provide some background from symplectic geometry about hamiltonian group actions. We also show that for a complex vector space endowed with a pseudo-Hermitean form we have a pseudo-Kähler structure on an open subset of the corresponding projective space. The associated symplectic structure is shown to be the Marsden-Weinstein reduction of the natural symplectic structure on the vector space. We also give a description of the corresponding moment map.

Section 8.3 starts with a brief survey on the representation theory of complex semisimple Lie algebras where the main accent lies on the action of the complex group on the orbit of the highest weight ray which is a complex flag manifold. To relate this to the real groups, we give a discription of the open orbits of a real form in a complex flag manifold and we show that if $\mathfrak{g}$ contains a compactly embedded Cartan algebra, then every open G-orbit in an associated complex flag manifold $G_{\mathbb{C}}/P$ can be realized as a G-orbit of a highest weight ray in the $G_{\mathbb{C}}$-orbit of the same ray which in turn lies in the projective space of an irreducible $\mathfrak{g}_{\mathbb{C}}$-module.

If, in addition, G is a Hermitean simple Lie group, then every open G-orbit turns out to carry an invariant pseudo-Kähler structure such that the action of G is Hamiltonian. The main tool to obtain this realization is the observation that every complex $\mathfrak{g}$-module admits an invariant pseudo-Hermitean form, so that the results from Section 8.2 apply.

In the next section we consider the compression semigroup $S \subseteq G_{\mathbb{C}}$ of an open G-orbits in a complex flag manifold $G_{\mathbb{C}}/P$. We show that whenever S has non-empty interior and is different from $G_{\mathbb{C}}$, then G is Hermitean, S is one of the two maximal complex Ol'shanskiĭ semigroups in $G_{\mathbb{C}}$, and the moment mapping maps the open orbit into a coadjoint orbit in $\mathfrak{g}^*$ which lies in an invariant cone. Moreover, for Hermitean groups the latter property characterizes those open orbits for which S has non-empty interior.

We use the results from Section 8.4 to obtain in Section 8.5 a simple description of the compression and contraction semigroups associated to a pseudo-Hermitean form on a complex vector space.

Finally we combine in Section 8.6 the results from Sections 8.1 and 8.4 to show that maximal Ol'shanskiĭ semigroups $G \exp(iW_{\max})$ are in fact maximal subsemigroups of $G_{\mathbb{C}}$.

8.1. Invariant control sets

Let G be a Lie group, X a G-manifold and $C \subseteq X$ non-empty. We consider the *compression semigroup*

$$\mathrm{compr}(C) := \{g \in G : g.C \subseteq C\}$$

which is a subsemigroup of G. It will turn out that for suitable C the semigroup $\mathrm{compr}(C)$ is maximal. Here suitable will in particular mean that C is an *invariant control set* for a subsemigroup S of G, i.e.,

$$(8.1) \qquad\qquad \overline{S.c} = \overline{C} \qquad \forall c \in C$$

and C is maximal with respect to this property. The semigroup $S \subseteq G$ will be called *accessible on* X if

$$\mathrm{int}(S.x) \neq \emptyset \quad \forall x \in X.$$

In the following proposition we collect a few facts about invariant control sets:

Proposition 8.1. *Let G be a topological group, S a subsemigroup of G and X a G-space.*

(i) *If $\emptyset \neq C \subseteq X$ is closed and satisfies (8.1), then it is an invariant control set.*

(ii) *If C_1 and C_2 are invariant control sets with non-empty intersection, then they agree.*

(iii) *If S is accessible, then every invariant control set is closed.*

(iv) *If $\bigcap_{x \in X} \overline{S.x}$ is non-empty, it is the unique invariant control set for S.*

(v) *Each compact S-invariant subset $Y \subseteq X$ contains an invariant control set. If S is accessible, then Y contains only finitely many invariant control sets.*

(vi) *If S has non-empty interior and C is an invariant control set for S, then the set $C_0 := (\mathrm{int}\, S).C$ is open and dense in C. It is S-invariant and satisfies*

$$(8.2) \qquad\qquad C_0 = \{c \in C : (\exists s \in \mathrm{int}\, S)\, s.c = c\}$$

and

$$(8.3) \qquad\qquad C_0 = (\mathrm{int}\, S).c \quad \forall c \in C_0.$$

(vii) *Suppose that G is connected and acts transitively on X. Moreover, let C be an invariant control set for S and C' an invariant control set for S^{-1}. If S has non-empty interior and $\mathrm{int}\, C \cap \mathrm{int}\, C' \neq \emptyset$, then $C = C' = X$.*

Proof.   (i) Let C' satisfy (8.1) and $C \subseteq C'$. Then for $x' \in C'$ and $x \in C$ we have $x' \in \overline{S.x} \subseteq \overline{C} = C$.

(ii) Let $x \in C_1 \cap C_2$ then $\overline{C}_1 = \overline{S.x} = \overline{C}_2$ so that $C_1 \cup C_2$ satisfies (8.1). Now the maximality shows $C_1 = C_2$.

(iii) Let C be an invariant control set and $x \in \overline{C}$. Then $S.x \subseteq \overline{C}$ so that we can find an $x' \in \text{int}(S.x) \cap C$. Now

$$\overline{C} = \overline{S.x'} \subseteq \overline{S.x} \subseteq \overline{C}$$

and again $C = \overline{C}$ by maximality.

(iv) Let $C := \bigcap_{x \in X} \overline{S.x}$ and $x \in C$. Then $C \subseteq \overline{S.x} \subseteq C$ since C is closed and S-invariant. Thus (i) implies that C is an invariant control set. But since C intersects all S-orbit closures, the uniqueness follows from (ii).

(v) The set of non-empty closed S-invariant subsets of Y is inductively ordered and hence contains a minimal element. Each of these minimal sets is an invariant control set. Now suppose that there are countably many different control sets $(C_n)_{n \in \mathbb{N}}$. Take a sequence $c_n \in C_n$, which we may assume to converge to $c \in Y$. Then choose $s \in S$ such that $s.c \in \text{int}(S.c)$. Since $s.c_n \to s.c$ we may also assume that $s.c_n \in \text{int}(S.c)$ for all n, so that

$$(\text{int } S).c_n \subseteq \text{int } C_n \subseteq C_n = \overline{S.(s.c_n)} \subseteq \overline{S.c}.$$

But then, for a fixed $n \in \mathbb{N}$, we find an $s' \in S$ such that $s'.c \in \text{int } C_n$ and, since $s'.c_m \to s'.c$, an $m_0 \in \mathbb{N}$ such that $s'.c_m \in \text{int } C_n$ for all $m \geq m_0$. This proves the claim in view of (ii).

(vi) It is clear that C_0 is open. The S-invariance follows from the fact that int S is a semigroup ideal. The density is a consequence of the S-invariance and (8.1). Thus it only remains to prove the identities (8.2) and (8.3). So let $c \in C$. If there exists $s \in \text{int } S$ with $s.c = c$, then clearly $c \in C_0$. Conversely, if $c \in C_0$, then $(\text{int } S^{-1}).c \cap C$ is relatively open and therefore intersects the dense subset $S.c$. Let $s.c = t^{-1}.c$ with $s \in S$ and $t \in \text{int } S$, then $c = ts.c$ and $ts \in \text{int } S$. To show (8.3), note that clearly $(\text{int } S).c \subseteq C_0$. Conversely, if $c_0 \in C_0$, then, as before, $(\text{int } S^{-1}).c_0 \cap C$ intersects $S.c$, i.e., $c_0 \in \text{int } S.c$.

(vii) If $C_0 = \text{int } S.C$ and $C_0' = \text{int } S^{-1}.C'$, then the assumption yields an $x \in C_0 \cap C_0'$. If $y \in C_0'$, then the equality (8.3) guarantees the existence of an element $g \in \text{int } S$ with $y = g.x$. This shows $y \in C_0$ and hence $C_0 \subseteq C_0'$. By symmetry we even have equality. Now the invariance properties of C_0 and C_0' say that C_0 is G-invariant, hence equal to X.   ∎

We now consider the situation, where G is a real reductive Lie group with Iwasawa decomposition KAN and $X = G/P$ with a parabolic subgroup P of G. This means that P contains a conjugate of $P_{\min} := MAN$, where M is the centralizer $M = Z_K(A)$ of A in K. We will assume that $MAN \subseteq P$.

Note that Proposition 8.1 shows that any subsemigroup S of G admits an invariant control set. In fact, if we assume that $(\text{int } S) \cap A \neq \emptyset$, then it even shows that the invariant control set is unique. To see this, note first that conjugating S if necessary we may assume that int S contains an element

$$a \in A^+ := \{a \in A: (\forall \lambda \in \Delta^+(\mathfrak{g}, \mathfrak{a}))\, \lambda(\log a) > 0\}.$$

If now $x_0 = 1P \in G/P$, then

$$\lim_{n \to \infty} a^n . x = x_0 \quad \forall x \in \overline{N} . x_0,$$

where $\overline{N} = \theta(N)$ for a Cartan involution θ leaving K and A invariant. Note that $\overline{N} . x_0$ is the open and dense Bruhat cell so that

$$x_0 \in \bigcap_{x \in G/P} \overline{S . x}$$

since S has non-empty interior and hence is accessible.

Suppose that G has compact center. Then the hypothesis made on S is always satisfied for *some* Iwasawa A as long as S has non-empty interior since then int S contains regular points of G so that int S intersects a Cartan subgroup H of G. If θ is a Cartan involution leaving $\mathfrak{h} = L(H)$ invariant, then

$$H = (H \cap K) \exp(\mathfrak{h} \cap \mathfrak{p}),$$

where $\mathfrak{g} = \mathfrak{k} + \mathfrak{p}$ is the corresponding Cartan decomposition and K the (compact!) group belonging to $\mathfrak{k}$. Now we set

$$S_k := \{ k \in H \cap K : k \exp(\mathfrak{h} \cap \mathfrak{p}) \cap \operatorname{int} S \neq \emptyset \}$$

which is an open subsemigroup of $H \cap K$ because H is abelian. But then Corollary 1.21 shows that S_k contains the identity component $(H \cap K)_0$ of $H \cap K$ and the Cartan decomposition of H given above shows that int S has to contain points in $\exp(\mathfrak{h} \cap \mathfrak{p})$. If we now choose a maximal abelian subspace $\mathfrak{a}$ of $\mathfrak{p}$ containing the logarithm of such a point, then the corresponding Iwasawa decomposition $G = KAN$ satisfies int $S \cap A \neq \emptyset$.

Thus we have shown the major part of the following proposition:

Proposition 8.2. *Let G be a reductive Lie group with compact center and S a subsemigroup of G with non-empty interior. Then for any parabolic subgroup P of G the space G/P contains a unique invariant control set*

$$C_S = \bigcap_{x \in G/P} \overline{S . x}$$

for S. Replacing S by a conjugate, we may assume that we have an Iwasawa decomposition $G = KAN$ of G such that P contains AN and int $S \cap A^+ \neq \emptyset$. Then, in particular, the base point x_0 is contained in $(C_S)_0$ and $(C_S)_0 = (\operatorname{int} S) . x_0$.

Proof. We only have to show the last assertion. But that follows immediately from $(A \cap \operatorname{int} S) . x_0 = \{x_0\}$ and Proposition 8.1(vi). ∎

The following observations will be useful later on.

Lemma 8.3. *Let V be a finite dimensional real vector space and $S \subseteq V$ an open subsemigroup. Then $(\mathbb{R}^+ \setminus \{0\})S$ is an open convex cone.* ∎

Lemma 8.4. *Let G be linear reductive and S a subsemigroup of G such that there exists a nilpotent element $X \in \mathfrak{g}$ with $\exp X \in \operatorname{int} S$. Then $S = G$.*

Proof. The Jacobson-Morozov Theorem ([Bou75, Ch. VIII, §11, no. 3, Prop. 2]) then says that we can find a subalgebra $\mathfrak{h}$ of $\mathfrak{g}$ isomorphic to $\mathfrak{sl}(2, \mathbb{R})$ containing X. In $\mathfrak{h}$ the element X lies on the zero set of the Killing form. If H is the analytic subgroup of G with $\mathbf{L}(H) = \mathfrak{h}$ and G is linear, then H is not simply connected and hence $H \cap \operatorname{int} S$ contains elements on compact one parameter groups (cf. Corollary 1.21). Therefore $\mathbf{1} \in \operatorname{int} S$ and $S = G$. ∎

Theorem 8.5. (San Martin's Theorem) *Let G be a reductive Lie group with compact center, $P_{\min}$ a minimal parabolic subgroup, and S a subsemigroup of G with non-empty interior. Then $C_S = G/P_{\min}$ if and only if $S = G$.*

Proof. We note first that $S = G$ trivially implies $C_S = G/P_{\min}$. For the converse consider the semigroups

$$A_S^0 := A \cap \operatorname{int} S$$

and

$$B_S^0 := B \cap \operatorname{int} S$$

for $B = AN$. If $\operatorname{pr}: AN \to A$ is the canonical projection, then

$$C^0 := (\mathbb{R}^+ \setminus \{0\}) \exp_A^{-1} A_S^0 \quad \text{and} \quad \widetilde{C}^0 := (\mathbb{R}^+ \setminus \{0\}) \exp_A^{-1} \left(\operatorname{pr}(B_S^0)\right)$$

are open convex cones in $\mathfrak{a}$.

Note that the normalizer $N_K(A)$ also normalizes $M = Z_K(A)$ so that $N_K(A) \subseteq N_K(M)$. We claim that $w.x_0 \in (C_S)_0$ with $w \in N_K(A)$ implies $w^{-1}.C^0 \subseteq \widetilde{C}^0$. In order to show this, we note first that, according to Proposition 8.1, we can find $s_1, s_2 \in \operatorname{int} S$ with $s_1.x_0 = w.x_0$ and $s_2.(w.x_0) = x_0$. In other words $w^{-1}s_1, s_2 w \in P_{\min}$. Writing

$$w^{-1}s_1 = m_1 a_1 n_1, \quad s_2 w = m_2 a_2 n_2$$

we calculate for $s \in A_S^0$

$$s_2 s s_1 = m_2 a_2 n_2 (w^{-1} s w) m_1 a_1 n_1$$
$$= m' a_2 w^{-1} s w a_1 n'$$

with $m' \in M$ and $n' \in N$. We set $s' = a_2 w^{-1} s w a_1 = w^{-1} a_2' s a_1' w$ and see that $M s' N \cap \operatorname{int} S \neq \emptyset$. But

$$\{m \in M : m \langle s' \rangle N \cap \operatorname{int} S \neq \emptyset\}$$

with $\langle s' \rangle = \{(s')^k : k \in \mathbb{N}\}$ is an open subsemigroup of M and hence by Corollary 1.21 equal to M since M is compact and connected. This proves $\langle s' \rangle \cap \operatorname{pr}(B_S^0) \neq \emptyset$. Consider $X \in C^0$. Then for $s = \exp X$ we have

$$\lim_{k \to \infty} \frac{\log(a_2' s^k a_1')}{k} = \lim_{k \to \infty} \left(\frac{\log a_1' + \log a_2'}{k} + \log s\right) = X.$$

Moreover, we have seen that $w^{-1}a_2's^ka_1'w \in \exp(\widetilde{C}_0)$ since $s^k \in A_S^0$. But then $w^{-1}.\log a_2's^ka_1' \in \widetilde{C}^0$ and

$$w^{-1}.X = w^{-1}.\lim_{k\to\infty}\frac{\log a_2's^ka_1'}{k} \in \overline{\widetilde{C}^0}.$$

Thus $w^{-1}.C^0 \subseteq \overline{\widetilde{C}^0}$ which implies $w^{-1}.C^0 \subseteq \widetilde{C}^0$ since $w^{-1}.C^0$ is open. This proves our claim.

Now suppose that $C_S = G/P_{\min}$. Then we have $w^{-1}.C^0 \subseteq \widetilde{C}^0$ for *all* $w \in N_K(A)$. This means that $X \in C^0$ implies

$$0 = \sum_{w \in W} w.X \in \widetilde{C}^0$$

whence $\widetilde{C}^0 = \mathfrak{a}$. In particular $0 \in \mathrm{pr}(B_S^0)$ which shows $(\mathrm{int}\,S) \cap N \neq \emptyset$. Now Lemma 8.4 applied to $\mathrm{Ad}\,G$ shows

$$\mathbf{1} \in (\mathrm{int}\,S)Z(G),$$

where $Z(G)$ is the center of G since the center of G acts trivially on $G/P_{\min}$. But $Z(G)$ is compact, so that the semigroup $Z(G)\cap\mathrm{int}\,S$ is a group and hence $\mathbf{1} \in \mathrm{int}\,S$ which again shows $S = G$. $\qquad\blacksquare$

With this theorem we can show that in a reductive Lie group with compact center maximal subsemigroups with non-empty interior are always of the form S_C for some closed subset $C \subseteq G/P_{\min}$ with $P_{\min}$ a minimal parabolic:

Theorem 8.6.$\qquad$*Let G be a reductive Lie group with compact center and S a maximal subsemigroup of G with non-empty interior and $C_S \subseteq G/P_{\min}$ the corresponding invariant control set. Then*

$$S = \mathrm{compr}(C_S) = \{g \in G : g.C_S \subseteq C_S\}.$$

Proof.$\qquad$We know from San Martin's Theorem (Theorem 8.5) that $C_S \neq G/P_{\min}$. The definition of C_S shows that $S \subseteq \mathrm{compr}(C_S)$ so that maximality shows the equality. $\qquad\blacksquare$

Note at this point that Theorem 8.6 is, in its present form, not of much practical use. What is needed to make it useful is a sufficiently general prescription to find sets in flag manifolds whose associated semigroups have non-empty interior. At present such a prescription is not available, but strikingly enough all known examples may described in a very similar fashion using the Kähler structure of complex flag manifolds and the associated moment map. But before we describe such examples, we show how in some cases one can use Theorem 8.5 to actually prove the maximality of a given semigroup. Sometimes for a semigroup S with non-empty interior acting on the flag manifold G/P, the invariant control set $C_{S^{-1}}$ satisfies

(8.4)$\qquad\qquad\qquad C_{S^{-1}} = (G/P)\backslash \mathrm{int}\,C_S.$

If now $S_{\max}$ is a maximal semigroup containing $\mathrm{compr}(C_S)$, then C_S is properly contained in $C := C_{S_{\max}}$ unless $\mathrm{compr}(C_S)$ is maximal which we assume not to be the case. But then

$$C \backslash C_S \subseteq \mathrm{int}\, C_{S^{-1}} \subseteq \mathrm{int}\, C_{S_{\max}^{-1}}.$$

Then

$$\mathrm{int}\,(S.(C \backslash C_S)) \subseteq \mathrm{int}\,(S_{\max}.(C \backslash C_S)) \subseteq \mathrm{int}\, C_{S_{\max}},$$

so that Proposition 8.1(vii) shows that $C = G/P$. If P is a minimal parabolic, then Theorem 8.5 shows at this point that $S_{\max} = G$, contradicting our hypotheses. This means that (8.4) for P minimal means that $\mathrm{compr}(C_S)$ is maximal in G. Unfortunately it turns out that (8.4) is much more likely to be true for maximal parabolics. Therefore the following elementary lemma is useful:

Lemma 8.7.   *Let* $\pi: E \to X$ *be a fibre bundle on which the Lie group* G *acts transitively by bundle automorphisms. If* S *is a subsemigroup of* G, *then the following statements are equivalent:*

(1) $S.e = E$ *for all* $e \in E$.

(2) $S.x = X$ *for all* $x \in X$ *and there exists* $x_0 \in X$ *such that* $\pi^{-1}(x_0) \subseteq S.e$ *for all* $e \in \pi^{-1}(x_0)$.

Proof.   The implication (1) $\Rightarrow$ (2) is trivial. For the converse fix $e_1, e_2 \in E$. Then we find s_1, s_2 and s in S such that $s_1.\pi(e_1) = x_0$, $s_2.x_0 = \pi(e_2)$, and $s.(s_1.e_1) = s_2^{-1}.f_2$. Thus $s_2 s s_1.e_1 = e_2$ which proves the claim.   ∎

Proposition 8.8.   *Let* S *be a subsemigroup of a connected Lie group* G *such that* $\mathrm{int}(S) \neq \varnothing$, $H \subseteq S$ *is subgroup of* S, *and* M *a homogeneous* G-*space. Suppose that the* H-*orbit* $\Omega \subseteq M$ *is invariant under* S. *Then* $\overline{\Omega}$ *is an invariant control set for every subsemigroup* S' *containing* S *which maps* $\overline{\Omega}$ *into itself.*

Proof.   In view of Proposition 8.1, we only have to show that the S'-orbit of every $p \in \overline{\Omega}$ is dense. Let us first suppose that $p \in \Omega$. Then the S'-orbit contains the H-orbit and therefore Ω. So we may assume that $p \in \partial\Omega$. Pick $s \in \mathrm{int}\, S$, so that $s.p \in \Omega$ (note that $\mathrm{int}(S).p$ is an open subset of $\overline{\Omega}$). Then

$$S'.p \supseteq S's.p = S'(Hs.p) = S'.\Omega \supseteq \Omega$$

shows that $S'.p$ is dense in $\overline{\Omega}$.   ∎

Corollary 8.9.   *Let* M *be a homogeneous* G-*space and* H *a closed subgroup such that the compression semigroup* $S = \mathrm{compr}(\Omega)$ *of an open* H-*orbit* Ω *in* M *has interior points. Then* $\overline{\Omega}$ *is an invariant control set for* S.

Proof.   Since $H \subseteq \mathrm{compr}(\Omega) = S$, the assumptions of Proposition 8.8 are satisfied.   ∎

Proposition 8.10.   *Let* G *be a connected Lie group,* $M = G/P$ *be a homogeneous* G-*space,* H *a closed subgroup such that there exist open* H-*orbits in* G/P, *and* S *a subsemigroup with non-empty interior such that* $H \subseteq S$. *Then the following assertions hold:*

(i) *The set of all elements in M such that the H-orbit is open and dense in M.*

(ii) *Every invariant control set for S is the closure of a union of open H-orbits.*

Proof. (i) We may assume that $\dim M > 0$. The closed subgroup H has at most countably many components. Therefore an H-orbit is a countable union of H_0-orbits. Thus, if there exists an open H-orbit, Ω say, then $\Omega = H.x_1$ is locally compact and therefore

$$\Omega \cong H/H^{x_1}$$

is a homogeneous space of H. It follows that the image of H_0 is open in Ω. Thus $H_0.x_1$ is open. Let $x_1 = g.x_0$, where $P = G^{x_0}$, i.e., x_0 is the base point in M. Then the condition that $H_0.x_1$ is open means that

$$\mathfrak{h} + \mathrm{Ad}(g)\mathfrak{p} = \mathfrak{g},$$

where $\mathfrak{h} = \mathbf{L}(H)$, $\mathfrak{p} = \mathbf{L}(P)$ and $\mathrm{Ad}(g)\mathfrak{p} = \mathbf{L}(G^{x_1})$.

Let $k = \dim \mathfrak{p}$ and S_k the space of all k-dimensional subspaces of $\mathfrak{g}$. The set of all elements V of S_k such that $V + \mathfrak{h} \neq \mathfrak{g}$ is an algebraic subvariety (we only need that it is analytic), i.e., it is locally defined as the set of zeros of finitely many polynomials $f_1, \ldots, f_n$. Let $h_i(g) := f_i(\mathrm{Ad}(g)\mathfrak{p})$. Since $\mathrm{Ad}(g)\mathfrak{p} + \mathfrak{h} = \mathfrak{g}$, there exists i_0 with $h_{i_0}(g) \neq 0$. Since the function h_{i_0} is analytic, its set of zeros is closed and has no interior points. For all g' with $h_{i_0}(g') \neq 0$ the H_0-orbit, and therefore the H-orbit, of $g.x_0$ is open in M.

(ii) Let C be an invariant control set for S and

$$U = \{x \in X : H.x \text{ is open}\}.$$

According to (i) U is open and dense so that for $C_0 = (\mathrm{int}\, S).C$ which is open and dense in C we have that $C_0 \cap U$ is open and dense in C_0. But $C_0 \cap U$ clearly is H-invariant, whence

$$C = \overline{C_0} = \overline{C_0 \cap U} = \overline{\bigcup_{c \in C_0 \cap U} H.c}$$

and the $H.c$ with $c \in C_0 \cap U$ are open. ∎

8.2. Moment maps and projective spaces

Let M be a real symplectic manifold, i.e., we have a non-degenerate closed 2-form ω on M. Contraction with this 2-form yields an isomorphism between the space $\mathcal{X}(M)$ of vector fields on M and the space $\Omega^1(M)$ of 1-forms on M:

$$\omega^\sharp : \Omega^1(M) \to \mathcal{X}(M)$$

with

$$\omega(\omega^\sharp(\alpha), v) = \alpha(v) \quad \forall v \in v \in T_m(M), \alpha \in T_m(M)^*.$$

$\omega^\sharp$ Using $\omega^\sharp$ we can associate a *Hamiltonian vector field* $\mathcal{X}_f \in \mathcal{X}(M)$ to each function $f \in C^\infty(M)$ via $\mathcal{X}_f = -\omega^\sharp(df)$. We obtain

(8.5) $$\omega(\mathcal{X}_f(m), v) = -df_m(v) \quad \forall m \in M, v \in T_m M.$$

We define the *Poisson bracket* of two functions $f_1, f_2 \in C^\infty(M)$ by

$$\{f_1, f_2\} := \omega(\mathcal{X}_{f_1}, \mathcal{X}_{f_2}).$$

Then $(C^\infty(M), \{ , \})$ is a Lie algebra and the map

$$j: C^\infty(M) \to \mathcal{X}(M)$$
$$f \mapsto \mathcal{X}_f$$

is a Lie algebra homomorphism (cf. [LM87]).

Now suppose that a Lie group G acts smoothly on the left on M by diffeomorphisms preserving the symplectic form:

$$\tau: G \times M \to M$$
$$(g, m) \mapsto g.m.$$

Then

$$\dot\tau(X)(m) := \frac{d}{dt}\Big|_{t=0} \exp(-tX).m$$

defines a homomorphism of Lie algebras

$$\dot\tau(X): \mathfrak{g} \to \mathcal{X}(M).$$

The action τ is called *Hamiltonian* if there exists a homomorphism $\varphi: \mathfrak{g} \to C^\infty(M)$ such that

$$j \circ \varphi = \dot\tau.$$

For a Hamiltonian group action τ with homomorphism φ one defines the *moment map* $\Phi: M \to \mathfrak{g}^*$ via

$$\langle \Phi(m), X \rangle = \varphi(X)(m) \quad \forall m \in M, X \in \mathfrak{g}.$$

The moment map is (τ, Ad^*)-equivariant (cf. [LM87]).

As a first example we consider the space $\mathbb{C}^N$ with the Hermitean metric

$$\widetilde{\omega}_J = \sum_{j=1}^N \varepsilon_j dz_j d\bar{z}_j = \sum_{j=1}^N \varepsilon_j\big((dx_j)^2 + (dy_j)^2 + i(dy_j \wedge dx_j)\big),$$

where $z_j = x_j + iy_j$ are the complex coordinate functions and $J = (\varepsilon_1, \ldots, \varepsilon_N)$ is a vector whose components are either 1 or -1. The symplectic form we want to use is two times the imaginary part of Ω:

$$\omega_J = 2 \sum_{j=1}^N \varepsilon_j dy_j \wedge dx_j = \frac{1}{i} \sum_{j=1}^N \varepsilon_j dz_j \wedge d\bar{z}_j.$$

For $f \in C^\infty(\mathbb{C}^N)$ the Hamiltonian vector field is given by

$$\mathcal{X}_f(z) = \frac{1}{2} \sum_{j=1}^N \varepsilon_j \Big(\frac{\partial f}{\partial x_j}(z)\frac{\partial}{\partial y_j} - \frac{\partial f}{\partial y_j}(z)\frac{\partial}{\partial x_j}\Big) = i \sum_{j=1}^N \varepsilon_j \Big(\frac{\partial f}{\partial \bar{z}_j}(z)\frac{\partial}{\partial z_j} - \frac{\partial f}{\partial z_j}(z)\frac{\partial}{\partial \bar{z}_j}\Big),$$

where

$$\frac{\partial}{\partial z_j} = \frac{1}{2}\Big(\frac{\partial}{\partial x_j} - i\frac{\partial}{\partial y_j}\Big), \quad \frac{\partial}{\partial \bar{z}_j} = \frac{1}{2}\Big(\frac{\partial}{\partial x_j} + i\frac{\partial}{\partial y_j}\Big).$$

The Poisson bracket is

$$\{f_1, f_2\} = i\sum_{j=1}^{N} \varepsilon_j \Big(\frac{\partial f_1}{\partial z_j}\frac{\partial f_2}{\partial \bar{z}_j} - \frac{\partial f_1}{\partial \bar{z}_j}\frac{\partial f_2}{\partial z_j}\Big).$$

Note that a smooth function $F{:}\mathbb{C}^N \to \mathbb{C}^N$ corresponds to a *real* vector field on $\mathbb{C}^N$ via

$$\mathcal{X}_F(z) = \sum_{j=1}^{N} \mathrm{Re}\,(F(z)_j)\frac{\partial}{\partial x_j} + \mathrm{Im}\,(F(z)_j)\frac{\partial}{\partial y_j}$$

$$= \sum_{j=1}^{N} F(z)_j\frac{\partial}{\partial z_j} + \overline{F(z)}_j\frac{\partial}{\partial \bar{z}_j}.$$

We identify J with the Hermitean form it determines and denote by $\mathrm{U}(J)$ the group of isometries of J. This group acts on $\mathbb{C}^N$ by matrix multiplication. Then for $X = (X_{rs})_{r,s=1,\ldots,N} \in \mathfrak{u}(J) := \mathrm{L}\,\big(\mathrm{U}(J)\big)$ we have

$$\varepsilon_s\overline{X}_{sr} = -\varepsilon_r X_{rs}$$

and the derived action can be written

$$\dot{\tau}(X)(z) = -X.z = -\Big(\sum_{s=1}^{N} X_{1s}z_s, \ldots, \sum_{s=1}^{N} X_{Ns}z_s\Big)$$

$$= -\sum_{r,s=1}^{N} \Big(X_{rs}z_s\frac{\partial}{\partial z_r} + \overline{X_{rs}}\bar{z}_s\frac{\partial}{\partial \bar{z}_r}\Big)$$

$$= -\sum_{r,s=1}^{N} \Big(X_{rs}z_s\frac{\partial}{\partial z_r} - \varepsilon_r\varepsilon_s X_{sr}\bar{z}_s\frac{\partial}{\partial \bar{z}_r}\Big).$$

Note that the identity function $\mathrm{id}{:}\mathbb{C}^N \to \mathbb{C}^N$ corresponds to the vector field

$$\mathcal{X}_{\mathrm{id}}(z) = \sum_{j=1}^{N} z_j\frac{\partial}{\partial z_j} + \bar{z}_j\frac{\partial}{\partial \bar{z}_j}.$$

We set

$$\varphi(X)(z) = \omega_J\big(\dot{\tau}(X)(z), z\big).$$

Then

$$\varphi(X)(z) = i\sum_{r,s} \varepsilon_r X_{rs}\bar{z}_r z_s$$

and

$$\frac{\partial\varphi(X)}{\partial z_r} = i\sum_{j} \varepsilon_j X_{jr}\bar{z}_j.$$

$$\frac{\partial \varphi(X)}{\partial \bar{z}_r} = i \sum_j \varepsilon_r X_{rj} z_j.$$

Therefore a quick calculation gives

$$\mathcal{X}_{\varphi(X)}(z) = \dot{\tau}(X)(z).$$

Next we show that φ is a homomorphism. To this end note that $[X, X']$ is given by the matrix

$$\left(\sum_j (X_{rj} X'_{js} - X'_{rj} X_{js}) \right)_{r,s=1,\dots,N}$$

so that

$$\varphi([X, X'])(z) = i \sum_{r,s,j} (X_{rj} X'_{js} - X'_{rj} X_{js}) \varepsilon_r \bar{z}_r z_s.$$

On the other hand

$$\{\varphi(X), \varphi(X')\}(z) =$$

$$= -i \sum_l \left(\left(i \sum_r X_{rl} \varepsilon_r \bar{z}_r \right) \left(i \sum_s X'_{ls} \varepsilon_l z_s \right) - \left(i \sum_s X_{ls} \varepsilon_l z_s \right) \left(i \sum_r X'_{rl} \varepsilon_r \bar{z}_r \right) \right) \varepsilon_l$$

$$= i \sum_{r,s,l} \varepsilon_r \left(X_{rl} X'_{ls} \bar{z}_r z_s - X'_{rl} X_{ls} \bar{z}_r z_s \right)$$

$$= \varphi([X, X'])(z).$$

Thus τ is a Hamiltonian action with moment map

$$\langle \Phi(z), X \rangle = i \sum_{r,s} X_{rs} \varepsilon_r \bar{z}_r z_s.$$

If we identify $\mathfrak{u}(J)$ with $\mathfrak{u}(J)^*$ via the trace form $\langle X, X' \rangle = \operatorname{tr}(XX')$ (cf. Lemma 8.11 below), then $\Phi(z) \in \mathfrak{u}(J)$ is given by the matrix

$$\left(\Phi(z)_{rs} \right)_{r,s=1,\dots,N} = i(\varepsilon_s \bar{z}_s z_r)_{r,s=1,\dots,N}.$$

Before we consider Marsden-Weinstein reductions of this example which will then yield symplectic structures on open domains in $\mathbb{P}(\mathbb{C}^N)$ we reformulate it in a coordinate free way.

Let V be a finite dimensional complex Hilbert space with respect to the scalar product $\langle \cdot, \cdot \rangle$. Moreover, let A be a selfadjoint operator defining the non-degenerate Hermitean form

$$J(v, w) := \langle v, Aw \rangle.$$

We denote the transpose of an operator with respect to J by $X^\sharp$, i.e.,

$$J(X.v, w) = J(v, X^\sharp.w) \qquad \forall v, w \in V.$$

Then the group of all J-preserving invertible transformations of V is given by

$$U(J) = \{g \in Gl(V) : g^\sharp = g^{-1}\}$$

and

$$u(J) = \{X \in gl(V) : X^\sharp = -X\}.$$

As before we identify the dual space of $u(J)$ with $u(J)$ via the trace form. Consider therefore the following mapping:

$$\psi : u(J) \to u(J)^*, \quad Y \mapsto (X \mapsto tr(XY)).$$

At this point we note that indeed $tr(XY) \in \mathbb{R}$ for $X, Y \in u(J)$ since $tr(X^\sharp) = tr(A^{-1}X^*A) = \overline{tr(X)}$, so that

$$tr(XY) = tr\left((-X)^\sharp(-Y)^\sharp\right) = tr\left((YX)^\sharp\right) = \overline{tr(YX)} = \overline{tr(XY)}.$$

Lemma 8.11.

(i) $\psi(gYg^{-1}) = Ad^*(g)\psi(Y)$ for all $g \in U(J), Y \in u(J)$.

(ii) ψ is a linear isomorphism.

Proof. (i)

$$\langle \psi(gYg^{-1}), X \rangle = tr(gYg^{-1}X) = tr(Yg^{-1}Xg) =$$
$$= \langle \psi(Y), Ad(g^{-1})X \rangle$$

(ii) It suffices to show that ψ is injective. Suppose that $\psi(Y) = 0$. Since J is defined by the Hermitean matrix A, we have that $X^\sharp = A^{-1}X^*A$. Hence $\psi(Y) = \psi(-Y^\sharp) = 0$ implies that

$$tr(XA^{-1}Y^*A) = tr\left((AX)(YA^{-1})^*\right) = 0$$

for all $X \in u(J)$. Since YA^{-1} is skew Hermitean and the same is true for AX, it follows that $YA^{-1} = 0$, hence that $Y = 0$. ∎

The group $U(J)$ acts by symplectic linear transformations on the symplectic vector space V. If we endow V with the constant symplectic form this linear action is Hamiltonian with moment map $\Phi : V \to u(J)^*$ given by

$$\langle \Phi(v), X \rangle = -\omega(X.v, v)$$

for $X \in u(J)$ and $v \in V$. The Hamiltonian vector field on V associated to an element $X \in u(J)$ is $v \mapsto -X.v$. Now it is easy to check that this moment map agrees with the one given in coordinates.

The next class of examples consists of open domains in projective spaces. Here we do the coordinate free version first. Let $\mathbb{P}(V)$ denote the projective space of V and $\Omega \subseteq \mathbb{P}(V)$ the set of all equivalence classes of non-isotropic vectors for J, i.e.,

$$\Omega = \{[v] \in \mathbb{P}(V) : J(v, v) \neq 0\}.$$

We want to endow Ω with a symplectic structure that is invariant under the unitary group $U(J)$ of all J-preserving invertible transformations of V.

Lemma 8.12.

$$\psi' : \Omega \to \mathfrak{u}(J), \quad [v] \mapsto \left(w \mapsto i\frac{J(w,v)}{J(v,v)}v \right)$$

is injective and equivariant with respect to the action of $\mathrm{U}(J)$.

Proof.   First we note that the operators $\psi'([v])$ are in $\mathfrak{u}(J)$. To see this, let $w, w' \in V$ and $[v] \in \Omega$. Then

$$J\big(w, \psi'([v])w'\big) = -i\frac{\overline{J(w',v)}}{J(v,v)}J(w,v)$$

$$= -i\frac{J(v,w')}{J(v,v)}J(w,v)$$

$$= -J\big(\psi'([v]).w, w'\big).$$

This proves that the image of ψ' is contained in $\mathfrak{u}(J)$. Suppose that $\psi'([v]) = \psi'([w])$. Then

$$iv = \psi'([v])(v) = \psi'([w])(v) \in \mathbb{C}^*.w$$

entails that $[v] = [w]$, so ψ' is injective.

To see that ψ' is equivariant, we only have to note that

$$g\psi'([v])g^{-1}.w = i\frac{J(g^{-1}.w,v)}{J(v,v)}g.v = i\frac{J(w,g.v)}{J(g.v,g.v)}g.v = \psi'(g.[v]).w.$$

$\blacksquare$

Let us calculate the composition of ψ and ψ'. For a vector $v \in V$ with $J(v,v) \neq 0$ we construct a basis $e_1 = v$, $e_2,\ldots,e_n$ such that $J(e_1,e_j) = 0$ for $j > 1$. Let e_j^* denote the elements of the dual basis. Then $e_1^*(w) = \frac{J(w,v)}{J(v,v)}$. So we find that

$$\langle \psi \circ \psi'([v]), X \rangle = \mathrm{tr}\left(X\psi'([v])\right)$$

$$= \sum_{j=1}^n e_j^*(X\psi'([v])e_j) = i\sum_{j=1}^n e_j^*(Xv)\frac{J(e_j,v)}{J(v,v)}$$

$$= i\frac{J(Xv,v)}{J(v,v)}\frac{J(e_1,v)}{J(v,v)} = i\frac{J(Xv,v)}{J(v,v)}.$$

Lemma 8.13.   *The set* Ω *consists at most of 2 orbits under the action of the group* $\mathrm{U}(J)$, *namely*

$$\Omega_+ = \{[v] : J(v,v) > 0\} \quad and \quad \Omega_- = \{[v] : J(v,v) < 0\}.$$

If J *is positive or negative definite, then* $\Omega = \mathbb{P}(V)$ *is a* $\mathrm{U}(J)$-*orbit.*

Proof.   This is an immediate consequence of Witt's Theorem. $\blacksquare$

Definition 8.14.   Let $\Phi_\Omega := \psi \circ \psi' : \Omega \to \mathfrak{u}(J)^*$. In view of Lemma 8.12, this mapping is injective, and, since Ω consists precisely of two $\mathrm{U}(J)$-orbits, its image, consisting of two coadjoint orbits, is a symplectic manifold. So we may define a symplectic structure on Ω as the pullback of the symplectic structure of these two orbits. To see this, recall that the injectivity of Φ_Ω implies that the corresponding 2-form is nondegenerate, and that the pullback of a closed form is closed. $\blacksquare$

The symplectic structure on Ω has the following properties:

Proposition 8.15. *The action of* $U(J)$ *on* Ω *is a Hamiltonian action and the moment mapping is given by*

$$\Phi_\Omega : \Omega \to \mathfrak{u}(J)^*, \quad [v] \mapsto \left(X \mapsto i\,\frac{J(X.v,v)}{J(v,v)} \right).$$

Proof.   Since the action of $U(J)$ on $\mathfrak{u}(J)^*$ is a Hamiltonian action and Φ_Ω is a $U(J)$-equivariant bijection, we only have to check that the formula for the moment mapping is correct. But for the coadjoint orbit the moment map simply is the inclusion in $\mathfrak{u}(J)$ and the moment map of the Hamiltonian action which is transported via the symplectic map Φ_Ω then is the composition of Φ_Ω with that inclusion. ∎

The construction of the symplectic forms on the open domains $\Omega_\pm$ in $\mathbb{P}^n(\mathbb{C})$ given above may, as we have mentioned already, be interpreted as a Marsden-Weinstein reduction of the form J on $\mathbb{C}^{n+1}$. The group action that has to be factored here is that of $T := Z\big(U(J)\big) \cong U(1)$. More precisely, we restrict τ to the torus T and in this way obtain a Hamiltonian action of T on $\mathbb{C}^{n+1}$ whose moment map Φ_T simply is the composition of Φ with the canonical projection $\mathfrak{u}(J)^* \to \mathfrak{t}^*$. Using the trace form again we find

$$\Phi_T(z) = i\Big(\sum_{j=0}^n \varepsilon_j |z_j|^2 \Big)\mathbf{1} \in \mathfrak{t}.$$

Now we can identify $\Omega_\pm$ with the orbit space

$$\Phi_T^{-1}(\pm i\mathbf{1})/T \cong \Omega_\pm.$$

Consider the commutative diagram

$$
\begin{array}{ccc}
H_\pm & := & \Phi_T^{-1}(\pm i\mathbf{1}) \xrightarrow{\ \pi_T\ } \Omega_\pm \\
 & & \ \ \downarrow \iota_\pm \qquad \nearrow \pi \\
 & & \mathbb{C}^{n+1}_{J_\pm}
\end{array}
$$

where for a complex vector space V with Hermitean form J we set

$$V_{J_+} := \{v \in V : J(v,v) > 0\} \quad \text{and} \quad V_{J_-} := \{v \in V : J(v,v) < 0\}.$$

Then the Marsden-Weinstein reduction (cf. [LM87]) shows that on $\Omega_\pm$ a symplectic form $\omega_\pm$ is uniquely defined by

$$\pi_T^* \omega_\pm = \iota_\pm^* \omega_J,$$

where ω_J is the symplectic form on $\mathbb{C}^{n+1}$ which is obtained from the imaginary part of J. We set

$$\|z\|_J := \sqrt{|J(z,z)|}.$$

Consider the projection $\pi_\pm : \mathbb{C}^{n+1}_\pm \to H_\pm$ given by

$$\pi_\pm(z) = \frac{z}{\|z\|_J} = (|J(z,z)|)^{-\frac{1}{2}}(z_0, \ldots, z_n).$$

Then $\pi = \pi_T \circ \pi_\pm$. Therefore we have

$$\pi^* \omega_\pm = \pi_\pm^* \pi_T^* \omega_\pm = \pi_\pm^* \iota_\pm^* \omega_J.$$

This means

$$\omega_\pm([z])(d\pi(z)v, d\pi(z)w) = \omega_J([z])((d\pi_\pm)(z)v, (d\pi_\pm)(z)w)$$

for $v, w \in T_z(\mathbb{C}^{n+1})$. But

$$dz_k\left((d\pi_\pm)\frac{\partial}{\partial z_j}\right) = \frac{\partial(z_k \circ \pi_\pm)}{\partial z_j} = -\frac{1}{2\|z\|_J^3}(\pm\varepsilon_j)\overline{z}_j z_k + \frac{1}{\|z\|_J}\delta_{jk}$$

$$dz_k\left((d\pi_\pm)\frac{\partial}{\partial \overline{z}_j}\right) = \frac{\partial(z_k \circ \pi_\pm)}{\partial \overline{z}_j} = -\frac{1}{2\|z\|_J^3}(\pm\varepsilon_j)z_j z_k$$

$$d\overline{z}_k\left((d\pi_\pm)\frac{\partial}{\partial z_j}\right) = \frac{\partial(\overline{z}_k \circ \pi_\pm)}{\partial z_j} = -\frac{1}{2\|z\|_J^3}(\pm\varepsilon_j)\overline{z}_j \overline{z}_k$$

$$d\overline{z}_k\left((d\pi_\pm)\frac{\partial}{\partial \overline{z}_j}\right) = \frac{\partial(\overline{z}_k \circ \pi_\pm)}{\partial \overline{z}_j} = -\frac{1}{2\|z\|_J^3}(\pm\varepsilon_j)z_j \overline{z}_k + \frac{1}{\|z\|_J}\delta_{jk}.$$

From this an easy calculation yields

$$\omega_\pm\left(d\pi\frac{\partial}{\partial z_j}, d\pi\frac{\partial}{\partial \overline{z}_m}\right) = \frac{1}{i\|z\|_J^2}\left(\varepsilon_j\delta_{jm} \mp \frac{(\varepsilon_j + \varepsilon_m)\overline{z}_j z_m}{2\|z\|_J^2}\right),$$

$$\omega_\pm\left(d\pi\frac{\partial}{\partial z_j}, d\pi\frac{\partial}{\partial z_m}\right) = \omega_\pm\left(d\pi\frac{\partial}{\partial \overline{z}_j}, d\pi\frac{\partial}{\partial \overline{z}_m}\right) = 0.$$

We introduce a complex projection on $T_z(\mathbb{C}^{n+1}_{J_\pm})_{\mathbb{C}}$ via the vector fields

$$\left(\frac{\partial}{\partial z_j}\right)_{\mathbb{P}} := \frac{\partial}{\partial z_j} - \sum_{l=0}^{n}\frac{\varepsilon_j\overline{z}_j z_l}{J(z,z)}\frac{\partial}{\partial z_l} = \frac{\partial}{\partial z_j} \mp \sum_{l=0}^{n}\frac{\varepsilon_j\overline{z}_j z_l}{\|z\|_J^2}\frac{\partial}{\partial z_l}$$

and

$$\left(\frac{\partial}{\partial \overline{z}_j}\right)_{\mathbb{P}} := \frac{\partial}{\partial \overline{z}_j} - \sum_{l=0}^{n}\frac{\varepsilon_j z_j\overline{z}_l}{J(z,z)}\frac{\partial}{\partial \overline{z}_l} = \frac{\partial}{\partial \overline{z}_j} \mp \sum_{l=0}^{n}\frac{\varepsilon_j z_j\overline{z}_l}{\|z\|^2}\frac{\partial}{\partial \overline{z}_l}$$

on $\mathbb{C}^{n+1}_{J_\pm}$. Since $d\pi(z)$ annihilates $\mathbb{C}z$, we have

$$\omega_\pm\left(d\pi\left(\frac{\partial}{\partial z_j}\right)_{\mathbb{P}}, d\pi\left(\frac{\partial}{\partial \overline{z}_m}\right)_{\mathbb{P}}\right) = \omega_\pm\left(d\pi\left(\frac{\partial}{\partial z_j}\right), d\pi\left(\frac{\partial}{\partial \overline{z}_m}\right)\right)$$

and similarly for the other products. A straightforward calculation now shows that

$$\pi^* \omega_\pm\big|_{z^\perp \times z^\perp} = \frac{1}{i\|z\|_J^2}\sum_{r=0}^{n}\varepsilon_r(dz_r \wedge d\overline{z}_r)\big|_{z^\perp \times z^\perp},$$

where $z^\perp = \text{span}\{(\frac{\partial}{\partial z_j})_{\mathbb{P}}, (\frac{\partial}{\partial \bar{z}_j})_{\mathbb{P}}; j = 0, \ldots, n)\}$. This fact is also expressed by saying that

$$(8.6) \qquad \omega_\pm = \frac{1}{i\|z\|_J^2} \sum_{r=0}^{n} \varepsilon_r (dz_r \wedge d\bar{z}_r)$$

in *homogeneous coordinates*.

Note that similarly to the case of ω_J also the pseudo-Kähler metric $\tilde{\omega}_J$ on $\mathbb{C}^{n+1}$ gives rise to pseudo-Kähler metrics $\tilde{\omega}_\pm$ on $\Omega_\pm$. In the positive definite case this metric simply is the *Fubini study metric*. In homogeneous coordinates the pseudo-Kähler metric is given by

$$(8.7) \qquad \tilde{\omega}_\pm = \frac{1}{\|z\|^2} \sum_{r=0}^{n} \varepsilon_r dz_r d\bar{z}_r.$$

In particular we see that $\omega_\pm$ is two times the imaginary part of $\tilde{\omega}_\pm$. This fact will be important for us since it helps to decide for which *complex* submanifolds of $\Omega_\pm$ the restriction of $\omega_\pm$ to this submanifold is again non-degenerate, i.e., a symplectic form. In the positive definite case this is true for *all* complex submanifolds.

Recall that we obtained the reduction from V to $\Omega_\pm$ via the action of the center of $\mathrm{U}(J)$. But then the general theory of commuting Hamiltonian actions shows how to find the moment map on $\Omega_\pm$. In fact, the Lie algebra homomorphism $\varphi: \mathfrak{u}(J) \to C^\infty(V)$ yields upon restriction, functions in $C^\infty(H_\pm)$ which factor to functions on $\Omega_\pm$. This shows that

$$\left(\Phi_{\mathbb{P}}(z)_{rs}\right)_{r,s=0,\ldots,n} = i\left(\frac{z_r \bar{z}_s}{J(z,z)}\right)_{r,s=0,\ldots,n},$$

where we use the identification of $\mathfrak{u}(J)$ and $\mathfrak{u}(J)^*$ via the trace form as before. Note that we now see that the symplectic structure on $\Omega_\pm$ and the Hamiltonian action of $\mathrm{U}(J)$ on it as defined in Definition 8.14 agree with the structure and action as defined from the Marsden Weinstein reduction.

Lemma 8.16. *Let $\tau: G \times M \to M$ be a Hamiltonian action with moment map Φ_M and N a G-invariant symplectic submanifold of M. Then the restriction τ_N of τ to N is Hamiltonian with moment map $\Phi_N = \Phi_M|_N$.*

Proof. The inclusion $N \to M$ is a symplectic map, hence the restriction map $r: C^\infty(M) \to C^\infty(N)$ preserves the Poisson bracket. Therefore $\varphi_N = r \circ \varphi_M: \mathfrak{g} \to C^\infty(N)$ is a homomorphism. Consider the restriction $TM|_N$ of the tangent bundle TM of M to N. Then $\mathcal{X}(N)$ is a subspace of the space of sections $\Gamma(TM|_N)$ for $TM|_N$. We have the following diagram

$$
\begin{array}{ccc}
C^\infty(M) & \xrightarrow{\;j_M\;} & \mathcal{X}(M) \\
\llap{$r\downarrow$}\qquad \searrow \Phi_M & \nearrow \dot{\tau} & \downarrow \\
& \mathfrak{g} & \Gamma(TM|_N) \\
\qquad \nearrow \Phi_N & \searrow \dot{\tau}_N & \uparrow \\
C^\infty(N) & \xrightarrow{\;j_N\;} & \mathcal{X}(N)
\end{array}
$$

The invariance of N shows that $\dot{\tau}(X)_n \in T_n(N)$ so that $\dot{\tau}(X)|_N = \dot{\tau}_N(X) \in \mathcal{X}(N)$. We have to show that the diagram commutes which would then imply

$$\langle \Phi_N(n), X \rangle = \varphi_N(X)(n) = \varphi_M(X)(n) = \langle \Phi_M(n), X \rangle.$$

Note that for all $n \in N$ we have $T_n N \oplus T_n N^\perp = T_n M$, where $\perp$ denotes the orthogonal complement with respect to the symplectic form ω, since N is symplectic. If ω_N is the restriction of ω to N and $v = v_1 + v_2 \in T_n M$ with $v_1 \in T_n N, v_2 \in T_n N^\perp$, then we calculate

$$\begin{aligned}
\omega(\mathcal{X}_{\varphi_M(X)}(n), v) &= \omega(\mathcal{X}_{\varphi_M(X)}(n), v_1) \\
&= -d\big(\varphi_M(X)\big)(n)(v_1) \\
&= -d\big(\varphi_N(X)\big)(n)(v_1) \\
&= \omega_N(\mathcal{X}_{\varphi_N(X)}(n), v_1) \\
&= \omega_M(\mathcal{X}_{\varphi_N(X)}(n), v)
\end{aligned}$$

since $\mathcal{X}_{\varphi_M(X)}(n)$ and $\mathcal{X}_{\varphi_N(X)}(n)$ are both contained in $T_n N$. Thus $\mathcal{X}_{\varphi_M(X)}(n) = \mathcal{X}_{\varphi_N(X)}(n)$ which implies the claim. ∎

8.3. Pseudo-unitary representations and orbits on flag manifolds

Complex semisimple Lie algebras

In this subsection $\mathfrak{g}_\mathbb{C}$ denotes a semisimple complex Lie algebra. We recall the basic facts from the finite dimensional representation theory of $\mathfrak{g}_\mathbb{C}$.

Let $\mathfrak{h}_\mathbb{C} \subseteq \mathfrak{g}_\mathbb{C}$ denote a Cartan subalgebra. For a linear functional λ on $\mathfrak{h}_\mathbb{C}$ we write

$$\mathfrak{g}_\mathbb{C}^\lambda := \{ X \in \mathfrak{g}_\mathbb{C} : (\forall E \in \mathfrak{h}_\mathbb{C})[E, X] = \lambda(E) X \}.$$

We set

$$\Lambda := \Lambda(\mathfrak{g}_\mathbb{C}, \mathfrak{h}_\mathbb{C}) := \{ \lambda \in \mathfrak{h}_\mathbb{C}^* \setminus \{0\} : \mathfrak{g}_\mathbb{C}^\lambda \neq \{0\} \}$$

and call this finite set the set of *roots* of $\mathfrak{g}_\mathbb{C}$ with respect to $\mathfrak{h}_\mathbb{C}$. A *positive system* Λ^+ of roots is a subset with the property that there exists $X_0 \in \mathfrak{h}_\mathbb{C}$ such that no root vanishes on $\mathfrak{h}_\mathbb{C}$, all roots are real on X_0, and

$$\Lambda^+ = \{ \lambda \in \Lambda : \lambda(X_0) > 0 \}.$$

Along with every positive system of roots comes a subalgebra

$$\mathfrak{b} = \mathfrak{b}(\Lambda^+) = \mathfrak{h}_\mathbb{C} \oplus \bigoplus_{\lambda \in \Lambda^+} \mathfrak{g}_\mathbb{C}^\lambda.$$

These subalgebras are called *Borel subalgebras* of $\mathfrak{g}_\mathbb{C}$ and a subalgebra $\mathfrak{p}$ containing a Borel subalgebra is called a *parabolic subalgebra*. We say that a parabolic subalgebra is associated with the positive system Λ^+ if it contains the Borel algebra $\mathfrak{b}(\Lambda^+)$.

A subset $\Upsilon = \{\alpha_1, \ldots, \alpha_l\} \subseteq \Lambda^+$ is called a *basis* of the positive system Λ^+ if Υ is a basis of $\mathfrak{h}_\mathbb{C}^*$ and every positive root $\lambda \in \Lambda^+$ is an integral linear combination of elements of Υ with non-negative coefficients. A basis always exists uniquely ([Bou75, Ch. VI, §1, no. 6, Th. 3], [Bou75, Ch. VIII, §2, no. 2, Th. 2]).

Let $\Sigma \subseteq \Upsilon$,

$$\Lambda_\Sigma := (\operatorname{span} \Sigma) \cap \Lambda \quad \text{and} \quad \mathfrak{s}_\Sigma := \mathfrak{h}_\mathbb{C} \oplus \bigoplus_{\lambda \in \Lambda_\Sigma} \mathfrak{g}_\mathbb{C}^\lambda.$$

Then $\mathfrak{s}_\Sigma$ is a reductive Lie algebra and its center is given by

$$Z(\mathfrak{s}_\Sigma) = \{X \in \mathfrak{h}_\mathbb{C} : (\forall \alpha \in \Sigma)\alpha(X) = 0\}.$$

([Bou75, Ch. VIII, §3, no. 1, Prop. 5]). Moreover, let

$$\mathfrak{n}_\Sigma := \bigoplus_{\lambda \in \Lambda^+ \setminus \Lambda_\Sigma} \mathfrak{g}_\mathbb{C}^\lambda.$$

Then $\mathfrak{n}_\Sigma$ is normalized by $\mathfrak{s}_\Sigma$,

$$\mathfrak{p}_\Sigma := \mathfrak{n}_\Sigma \rtimes \mathfrak{s}_\Sigma$$

is a parabolic subalgebra, and every parabolic subalgebra associated with Λ^+ is obtained in this way ([Bou75, Ch. VIII, §3, no. 4]).

The *Weyl group* $W = W_{\mathfrak{g}_\mathbb{C}} = W(\mathfrak{g}_\mathbb{C}, \mathfrak{h}_\mathbb{C})$ is the group generated by the reflections s_λ at the hyperplanes $\ker \lambda$ with respect to the restriction of the Cartan Killing form to $\mathfrak{h}_\mathbb{C}$.

Highest weight modules.

Let V be a $\mathfrak{g}_\mathbb{C}$-module. For a linear functional λ on $\mathfrak{h}_\mathbb{C}$ we write

$$V^\lambda := \{v \in V : (\forall X \in \mathfrak{h}_\mathbb{C})X.v = \lambda(X)v\}.$$

We set

$$\mathcal{P}_V := \{\lambda \in \mathfrak{h}_\mathbb{C}^* : V^\lambda \neq \{0\}\}$$

and call this set the set of *weights* of $\mathfrak{g}_\mathbb{C}$ with respect to $\mathfrak{h}_\mathbb{C}$. An element $\omega \in \mathcal{P}_V$ is called a *highest weight* with respect to the positive system Λ^+ if

$$(\omega + \Lambda^+) \cap \mathcal{P}_V = \emptyset.$$

Next let κ denote the Cartan Killing form of $\mathfrak{g}_\mathbb{C}$ and $(\cdot, \cdot)$ the bilinear form on the dual $\mathfrak{g}_\mathbb{C}^*$ of $\mathfrak{g}_\mathbb{C}$ induced by κ. Then $(\cdot, \cdot)$ is positive definite and real on $\operatorname{span}_\mathbb{R} \Lambda$ and for every root $\lambda \in \Lambda$ there exists an element $\check{\lambda} \in \mathfrak{h}_\mathbb{C}$ such that

$$\mu(\check{\lambda}) = \frac{2(\lambda, \mu)}{(\lambda, \lambda)} \qquad \forall \mu \in \mathfrak{h}_\mathbb{C}^*$$

We write $\check{\mathcal{R}}$ for the abelian subgroup of $\mathfrak{h}_{\mathbb{C}}$ generated by $\check{\Lambda}$, define the *weight lattice*

$$\mathcal{P} := \{\mu \in \mathfrak{h}_{\mathbb{C}}^* : \mu(\check{\mathcal{R}}) \subseteq \mathbb{Z}\},$$

and set

$$\mathcal{P}^+ := \mathcal{P}^+(\Lambda^+) := \{\mu \in \mathcal{P} : (\forall \alpha \in \Lambda^+)\langle \mu, \check{\alpha} \rangle \in \mathbb{N}_0\}.$$

Note that a basis of $\mathcal{P}$ is given by

$$\{\omega_\alpha : \alpha \in \Upsilon\}, \quad \text{where} \quad \omega_\alpha(\check{\beta}) = \begin{cases} 0 & \text{if } \beta \neq \alpha \\ 1 & \text{if } \beta = \alpha. \end{cases}$$

Then $\mathcal{P}^+ = \sum_{\alpha \in \Upsilon} \mathbb{N}_0 \omega_\alpha$.

Proposition 8.17. *Let V be a finite dimensional $\mathfrak{g}_{\mathbb{C}}$-module and $\Lambda^+ \subseteq \Lambda$ a positive system. Then the following assertions hold:*

(i) $\mathcal{P}_V \subseteq \mathcal{P}$.

(ii) $V = \bigoplus_{\mu \in \mathcal{P}_V} V^\mu$.

(iii) *If V is irreducible, then $\mathcal{P}_V \cap \mathcal{P}^+$ contains a highest weight with respect to Λ^+.*

(iv) *For every $\lambda \in \mathcal{P}^+$ there exists, up to isomorphy, a unique irreducible $\mathfrak{g}_{\mathbb{C}}$-module called V_λ such that λ is a highest weight with respect to Λ^+ in $\mathcal{P}_{V_\lambda}$.*

Proof. (i), (ii) ([Bou75, Ch. VIII, §7, no. 1, Prop. 1])
(iii), (iv) ([Bou75, Ch. VIII, §7, no. 2]). ∎

Next we consider a connected semisimple Lie group $G_{\mathbb{C}}$ with $\mathbf{L}(G_{\mathbb{C}}) = \mathfrak{g}_{\mathbb{C}}$. If $\mathfrak{b} = \mathfrak{b}(\Lambda^+) \subseteq \mathfrak{g}_{\mathbb{C}}$ is a Borel subalgebra, then the subgroup

$$B = B(\Lambda^+) := \langle \exp \mathfrak{b} \rangle$$

of $G_{\mathbb{C}}$ is called the corresponding *Borel subgroup* of $G_{\mathbb{C}}$. A *parabolic subgroup* P of $G_{\mathbb{C}}$ is a subgroup containing B. Note that this is equivalent to the condition that $\mathfrak{p} := \mathbf{L}(P)$ is a parabolic subalgebra of $\mathfrak{g}_{\mathbb{C}}$. Essential properties of parabolic subgroups are connectedness and that they are self-normalizing ([Ho81, p.198], [OV91, p.122]).

Let V be an irreducible $\mathfrak{g}_{\mathbb{C}}$-module and suppose that the representation of $\mathfrak{g}_{\mathbb{C}}$ integrates to a representation of $G_{\mathbb{C}}$ (this is always the case if $G_{\mathbb{C}}$ is simply connected). We write $\mathbb{P}(V)$ for the projective space of V. Then the representation of $G_{\mathbb{C}}$ on V induces an action of $G_{\mathbb{C}}$ on $\mathbb{P}(V)$ defined by

$$g.[v] = [g.v] \qquad \forall g \in G_{\mathbb{C}}, v \in V \setminus \{0\},$$

where $V \setminus \{0\} \to \mathbb{P}(V), v \mapsto [v]$ is the quotient mapping.

Proposition 8.18. *Let $\omega \in \mathcal{P}_V$ be a highest weight with respect to the positive system Λ^+ and $v_\omega \in V^\omega$ a highest weight vector. Then the following assertions hold:*

(i) *The stabilizer of $[v_\omega] \in \mathbb{P}(V)$ is a parabolic subgroup P_ω associated with Λ^+.*

(ii) *Let* $\omega = \sum_{\alpha \in \Upsilon} n_\alpha \omega_\alpha$. *Then* $\mathfrak{p}_\omega := L(P_\omega) = \mathfrak{p}_\Sigma$ *with*

$$\Sigma = \{\alpha \in \Upsilon : n_\alpha = 0\}.$$

(iii) *If* $\beta, \beta' \in \mathcal{P}_V$ *with* $G_{\mathbb{C}}.[v_\beta] = G_{\mathbb{C}}.[v_{\beta'}]$, *then* $\beta \in \mathcal{W}.\lambda$. *In particular if* β *is the highest weight, then* β' *is an extremal weight.*

Proof. (i) Let $\lambda \in \Lambda^+$. For $X \in \mathfrak{g}_{\mathbb{C}}^\lambda$ we have that $X.v_\omega \in V^{\omega + \lambda} = \{0\}$. Hence $\mathfrak{n}_\Upsilon.v_\omega = \{0\}$. It follows that v_ω is a common eigenvector for the Borel subgroup $B = B(\Lambda^+)$. Thus B fixes the point $[v_\omega]$ in the projective space. This means that the stabilizer of $[v_\omega]$ is a subgroup which contains B, hence parabolic.

(ii) It follows from [Bou75, Ch. VIII, §7, no. 2, Prop. 3] that $\mathfrak{g}_{\mathbb{C}}^{-\lambda} \subseteq \mathfrak{p}_\omega$ holds for $\lambda \in \Lambda^+$ if and only if

(8.8) $$\omega(\check{\lambda}) = 0.$$

Let $\omega = \sum_{\alpha \in \Upsilon} n_\alpha \omega_\alpha$ and $\lambda \in \Upsilon$. Then (8.8) means that $n_\lambda = 0$.

(iii) According to our assumption there exists $g \in G_{\mathbb{C}}$ such that $[v_{\beta'}] = g.[v_\beta]$. Hence the stabilizer $P_{\beta'}$ of $[v_{\beta'}]$ satisfies $P_{\beta'} = g P_\beta g^{-1}$. Since $\mathrm{Ad}(g)\mathfrak{t}_{\mathbb{C}} \subseteq \mathfrak{p}_{\beta'} := L(P_{\beta'})$ is a Cartan algebra, there exists $p \in P_{\beta'}$ such that $\mathrm{Ad}(p)\,\mathrm{Ad}(g)\mathfrak{t}_{\mathbb{C}} = \mathfrak{t}_{\mathbb{C}}$. Now $pg \in N_{G_{\mathbb{C}}}(\mathfrak{t}_{\mathbb{C}})$. Hence $\gamma := \mathrm{Ad}(pg)\big|_{\mathfrak{t}_{\mathbb{C}}} \in \mathcal{W}$ satisfies $\gamma.\beta = \beta'$ since $g.v_\beta \subseteq \mathbb{C}v_{\beta'}$. ∎

The complex manifolds $G_{\mathbb{C}}/P$, where $P \subseteq G_{\mathbb{C}}$ is a parabolic subgroup are called *complex flag manifolds* because they are manifolds of flags in $\mathbb{C}^n$ if $G_{\mathbb{C}} = \mathrm{Sl}(n, \mathbb{C})$ and P is the stabilizer of a flag in $\mathbb{C}^n$.

Since every parabolic subalgebra $\mathfrak{p} \subseteq \mathfrak{g}_{\mathbb{C}}$ associated with $\mathfrak{b}(\Lambda^+)$ occurs as some $\mathfrak{p}_\Sigma$ for a subset $\Sigma \subseteq \Upsilon$, let us consider the weight

$$\omega_\Sigma := \sum_{\alpha \notin \Sigma} \omega_\alpha$$

and the corresponding highest weight module V. Then the preceding proposition shows that $P = P_\Sigma = P_{\omega_\Sigma}$ arises as the stabilizer of a highest weight vector $[v_{\omega_\Sigma}]$ in $\mathbb{P}(V)$. Thus we have obtained a realization of the flag manifold $G_{\mathbb{C}}/P$ as a compact submanifold of the projective space $\mathbb{P}(V)$. This realization will turn out to be crucial for the investigation of their structure and the orbits of real forms of $G_{\mathbb{C}}$ on $G_{\mathbb{C}}/P$.

Real forms and open orbits

If G is a connected real Lie group, then there exists a complex Lie group $G_{\mathbb{C}}$ together with a Lie group homomorphism $\eta: G \to G_{\mathbb{C}}$ with the universal property that for every homomorphism $\alpha: G \to H$ of G into a complex Lie group H there exists a unique holomorphic homomorphism $\alpha_{\mathbb{C}}: G_{\mathbb{C}} \to H$ such that $\alpha_{\mathbb{C}} \circ \eta = \alpha$. In the case where η is injective, $\alpha_{\mathbb{C}}$ can be interpreted as a holomorphic extension of α to the complex group $G_{\mathbb{C}}$. Moreover, there exists an antiholomorphic involutive automorphism σ of $G_{\mathbb{C}}$ such that $\eta(G)$ is an open subgroup of $G_{\mathbb{C}}^\sigma := \{g \in G_{\mathbb{C}} : \sigma(g) = g\}$ and $\eta(G)$ is closed in $G_{\mathbb{C}}$. The automorphism $d\sigma$ of $L(G_{\mathbb{C}}) \cong \eta(\mathfrak{g})_{\mathbb{C}}$

coincides with complex conjugation on this Lie algebra. So we also write $\overline{X} = d\sigma(1)X$. (cf. [HiNe91, III.9.22]).

Note that every finite dimensional representation $\pi: G \to \mathrm{Gl}(V)$ may be interpreted as a homomorphism into the complex group $\mathrm{Gl}(V_{\mathbb{C}})$ so that it factors to a representation of the group $\eta(G)$. Hence the finite dimensional representation theory of the groups G and $\eta(G)$ is the same. Thus we may assume, and we will do so in the following, that η is injective on G, i.e., $\eta(G) \cong G$.

For the remainder of this section we assume that G is a connected semisimple Lie group. Then the condition that η is injective means exactly that G is linear, i.e., isomorphic to a subgroup of a general linear group (cf. [HiNe91]).

We are interested in orbits of G on complex flag manifolds of the type $G_{\mathbb{C}}/P$, where $P \subseteq G_{\mathbb{C}}$ is a parabolic subgroup. Let $X := G_{\mathbb{C}}/P$ be such a flag manifold. To see that there exists a dense open subset $X' \subseteq X$ such that the G-orbit through every element of X' is open, it suffices to find at least one element in X having an open G-orbit since the condition that the G-orbit of a point in X is open is algebraic in nature so the set of points satisfying it is Zariski open (cf. Proposition 8.10).

Let $\mathfrak{g} = \mathbf{L}(G)$ and $\mathfrak{g} = \mathfrak{k} + \mathfrak{p}_0$ a Cartan decomposition of $\mathfrak{g}$. We pick a Cartan subalgebra $\mathfrak{t} \subseteq \mathfrak{k}$. Then $Z_{\mathfrak{g}}(\mathfrak{t})$ is a subalgebra of $\mathfrak{g}$ and

$$Z_{\mathfrak{g}}(\mathfrak{t}) = Z_{\mathfrak{k}}(\mathfrak{t}) + \big(Z_{\mathfrak{g}}(\mathfrak{t}) \cap \mathfrak{p}_0\big) = \mathfrak{t} + \mathfrak{a}.$$

Then $[\mathfrak{a}, \mathfrak{a}] \subseteq [\mathfrak{p}_0, \mathfrak{p}_0] \cap Z_{\mathfrak{g}}(\mathfrak{t}) = Z_{\mathfrak{k}}(\mathfrak{t}) = \mathfrak{t}$ and for $X, Y \in \mathfrak{a}$ and $Z \in \mathfrak{t}$ we find that

$$0 = \kappa(X, 0) = \kappa(X, [Y, Z]) = \kappa([X, Y], Z).$$

Hence $[X, Y] \in \mathfrak{t}^{\perp} \cap \mathfrak{t} = \{0\}$ since the restriction of κ to $\mathfrak{t}$ is negative definite. This shows that $\mathfrak{a}$ is an abelian subspace of $\mathfrak{p}_0$ and therefore $\mathfrak{h} := \mathfrak{t} + \mathfrak{a}$ is a Cartan subalgebra of $\mathfrak{g}$. Such Cartan subalgebras are called *maximally compact*. Then $\mathfrak{h}_{\mathbb{C}}$ is a Cartan subalgebra of $\mathfrak{g}_{\mathbb{C}}$ and we have a system of roots Λ. Moreover, since $\mathfrak{h}_{\mathbb{C}} = Z_{\mathfrak{g}_{\mathbb{C}}}(\mathfrak{t})$, we see that no root vanishes on $\mathfrak{t}$.

Let σ denote the complex conjugation of $\mathfrak{g}_{\mathbb{C}}$ with respect to $\mathfrak{g}$. Then an easy calculation shows that

$$\sigma(\mathfrak{g}_{\mathbb{C}}^{\lambda}) = \mathfrak{g}_{\mathbb{C}}^{\overline{\sigma^* \lambda}},$$

where $\overline{\sigma^* \lambda}(X) = \overline{\lambda(\sigma X)}$. We pick a regular element $X_0 \in i\mathfrak{t} \subseteq i\mathfrak{t} + \mathfrak{a}$. Since all roots are real valued on $i\mathfrak{t} + \mathfrak{a}$, we see that

$$\Lambda^+ := \{\lambda \in \Lambda : \lambda(X_0) > 0\}$$

is a positive system of roots. Moreover, $\sigma(X_0) = -X_0$ entails that

(8.9) $$\overline{\sigma^* \lambda} \in -\Lambda^+ \qquad \forall \lambda \in \Lambda^+,$$

i.e., $\overline{\sigma^*}$ interchanges the positive and the negative system of roots. Let $\mathfrak{p} \subseteq \mathfrak{g}_{\mathbb{C}}$ be any parabolic subalgebra associated with Λ^+. Then (8.9) entails that $\sigma(\mathfrak{p}) + \mathfrak{p} = \mathfrak{g}_{\mathbb{C}}$. On the other hand we know that $\mathfrak{p} + \sigma(\mathfrak{p}) \subseteq \mathfrak{p} + \mathfrak{g}$. Thus we have shown that $\mathfrak{p} + \mathfrak{g} = \mathfrak{g}_{\mathbb{C}}$ holds for all parabolic subalgebra associated with Λ^+. Applying this to the action of G on $G_{\mathbb{C}}/P$, and using the remarks from above, we obtain the following theorem:

Theorem 8.19. *Let G be a real form of $G_{\mathbb{C}}$ and $P \subseteq G_{\mathbb{C}}$ a parabolic subgroup. Then the set of all elements of $G_{\mathbb{C}}/P$ with open G-orbits is an open dense subset.* ∎

Wolf's analysis of open orbits in complex flag manifolds

Lemma 8.20. *If $P_1 \subseteq P_2$ are parabolic subgroups of $G_{\mathbb{C}}$, $X_i = G_{\mathbb{C}}/P_i$ the corresponding flag manifolds, $x_i \in X_i$, and $\pi: X_1 \to X_2$ the natural projection, then the following assertions hold:*

(i) *If $G.x_1$ is open in X_1, then $\pi(G.x_1) = G.\pi(x_1)$ is open in X_2.*

(ii) *If $G.x_2$ is open in X_2, then $\pi^{-1}(G.x_2)$ contains an open G orbit on X_1.*

Proof. (cf. [Wo69]) (i) This is a consequence of the fact that π is an open mapping.

(ii) Since $G.x_2$ is open, the same holds for the inverse image under π. On the other hand, the set of all points in X_1 with an open G-orbit is dense. Whence there exists $x_0 \in \pi^{-1}(G.x_2)$ such that $G.x_0$ is open. ∎

Lemma 8.21. *Let $X = G_{\mathbb{C}}/P$ be a complex flag manifold, $x \in X$, and $\mathfrak{p}_x := \mathfrak{g}_{\mathbb{C}}^x$ the isotropy subalgebra in x. Then the real isotropy subalgebra $\mathfrak{g}^x$ is a real form of $\mathfrak{p}_x \cap \sigma(\mathfrak{p}_x)$ and contains a Cartan subalgebra of $\mathfrak{g}$. Conversely, if $\mathfrak{h} \subseteq \mathfrak{g}$ is a Cartan subalgebra in $\mathfrak{g}$, then $\mathfrak{h} \subseteq \mathfrak{g} \cap \mathfrak{p}_x$ for some $x \in X$.*

Proof. (cf. [Wo69, 2.6]) The subalgebra $\mathfrak{g} \cap \mathfrak{p}_x$ is the isotropy subalgebra of G at x, and

$$\mathfrak{g} \cap \mathfrak{p}_x = (\mathfrak{g} \cap \mathfrak{p}_x) \cap (\mathfrak{g} \cap \sigma(\mathfrak{p}_x)) = \mathfrak{g} \cap (\mathfrak{p}_x \cap \sigma(\mathfrak{p}_x)).$$

Since $Y \in \mathfrak{p}_x \cap \sigma(\mathfrak{p}_x)$ implies $Y + \sigma(Y), i(Y - \sigma(Y)) \in \mathfrak{g} \cap \mathfrak{p}_x$ we see that $\mathfrak{g} \cap \mathfrak{p}_x$ is a real form of $\mathfrak{p}_x \cap \sigma(\mathfrak{p}_x)$.

Choose a Borel subalgebra $\mathfrak{b} \subseteq \mathfrak{p}_x$. Then $\mathfrak{b} \cap \sigma(\mathfrak{b})$ contains a Cartan subalgebra of $\mathfrak{g}_{\mathbb{C}}$ ([Bou75, Ch. VIII, §3, no. 3, Prop. 10]). It follows in particular that the σ-stable subalgebra $\mathfrak{b} \cap \sigma(\mathfrak{b})$ has full rank in $\mathfrak{g}_{\mathbb{C}}$. Pick a Cartan subalgebra $\mathfrak{h}$ in the real form $\mathfrak{b} \cap \mathfrak{g}$ of $\mathfrak{b} \cap \sigma(\mathfrak{b})$. Then $\mathfrak{h}$ must be a Cartan subalgebra in $\mathfrak{g}$.

For the converse, let $\mathfrak{h} \subseteq \mathfrak{g}$ be a Cartan subalgebra and $\mathfrak{h}_{\mathbb{C}} \subseteq \mathfrak{g}_{\mathbb{C}}$ its complexification. Choose a positive system $\Lambda^+ \subseteq \Lambda(\mathfrak{g}_{\mathbb{C}}, \mathfrak{h}_{\mathbb{C}})$ such that $\mathfrak{p} = \mathbf{L}(P)$ is conjugate to $\mathfrak{p}_\Sigma$ for some subset Σ of a basis $\Upsilon \subseteq \Lambda^+$. Then $\mathfrak{p}_\Sigma = \mathfrak{p}_y$ for some $y \in X$, and $\mathfrak{h} \subseteq \mathfrak{g} \cap \mathfrak{p}_y$. ∎

Theorem 8.22. *Let $X = G_{\mathbb{C}}/P$ be a complex flag manifold, $G \subseteq G_{\mathbb{C}}$ a real form, $x \in X$, and P_x the stabilizer of x in $G_{\mathbb{C}}$. Then the G-orbit of $x \in X$ is open iff there exists*

(1) *a maximally compact Cartan subalgebra $\mathfrak{h} \subseteq \mathfrak{g} \cap \mathfrak{p}_x$, and*

(2) *a positive system $\Lambda^+ \subseteq \Lambda(\mathfrak{g}_{\mathbb{C}}, \mathfrak{h}_{\mathbb{C}})$, a basis $\Upsilon \subseteq \Lambda^+$, and a subset $\Sigma \subseteq \Upsilon$ such that $\sigma.\Lambda^+ = -\Lambda^+$ and $\mathfrak{p}_x = \mathfrak{p}_\Sigma$.*

Proof. (cf. [Wo69]) The sufficiency of these condition has already been shown above in the proof of Theorem 8.19.

Suppose for a moment that P_x is a Borel subgroup of $G_\mathbb{C}$. Let $\mathfrak{h} \subseteq \mathfrak{p}_x \cap \mathfrak{g}$ be a Cartan subalgebra of $\mathfrak{g}$ (Lemma 8.21). Then there exists a unique positive system $\Lambda^+ \subseteq \Lambda(\mathfrak{g}_\mathbb{C}, \mathfrak{h}_\mathbb{C})$ such that $\mathfrak{p}_x = \mathfrak{b}(\Lambda^+)$. If $G.x$ is open, then

$$\dim_\mathbb{R} G.x = \dim_\mathbb{R} X = 2 \dim_\mathbb{C} X = 2|\Lambda^+|$$

together with

$$\begin{aligned}
\dim_\mathbb{R} G.x &= \dim_\mathbb{R} G - \dim_\mathbb{R}(G \cap P_x) \\
&= \dim_\mathbb{C} G_\mathbb{C} - \dim_\mathbb{C}(P_x \cap \sigma P_x) \\
&= (\dim_\mathbb{C} \mathfrak{h}_\mathbb{C} + 2|\Lambda^+|) - (\dim_\mathbb{C} \mathfrak{h}_\mathbb{C} + 2|\Lambda^+ \cap \sigma \Lambda^+|)
\end{aligned}$$

implies $\Lambda^+ \cap \sigma\Lambda^+ = \emptyset$, i.e., $\sigma.\Lambda^+ = -\Lambda^+$. Suppose that $\mathfrak{h}$ is not maximally compact. Then there exists a root λ vanishing on $\mathfrak{t}$, where $\mathfrak{h} = \mathfrak{t} + \mathfrak{a}$ is the decomposition into compact and vector part. Whence $\sigma.\lambda = \lambda$ and $\sigma.\Lambda^+ = -\Lambda^+$ is impossible.

Now in general choose a Borel subgroup $B \subseteq P_x$, let $Y := G_\mathbb{C}/B$ denote the corresponding flag manifold, and $\pi: Y \to X$ the natural projection. Suppose that $G.x$ is open in X. Then, in view of Lemma 8.20(ii), there exists an open G-orbit in $\pi^{-1}(G.x)$. Thus we find $y \in \pi^{-1}(x)$ such that $G.y$ is open in Y. Then $B_y \subseteq P_x$. Choose a Cartan subalgebra $\mathfrak{h} \subseteq \mathfrak{b}_y \cap \mathfrak{g}$ and a positive system in $\Lambda(\mathfrak{g}_\mathbb{C}, \mathfrak{h}_\mathbb{C})$ such that $\mathfrak{b}_y = \mathfrak{b}(\Lambda^+)$. As we have already seen, $\mathfrak{h}$ is maximally compact in $\mathfrak{g}$ and $\sigma.\Lambda^+ = -\Lambda^+$ because $G.y$ is open in Y. Now $\mathfrak{p}_x = \mathfrak{p}_\Sigma$ for some subset Σ in a basis $\Upsilon \subseteq \Lambda^+$ because $\mathfrak{b}_y \subseteq \mathfrak{p}_x$. ∎

Now we are ready to give an explicit parametrization of the set of open G-orbits in a flag manifold $X = G_\mathbb{C}/P$. To do this, we choose a Borel subgroup $B \subseteq P$ and have a look at the open G-orbits in $Y = G_\mathbb{C}/B$.

Theorem 8.23.   *Let $\mathfrak{h} = \mathfrak{t} + \mathfrak{a} \subseteq \mathfrak{g}$ be a maximally compact Cartan subalgebra, $\mathcal{W}_\mathfrak{t}$ the Weyl group of $\mathfrak{t}$ with respect to its Cartan subalgebra $\mathfrak{t}$, and $\mathcal{D} := \{D_1, \ldots, D_m\}$ the chambers in it cut out by the roots in $\Lambda = \Lambda(\mathfrak{g}_\mathbb{C}, \mathfrak{h}_\mathbb{C})$. Then the following assertions hold:*

(i) *For $D_i \in \mathcal{D}$ let*

$$\Lambda_i^+ := \{\lambda \in \Lambda : (\forall X \in D_i)\lambda(X) \geq 0\}$$

denote the corresponding positive system and $\mathfrak{b}_i := \mathfrak{b}(\Lambda_i^+)$ the associated Borel algebra. Let $y_i \in Y$ with $B^{y_i} = B_i$. Then the open G-orbits on Y are just the $G.y_i$ and $G.y_i = G.y_j$ iff some element of $\mathcal{W}_\mathfrak{t}$ sends D_i to D_j.

(ii) *$\mathcal{W}_\mathfrak{t}$ acts simply on $\mathcal{D}$.*

(iii) *There are precisely $m/\|\mathcal{W}_\mathfrak{t}\|$ distinct open G-orbits on Y.*

Proof.   (i) Let $G.y$ be an open orbit on Y. Then Theorem 8.22 provides a maximally compact Cartan subalgebra $\mathfrak{h}' \subseteq \mathfrak{g} \cap \mathfrak{b}_y$ and since two such Cartan algebras are conjugate ([Bou75, Ch. VII, §2, Th. 1]), we find $g \in G$ such that $\mathfrak{h} = \mathrm{Ad}(g)\mathfrak{h}'$. Thus, replacing y by $g.y$, we may assume that $\mathfrak{h} \subseteq \mathfrak{b}_y$. Now Theorem 8.22 provides a positive system $\Lambda^+ \subseteq \Lambda(\mathfrak{g}_\mathbb{C}, \mathfrak{h}_\mathbb{C})$ such that $\mathfrak{b}_y = \mathfrak{b}(\Lambda^+)$ and $\sigma.\Lambda^+ = -\Lambda^+$. Let $X_0 \in i\mathfrak{t} + \mathfrak{a}$ be a regular element such that

$$\Lambda^+ = \{\lambda \in \Lambda : \lambda(X_0) > 0\}.$$

Write $X_0 = X_t + X_\mathfrak{a}$ with $X_t \in it$ and $X_\mathfrak{a} \in \mathfrak{a}$, so

$$(\sigma.\lambda)(X_0) = -\lambda(X_t) + \lambda(X_\mathfrak{a}) \qquad \forall \lambda \in \Lambda.$$

Let $\lambda \in \Lambda^+$. Then

$$(\sigma.\lambda)(X_0) = -\lambda(X_t) + \lambda(X_\mathfrak{a}) < 0 < \lambda(X_0) = \lambda(X_t) + \lambda(X_\mathfrak{a}).$$

Thus $\lambda(X_t) > 0$. Whence X_t is regular and $\Lambda^+ = \Lambda_i^+$ for the chamber containing X_t, so $\mathfrak{b}_y = \mathfrak{b}_i$ and we have shown that every open G-orbit is one of the $G.y_i$. That every orbit $G.y_i$ is open follows immediately from Theorem 8.22 since $\sigma(D_i) = -D_i$ for $i = 1, \ldots, m$.

Let $\gamma \in \mathcal{W}_t$ send D_i to D_j. Then there exists $k \in K = \exp \mathfrak{k}$ such that $\gamma = \mathrm{Ad}(\mathfrak{k})\,|_t$. Since $\mathfrak{h} = Z_\mathfrak{g}(t)$, the operator $\mathrm{Ad}(k)$ leaves $\mathfrak{h}$ invariant and it is immediate that $\mathrm{Ad}(k)\mathfrak{b}_i = \mathfrak{b}_j$. Hence $k.y_i = y_j$, so $G.y_i = G.y_j$.

Conversely, suppose that $G.y_i = G.y_j$, i.e., $g.y_i = y_j$ for some $g \in G$. Then $\mathrm{Ad}(g)\mathfrak{b}_i = \mathfrak{b}_j$ entails that

$$\mathrm{Ad}(g)\mathfrak{h} = \mathrm{Ad}(g)(\mathfrak{b}_i \cap \mathfrak{g}) = \mathfrak{b}_j \cap \mathfrak{g} = \mathfrak{h}.$$

Moreover, $\exp t$ is the maximal compact subgroup of $\exp \mathfrak{h}$, hence a characteristic subgroup, so t is also invariant under $\mathrm{Ad}(g)$ and thus $\mathrm{Ad}(g)|_t \in \mathcal{W}_t$. This completes the proof of (i).

(ii) Let $\gamma \in \mathcal{W}_t$ with $\gamma(D_i) = D_i$. Since $\mathcal{W}_t \subseteq \mathcal{W}_{\mathfrak{g}_\mathbb{C}}$ and $D_i = D \cap it$ for some Weyl chamber D with respect to Λ, it follows that $\gamma(D) = D$, so $\gamma = 1$ because $\mathcal{W} = \mathcal{W}_{\mathfrak{g}_\mathbb{C}}$ acts simply transitive on the Weyl chambers ([Bou68, Ch. V, §4, no. 4, Cor. 1]).

(iii) In view of (ii), the action of $\mathcal{W}_t$ partitions the set of open G-orbits $G.y_i$ into $m/|\mathcal{W}_t|$ distinct ones. ∎

Corollary 8.24. *Let* $Y = G_\mathbb{C}/B$ *a complex flag manifold, where* B *is a Borel subgroup. Let* G *be a real form of* $G_\mathbb{C}$ *whose maximal compact subgroup* K *has full rank. Then there are precisely*

$$|\mathcal{W}|/|\mathcal{W}_t|$$

distinct open G-*orbits on* Y. ∎

Note that in the example $G = \mathrm{Sl}(3, \mathbb{R})$ and $G_\mathbb{C} = \mathrm{Sl}(3, \mathbb{C})$ the complex Weyl group contains 6 elements, $|\mathcal{W}_t| = 2$, and $m = 2$, so that there is only one open G-orbit on $Y = G_\mathbb{C}/B$.

For the application we have in mind we are mainly interested in the case where the assumptions of Corollary 8.24 are satisfied. Let us in addition assume that we have a realization of the flag manifold X as a $G_\mathbb{C}$-orbit $G_\mathbb{C}.[v_\omega]$ of a highest weight ray in $\mathbb{P}(V)$, where V is a highest weight module and $\mathfrak{h} = t$ is a compactly embedded Cartan subalgebra in $\mathfrak{g}$. Then Theorem 8.22 shows that the G-orbit of $[v_\omega]$ is open. Let $B \subseteq P$ be a Borel subgroup, $Y = G_\mathbb{C}/B$, and $\pi \colon Y \to X$ the projection. Since every open G-orbit on X is covered by an open G-orbit in Y (Lemma 8.20), Corollary 8.24 shows that every open G-orbit on X contains a Weyl group translate of $[v_\omega]$. Since the Weyl group translates correspond to the extremal weights in $\mathcal{P}(V)$ ([Bou75, Ch. VIII, §7, no. 2, Prop. 5]), we have proved the following proposition.

Proposition 8.25.   *Suppose that* rank G = rank K *and that the complex flag manifold is realized as a $G_{\mathbb{C}}$-orbit of a highest weight vector in* $\mathbb{P}(V)$. *Then the following assertions hold:*

(i) *Every open G-orbit in X contains an element $[v_\lambda]$, where $\lambda \in \mathcal{P}_V$ is an extremal weight.*

(ii) *Every G-orbit of an extremal weight ray is open in X.*   ■

Pseudo-unitary representations

A pair (V, J) of a finite dimensional complex vector space V and a nondegenerate Hermitean form J on V is called a finite dimensional *Krein space*. By abuse of language we also write V for the pair (V, J). For an operator $A \in \text{gl}(V)$ we write $A^\sharp$ for the adjoint, i.e.,

$$J(A.v, w) = J(v, A^\sharp.w) \qquad \forall v, w \in V.$$

Then

$$U(J) := \{g \in \text{Gl}(V) : g^\sharp = g^{-1}\}$$

is called the *pseudo-unitary group* of the Hermitean form J, and

$$\mathfrak{u}(J) := \{X \in \text{gl}(V) : X^\sharp = -X\}$$

is its Lie algebra (cf. Section 8.2).

Let G be a Lie group. A representation $\pi \colon G \to \text{Gl}(V)$ is said to be *pseudo-unitary*, or more precisely *J-unitary* if the image $\pi(G)$ is contained in $U(J)$. If G is connected, then this is clearly equivalent to the condition that the derived representation $d\pi \colon \mathfrak{g} = \mathbf{L}(G) \to \text{gl}(V)$ maps $\mathfrak{g}$ into $\mathfrak{u}(J)$. A pseudo-unitary representation is said to be unitary if J is positive definite.

Proposition 8.26.   *Let G be a connected Lie group, V a pseudo-unitary G-module defined by the representation π, and $\pi_{\mathbb{C}}$ the holomorphic extension to $G_{\mathbb{C}}$. Then the following assertions hold:*

(i) $\pi_{\mathbb{C}}(g)^\sharp = \pi_{\mathbb{C}}(\sigma(g)^{-1})$ *for all $g \in G_{\mathbb{C}}$.*

(ii) $d\pi_{\mathbb{C}}(X)^\sharp = d\pi_{\mathbb{C}}(-\overline{X})$ *for all $X \in \mathfrak{g}_{\mathbb{C}}$.*

Proof.   (i) For the elements of G this follows from the J-unitarity of the representation π. Hence

$$\pi_{\mathbb{C}}(g) = \pi_{\mathbb{C}}(\sigma(g)^{-1})^\sharp$$

holds for all $g \in G$. Both define holomorphic extensions of π to $G_{\mathbb{C}}$ so they have to coincide on $G_{\mathbb{C}}$ by the uniqueness of extension.

(ii) This is the infinitesimal version of (i).   ■

Proposition 8.27.   *Let (V, J) be an irreducible pseudo-unitary module of the Lie group G and J' another G-invariant Hermitean form on V. Then J' is proportional to J.*

Proof.   Since J is non-degenerate, there exists a J-Hermitean operator $A \in \mathrm{Gl}(V)$ such that

$$J'(v,w) = J(Av,w) \qquad \forall v,w \in V.$$

For $g \in G$ this leads to

$$J(Ag.v,w) = J'(g.v,w) = J'(v,g^{-1}.w) = J(Av,g^{-1}.w) = J(gAv,w),$$

i.e., $gA = Ag$. Now Schur's Lemma shows that $A \in \mathbb{C}\,\mathrm{id}_V$ and $A \in \mathbb{R}\,\mathrm{id}_V$ follows from $A^{\sharp} = A$.   ∎

The preceding proposition shows that invariant Hermitean forms on irreducible G-modules are essentially unique whenever they exist. That this is not always true can be seen from the following example.

Let $V = \mathbb{C}^3$ be considered as an irreducible complex module of the real group $G = \mathrm{Sl}(3,\mathbb{R})$. Suppose that V can be made into a pseudo-unitary G-module with respect to J. Since G is not compact, J must be indefinite. Thus, up to sign, the signature of J must be $(+,+,-)$. Then $\mathrm{U}(J) \cong \mathrm{U}(2,1)$ and therefore we have an embedding of G into $\mathrm{SU}(2,1)$. Such an embedding cannot exist because the split rank of $\mathrm{Sl}(3,\mathbb{R})$ is 2 and the split rank of $\mathrm{SU}(2,1)$ is 1.

Pseudo-unitarizability of representations

A well known result in the representation theory of compact groups is Weyl's trick:

Proposition 8.28.   (Weyl's trick) *Let V be a finite dimensional complex representation of the compact group K, then there exists a positive definite Hermitean form J on V turning V into a unitary K-module.*   ∎

In this section we extend this result to representations of *quasihermitean semisimple* Lie algebras. Recall that a real semisimple Lie algebra $\mathfrak{g}$ is said to be quasihermitean if its simple ideals are either compact or Hermitean. The Hermitean Lie algebras are those where the centralizer of the center of a maximal compactly embedded subalgebra $\mathfrak{k}$ coincides with $\mathfrak{k}$. It follows in particular that $Z(\mathfrak{k}) \neq \{0\}$ if $\mathfrak{g}$ is not compact.

We recall a few basic facts about Hermitean (simple) Lie algebras: If $\mathfrak{t}$ is a compactly embedded Cartan subalgebra of $\mathfrak{g}$, then any root $\alpha \in \Lambda(\mathfrak{g}_{\mathbb{C}}, \mathfrak{t}_{\mathbb{C}})$ is either compact, i.e., its root space is contained in $\mathfrak{k}_{\mathbb{C}}$, where $\mathfrak{k}$ is the unique maximal compactly embedded subalgebra of $\mathfrak{g}$ containing $\mathfrak{t}$ (cf. [HHL89, A.2.40]), or else in $(\mathfrak{p}_0)_{\mathbb{C}}$, where $\mathfrak{g} = \mathfrak{k} + \mathfrak{p}_0$ is a Cartan decompostion. Any base of $\Lambda(\mathfrak{g}_{\mathbb{C}}, \mathfrak{t}_{\mathbb{C}})$ contains precisely one non-compact root (cf. [HiNe92, 5.1]). The reflections associated to the compact roots generate a Weyl group $\mathcal{W}_{\mathfrak{k}}$ which coincides with the Weyl group for the pair $(\mathfrak{k}, \mathfrak{t})$. The minimal cone $C_{\min}$ in $\mathfrak{t}$ is generated by the elements $i[\overline{Z}, Z] = -i\mathbb{R}^+ \check{\beta}$, where β is a non-compact root and $Z \in \mathfrak{g}_{\mathbb{C}}^{\beta} \setminus \{0\}$. The dual cone $C_{\min}^*$ is the union of the $\mathcal{W}_{\mathfrak{k}}$-translates of the closure of the positive Weyl chamber.

Lemma 8.29. *Let $\mathfrak{g}$ be a quasihermitean semisimple Lie algebra and $\mathfrak{k} \subseteq \mathfrak{g}$ maximal compactly embedded, then there exists $Z \in Z(\mathfrak{k})$ such that $e^{\pi \operatorname{ad} Z}$ is a Cartan involution of $\mathfrak{g}$.*

Proof. We clearly may assume that $\mathfrak{g}$ is simple. If $\mathfrak{g}$ is compact, the assertion is trivial. If $\mathfrak{g}$ is Hermitean, then $\dim Z(\mathfrak{k}) = 1$ ([Hel78, p.382]) and there exists $Z \in Z(\mathfrak{k})$ such that $\operatorname{ad} Z_0$ defines a complex structure on the complement of $\mathfrak{k}$ ([Hel78, p.382]). Then $e^{\operatorname{ad} \pi Z_0}$ is a Cartan involution of $\mathfrak{g}$.   ∎

Theorem 8.30. (The Pseudo-unitary Trick) *Let G be a connected quasihermitean semisimple Lie group and V a complex G-module. Then there exists a Hermitean form J on V turning V into a pseudo-unitary G-module.*

Proof. Since V is a semisimple G-module by Weyl's Theorem ([Bou71a, Ch. I, §6, no. 2, Th. 2]), we may assume that V is irreducible.

Let $G_{\mathbb{C}}$ denote the complexification of G and $U \subseteq G_{\mathbb{C}}$ a compact real form. On the level of Lie algebras this means that we have a Cartan decomposition $\mathfrak{g} = \mathfrak{k} + \mathfrak{p}_0$, and $\mathfrak{u} = \mathfrak{k} + i\mathfrak{p}_0$. Since the representation $\pi \colon G \to \operatorname{Gl}(V)$ extends to a holomorphic representation $\pi_{\mathbb{C}} \colon G_{\mathbb{C}} \to \operatorname{Gl}(V)$, we consider V as a $G_{\mathbb{C}}$-module. Using Weyl's trick (Theorem 8.28) we find a positive definite Hermitean form J' on V such that V is a unitary U-module with respect to J'. Let X^* denote the adjoint with respect to J' and θ the Cartan involution of $\mathfrak{g}$. Then the J'-unitarity of V entails that

$$d\pi_{\mathbb{C}}(X)^* = -d\pi_{\mathbb{C}}(\overline{\theta X}).$$

Let Z be as in Lemma 8.29. We set $A := \pi\big(\exp(\pi Z)\big)$. Since

$$\operatorname{Ad}\big(\exp(2\pi Z)\big) = e^{2\pi \operatorname{ad} Z} = \operatorname{id}_{\mathfrak{g}},$$

it follows that $\exp(2\pi Z) \in Z(G)$, so that $A^2 \in \pi\big(Z(G)\big)$ commutes with $\pi(G)$ and therefore with $\pi_{\mathbb{C}}(G_{\mathbb{C}})$. Now the assumption that V is irreducible, together with Schur's Lemma, shows that $A^2 \in \mathbb{C}^* \operatorname{id}_V$. On the other hand we know that A^2 is J'-unitary, hence $A^2 = z \operatorname{id}_V$ with $|z| = 1$. Let $\gamma \in \mathbb{C}^*$ with $|\gamma| = 1$ and $\gamma^2 A^2 = \operatorname{id}_V$. Set $B := \gamma A$. Then B is J'-unitary and $B^2 = \operatorname{id}_V$, hence $B^* = B^{-1} = B$. Thus B is also J'-Hermitean, so

$$J(v, w) := J'(Bv, w)$$

defines a non-degenerate Hermitean form on V.

We claim that V is a pseudo-unitary G-module with respect to this form. It is clear that the adjoint with respect to J is given by

$$X^{\sharp} = B^{-1} X^* B = A^{-1} X^* A.$$

Therefore, for $X \in \mathfrak{g}$,

$$d\pi(X)^{\sharp} = A^{-1} d\pi(X)^* A = -\pi\big(\exp(-\pi Z)\big) d\pi(\theta X) \pi\big(\exp(\pi Z)\big)$$
$$= -d\pi(e^{-\pi \operatorname{ad} Z} \theta X) = -d\pi(X)$$

proves the claim (Proposition 8.26).   ∎

Moment mappings

In this subsection we combine the results on open G-orbits in complex flag manifolds with the theory of pseudo-unitary representations.

Let G denote a connected Lie group with $L(G) = \mathfrak{g}$ and $\pi: G \to Gl(V)$ be a pseudo-unitary representation on (V, J), i.e., a representation such that $\pi(G)$ leaves J invariant. Let $d\pi: \mathfrak{g} \to gl(V)$ be the derived representation. Then $d\pi(\mathfrak{g})$ consists of skew-Hermitian operators with respect to J and $d\pi(i\mathfrak{g})$ consists of Hermitean operators. We have a morphism of Lie groups $\pi: G \to U(J)$. Since we have a Hamiltonian action of $U(J)$ on

$$\Omega = \{[v] \in \mathbb{P}(V): J(v, v) \neq 0\},$$

we get a Hamiltonian action of G on Ω by pulling back via

$$G \times \Omega \to \Omega, \qquad (g, [v]) \mapsto [\pi(g).v].$$

The corresponding moment mapping is given by $d\pi^* \circ \Phi$, namely

$$\Phi_\pi : \Omega \to \mathfrak{g}^*, \quad [v] \mapsto \left(X \mapsto i\frac{J(\pi(X).v, v)}{J(v, v)} \right).$$

Let now G be connected semisimple such that the mapping $\eta: G \to G_{\mathbb{C}}$ is injective and $K \subseteq G$ is a maximal compact subgroup. We assume that $\operatorname{rank} K = \operatorname{rank} G$.

Further assume that V be an irreducible pseudo-unitary G-module with respect to the Hermitean form J. Moreover, let $\mathfrak{t} \subseteq \mathfrak{k}$ be a compactly embedded Cartan subalgebra of $\mathfrak{g}$, $\Lambda^+ \subseteq \Lambda$ a positive system, $\lambda \in \mathcal{P}$ the highest weight of V, and $\mathfrak{p} = \mathfrak{p}_\lambda$ the corresponding parabolic subgroup.

Then $G_{\mathbb{C}}/P$ can be realized as the $G_{\mathbb{C}}$-orbit of the highest weight ray $[v_\lambda] \in \mathbb{P}(V)$. The decomposition of V into weight spaces is J-orthogonal because the $\mathfrak{t}_{\mathbb{C}}$-weight spaces are the same as the $i\mathfrak{t}$-weight spaces and $i\pi(\mathfrak{t})$ consists of J-Hermitean operators. Whence the restriction of J to the weight spaces is non-degenerate. Since the multiplicities of the extreme weights are 1 ([Bou75, Ch. VIII, §7, no. 2, Prop. 5]), we find that $J(v_\lambda, v_\lambda) \neq 0$ whenever v_λ is a weight vector for an extreme weight.

This shows that Ω contains all the rays corresponding to extreme weights. In view of Proposition 8.25, this means that every open G-orbit in X is contained in Ω.

Since every such orbit contains a weight ray, let v_λ be a weight vector for an extremal weight. Then the moment mapping Φ satisfies

$$\Phi([v_\lambda])(X) = i\frac{J(X.v_\lambda, v_\lambda)}{J(v_\lambda, v_\lambda)} = i\lambda(X) \qquad \forall X \in \mathfrak{t}.$$

Whence $\Phi([v_\lambda]) = i\lambda$ and the open G-orbit through $[v_\lambda]$ is mapped by Φ onto the coadjoint orbit of the functional $i\lambda \in \mathfrak{t}^* \cong [\mathfrak{t}, \mathfrak{g}]^\perp \subseteq \mathfrak{g}^*$.

Let ω denote the symplectic structure on Ω defined by the imaginary part of the pseudo-Kähler structure. Further, let $M := G.[v_\lambda]$ denote the G-orbit in Ω, and $\alpha\colon M \to \Omega$ the inclusion mapping. Then $\alpha^*\omega$ coincides with the pull-back of the symplectic structure on the coadjoint orbit $\mathcal{O}_{i\lambda}$ to M (cf. [LM87]). Whence $\alpha^*\omega$ is non-degenerate and therefore a symplectic structure on M, so Ω even induces the structure of a pseudo-Kähler manifold on M if M is endowed with the complex structure inherited by X.

This is a direct way to get the pseudo-Kähler structure on open G-orbits in complex flag manifolds under the assumption that G is quasihermitean, or, more generally, the flag manifold X embeds into the projective space of a pseudo-unitary G-module. We will describe a more general version below.

Let $V = \mathbb{C}^3$, $G = \mathrm{SU}(2,1)$, and $G_\mathbb{C} = \mathrm{SO}(3,\mathbb{C})$ with the natural action on V. Then

$$J(z,z) = |z_1|^2 + |z_2|^2 - |z_3|^2$$

defines a Hermitean form on V such that V is a pseudo-unitary G-module. Let $\Upsilon = \{\alpha_1, \alpha_2\}$ be a basis of a positive system of roots, where α_1 is non-compact. Then $\omega_2 := \omega_{\alpha_2}$ is a highest weight for V. The other extreme weights are

$$s_{\alpha_2}(\omega_2) \quad \text{and} \quad -\omega_1 := -\omega_{\alpha_1}.$$

The G-orbit of $[v_{-\omega_1}] = [(0,0,1)]$ is isomorphic to the Hermitean symmetric space $G/K = \mathrm{SU}(2,1)/\mathrm{SU}(2)$, and the open orbit through $[v_{\omega_2}]$ is a pseudo-Kähler manifold. The corresponding parabolic is

$$\mathfrak{p}_{\omega_2} = \mathfrak{p}_{\{\alpha_1\}} = \mathfrak{t}_\mathbb{C} \oplus \mathfrak{g}_\mathbb{C}^{\alpha_1} \oplus \mathfrak{g}_\mathbb{C}^{-\alpha_1} \oplus \mathfrak{g}_\mathbb{C}^{\alpha_2} \oplus \mathfrak{g}_\mathbb{C}^{\alpha_1+\alpha_2}.$$

Pseudo-Kähler structures on open G-orbits

In this subsection we describe pseudo-Kählerian structures in a more general context than we did in the previous subsection. Let $G_\mathbb{C}$ denote a complex connected semisimple Lie group and G a real form of $G_\mathbb{C}$ with the property that $\mathfrak{g} = \mathbf{L}(G)$ contains a compactly embedded Cartan subalgebra $\mathfrak{t}$, i.e., $\mathrm{rank}(G) = \mathrm{rank}(K)$, where $K = \exp \mathfrak{k}$ for a maximal compactly embedded subalgebra $\mathfrak{k}$.

Let $X = G_\mathbb{C}/P$ be a complex flag manifold, y_0 the base point, and suppose that $G.y_0$ is open in X. Then $\mathfrak{p} = \mathbf{L}(P)$ may be written as $\mathfrak{p} = \mathfrak{p}_\Sigma$, where $\Sigma \subseteq \Upsilon$, and Υ is a basis in a positive system $\Lambda \subseteq \Lambda(\mathfrak{t}_\mathbb{C}, \mathfrak{g}_\mathbb{C})$, where $\mathfrak{t} \subseteq \mathfrak{g}$ is a compactly embedded Cartan subalgebra.

The involution σ of $\mathfrak{g}_\mathbb{C}$ over $\mathfrak{g}$ acts on Λ by $\sigma.\lambda = -\lambda$. Whence

$$\mathfrak{s}_\Sigma = \mathfrak{p} \cap \bar{\mathfrak{p}} \quad \text{and} \quad \mathfrak{p} + \bar{\mathfrak{p}} = \mathfrak{g}_\mathbb{C}$$

as we have already seen in the preceding subsections. Suppose that X is realized as a G-orbit of a highest weight ray $[v_\omega]$ for $\omega \in \mathcal{P}$. Write

$$\omega = \sum_{\alpha \in \Upsilon} n_\alpha \omega_\alpha.$$

Then, for a root $\beta = \sum_{\alpha \in \Upsilon} m_\alpha \alpha$ we find that

$$(\omega, \beta) = \sum_{\alpha \in \Upsilon} n_\alpha m_\alpha (\omega_\alpha, \alpha) = \frac{1}{2} \sum_{\alpha \in \Upsilon} n_\alpha m_\alpha (\alpha, \alpha),$$

so $\omega^\perp \cap \Upsilon = \Sigma$ and $\omega^\perp \cap \Lambda$ is the set of roots of the reductive algebra $\mathfrak{s}_\Sigma$ (cf. Proposition 8.18). Let $X_\omega \in \mathfrak{t}$ be an element representing the linear function ω via the Cartan Killing form. Then the stabilizer of ω with respect to the coadjoint representation is given by

$$(8.10) \qquad \mathfrak{g}^\omega = Z_{\mathfrak{g}}(X_\omega) = \mathfrak{g} \cap Z_{\mathfrak{g}_{\mathbb{C}}}(X_\omega) = \mathfrak{g} \cap \left(\mathfrak{t}_{\mathbb{C}} \oplus \bigoplus_{\beta(X_\omega)=0} \mathfrak{g}_{\mathbb{C}}^\beta \right) = \mathfrak{g} \cap \mathfrak{s}_\Sigma.$$

A subalgebra $\mathfrak{b}$ of $\mathfrak{g}_{\mathbb{C}}$ is called *isotropic* with respect to ω if the complex extension $\omega_{\mathbb{C}}$ vanishes on $[\mathfrak{b}, \mathfrak{b}]$. It is called a *complex polarization* if it is maximally isotropic and satisfies $\mathfrak{b} \cap \mathfrak{g} = \mathfrak{g}^\omega$. Counting dimensions it is not hard to see that an isotropic subalgebra $\mathfrak{b}$ of $\mathfrak{g}_{\mathbb{C}}$ satisfying $\mathfrak{b} \cap \mathfrak{g} = \mathfrak{g}^\omega$ is maximally isotropic if and only if

$$\mathfrak{b} \cap \overline{\mathfrak{b}} = \mathfrak{g}_{\mathbb{C}}^\omega \quad \text{and} \quad \mathfrak{b} + \overline{\mathfrak{b}} = \mathfrak{g}_{\mathbb{C}}.$$

Thus to see that $\mathfrak{p}$ is a complex polarization for ω, it remains to check that $\mathfrak{p}$ is isotropic.

$$\omega_{\mathbb{C}}([\mathfrak{p}, \mathfrak{p}]) = \omega_{\mathbb{C}}([\mathfrak{s}_\Sigma, \mathfrak{s}_\Sigma] + \mathfrak{n}_\Sigma) = \omega_{\mathbb{C}}([\mathfrak{s}_\Sigma, \mathfrak{s}_\Sigma] \cap \mathfrak{t}_{\mathbb{C}}) = 0$$

by (8.10).

The following theorem shows how complex polarizations give rise to complex structures (cf. [Ki76, p.203]).

Theorem 8.31. *Let G be a connected Lie group with Lie algebra $\mathfrak{g} := \mathbf{L}(G)$. For $\omega \in \mathfrak{g}^*$ and G^ω the stabilizer of ω under the coadjoint action we write $\mathfrak{g}^\omega := \mathbf{L}(G^\omega)$. Let $\mathfrak{b} \subseteq \mathfrak{g}_{\mathbb{C}}$ be an $\operatorname{Ad} G^\omega$-invariant complex polarization. Then the identification of $T_\omega(\mathcal{O}_\omega) \cong T_{1G^\omega}(G/G^\omega) \cong \mathfrak{g}/\mathfrak{g}^\omega$ with the complex vector space $\mathfrak{g}_{\mathbb{C}}/\overline{\mathfrak{b}}$ yields a complex structure on $G/G^\omega \cong \mathcal{O}_\omega$.* ∎

It follows that the coadjoint orbit $\mathcal{O}_\omega$ carries an invariant pseudo-Kähler structure whose pull-back from the tangent space in the base point to the subalgebra $\mathfrak{p}$ is given by

$$C(X, Y) = \omega_{\mathbb{C}}([iX, \overline{Y}]).$$

Note that the radical of this Hermitean form on $\mathfrak{p}$ is precisely $\mathfrak{s}_\Sigma = \mathfrak{p} \cap \sigma(\mathfrak{p})$. As we have seen in the previous subsection, in those cases which are of interest to us, these pseudo-Kähler structures come up more naturally by the realizations of the flag manifolds in projective spaces.

8.4. Compression semigroups of open G-orbits

In this section we consider the following problem. Let G be a linear simple Lie group, $G_{\mathbb{C}}$ its complexification, $M = G_{\mathbb{C}}/P$ a complex flag manifold and $\mathcal{O} \subseteq M$ an open G-orbit. Then we are interested in the semigroup

$$\text{compr}(\mathcal{O}) := \{g \in G_{\mathbb{C}} : g.\mathcal{O} \subseteq \mathcal{O}\}.$$

For a parabolic subgroup $P \subseteq G_{\mathbb{C}}$ let

$$S(P) := \{g \in G_{\mathbb{C}} : gGP \subseteq GP\}.$$

These are subsemigroups of $G_{\mathbb{C}}$ containing G and it is immediate that

$$S(P) = \text{compr}(\mathcal{O})$$

whenever $P = \{g \in G_{\mathbb{C}} : g.x = x\}$ for a point $x \in \mathcal{O}$.

We will show that these semigroups have non-empty interior different from $G_{\mathbb{C}}$ iff G is Hermitean and the moment mapping of $\mathcal{O}$ maps $\mathcal{O}$ into a coadjoint orbit lying in a pointed invariant cone. Note that in this case $\mathcal{O}$ always carries a natural symplectic structure such that the action of G on $\mathcal{O}$ is a Hamiltonian action, so that it makes sense to talk about a moment mapping.

First we apply the results on open G-orbits to obtain some more precise information on the situation. Let $\mathfrak{g} = \mathfrak{k} + \mathfrak{p}_0$ be a Cartan decomposition, $\mathfrak{t} \subseteq \mathfrak{k}$ a Cartan subalgebra, and $\mathfrak{h} = \mathfrak{t} + \mathfrak{a}$ the corresponding Cartan subalgebra of $\mathfrak{g}$. Then $\mathfrak{g}_{\mathbb{C}} = (\mathfrak{k} + i\mathfrak{p}_0) + (i\mathfrak{k} + \mathfrak{p}_0)$ is a Cartan decomposition of $\mathfrak{g}_{\mathbb{C}}$ and $\mathfrak{h}' := i\mathfrak{t} + \mathfrak{a}$ is a maximal abelian subalgebra of $i\mathfrak{k} + \mathfrak{p}_0$.

To study our semigroup $S := S(P)$ we also need some knowledge on the G-double coset decomposition of $G_{\mathbb{C}}$.

Proposition 8.32.   *Let G be semisimple and $\mathfrak{t}_1, \ldots, \mathfrak{t}_n$ representatives for the conjugacy classes of Cartan subalgebras in $\mathfrak{g}$. Then the set*

$$\bigcup_{i=1}^{n} G N_{G_{\mathbb{C}}}(\mathfrak{t}_i) G$$

contains an open dense subset of $G_{\mathbb{C}}$.

Proof.   For $g \in G_{\mathbb{C}}$ we set $g^* := \sigma(g)^{-1}$, where σ denotes complex conjugation on $G_{\mathbb{C}}$. Let $G_{\mathbb{C}}^{\text{reg}}$ denote the set of regular elements in $G_{\mathbb{C}}$,

$$G_{\mathbb{C}}' := \{g \in G_{\mathbb{C}} : gg^* \in G_{\mathbb{C}}^{\text{reg}}\},$$

and $T_i := Z_{G_{\mathbb{C}}}(\mathfrak{t}_i)$ the Cartan subgroups corresponding to the Cartan subalgebras $\mathfrak{t}_{i\mathbb{C}} \subseteq \mathfrak{g}_{\mathbb{C}}$. Then it follows from [OM80, p.400] that the open set $G_{\mathbb{C}}'$ is a union of the finitely many open sets

$$G_{\mathbb{C}}^i := G H_i' G,$$

where
$$H_i := \{g \in G_{\mathbb{C}}: gg^* \in T_i\} \quad \text{and} \quad H_i' := H_i \cap G_{\mathbb{C}}'.$$

To obtain a better description of the representatives of the double cosets, we need to shrink the sets H_i.

So let $g = hh^* \in T_i \cap G_{\mathbb{C}}^{\text{reg}}$. Then $\text{Ad}(g)$ fixes $t_{i\mathbb{C}}$ pointwise. Since
$$\text{Ad}(h^*) = \sigma \circ \text{Ad}(h)^{-1} \circ \sigma,$$

it follows that $\text{Ad}(h)^{-1}$, and therefore $\text{Ad}(h)$, commutes with σ on $t_{i\mathbb{C}}$. Whence $\text{Ad}(h)(t_i)$ is a Cartan subalgebra of $\mathfrak{g}$ and we find $h' \in G$ and $j \in \{1,\ldots,n\}$ with
$$\text{Ad}(h')\,\text{Ad}(h)t_i = t_j.$$

Then $i = j$ by Corollary 2.4 in [Ro72]. Hence
$$(h'h)(h'h)^* = h'(hh^*)h'^{-1} = h'gh'^{-1}.$$

We conclude that
$$GH_i'G \subseteq GN_{G_{\mathbb{C}}}(t_i)G$$
and the assertion follows.   ∎

Let $S^0 := \text{int}\,S$ and suppose that this semigroup is non-empty. Then the preceding proposition shows that there exists a Cartan subalgebra $t_j \subseteq \mathfrak{g}$ such that $S^0 \cap N_{G_{\mathbb{C}}}(t_j) \neq \emptyset$. Let $\mathfrak{a}_j \subseteq it_j$ denote the vector part of it_j. Then an application of Corollary 1.20 entails that $S^0 \cap \exp \mathfrak{a}_j \neq \emptyset$ because $\exp(t_j) \subseteq S$ and therefore $S^0 \exp(t_j) = S^0$. The subspace $i\mathfrak{a}_j \subseteq \mathfrak{g}$ is abelian and compactly embedded. Hence there exists $g \in G$ such that $\text{Ad}(g)i\mathfrak{a}_j \subseteq t$ (cf. Proposition 7.3). Then $\text{Ad}(g)\mathfrak{a}_j \subseteq it$ and consequently $S^0 \cap \exp(it) \neq \emptyset$.

Let $C := \exp|_{it}^{-1}(S^0)$. The semigroup C is an open subsemigroup of it which is invariant under the Weyl group $\mathcal{W}_t$. Let $c \in C$. Then
$$c_0 := \sum_{\gamma \in \mathcal{W}_t} \gamma(c) \in C$$

is a fixed point under $\mathcal{W}_t$. There are two possible cases:

Case (1): $c_0 = 0$. Then $0 \in C$ and $1 \in \text{int}\,S$. This means that $S = G_{\mathbb{C}}$ because $G_{\mathbb{C}}$ is connected.

Case (2): $c_0 \neq 0$. Then c_0 is a non-zero $\mathcal{W}_t$-invariant element in it. It follows that $i\mathbb{R}c_0 \subseteq Z(\mathfrak{k})$, and in particular that $Z(\mathfrak{k}) \neq \{0\}$. Whence $\mathfrak{g}$ is a Hermitean simple Lie-algebra (cf. [Hel78, p.382]) and $S^0 \cap \exp(iZ(\mathfrak{k})) \neq \emptyset$.

We collect the results of this discussion in the following proposition:

Proposition 8.33.   *Let G be a linear simple Lie group, $G_{\mathbb{C}}$ its complexification, $M = G_{\mathbb{C}}/P$ a complex flag manifold, $\mathcal{O} \subseteq M$ an open G-orbit, and*
$$S = \text{compr}(\mathcal{O}) := \{g \in G_{\mathbb{C}} : g.\mathcal{O} \subseteq \mathcal{O}\}.$$

Assume that $S^0 \neq \emptyset$. Then either

(1) *$S = G_{\mathbb{C}}$ and G acts transitively on $G_{\mathbb{C}}/P$ or*

(2) *G is Hermitean and $S^0 \cap \exp(iZ(\mathfrak{k})) \neq \emptyset$.*   ∎

Lemma 8.34. *If $\mathcal{O}$ is an open subset of a locally compact G-space M, then the semigroup $S = \mathrm{compr}(\mathcal{O})$ is closed in G.*

Proof.   First we note that the action of G on the space M extends to a continuous action on the space $\mathcal{F}(M)$ of all closed subsets of M by

$$G \times \mathcal{F}(M) \to \mathcal{F}(M), \quad (g, F) \mapsto g.F$$

([HiNe92, I.5]). On the other hand, for a closed subset $F \subseteq M$, the set

$$\uparrow F = \{E \in \mathcal{F}(M) : F \subseteq E\}$$

is closed (Proposition 4.1(iii)). It follows that the set

$$\{g \in G : g(M \setminus \mathcal{O}) \supseteq (M \setminus \mathcal{O})\}$$

is closed. But this set is exactly the semigroup S.   ∎

Lemma 8.35.   *Let $g \in \mathrm{Gl}(V)$ be diagonalizable with positive real eigenvalues, and $[v] \in \mathbb{P}(V)$. Then*

$$[v_0] := \lim_{n \to \infty} g^n.[v]$$

exists and $g.v_0 = \lambda_{\max} v_0$, where $\lambda_{\max}$ is the largest eigenvalue of g on the smallest g-invariant subspace of V containing v.

Proof.   We may assume that the smallest g-invariant subspace of V containing v coincides with V. Let

$$\lambda_0 < \lambda_1 < \ldots < \lambda_k = \lambda_{\max}$$

denote the different eigenvalues of g on V. Then

$$v = \sum_{i=0}^{k} v_i \quad \text{with} \quad g.v_i = \lambda_i v_i$$

and therefore

$$g^n.[v] = [g^n.v] = \left[\sum_{i=0}^{k} g^n.v_i \right]$$

$$= \left[\sum_{i=0}^{k} \lambda_i^n.v_i \right] = \left[\sum_{i=0}^{k} \left(\frac{\lambda_i}{\lambda_{\max}} \right)^n .v_i \right] \to [v_k].$$

                                                                                ∎

Let us return to the setting where G is simple Hermitean and

$$S^0 \cap \exp(iZ(\mathfrak{k})) \neq \emptyset.$$

Then S is a subsemigroup of $G_{\mathbb{C}}$ with non-empty interior containing G and $C_S = \overline{\mathcal{O}}$ is the corresponding unique invariant control set on M (Proposition 8.2, Corollary 8.9). This situation will arise again in Section 8.6, so we need some general results dealing with this situation.

Definition 8.36. (a) Let Y be a diagonalizable endomorphism of the complex vector space V with real eigenvalues. An element $v \in V$ is said to be *generic* with respect to Y if the smallest Y-invariant subspace containing v contains eigenvectors for the maximal and minimal eigenvalues of Y.

(b) If V is a finite dimensional module of the Lie algebra $\mathfrak{g}_{\mathbb{C}}$ and $\mathcal{P}_V$ the corresponding set of weights, then we say that an element $X \in it$ is *weight separating* if it separates the set of all weights, i.e., if the values $\alpha(X), \alpha \in \mathcal{P}_V$ are pairwise distinct. ∎

Lemma 8.37. *Let V be an irreducible finite dimensional $G_{\mathbb{C}}$-module, $v_\lambda \in V$ a highest weight vector, $\mathcal{O} \subseteq G_{\mathbb{C}}.[v_\lambda]$ an open G-orbit, and $Y \in \exp(it)$. Then there exists a vector $v \in V$ which is generic with respect to Y such that $[v] \in \mathcal{O}$.*

Proof. The fact that V is a simple $G_{\mathbb{C}}$-module entails that V is spanned by the set $\{v \in V : [v] \in G_{\mathbb{C}}.[v_\lambda]\}$, and, by the analyticity of the orbit mapping, it is even spanned by $\{v \in V : [v] \in U\}$ for every open subset $U \subseteq G_{\mathbb{C}}.[v_\lambda]$. This applies in particular to the G-orbit $\mathcal{O}$. Let $v = \sum v_\alpha$ denote the decomposition of a vector $v \in V$ into Y-eigenvectors, where v_α is an eigenvector with eigenvalue α. Write $\lambda_{\min}$ and $\lambda_{\max}$ for the minimal and maximal eigenvalue. Then, since $\mathcal{O}$ spans V, we first find $[v'] \in \mathcal{O}$ with $v'_{\lambda_{\max}} \neq 0$. We note that the complement of this set is an analytic set, hence nowhere dense. So we even find $[v] \in \mathcal{O}$ with $v_{\lambda_{\max}} \neq 0$ and $v_{\lambda_{\min}} \neq 0$. Now the smallest Y-invariant subspace containing v also contains $v_{\lambda_{\max}}$ and $v_{\lambda_{\min}}$. ∎

Lemma 8.38. *Let $S \subseteq G_{\mathbb{C}}$ be a subsemigroup with non-empty interior containing G, M a complex flag manifold realized as a highest weight orbit $G_{\mathbb{C}}.[v_\lambda]$, $C_S \subseteq M$ the invariant control set for S, and $\mathcal{P}_{C_S}$ the set of all extremal weights α with $[v_\alpha] \in C_S$. Then $\mathcal{P}_{C_S}$ has the following properties:*

(i) $\mathcal{W}_{\mathfrak{t}}.\mathcal{P}_{C_S} = \mathcal{P}_{C_S}$.

(ii) *If $X \in it$ is weight separating with $\exp X \in S$, then*

$$\alpha(X) = \max\{\beta(X) : \beta \in \mathcal{P}_V\}$$

implies that $\alpha \in \mathcal{P}_{C_S}$.

(iii) *If $X \in it$ is weight separating with $\exp X \in S$, $\mu \in \mathcal{P}_{C_S}$, $\alpha \in \Lambda$, and $s_\alpha \in \mathcal{W}$ is the corresponding reflection, then $s_\alpha(\mu)(X) > \mu(X)$ implies that $s_\alpha(\mu) \in \mathcal{P}_{C_S}$.*

(iv) $\mathcal{P}_{C_S} = \{\alpha \in \mathcal{P}_V : [v_\alpha] \in \operatorname{int} C_S\}$.

Proof. (i) Let $\gamma \in \mathcal{W}_{\mathfrak{t}} \cong N_K(\mathfrak{t})/Z_K(\mathfrak{t})$. Then there exists $k \in K$ such that $\operatorname{Ad}(k)|_{\mathfrak{t}_{\mathbb{C}}} = \gamma$. It follows that

$$k.[v_\alpha] = [k.v_\alpha] = [v_{\alpha \circ \gamma^{-1}}] = [v_{\gamma.\alpha}].$$

(ii) In view of Proposition 8.10(ii), the invariant control set C_S contains at least one open G-orbit $\mathcal{O}$. Using Lemma 8.37, we find $[v] \in \mathcal{O}$ such that v is generic for X. Then, according to Lemma 8.35,

$$[v'] := \lim_{n \to \infty} \exp(X)^n.[v] \in \overline{\mathcal{O}} \subseteq C_S$$

exists in $\mathbb{P}(V)$ and v' is an eigenvector of $\exp(X)$ for the maximal eigenvalue $e^{\alpha(X)}$, hence a weight vector of weight α for $\mathfrak{t}_{\mathbb{C}}$ because X is weight separating. Finally

$$\alpha(X) = \max\{\beta(X) : \beta \in \mathcal{P}_V\}$$

and the weight separating property of X show that α is extremal.

(iii) Let $G_{\mathbb{C}}(\alpha)$ denote the analytic subgroup of $G_{\mathbb{C}}$ with

$$\mathfrak{g}_{\mathbb{C}}(\alpha) := \mathbf{L}\left(G_{\mathbb{C}}(\alpha)\right) = \mathfrak{g}_{\mathbb{C}}^{\alpha} + \mathfrak{g}_{\mathbb{C}}^{-\alpha} + [\mathfrak{g}_{\mathbb{C}}^{\alpha}, \mathfrak{g}_{\mathbb{C}}^{-\alpha}] \cong \mathfrak{sl}(2,\mathbb{C}).$$

Further let W denote the smallest $\mathfrak{g}_{\mathbb{C}}(\alpha)$-submodule containing v_{μ}. This module is irreducible ([Bou75, Ch. VIII, §7, no. 2, Prop. 3]) with highest weight vector v_{μ} and lowest weight vector $v_{\mu'}$, where $\mu' := s_{\alpha}(\mu)$.

It follows that the $G_{\mathbb{C}}(\alpha)$-orbit M_{α} of $[v_{\mu}]$ contains exactly two weight rays, namely $[v_{\mu}]$ and $[v_{\mu'}]$ (cf. [Bou75, Ch. VIII, §7, no. 2, Prop. 5]). The orbits of these elements under the group $G_{\alpha} := \langle \exp\left(\mathfrak{g}_{\mathbb{C}}(\alpha) \cap \mathfrak{g}\right)\rangle$ are relatively open in M_{α} (Proposition 8.25).

Since $\mathfrak{g}_{\mathbb{C}}(\alpha)$ is invariant under $\mathrm{Ad}(\exp it)$, it follows that

$$\exp(X).M_{\alpha} = \exp(X).\left(G_{\mathbb{C}}(\alpha).[v_{\mu}]\right) = G_{\mathbb{C}}(\alpha).\left((\exp X).[v_{\mu}]\right) = G_{\mathbb{C}}(\alpha).[v_{\mu}] = M_{\alpha}.$$

On the other hand the orbit $G_{\alpha}.v_{\mu}$ spans W, so it contains a generic vector v for X on W (Lemma 8.37). Note that $G_{\alpha}.[v_{\mu}] \subseteq G.[v_{\mu}] \subseteq C_S$ since $\mu \in \mathcal{P}_{C_S}$. Now our assumption $\mu'(X) > \mu(X)$ shows that the maximal eigenvalue of X on W is $\mu'(X)$. Whence

$$[v_{\mu'}] = \lim_{n \to \infty} \exp(X)^n.[v] \in \overline{G.[v]} \subseteq C_S$$

(Lemma 8.35), so that $\mu' \in \mathcal{P}_{C_S}$.

(iv) Since every G-orbit of an extremal weight ray is open by Proposition 8.25, and C_S is the closure of a union of open G-orbits (Proposition 8.10(ii)), the condition $[v_{\alpha}] \in C_S$ even implies that $[v_{\alpha}] \subseteq \mathrm{int}\, C_S$. ∎

We apply these results in the special case where G is simple Hermitean and the interior of S intersects $\exp\left(iZ(\mathfrak{k})\right)$ non-trivially. We fix an element $Z_k \in S^0 \cap \exp\left(iZ(\mathfrak{k})\right)$ and consider a realization of the flag manifold $M = G_{\mathbb{C}}/P$ as a $G_{\mathbb{C}}$-orbit of a highest weight ray $[v_{\lambda}]$ in a highest weight module V of $G_{\mathbb{C}}$. Let $\mathcal{P}_V$ denote the corresponding set of weights. Then the extreme points of the convex hull of $\mathcal{P}_V$ consists precisely of the Weyl group orbit $W.\lambda$ of the highest weight λ ([Bou75, Ch. VIII, §7, no. 2, Prop. 5]).

Choose a weight $\alpha \in \mathcal{P}_V$ such that $\alpha(Z_k)$ is maximal. Then there exists a weight separating element $Z_{\alpha} \in it$ arbitrarily close to Z_k such that

$$\alpha(Z_{\alpha}) = \max\{\beta(Z_{\alpha}) : \beta \in \mathcal{P}_V\},$$

and $\exp(Z_{\alpha}) \in S^0$. Now Lemma 8.38(iii), (iv) yield

$$[v_{\alpha}] \in \mathrm{int}\, C_S = \mathcal{O}.$$

So we have shown that $\mathcal{O}$ contains every weight ray $[v_{\alpha}]$, where $\alpha(Z_k)$ is maximal. To evaluate this condition, we need the following lemma.

Lemma 8.39. *Let $\Lambda^+ \subseteq \Lambda$ be a $\mathfrak{k}$-adapted positive system. Then the following assertions hold:*

(i) *Let $\mu \in C^*_{\min}$. Then*
$$W.\mu \cap C^*_{\min} = W_{\mathfrak{k}}.\mu.$$

(ii)
$$W_{\mathfrak{k}} = \{\gamma \in W : \gamma.iC^*_{\min} = iC^*_{\min}\}.$$

(iii) *If $Z \in iZ(\mathfrak{k})$ such that $\alpha(Z) > 0$ holds for the positive non-compact roots and $\mu \in (it)^*$, then*
$$i\gamma.\mu \in C^*_{\min} \iff (\gamma.\mu)(Z) = \max\{(\gamma'.\mu)(Z) : \gamma' \in W\}.$$

Proof.   (i) That the right hand side is contained in the left hand side follows from the invariance of $C_{\min}$ under the small Weyl group $W_{\mathfrak{k}}$. Suppose that $\gamma \in W$ with $\gamma.\mu \in C^*_{\min}$. Then $(\gamma.\mu)(i\check\alpha) \leq 0$ for all $\alpha \in \Lambda_p^+$. Thus there exists $\gamma' \in W_{\mathfrak{k}}$ such that
$$((\gamma'\gamma).\mu)(i\check\alpha) \leq 0 \qquad \forall \alpha \in \Lambda^+.$$
On the other hand there exists $\gamma'' \in W_{\mathfrak{k}}$ with
$$(\gamma''.\mu)(i\check\alpha) \leq 0 \qquad \forall \alpha \in \Lambda^+.$$
Thus $(\gamma'\gamma).\mu = \gamma''.\mu$ ([Bou68, Ch. V, §3, no. 3.3, Th. 2]) and therefore
$$\gamma.\mu = (\gamma')^{-1}\gamma''.\mu \in W_{\mathfrak{k}}.\mu.$$

(ii) That $W_{\mathfrak{k}}$ leaves $C^*_{\min}$ invariant is clear. To prove the assertion, pick $\mu \in C^*_{\min}$ such that the stabilizer of μ in W is trivial. Suppose that $\gamma \in W$ leaves the cone $C^*_{\min}$ invariant. Then
$$\gamma.\mu \in C^*_{\min} \subseteq W_{\mathfrak{k}}.\mu$$
shows that $\gamma \in W_{\mathfrak{k}}$.

(iii) Let $\alpha \in it^* \subseteq \mathfrak{t}_{\mathbb{C}}^*$ such that
$$(8.11) \qquad\qquad \alpha(Z) = \max\{(\gamma.\alpha)(Z) : \gamma \in W\}.$$
Pick a positive non-compact root β and consider the reflection s_β at the hyperplane $\ker \beta$. Then
$$s_\beta(\alpha) = \alpha - \frac{2(\alpha, \beta)}{(\beta, \beta)}\beta = \alpha - \alpha(\check\beta)\beta.$$
Since $s_\beta(\alpha)(Z) \leq \alpha(Z)$, this means that
$$(8.12) \qquad\qquad \alpha(\check\beta)\beta(Z) \geq 0.$$

Now (8.12) shows that α is non-negative on all the elements $[X, \overline{X}]$ for $X \in \mathfrak{g}_{\mathbb{C}}^\beta$, $\beta \in \Lambda_p^+$. In view of Lemma 7.7, this means that
$$i\alpha \in C^*_{\min},$$
since $\check\beta \in \mathbb{R}^+[X, \overline{X}]$ (cf. definition of $C_{\min}$).

If, conversely, $i\alpha \in C^*_{\min}$, then $(\gamma.\alpha)(Z) = \alpha(Z)$ holds for all $\gamma \in W_{\mathfrak{k}}$. So, as in (i), we may assume that $\alpha(\check\beta) \geq 0$ holds for all $\beta \in \Lambda^+$. This means that α is contained in the positive Weyl chamber corresonding to Λ^+. It follows that
$$W.\alpha \subseteq \alpha - \sum_{\beta \in \Lambda^+} \mathbb{R}^+\beta$$
(cf. [Hel84, p.459]), so that the assertion follows from $\beta(Z) = 0$ if β is a compact root and $\beta(Z) > 0$ if β is non-compact. ∎

If $\alpha \in \mathcal{P}_V$ is such that $\alpha(Z_k)$ is maximal among the Weyl group translates of α, then $\alpha(Z_k) \geq 0$ and Lemma 8.39(iii) yields that

$$i\alpha \in C_{\min}^*.$$

Thus there exists $\gamma \in \mathcal{W}_{\mathfrak{k}}$ such that $(\gamma.\alpha)(\check{\beta}) \geq 0$ for all $\beta \in \Upsilon_k$ since $i\alpha \in C_{\min}^*$ and the $\mathcal{W}_{\mathfrak{k}}$-translates of the positive Weyl chamber cover $C_{\min}^*$ ([Bou68, Ch. V, §4, no. 4, Cor. 1]). Thus $\lambda = \gamma.\alpha$ is a highest weight for V with respect to the positive system Λ^+. Pick $g \in G$ such that $\gamma = \mathrm{Ad}(g)|_{\mathfrak{t}_{\mathbb{C}}}$. Then

$$g.[v_\alpha] = [v_{\gamma.\alpha}]$$

entails that $\mathcal{O} = G.[v_\lambda]$. If we let $M \subseteq \mathbb{P}(V)$ be the $G_{\mathbb{C}}$-orbit containing all the weight rays for the extremal weights, then we just have shown:

Lemma 8.40. *If $G_{\mathbb{C}} \neq S^0 \neq \emptyset$ and $M = G_{\mathbb{C}}.[v_\lambda]$, then there exists a $\mathfrak{k}$-adapted positive system Λ^+ such that*

$$i\lambda \in C_{\min}(\Lambda^+)^*$$

is a highest weight, and

$$\mathcal{O} = G.[v_\lambda]$$

holds for a highest weight vector v_λ in V with respect to Λ^+.   ∎

Lemma 8.41. *Let $\mathcal{O} = G.x$, P_x the stabilizer of x in $G_{\mathbb{C}}$ and $B \subseteq P_x$ a Borel subgroup. Then*

$$S(B) \subseteq S(P_x),$$

and equality holds if $B = B(\Lambda^+)$ for a $\mathfrak{k}$-adapted positive system and $P_x \cap G$ is compact.

Proof.   Let $g \in S(B)$. Then $gGB \subseteq GB$ and therefore $gG \subseteq GB \subseteq GP_x$ entails

$$gGP_x \subseteq (GP_x)P_x = GP_x,$$

i.e., $g \in S(P_x)$.

Now suppose that $B = B(\Lambda^+)$ for a $\mathfrak{k}$-adapted positive system, and that $P_x \cap G$ is compact. Let Υ_k denote the set of all compact roots in Υ. Then $\Upsilon = \Upsilon_k \cup \{\delta\}$ and the parabolic $P := P_x$ is a parabolic containing B. We also write N for the commutator group of B. We claim that $GB = GP$. To see this, write $\mathfrak{p} := \mathbf{L}(P)$ as $\mathfrak{p}_\Sigma$ for a subset $\Sigma \subseteq \Upsilon$. Then the compactness of $P \cap G$ is equivalent to $\Sigma \subseteq \Upsilon_k$. If this condition is satisfied, then

$$P \cong N_P \rtimes (P \cap G)_{\mathbb{C}},$$

where $N_P \subseteq N$ is the unipotent radical of P. Since $(P \cap G)_{\mathbb{C}}$ is a reductive complex group and $P \cap G$ is a compact real form, we obtain an Iwasawa decomposition of this group by

$$(P \cap G)_{\mathbb{C}} = (P \cap G)\exp(i\mathfrak{t})((P \cap G)_{\mathbb{C}} \cap N).$$

Whence

$$GP = G(P \cap G)\exp(i\mathfrak{t})N = G\exp(i\mathfrak{t})N = G\exp(\mathfrak{t}_{\mathbb{C}})N = GB.$$

It follows in particular that $GP = GB$, thus $S(P) = S(B)$ is clear.   ∎

In view of the preceding lemmas, we can expect to get a good deal of information on the compression semigroups on general flag manifolds from the compression semigroups on maximal flag manifolds, i.e., quotients $G_{\mathbb{C}}/B$, where $B \subseteq G_{\mathbb{C}}$ is a Borel subgroup. Moreover, we know that it suffices to consider Borel subgroups $B = B(\Lambda^+)$, where Λ^+ is a $\mathfrak{k}$-adapted positive system. In this case we even have another possibility to simplify matters. Let $P_k := P_{\Upsilon_k}$ denote the maximal parabolic subgroup containing B with the property that its intersection with G is compact. Then, by Lemma 8.40, we have that $S(B) = S(P_k) \subseteq S(P_x)$.

So our next task is to analyse the semigroup $S_k := S(P_k)$. First we will show that S_k contains the maximal Ol'shanskiĭ semigroup $S_{\max}$ (cf. Lemma 7.25).

Theorem 8.42.   (Paneitz's Convexity Theorem) *Let $\mathfrak{g}$ be a Hermitean simple Lie-algebra, $\mathfrak{k} \subseteq \mathfrak{g}$ a compactly embedded Cartan algebra, and $\Lambda^+ \subseteq \Lambda(\mathfrak{g}_{\mathbb{C}}, \mathfrak{k}_{\mathbb{C}})$ a positive $\mathfrak{k}$-adapted system. Further let $p: \mathfrak{g}^* \to \mathfrak{k}^*$ denote the restriction mapping. Then*

$$p\big(\operatorname{Ad}(G)^*.\omega\big) \subseteq \operatorname{conv}(\mathcal{W}_{\mathfrak{k}}.\omega) + C^*_{\max} \subseteq C^*_{\min} \qquad \forall \omega \in C^*_{\min}$$

and equality holds for $\omega \in \operatorname{int} C^*_{\min}$.

Proof.   For $\omega \in \operatorname{int} C^*_{\min}$ this follows from [Pa84, p.224]. Let $\omega \in C^*_{\min}$ be arbitrary. Then there exists a sequence $\omega_n \in \operatorname{int} C^*_{\min}$ with $\omega_n \to \omega$. For $g \in G$ we now have

$$\operatorname{Ad}^*(g).\omega = \lim_{n \to \infty} \operatorname{Ad}^*(g).\omega_n$$

and

$$\operatorname{Ad}^*(g).\omega_n \in \operatorname{conv}(\mathcal{W}_{\mathfrak{k}}.\omega_n) + C^*_{\max}.$$

Hence we find for every $\gamma \in \mathcal{W}_{\mathfrak{k}}$ a real number $\mu_\gamma^n \in [0,1]$ such that $\sum_{\gamma \in \mathcal{W}_{\mathfrak{k}}} \mu_\gamma^n = 1$ and $\alpha_n \in C^*_{\max}$ with

$$\operatorname{Ad}^*(g).\omega_n = \sum_{\gamma \in \mathcal{W}_{\mathfrak{k}}} \mu_\gamma^n \gamma.\omega_n + \alpha_n.$$

After passing to a subsequence we may assume that $\mu_\gamma^n \to \mu_\gamma$ and, since the sequence α_n lies in a compact set, that $\alpha_n \to \alpha$. Then

$$\operatorname{Ad}^*(g).\omega_n \to \sum_{\gamma \in \mathcal{W}_{\mathfrak{k}}} \mu_\gamma \gamma.\omega + \alpha \in \operatorname{conv}(\mathcal{W}_{\mathfrak{k}}.\omega) + C^*_{\max}.$$

This proves that $\operatorname{Ad}^*(g).\omega \in \operatorname{conv}(\mathcal{W}_{\mathfrak{k}}.\omega) + C^*_{\max}$.   ∎

Corollary 8.43.   *Suppose that, under the assumptions of Theorem 8.42, the functional ω is contained in $C^*_{\min}$ and $\mathcal{O}_\omega$ is the coadjoint orbit through ω. Then $\operatorname{wedge}(\mathcal{O}_\omega)$ is a pointed and generating invariant cone in $\mathfrak{g}^*$ and*

$$p\big(\operatorname{wedge}(\mathcal{O}_\omega)\big) \subseteq C^*_{\min}.$$

Proof.   That $W_\omega := \operatorname{wedge}(\mathcal{O}_\omega)^*$ is invariant follows from the $\operatorname{Ad}(G)^*$-invariance of the generating set $\mathcal{O}_\omega$. Since $\mathfrak{g}^*$ is a simple $\operatorname{Ad}(G)^*$-module, it follows that W_ω is pointed because W_ω^* is different from $\{0\}$.

On the other hand, according to Paneitz's Convexity Theorem, W_ω contains $C_{\min}$. Since $C_{\min}$ has non-empty interior with respect to $\mathfrak{k}$, it follows that W_ω contains $W_{\min}$, so it is generating, too (Lemma 7.25). Thus W_ω^* is pointed.   ∎

Corollary 8.44. *Let Λ^+ be a $\mathfrak{k}$-adapted positive system, $\lambda \in \mathcal{P}^+$ such that $i\lambda \in C_{\min}(\Lambda^+)^*$, V_λ a corresponding highest weight module, and $\mathcal{O} := G.[v_\lambda] \subseteq G_{\mathbb{C}}.[v_\lambda]$ the corresponding open G-orbit. Then $\beta \in \mathcal{P}_V$ and $[v_\beta] \in \mathcal{O}$ implies that $i\beta \in \mathcal{W}_{\mathfrak{t}}.i\lambda \subseteq C_{\min}(\Lambda^+)^*$.*

Proof. Since $[v_\beta] = G.[v_\lambda]$, there exists $g \in G$ with $g.[v_\lambda] = [v_\beta]$.

To prepare the application of Paneitz's Convexity Theorem, we first use Theorem 8.30 to find a G-invariant pseudo-Hermitean structure J on V such that the corresponding moment mapping is given by

$$\Phi : \Omega \to \mathfrak{g}^*, \qquad [v] \mapsto \left(X \mapsto i\frac{J(X.v, v)}{J(v, v)} \right),$$

where $\Omega = \{[v] \in \mathbb{P}(V) : J(v, v) \neq 0\}$. Then $\Phi([v_\lambda]) = i\lambda$ and

$$i\beta = \Phi([v_\beta]) = \mathrm{Ad}^*(g)\Phi([v_\lambda]) = \mathrm{Ad}^*(g).i\lambda$$

because Φ is equivariant. Let $p : \mathfrak{g}^* \to \mathfrak{t}^*$ denote the restriction mapping. Then, according to Paneitz's Convexity Theorem,

$$i\beta = p(i\beta) \in p\big(\mathrm{Ad}^*(G).i\lambda\big) \subseteq \mathrm{conv}(\mathcal{W}_{\mathfrak{t}}.i\lambda) + C_{\max}^* \subseteq C_{\min}^*.$$

Since, according to Proposition 8.18(iii), $i\beta \in \mathcal{W}.i\lambda$, we conclude with Lemma 8.39 that

$$i\beta \in \mathcal{W}.i\lambda \cap C_{\min}^* = \mathcal{W}_{\mathfrak{t}}.i\lambda.$$

                                                   ∎

Corollary 8.45. *Let $\omega \in \mathrm{int}\, C_{\min}^*$ and $X \in \mathrm{int}\, C_{\min}$. Then the function*

$$H_X : \mathcal{O}_\omega \to \mathbb{R}, \qquad \nu \mapsto \langle \nu, X \rangle$$

is proper and positive, where the closed orbit $\mathcal{O}_\omega$ carries its manifold topology.

Proof. Let $W_\omega := \mathrm{wedge}(\mathcal{O}_\omega)^*$. In view of Corollary 8.43, this cone is pointed and generating. Moreover, since $\mathrm{Ad}(G) = \mathrm{Aut}(\mathfrak{g})_0$ is closed and consists of linear mappings with determinant 1, the group $\mathrm{Ad}(G)^*$ acts as a closed group of automorphisms of W_ω^* with determinant 1. Whence the orbit $\mathcal{O}_\omega$ is closed in W_ω^* because $\omega \in \mathrm{int}\, \mathcal{O}_\omega$ (Proposition 1.12). Since $X \in \mathrm{int}\, W_\omega$, the function $\alpha \mapsto \alpha(X)$ is a proper and non-negative function on W_ω^*. Hence the function H_X is proper on $\mathcal{O}_\omega$.   ∎

To apply Corollary 8.45 to the semigroup $S(P_k)$, we fix a realization of the flag manifold $G_{\mathbb{C}}/P_k$ as an orbit of a highest weight ray $[v_{\omega_\delta}]$, where ω_δ is the fundamental weight belonging to the non-compact root $\delta \in \Upsilon$. Using Theorem 8.30, we find a G-invariant pseudo-Hermitean structure J on V such that the corresponding moment mapping is given by

$$\Phi : \Omega \to \mathfrak{g}^*, \qquad [v] \mapsto \left(X \mapsto i\frac{J(X.v, v)}{J(v, v)} \right),$$

where $\Omega = \{[v] \in \mathbb{P}(V) \colon J(v,v) \neq 0\}$. It follows in particular that $\Phi([v_\alpha]) = i\alpha$ holds for every weight $\alpha \in \mathcal{P}_V$.

Let $X \in \operatorname{int} W_{\max} \cap \mathfrak{t}$ and consider the function H_X on $\mathfrak{g}^*$ defined by evaluation in X. Then

$$H_X\big(\operatorname{Ad}^*(g).i\omega_\delta\big) = i\omega_\delta\big(\operatorname{Ad}(g)^{-1}.X\big) = \Phi([v_{\omega_\delta}])\big(\operatorname{Ad}(g)^{-1}.X\big) = \Phi([g.v_{\omega_\delta}])(X).$$

Thus $F_X([v]) := \Phi([v])(X)$ defines a proper positive function on the open G-orbit $G.[v_{\omega_\delta}]$ in $G_\mathbb{C}.[v_{\omega_\delta}]$ (Corollary 8.45).

We claim that $\exp(\mathbb{R}^+ iX) \subseteq S(P_k)$. Since F_X is proper, we know that

$$\lim_{p \to \partial \mathcal{O}} F_X(p) = \infty,$$

and hence it suffices to show that, for $t \geq 0$, we have that

$$F_X(\exp itX.p) \leq F_X(p) \qquad \forall p \in G.[v_{\omega_\delta}].$$

Let $\mathcal{X}$ denote the vector field on $\mathbb{P}(V)$ defined by

$$\mathcal{X}(q) := \frac{d}{dt}\bigg|_{t=0} \exp(-tX).q.$$

Then the restriction of $\mathcal{X}$ to $G.[v_{\omega_\delta}]$ corresponds via the G-equivariant map Φ to the vector field $\mathcal{X}'$ on $\mathcal{O} := \mathcal{O}_{i\omega_\delta}$ which satisfies

$$\mathcal{X}'(\beta) = \frac{d}{dt}\bigg|_{t=0} \operatorname{Ad}^*\big(\exp(-tX)\big).\beta = -\operatorname{ad}^*(X)(\beta) = \beta \circ \operatorname{ad} X.$$

Since the canonical symplectic form on $\mathcal{O}$ satisfies

$$\omega(\beta)(\beta \circ \operatorname{ad} Y, \beta \circ \operatorname{ad} Z) = \beta([Y,Z]) \qquad \forall Y, Z \in \mathfrak{g},$$

we find that

$$\omega(\beta)\big(\mathcal{X}'(\beta), \beta \circ \operatorname{ad} Y\big) = \beta([X,Y]) = -\langle \beta \circ \operatorname{ad} Y, X\rangle = -dH_X(\beta)(\beta \circ \operatorname{ad} Y).$$

Hence $\mathcal{X}'$ is the Hamiltonian vector field on $\mathcal{O}$ corresponding to the function H_X.

Let ω denote the 2-form defining the symplectic structure on $\mathcal{O}$ and I the tensor field defining the complex structure on $\mathcal{O}$. Then, for $\Phi(p) = \beta = i \operatorname{Ad}^*(g)\omega_\delta$,

$$dF_X(p)\big(-i\mathcal{X}(p)\big) = dH_X(\beta)\big(-I(\beta)\mathcal{X}'(\beta)\big)$$
$$= \omega(\beta)\big(\mathcal{X}'(\beta), I(\beta)\mathcal{X}'(\beta)\big).$$

To see that this expression is always negative, in view of the G-invariance of the symplectic form ω and the complex structure I, it suffices to see that the bilinear form

$$(X, X') \mapsto \omega(i\omega_\delta)(i\omega_\delta \circ \operatorname{ad} X, Ii\omega_\delta \circ \operatorname{ad} X')$$

is negative definite on $\mathfrak{p}$. Since

$$T_{\omega_\delta}(\mathcal{O}) \cong T_{[v_{\omega_\delta}]}(G.[v_{\omega_\delta}]) \cong \mathfrak{g}_{\mathbb{C}}/\mathfrak{p}_k,$$

this space inherits its complex structure from the subalgebra $\mathfrak{p}_{\mathbb{C}}^-$ spanned by the negative non-compact root spaces. So we can use Lemma 7.7(ii) to find for $\overline{Z} \in \mathfrak{g}_{\mathbb{C}}^{-\lambda}$, $\lambda \in \Lambda_p^+$, and $X = \overline{Z} + Z \in \mathfrak{g}$, we have that

$$I(i\omega_\delta)(i\omega_\delta \circ \operatorname{ad} X) = i\omega_\delta \circ \operatorname{ad}(i\overline{Z} - iZ)$$
$$= i\omega_\delta \circ \operatorname{ad}(i(\overline{Z} - Z)).$$

Now

$$\omega(i\omega_\delta)(i\omega_\delta \circ \operatorname{ad} X, Ii\omega_\delta \circ \operatorname{ad} X) = i\omega_\delta([\overline{Z} + Z, i(\overline{Z} - Z)]) = -i\omega_\delta(2i[\overline{Z}, Z]).$$

Since

$$C_{\min} = i \operatorname{wedge}\{[\overline{Z}, Z] : Z \in \mathfrak{g}_{\mathbb{C}}^\lambda, \lambda \in \Lambda_p^+\}$$

and $i\omega_\delta \in C_{\min}^*$, we see that

$$dF_X(p)(-i\mathcal{X}(p)) \le 0 \qquad \forall p \in G.[v_{\omega_\delta}].$$

Whence the functions

$$t \mapsto F_X(\exp(itX).p)$$

are decreasing for $X \in \operatorname{int} W_{\max} \cap \mathfrak{t}$.

Thus we have shown that $iW_{\max} \cap \mathfrak{t} \subseteq \mathbf{L}(S)$ since $\mathbf{L}(S)$ is closed. Note that $\mathbf{L}(S)$ is well defined by the closedness of S (Lemma 8.34). The fact that $G \subseteq H(S)$ entails that

$$\mathbf{L}(S) \cap i\mathfrak{g}$$

is a wedge invariant under $\operatorname{Ad}(G)$. Hence $iW_{\max} \subseteq \mathbf{L}(S)$ since invariant cones are determined by their intersection with a compactly embedded Cartan algebra (Theorem 7.27, and so

$$S_{\max} = G \exp(iW_{\max}) \subseteq S.$$

Whence the preceding discussion, together with Lemma 8.40, proves the following proposition.

Proposition 8.46.   *If $\emptyset \ne S(P) \ne G_{\mathbb{C}}$ and $G_{\mathbb{C}}/P = G_{\mathbb{C}}.[v_\lambda]$ is a realization as an orbit of a highest weight ray with respect to a $\mathfrak{t}$-adapted positive system Λ^+, then*

$$S_{\max} = G \exp(iW_{\max}) \subseteq S(P) = \operatorname{compr}(G.[v_\lambda]).$$

∎

We want to show that $S_{\max} = S(P)$ holds in all these cases. Since $S_{\max}$ is a subsemigroup of $G_{\mathbb{C}}$ with dense interior, the following lemma may be helpful.

Lemma 8.47. *A subsemigroup $S \subseteq G_\mathbb{C}$ with dense interior containing G is completely determined by its intersections with the groups $N_{G_\mathbb{C}}(\mathfrak{t}_i)$, where $\mathfrak{t}_1, \ldots, \mathfrak{t}_n$ is a set of representatives for the set of conjugacy classes of Cartan subalgebras of $\mathfrak{g}$. More precisely*

$$S \cap G_\mathbb{C}' = \bigcup_{i=1}^{n} G\big(S \cap N_{G_\mathbb{C}}(\mathfrak{t}_i)'\big)G$$

Proof. Recall the definition of the subset $G_\mathbb{C}'$ from the proof of Proposition 8.32. Since $G_\mathbb{C}'$ is open and dense in $G_\mathbb{C}$ and int S is open and dense in S, it follows that $S \cap G_\mathbb{C}'$ is dense in S. On the other hand $G \subseteq H(S)$ entails that

$$S \cap \big(GN_{G_\mathbb{C}}(\mathfrak{t}_i)G\big) = G\big(S \cap N_{G_\mathbb{C}}(\mathfrak{t}_i)\big)G.$$

∎

Proposition 8.48. *Let $\mathfrak{g}$ be simple Hermitean, $\mathfrak{t} \subseteq \mathfrak{g}$ be a compactly embedded Cartan algebra, Λ^+ a $\mathfrak{t}$-adapted positive system, and $C_{\max}$ the corresponding maximal cone. Suppose that $S \subseteq G_\mathbb{C}$ is a subsemigroup with dense interior containing G. If*

$$S \cap \exp(i\mathfrak{t}) \subseteq \exp(iC_{\max} \cup -iC_{\max}),$$

then

$$S \subseteq \overline{GN_{G_\mathbb{C}}(\mathfrak{t})G}.$$

Proof. In view of Lemma 8.47, we have to show that

$$\operatorname{int}(S) \cap N_{G_\mathbb{C}}(\mathfrak{t}') = \emptyset$$

for every Cartan subalgebra $\mathfrak{t}' \subseteq \mathfrak{g}$ which is not conjugate to $\mathfrak{t}$.

As before, let $\mathfrak{k} \subseteq \mathfrak{g}$ be the unique maximal compactly embedded subalgebra containing $\mathfrak{t}$, and pick a Cartan subalgebra $\mathfrak{t}' \subseteq \mathfrak{g}$ not conjugate to $\mathfrak{t}$. Using [Ro72, 1.3], we may assume that $\mathfrak{t}'$ is invariant under the Cartan involution determined by $\mathfrak{k}$, i.e.,

$$\mathfrak{t}' = (\mathfrak{t}' \cap \mathfrak{k}) + (\mathfrak{t}' \cap \mathfrak{p}_0),$$

where $\mathfrak{p}_0$ is the orthogonal complement of $\mathfrak{k}$ with respect to the Cartan Killing form. Moreover, since all compactly embedded Cartan algebras are conjugate [Ro72, 1.4], and $\mathfrak{t}' \cap \mathfrak{k}$ may be extended to a Cartan subalgebra of $\mathfrak{k}$, we even may assume that $\mathfrak{t}' \cap \mathfrak{k} \subseteq \mathfrak{t}$. In particular $\mathfrak{t}' \cap \mathfrak{p}_0 \neq \{0\}$.

Now we consider the group $N' := N_{G_\mathbb{C}}(\mathfrak{t}')$. Its Lie algebra coincides with $\mathfrak{t}'_\mathbb{C}$ and $A' := \exp\big(i(\mathfrak{t}' \cap \mathfrak{k}) + (\mathfrak{t}' \cap \mathfrak{p}_0)\big)$ is a normal subgroup such that N'/A' is compact. Let $S' := \operatorname{int}(S) \cap N'$ and suppose that $S' \neq \emptyset$. Then $S'A'/A'$ is an open subsemigroup of the compact group N'/A' and therefore it is a subgroup (Corollary 1.21). It follows in particular that it contains the unit element. This means that $A' \cap S' \neq \emptyset$. Since $\exp(\mathfrak{t}' \cap \mathfrak{p}_0) \subseteq G \subseteq S$, it even follows that

$$S' \cap \exp\big(i(\mathfrak{t}' \cap \mathfrak{k})\big) = \operatorname{int}(S) \cap \exp\big(i(\mathfrak{t}' \cap \mathfrak{k})\big) \neq \emptyset.$$

Let $X \in i(\mathfrak{t}' \cap \mathfrak{k}) \subseteq i\mathfrak{t}$. If there exists no non-compact root vanishing on X, then

$$\mathfrak{t}' \cap \mathfrak{p}_0 \subseteq Z_\mathfrak{g}(X) \subseteq \mathfrak{k}$$

yields a contradiction. Therefore we find an element $X \in it \cap it'$ and a non-compact root α such that $\alpha(X) = 0$ and

$$\exp(X) \in \text{int}(S).$$

This is impossible because

$$S \cap \exp(it) \subseteq \exp(iC_{\max} \cup -iC_{\max}).$$

∎

We want to apply Proposition 8.48 to study our compression semigroup $S(P)$. So we first have to show that

$$S \cap \exp(it) = \exp(iC_{\max})$$

because we know already that $\exp(iC_{\max}) \subseteq S$.

Let us return to the realization of the complex flag manifold $G_{\mathbb{C}}/P$ as an orbit of a highest weight ray $[v_\lambda]$ in the projective space $\mathbb{P}(V)$ of a highest weight module.

Let $X \in it \setminus iC_{\max}$ be weight separating and suppose that $\exp(X) \in \text{int}(S)$. Then there exists a positive non-compact root α with $(i\alpha)(-iX) = \alpha(X) < 0$. Since $\lambda \in \mathcal{P}^+$, we have that $\langle \lambda, \check{\alpha} \rangle \geq 0$. Hence

$$s_\alpha(\lambda)(X) = \lambda(X) - \langle \lambda, \check{\alpha} \rangle \alpha(X) > \lambda(X).$$

So, using Lemma 8.38(iii) and (iv), we see that $[v_\lambda] \in \mathcal{O}$ implies that $[v_{s_\alpha.\lambda}] \in \mathcal{O}$. Let $\beta := s_\alpha.\lambda$. Then

$$\beta \in \mathcal{W}_t.\lambda$$

by Corollary 8.44. Let δ denote the non-compact base root and $\omega := \omega_\delta$ the corresponding fundamental weight. Then ω is invariant under the Weyl group $\mathcal{W}_t$, and

$$(8.13) \qquad\qquad \langle \omega, \check{\mu} \rangle = \frac{2(\omega, \mu)}{(\mu, \mu)} \geq 1$$

for all $\mu \in \Lambda_p^+$. Since β and λ are $\mathcal{W}_t$-conjugate, we have

$$\langle \beta, \omega \rangle = \langle \lambda, \omega \rangle.$$

On the other hand

$$\beta = s_\alpha(\lambda) = \lambda + \langle \lambda, \check{\alpha} \rangle \alpha$$

entails that $\langle \alpha, \omega \rangle = 0$, contradicting (8.13).

So we have proved that $S(P) \cap \exp(it) = \exp iC_{\max}$. Now Proposition 8.47 shows that

$$S(P) \subseteq \overline{G N_{G_{\mathbb{C}}}(t) G}.$$

Let $s \in S \cap N_{G_{\mathbb{C}}}(t)$. Then $s.[v_\lambda] = [v_{\gamma.\lambda}]$, where $\gamma \in \mathcal{W}$ is the element of the big Weyl group represented by s, i.e., $\gamma = \text{Ad}(s)|_{t_{\mathbb{C}}}$. Now, using Corollary 8.44, we find that $\gamma.i\lambda \in \mathcal{W}_t.i\lambda$. This means that $s.[v_\lambda] \in N_G(t).[v_\lambda]$. The same argument applies to every other weight vector in $\mathcal{W}_t.[v_\lambda]$. Thus

$$\gamma.(\mathcal{W}_t.i\lambda) \subseteq \mathcal{W}_t.i\lambda.$$

Let $\beta := \sum_{w \in \mathcal{W}_t} w.i\lambda \in \operatorname{int} C_{\min}^*$. Then $\beta \neq 0$ and $\gamma.\beta = \beta$. It follows that γ preserves the set of $\mathcal{W}$-Weyl chambers containing $\mathbb{R}^+\beta$. Since the small Weyl group $\mathcal{W}_t$ acts simply transitive on this set of Weyl chambers and $\mathcal{W}$ acts simply transitive on the set of all Weyl chambers, it follows that $\gamma \in \mathcal{W}_t$ (cf. Lemma 8.39(ii)). Whence $s \in S \cap N_{G_{\mathbb{C}}}(t)$ is represented by an element in $\mathcal{W}_t$, so that

$$S \cap N_{G_{\mathbb{C}}}(t) \subseteq N_G(t)Z_{G_{\mathbb{C}}}(t) = N_G(t)\exp(t_{\mathbb{C}}) \subseteq G\exp(it).$$

For $s \in S \cap N_{G_{\mathbb{C}}}(t)$ this implies the existence of $g \in G$ with $gs \in \exp(it) \cap S = \exp(iC_{\max})$. So $S \cap N_{G_{\mathbb{C}}}(t) \subseteq S_{\max}$ and since $G(S \cap N_{G_{\mathbb{C}}}(t))G$ is dense in S, we conclude that $S \subseteq S_{\max}$.

Theorem 8.49. (Main Theorem on Compression Semigroups) *Let P be a parabolic in the complexification $G_{\mathbb{C}}$ of the real linear simple Lie group G and suppose that $\emptyset \neq \operatorname{int} S(P) \neq G_{\mathbb{C}}$. Then G is Hermitean and $S(P)$ is one of the two maximal Ol'shanskiĭ semigroups*

$$G\exp(iW_{\max}) \quad and \quad G\exp(-iW_{\max}).$$

Moreover, if G is supposed to be Hermitean and $\mathcal{O}$ is an open G-orbit in the complex flag manifold $G_{\mathbb{C}}/P$, then $\mathcal{O}$ carries a natural pseudo-Kähler structure such that the action of G on $\mathcal{O}$ is a Hamiltonian action and $\operatorname{compr}(\mathcal{O})$ has inner points iff its image under the moment mapping is contained in a pointed invariant cone.

Proof. It only remains to prove the second statement. We choose a realization of $G_{\mathbb{C}}/P$ as an orbit of a highest weight ray $[v_\lambda]$ in the projective space $\mathbb{P}(V)$ of a highest weight module V. The existence of the pseudo-Kähler structure follows from the section on moment mappings (cf. (8.6) and (8.7) as well as the discussion after Theorem 8.30). According to Proposition 8.25 there exists an extremal weight $\alpha \in \mathcal{P}_V$ and a weight vector v_α such that $[v_\alpha] \in \mathcal{O}$.

Suppose that the semigroup $S := S(P_{[v_\alpha]}) = \operatorname{compr}(\mathcal{O})$ has non-empty interior. We claim that $S \neq G_{\mathbb{C}}$. Let us assume that this is false, i.e., that $S = G_{\mathbb{C}}$. Then G is transitive on $G_{\mathbb{C}}/P$. If $G = KAN$ is an Iwasawa decomposition, then there exists a Borel subgroup $B \subseteq G_{\mathbb{C}}$ such that $AN \subseteq B$. On the other hand there exists a fixed point y of B on $G_{\mathbb{C}}/P$ because there exists a parabolic conjugate to P which contains B. Hence $G.y = K.y = G_{\mathbb{C}}/P$ and therefore K acts transitively on $G_{\mathbb{C}}/P$. It follows in particular that $K.v_\alpha$ contains weight vectors for all extremal weights. Let $Z \in Z(\mathfrak{k}) \setminus \{0\}$ and $v_\beta = k.v_\alpha$. Then

$$\beta(Z)v_\beta = Z.v_\beta = Z.(k.v_\alpha) = k.(Z.v_\alpha) = \alpha(Z)k.v_\alpha = \alpha(Z)v_\beta.$$

Thus all extremal weights take the same value on Z. It follows that Z acts as a scalar multiple of the identity on V, so $Z(K) = \exp(\mathbb{R}Z)$ acts trivially on $\mathbb{P}(V)$.

This contradicts the fact that $G_{\mathbb{C}}$ is simple because the effectivity kernel for the action of $G_{\mathbb{C}}$ on X is a normal subgroup containing $Z(K)$. We conclude that $S \neq G_{\mathbb{C}}$.

Now Lemma 8.40 yields that the image of $\mathcal{O}$ under the moment mapping $\Phi : \mathcal{O} \to \mathfrak{g}^*$ is a coadjoint orbit of an element in $C_{\min}(\Lambda^+)^*$ for a $\mathfrak{k}$-adapted positive system Λ^+. So Corollary 8.43 shows that $\Phi(\mathcal{O})$ is contained in a pointed invariant cone.

Suppose, conversely, that $\Phi(\mathcal{O})$ is contained in a pointed invariant cone. In view of Proposition 8.25, the coadjoint orbit $\Phi(\mathcal{O})$ is the orbit through a functional $i\lambda$, where λ is a highest weight of an irreducible representation of $G_{\mathbb{C}}$. Now the condition that the wedge wedge$(\mathcal{O}_{i\lambda})$ is pointed entails that

$$i\lambda \in C^*_{\min}(\Lambda^+)$$

with respect to a $\mathfrak{k}$-adapted positive system Λ^+ (Lemma 7.25). Now Proposition 8.46 yields the inclusion $S_{\max}(\Lambda^+) \subseteq S$. It follows in particular that S has dense interior. ∎

8.5. Contraction semigroups for indefinite forms

Let V be a finite dimensional real vector space endowed with a non-degenerate symmetric bilinear form J or a complex vector space endowed with a pseudo-Hermitean form J. We recall the definitions of the associated semigroups from Section 2.6. We have the two compression semigroups

$$\operatorname{compr}(\Omega_+) := \{ s \in \operatorname{Gl}(V) : s.\Omega_+ \subseteq \Omega_+ \},$$

and

$$\operatorname{compr}(\Omega_-) := \{ s \in \operatorname{Gl}(V) : s.\Omega_- \subseteq \Omega_- \},$$

the contraction semigroup,

$$\mathcal{C}(J) := \{ s \in \operatorname{Gl}(V) : (\forall v \in V) J(s.v, s.v) \leq J(v, v) \},$$

and the expansion semigroup

$$\mathcal{E}(J) := \{ s \in \operatorname{Gl}(V) : (\forall v \in V) J(s.v, s.v) \geq J(v, v) \}.$$

The following inclusion are trivial:

$$\mathcal{E}(J) \subseteq \operatorname{compr}(\Omega_+) \quad \text{and} \quad \mathcal{C}(J) \subseteq \operatorname{compr}(\Omega_-).$$

The same definitions also apply when V is a complex vector space endowed with a non-degenerate Hermitean form.

In Section 2.6 we have seen that, if J is positive definite, the euclidean contraction semigroup

$$\mathcal{C}(J) = \mathrm{O}(J) \exp(C),$$

has two components, namely those containing the two components of the group of units $O(J)$, and that the component $\mathcal{C}(J)^+$ of the identity is a Lie semigroup. The same arguments as in Section 2.6 apply to the complex case, where J is a Hermitean form, i.e., where J is positive definite. In this case we have shown that $\mathcal{C}(J)$ is a Lie semigroup with

$$\mathcal{C}(J) = U(J)\exp(C),$$

where C is the cone of Hermitean matrices with non-positive spectrum.

In this section we will use the results of the preceding sections to show that one has similar decompositions for general expansion and contraction semigroups (cf. [BK79]).

The complex case

We start with the complex case. Let $V = \mathbb{C}^n$. Since every pseudo-Hermitean form ist equivalent to one of the standard forms given by

$$J_{p,q}(x,y) = \sum_{i=1}^{p} x_i\overline{y_i} - \sum_{i=p+1}^{p+q} x_i\overline{y_i},$$

we may assume that $J = J_{p,q}$ for a pair (p,q) with $p + q = n$.

Let $\mathfrak{g} = \mathfrak{u}(p,q) = \mathfrak{u}(J_{p,q})$. Then V is a highest weight module of $\mathfrak{g}_{\mathbb{C}} = \mathrm{gl}(n,\mathbb{C})$ with respect to the natural action.

The Lie algebra $\mathfrak{g}$ is given by

$$\mathfrak{u}(p,q) = \left\{ \begin{pmatrix} B & A \\ A^* & C \end{pmatrix} : B \in \mathfrak{u}(p), C \in \mathfrak{u}(q) \right\},$$

a maximal compactly embedded subalgebra is

$$\mathfrak{k} = \left\{ \begin{pmatrix} B & 0 \\ 0 & C \end{pmatrix} : B \in \mathfrak{u}(p), C \in \mathfrak{u}(q) \right\},$$

and

$$\mathfrak{t} = \left\{ \begin{pmatrix} B & 0 \\ 0 & C \end{pmatrix} : B, C \text{ diagonal} \right\}$$

is a compactly embedded Cartan algebra of $\mathfrak{g}$. Note that $\mathfrak{t}_{\mathbb{C}}$ consists exactly of the diagonal matrizes in $\mathrm{gl}(n,\mathbb{C})$. Define the functionals ε_i by

$$\varepsilon_i \begin{pmatrix} B & 0 \\ 0 & C \end{pmatrix} = \begin{cases} b_i, & \text{for } i = 1,\ldots,p \\ c_{i-p}, & \text{for } i = p+1,\ldots,n \end{cases}.$$

In the following we write $\mathrm{diag}(t_1,\ldots,t_n)$ for the diagonal matrix with entries $t_1,\ldots,t_n$. Then, according to [Pa81], the minimal and maximal invariant cone in $\mathfrak{su}(p,q)$ are determined by the following cones in $\mathfrak{t} \cap \mathfrak{su}(p,q)$:

$$C_{\min} = \{c = \mathrm{diag}(i\lambda_1,\ldots,i\lambda_p,i\sigma_1,\ldots,i\sigma_q) : \mathrm{tr}\, c = 0, (\forall i,j)\lambda_i \leq 0 \leq \sigma_j\}$$

and

$$C_{\max} = \{c = \operatorname{diag}(i\lambda_1,\ldots,i\lambda_p,i\sigma_1,\ldots,i\sigma_q) : \operatorname{tr} c = 0, (\forall i,j)\sigma_j - \lambda_i \geq 0\}.$$

Moreover,

$$W_{\min} = \{X \in \mathfrak{g} : (\forall v \in \mathbb{C}^n)J(v,iXv) \geq 0\}$$
$$= \{X \in \mathfrak{g} : iX \in \mathbf{L}\left(\mathcal{E}(J_{p,q})\right)\}$$

and

$$W_{\max} = \{X \in \mathfrak{g} : (\forall v \in \mathbb{C}^n, J(v,v) = 0)J(v,iXv) \geq 0\}$$
$$= \{X \in \mathfrak{g} : iX \in \mathbf{L}\left(\operatorname{compr}(\Omega^+)\right)\},$$

where the second equality follows from an easy application of the general version of the Invariance Theorem for Vector Fields (cf. [HHL89, I.5.17]).

The element $Z \in iZ(\mathfrak{k})$ is given by

$$Z = \frac{1}{p+q}\operatorname{diag}(q,\ldots,q,-p,\ldots,-p).$$

The highest weight is

$$\lambda = \omega_1 = \varepsilon_1 - \frac{1}{p+q}\sum_{i=1}^{n}\varepsilon_i$$

with highest weight vector $e_1 = (1,0,\ldots,0)$, and the lowest weight is

$$\mu = \omega_n = \varepsilon_n - \frac{1}{p+q}\sum_{i=1}^{n}\varepsilon_i$$

with lowest weight vector $e_n = (0,\ldots,0,1)$.

Next we consider the action of the group $\operatorname{Sl}(n,\mathbb{C})$ on the projective $\mathbb{P}(\mathbb{C}^n)$. Recall that $\Omega_+ := \{[v] \in \mathbb{P}(\mathbb{C}^n) : J(v,v) > 0\}$ and $\Omega_- := \{[v] \in \mathbb{P}(\mathbb{C}^n) : J(v,v) < 0\}$. By Witt's Theorem, these are two open orbits of $\operatorname{SU}(p,q)$ on the projective space $\mathbb{P}(\mathbb{C}^n)$.

Using Theorem 8.49, we see that the maximal Ol'shanskiĭ semigroup

$$S_{\max} = \operatorname{SU}(p,q)\exp(iW_{\max})$$

coincides with $\operatorname{compr}(\Omega_+) \cap \operatorname{Sl}(n,\mathbb{C})$. Similarly $S_{\max}^{-1} = \operatorname{compr}(\Omega_-) \cap \operatorname{Sl}(n,\mathbb{C})$. Thus

$$\operatorname{compr}(\Omega^+) = \operatorname{U}(p,q)(\mathbb{R}^*\mathbf{1})\exp(iW_{\max})$$

is the full semigroup of compressions.

To see what the expansions are, note that

$$\mathcal{E}(J_{p,q}) \subseteq \operatorname{compr}(\Omega^+)$$

and that $\operatorname{U}(p,q) \subseteq \mathcal{E}(J_{p,q})$. It follows that

(8.14) $\mathcal{E}(J_{p,q}) = \operatorname{U}(p,q)(\mathcal{E}(J_{p,q}) \cap \exp(iW_{\max} + \mathbb{R}\mathbf{1})) \supseteq \operatorname{U}(p,q)\exp(W),$

where
$$W := \{X \in i\mathfrak{u}(p,q) : (\forall v \in \mathbb{C}^n) J(v, Xv) \geq 0\}$$
$$= i\mathfrak{u}(p,q) \cap L\left(\mathcal{E}(J_{p,q})\right)$$

Since $\mathcal{E}(J_{p,q})$ has dense interior, the same holds for its intersection with $\exp(iW_{\max} + \mathbb{R}\mathbf{1})$. Let

$$s = \exp X \in \text{int} \, \mathcal{E}(J_{p,q}) \cap \exp(iW_{\max} + \mathbb{R}\mathbf{1}).$$

Then $s = \exp(X' + d\mathbf{1})$ with $X' \in \text{int} \, W_{\max}$ and X' is conjugate under $SU(p,q)$ to an element in it (Proposition 7.10). So we may assume that $X' \in it$. Write $X = \text{diag}(x_1, \ldots, x_p, y_1, \ldots, y_q)$. Then

$$e^X = \text{diag}(e^{x_1}, \ldots, e^{x_p}, e^{y_1}, \ldots, e^{y_q}) \in \mathcal{E}(J_{p,q})$$

entails that $e^{x_i} \geq 1$ for $i = 1, \ldots, p$ and $e^{y_j} \leq 1$ for $j = 1, \ldots, q$. Whence

$$X \in C := W \cap it = \{\text{diag}(\lambda_1, \ldots, \lambda_p, \sigma_1, \ldots, \sigma_q) : (\forall i, j)\lambda_i \geq 0 \geq \sigma_j\}.$$

This proves that
$$\mathcal{E}(J_{p,q}) = U(p,q) \exp W.$$

Finally, using Proposition 7.10, we obtain the following theorem:

Theorem 8.50.   *The full expansion and contraction semigroups of the pseudo-Hermitean form $J_{p,q}$ on $\mathbb{C}^n$ are given by*

$$\mathcal{E}(J_{p,q}) = U(p,q) \exp(W) \quad and \quad \mathcal{C}(J_{p,q}) = U(p,q) \exp(-W),$$

where $W \subseteq \mathfrak{u}(p,q)$ is the invariant cone determined by

$$C = \{\text{diag}(\lambda_1, \ldots, \lambda_p, \sigma_1, \ldots, \sigma_q) : (\forall i, j)\lambda_i \geq 0 \geq \sigma_j\}.$$

In particular both semigroups are Lie semigroups,

$$\text{int} \, \mathcal{E}(J_{p,q}) = U(p,q) \exp(\text{int} \, C) \, U(p,q),$$

and

$$\text{int} \, \mathcal{C}(J_{p,q}) = U(p,q) \exp(-\text{int} \, C) \, U(p,q).$$

∎

The real case

Let $V = \mathbb{R}^n$. Since every non-degenerate symmetric bilinear form ist equivalent to one of the standard forms given by

$$J_{p,q}(x,y) = \sum_{i=1}^{p} x_i y_i - \sum_{i=p+1}^{p+q} x_i y_i,$$

we may assume that $J = J_{p,q}$ for a pair (p,q) with $p + q = n$.

We consider the extension of J to the corresponding pseudo-Hermitean form $J_{\mathbb{C}}$ on $\mathbb{C}^n$.

Then

$$J(x + ix', x + ix') = J(x, x) + J(x', x')$$

so that every element in $\mathcal{C}(J)$ yields a contraction on $\mathbb{C}^n$, and conversely, every contraction in $\mathcal{C}(J_{\mathbb{C}})$ which leaves $\mathbb{R}^n$ invariant is a contraction on $\mathbb{R}^n$. So

$$\mathcal{C}(J) = \mathcal{C}(J_{\mathbb{C}}) \cap \mathrm{Gl}(n, \mathbb{R}) \quad \text{and} \quad \mathcal{E}(J) = \mathcal{E}(J_{\mathbb{C}}) \cap \mathrm{Gl}(n, \mathbb{R}).$$

If W is the invariant cone in $\mathfrak{u}(p,q)$ from above, and $V := W \cap \mathfrak{gl}(n, \mathbb{R})$, then we conclude with

$$\mathrm{U}(p,q) \cap \mathrm{Gl}(n, \mathbb{R}) = \mathrm{O}(p,q)$$

that the following theorem is true.

Theorem 8.51. *The full expansion and contraction semigroups of the form $J_{p,q}$ on $\mathbb{R}^{p+q}$ are given by*

$$\mathcal{E}(J_{p,q}) = \mathrm{O}(p,q) \exp V \quad \text{and} \quad \mathcal{C}(J_{p,q}) = \mathrm{O}(p,q) \exp(-V).$$

In particular the components of $\mathbf{1}$ in both semigroups are Lie semigroups,

$$\operatorname{int} \mathcal{E}(J_{p,q}) = \mathrm{U}(p,q) \exp(\operatorname{int} C) \mathrm{U}(p,q),$$

and

$$\operatorname{int} \mathcal{C}(J_{p,q}) = \mathrm{U}(p,q) \exp(-\operatorname{int} C) \mathrm{U}(p,q),$$

where C is the cone of diagonal matrices in V. Note that these semigroups have as many connected components as $\mathrm{O}(p,q)$, i.e., 4 if $1 < p < n$ and 2 otherwise. ∎

8.6. Maximality of complex Ol'shanskiĭ semigroups

In this section $\mathfrak{g}$ denotes a Hermitean simple Lie algebra, $\mathfrak{g}_{\mathbb{C}}$ its complexification, $G_{\mathbb{C}}$ a corresponding complex connected Lie group, and $G := \langle \exp \mathfrak{g} \rangle \subseteq G_{\mathbb{C}}$ the corresponding simple Hermitean Lie group. Let $W_{\max} \subseteq \mathfrak{g}$ be one of the two maximal invariant cones (cf. Lemma 7.25). We have already seen in Theorem 8.49 how the *maximal Ol'shanskiĭ semigroup*

$$S_{\max} := G \exp(iW_{\max})$$

arises as the compression semigroup of open G-orbits in complex flag manifolds. Next we want to show that this semigroup is a *maximal subsemigroup* of the group $G_{\mathbb{C}}$ (cf. Chapter 6).

We fix a compactly embedded Cartan algebra $\mathfrak{t} \subseteq \mathfrak{g}$ and set $T := \exp \mathfrak{t}$, $A := \exp i\mathfrak{t}$, and $T_{\mathbb{C}} := \exp \mathfrak{t}_{\mathbb{C}}$.

Let $S \supseteq S_{\max}$ denote a proper maximal subsemigroup of $G_{\mathbb{C}}$ (Proposition 6.7). We have to show that $S = S_{\max}$. The first step to this result is to see that the abelian semigroup $A_S := A \cap S$ coincides with $A_{S_{\max}} = \exp(iC_{\max})$.

Let $\Lambda = \Lambda(\mathfrak{g}_{\mathbb{C}}, \mathfrak{t}_{\mathbb{C}})$ be a root system of $\mathfrak{g}_{\mathbb{C}}$ and $\Lambda^+ \subseteq \Lambda$ a $\mathfrak{k}$-adapted system such that

$$C_{\max} = C_{\max}(\Lambda^+).$$

Further let $B := B(\Lambda^+)$ denote the corresponding Borel subgroup. This is a minimal parabolic subgroup of the real simple Lie group $G_{\mathbb{C}}$. So, in view of San Martin's Theorem (Theorem 8.5), the semigroup S does not act transitively on $M := G_{\mathbb{C}}/B$.

Let $C_S \subseteq M$ denote the unique invariant control set for S (Proposition 8.2). We recall that Proposition 8.10 shows that C_S is the closure of a union of open G-orbits in M. By Corollary 8.24 the set of open G-orbits in M may be parametrized by the coset space $\mathcal{W}/\mathcal{W}_{\mathfrak{k}}$ of the big Weyl group $\mathcal{W} = \mathcal{W}(\mathfrak{g}_{\mathbb{C}}, \mathfrak{t}_{\mathbb{C}})$ modulo the small Weyl group $\mathcal{W}_{\mathfrak{k}} = \mathcal{W}(\mathfrak{k}_{\mathbb{C}}, \mathfrak{t}_{\mathbb{C}})$.

To make this parametrization more explicit, recall that M is isomorphic to an orbit $G_{\mathbb{C}}.[v_\lambda]$, where v_λ is a highest weight vector. In our case the corresponding highest weight module may be chosen to be the adjoint module $V_\lambda = \mathfrak{g}_{\mathbb{C}}$. Then the highest weight vector v_λ corresponds to a highest root vector in the Lie algebra $\mathfrak{g}_{\mathbb{C}}$ and λ is the *highest root*. Now the open G-orbits are the orbits through the extremal root vectors (Proposition 8.25), and two G-orbits through extremal root vectors coincide if and only if the corresponding roots are conjugate under $\mathcal{W}_{\mathfrak{k}}$.

Our strategy to show that $A_S = \exp(iC_{\max})$ is as follows. Suppose that A_S is bigger. Then we will see that this implies that C_S contains all the extremal root rays, whence $C_S = M$, so that $S = G_{\mathbb{C}}$, a contradiction.

The following Lemma is a key ingredient in the proof of $A_S = \exp(iC_{\max})$. We recall from Definition 8.36(b) that a root separating element X in $i\mathfrak{t}$ is an element such that the values $\alpha(X)$, $\alpha \in \Lambda$ are pairwise distinct.

Lemma 8.52. *Let* $\lambda \in \Lambda^+$ *denote the highest root and* $X \in i(\mathfrak{t} \setminus C_{\max})$ *a root separating element. Further let* $\mathcal{S} \subseteq \Lambda$ *be the smallest subset with the following properties:*

(1) $\lambda \in \mathcal{S}$.

(2) $\mathcal{W}_{\mathfrak{k}}.\mathcal{S} = \mathcal{S}$.

(3) *If* $\mu \in \mathcal{S}$, $\alpha \in \Lambda$, *and* $s_\alpha(\mu)(X) > \mu(X)$, *then* $s_\alpha(\mu) \in \mathcal{S}$.

Then $\mathcal{S} = W.\lambda$.

Proof. Since $X \notin iC_{\max}$, there exists a positive non-compact root α such that $\alpha(X) < 0$. Let Υ denote a basis of the positive system Λ^+, δ the unique non-compact root in Υ, and $\Upsilon_k := \Upsilon \setminus \{\delta\}$ the set of compact base roots.

Next we find $\gamma \in \mathcal{W}_{\mathfrak{k}}$ such that $\langle \Upsilon_k, \gamma.X \rangle \subseteq \mathbb{R}^+$, so that

$$\langle \alpha, X \rangle = \langle \gamma.\alpha, \gamma.X \rangle < 0$$

entails that $\delta(\gamma.X) < 0$ because $\gamma.\alpha$ is a sum of δ and other positive compact roots. It follows that $\delta' := \gamma^{-1}.\delta$ satisfies $\delta'(X) < 0$. The non-compact root δ' is member of the basis $\gamma^{-1}.(\Upsilon)$ of the root system. Whence $\mathcal{W}$ is generated by the set

$$\{s_{\delta'}\} \cup \{s_\mu : \mu \in \gamma^{-1}.\Upsilon_k\} \subseteq \{s_{\delta'}\} \cup \mathcal{W}_{\mathfrak{k}}.$$

To show that $S = W.\lambda$, let us assume that this is false. We consider the ordering $\preceq$ in the vector space (it^*) defined by the simplicial cone spanned by $\gamma^*(\Upsilon)$. Let $\mu \in W.\lambda \setminus S$ be a maximal element in this ordering. Since $\widetilde{\gamma}.\mu \notin S$ for all $\widetilde{\gamma} \in W_t$, it follows that μ is maximal in the orbit $W_t.\mu$. Using

$$s_\beta(\mu) = \mu - 2\frac{(\beta, \mu)}{(\beta, \beta)}\beta,$$

we conclude that $(\beta, \mu) \geq 0$ holds for all $\beta \in \gamma^{-1}.\Lambda_k^+$.

If $(\delta', \mu) \geq 0$ holds, too, then μ is the unique element in $W.\lambda$ which is contained in the positive chamber defined by $\gamma^{-1}.\Upsilon$. Thus $\mu = \gamma^{-1}.\lambda \in W_t.\lambda \subseteq S$, a contradiction. So $(\delta', \mu) < 0$, therefore $\mu \preceq s_{\delta'}(\mu)$. Hence $s_{\delta'}(\mu) \in S$ by maximality of μ. Now, as a consequence of (3),

$$\mu = s_{\alpha'}\big(s_{\alpha'}(\mu)\big) \in S$$

because

$$s_{\delta'}(\mu)(X) = \mu(X) - 2\frac{(\delta', \mu)}{(\delta', \delta')}\delta'(X) < \mu(X).$$

This contradiction yields that the set S must be equal to $W.\lambda$. ∎

Lemma 8.53. $A_S := S \cap A = \exp(iC_{\max})$.

Proof.   Suppose that this is false. Since $1 \in \overline{\text{int}\, S \cap A}$, the interior of A_S with respect to A is dense in A_S, so that there exists a root separating element $X \in it$ such that $\exp(X) \in S$. Moreover

$$C_S \supseteq C_{S_{\max}} = G.[v_\lambda]$$

(Proposition 8.2, Corollary 8.9, Theorem 8.49) because $G_{\mathbb{C}}$ is a simple Lie group with compact center.

Now Lemmas 8.38 and 8.52 show that the set

$$\mathcal{P}_{C_S} := \{\alpha \in \mathcal{P}_V : [v_\alpha] \in C_S\}$$

must coincide with $W.\lambda$.

In view of Proposition 8.25, this means that C contains every open G-orbit in M, so the closed set C is dense and therefore it coincides with M. Finally San Martin's Theorem (Theorem 8.5) yields the contradiction $S = G_{\mathbb{C}}$. ∎

We have just shown that $S \cap \exp(it) = \exp(iC_{\max})$. So Proposition 8.48 tells us that

$$S \subseteq \overline{GN_{G_{\mathbb{C}}}(t)G}.$$

Let $s \in S \cap N_{G_{\mathbb{C}}}(t)$. Then the finiteness of the Weyl group W shows that there exists $n \in \mathbb{N}$ such that $\text{Ad}(s^n)|_t = \text{id}_t$. Let $\gamma := \text{Ad}(s)|_t$ denote the element of the Weyl group represented by s. Then

$$s\exp(iC_{\max})s^{n-1} = \exp(\gamma.iC_{\max})s^n \subseteq S.$$

On the other hand we know that $s^n \in Z_{G_\mathbb{C}}(t) = \exp \mathfrak{t}_\mathbb{C}$, so that there exists $t \in T$ and $Y \in it$ such that $s^n = t \exp(Y)$. It follows that

$$\exp(\gamma.iC_{max})\exp(Y) = \exp(\gamma.iC_{max} + Y) \subseteq S \cap (TA) = T(S \cap A) = T\exp(iC_{max}).$$

Whence $\gamma.iC_{max} + Y \subseteq iC_{max}$. This shows that $\gamma.iC_{max} \subseteq iC_{max}$, and, since $C_{max} = C_{min}^\star$, Lemma 8.39 entails that $\gamma \in \mathcal{W}_t$. Consequently

$$s \in \left(N_K(t)\mathcal{T}_\mathbb{C}\right) \cap S = N_K(t)(\mathcal{T}_\mathbb{C} \cap S) = N_K(t)TA_S = N_K(t)\exp(iC_{max}) \subseteq S_{max}.$$

Theorem 8.54.   (Maximality Theorem for Complex Ol'shanskiĭ Semigroups) *The maximal Ol'shanskiĭ semigroup $S_{max} = G\exp(iW_{max})$ is maximal in $G_\mathbb{C}$.*

Proof.   From the preceding discussion we infer that

$$S \subseteq \overline{GN_{G_\mathbb{C}}(t)G}$$

and that $S \cap N_{G_\mathbb{C}}(t) \subseteq S_{max}$. Therefore the density of

$$\operatorname{int} S \cap GN_{G_\mathbb{C}}(t)G = G\big(\operatorname{int} S \cap N_{G_\mathbb{C}}(t)\big)G \subseteq GS_{max}G = S_{max}$$

in S shows that $S \subseteq S_{max}$, i.e., $S = S_{max}$.   ∎

Notes

   The invariant control sets are borrowed from control theory, in particular from San Martin's paper [SM91]. The material on the momentum maps for the projective spaces is standard in symplectic geometry even though it is hard to find an explicit reference. The case of indefinite metrics also draws on the treatments given by representation theorists ([Wi89], [AL91], cf. also [Ne92d]). Wolf's analysis of open orbits appeared in [Wo69]. For the pseudo-unitary trick see also [Ne92d]. Compression semigroups for open orbits have been considered by Ol'shanskiĭ ([Ols82a]) and Stanton ([St86]). They also give arguments for the Main Theorem on Compression Semigroups in the case, where the open orbit is a Hermitian symmetric space. The general result is new. The contraction and expansion semigroups for indefinite forms have been described by Brunet and Kramer in [BK79]. The Maximality Theorem for Complex Ol'shanskiĭ Semigroups is also new.

9. Representation theory

In this chapter we will see how closely complex Ol'shanskiı semigroups are linked with the theory of unitary representations of Lie groups. The plan of the chapter is as follows. In Section 9.1 we describe some interesting properties of the semigroup of contractions of an infinite dimensional Hilbert space. These properties should be compared with the finite dimensional case and the case of indefinite metrics (Sections 2.6 and 8.5). Section 9.3 contains the main link between semigroup and group representations, the analytic extension process. These general results are prepared in Section 9.2 which contains the abelian case, i.e., the extension of unitary representations of the real line $\mathbb{R}$ to holomorphic contraction representations of the upper half-plane.

It is shown that unitary group representations satisfying a certain positivity condition are in one-to-one correspondence with holomorphic contraction representations of certain complex Ol'shanskiı semigroups. An interesting and also important class of examples where this process applies are the holomorphic discrete series representations of Hermitean simple Lie groups which are discussed in Section 9.4. For these groups, according to [Ol82a], the holomorphic discrete series representations may also be realized on Hardy spaces of holomorphic functions on Ol'shanskiı semigroups. The Hardy space construction which works in general is described in Section 9.5. A particularly interesting example of a complex Ol'shanskiı semigroup which is also linked to many interesting applications in mathematical physics is the *metaplectic semigroup* or *Howe's oscillator semigroup* (cf. [Ho88], [Hi89], [Fo89]). This semigroup and its principal realizations by integral operators is briefly discussed in Section 9.6. The final Section 9.7 contains the relation with the representation theory of real Ol'shanskiı semigroups, the Theorem of Lüscher-Mack. Here we present a global version of this theorem appropriate for our setting.

9.1. Involutive semigroups

For the following we recall that a *semitopological semigroup* is a topological space with an associative multiplication which is separately continuous in both arguments. An *involutive semigroup* S is a semitopological semigroup together with a continuous involutive antiautomorphism

$$* : S \to S, \qquad s \mapsto s^*.$$

There are some very natural examples for this concept.

(1) Every topological group G with $g^* = g^{-1}$.

(2) Every symmetric group (G, τ) (cf. Section 7.3) with $g^* = \tau(g)^{-1}$.

(3) Ol'shanskiĭ semigroups $\Gamma(C)$ with $s^* = \tau(s)^{-1}$. These are closed subsemigroups S of G with $G_0^\tau \subseteq S$ and $\tau(S)^{-1} = S$.

In this section $\mathcal{H}$ denotes a Hilbert space. We write $B(\mathcal{H})$ for the algebra of bounded operators on $\mathcal{H}$, $C(\mathcal{H})$ for the semigroup of contractions, and $U(\mathcal{H})$ for the group of unitary operators. If nothing else is said, we endow these semigroups with the weak operator topology. For $A \in B(\mathcal{H})$ we write A^* for the adjoint operator. Note that these semigroups are invariant under this operation. We also write $S(\mathcal{H})$ for the subspace of symmetric operators in $B(\mathcal{H})$.

In this first section we describe some properties of the semigroup $C(\mathcal{H})$ which contrast the finite dimensional case.

Lemma 9.1. *The semigroup $C(\mathcal{H})$ is compact.*

Proof.   [Bou67, EVT, Ch. IV, §2]   ∎

Proposition 9.2.

  (i) *The involution $A \mapsto A^*$ is continuous on $B(\mathcal{H})$ with respect to the weak operator topology.*

 (ii) *If $K \subseteq B(\mathcal{H})$ is a bounded subset, then the multiplication mapping $K \times K \to B(\mathcal{H})$ is strongly continuous.*

(iii) *If $v_i \to v$ is $\mathcal{H}$ converges weakly and $\|v_i\| \to \|v\|$, then $v_i \to v$.*

 (iv) *On the sphere $S := \{x \in \mathcal{H} : \|x\| = 1\}$ the weak topology coincides with the norm topology.*

Proof.   (i) Let $A_i \to A$ and $x, y \in H$. Then

$$\langle x, A_i^* y \rangle = \langle A_i x, y \rangle \to \langle A x, y \rangle = \langle x, A^* y \rangle.$$

(ii) Let $(A_i, B_i) \to (A, B)$ in $K \times K$ with respect to the strong operator topology, and suppose that $\|C\| \leq M$ for all $C \in K$. Pick $x \in \mathcal{H}$. Then

$$\|A_i B_i x - A B x\| \leq M \|B_i x - B x\| + \|A_i B x - A B x\| \to 0.$$

Hence $A_i B_i \to A B$.

(iii) Let $v = \lim v_i$ with respect to the weak topology such that $\|v_i\| \to \|v\|$. Then

$$\|v - v_i\|^2 = \|v\|^2 + \|v_i\|^2 - 2 \operatorname{Re}\langle v, v_i \rangle$$
$$= 2(1 - \langle v, v_i \rangle) \to \|v\|^2 + \|v\|^2 - 2 \operatorname{Re}\langle v, v \rangle = 0.$$

(iv) This is an immediate consequence of (iii).   ∎

Corollary 9.3.   *$C(\mathcal{H})$ is a compact involutive semitopological semigroup.*   ∎

Corollary 9.4. *On the unitary group $U(\mathcal{H})$ the strong operator topology coincides with the weak operator topology and $U(\mathcal{H})$ is a topological group with respect to this topology.*

Proof. Let $A = \lim A_i$ with respect to the weak operator topology. If $x \in \mathcal{H}$ with $\|x\| = 1$, then $A_i x$ converges weakly and is contained in the unit spere. Now the strong convergence follows from Lemma 9.2(iv). Hence $A = \lim A_i$ in the strong operator topology. That $U(\mathcal{H})$ is a topological group with this topology follows from the continuity of $A \mapsto A^*$ (Proposition 9.2(i)) and from the strong continuity of the multiplication on $U(\mathcal{H})$ (Proposition 9.2(ii)). ∎

Proposition 9.5. *If $\dim \mathcal{H} = \infty$, then $U(\mathcal{H})$ is dense in $C(\mathcal{H})$.*

Proof. Let $A \in C(\mathcal{H})$ and $(e_i)_{i \in I}$ denote an orthonormal basis in H. The topology on $C(\mathcal{H})$ is defined by the functions

$$C(\mathcal{H}) \to \mathbb{C}, \quad B \mapsto \langle e_i, Be_j \rangle.$$

For a finite subset $E \subseteq I$ let P_E be the projection onto $\mathrm{span}\{e_i : i \in E\}$. Then the net

$$\{P_E A P_E : E \subseteq I \text{ finite}\}$$

converges to A. So we may assume that there exists a finite subset $E \subseteq I$ with $A = P_E A P_E$. We choose a countable subset $F = \{e_n : n \in \mathbb{N}\} \subseteq I$ such that $E = \{1, \ldots, n\}$. We define the sequence $A_m \in C(\mathcal{H})$ by the relation

$$A_m = P_{E_m} A_m P_{E_m}$$

for every finite subset $E_m := \{1, \ldots, n, n+m, \ldots, 2n+m-1\}$ and by the prescription of the matrix

$$\begin{pmatrix} C & -(1-CC^*)^{\frac{1}{2}} \\ (1-CC^*)^{\frac{1}{2}} & C^* \end{pmatrix}$$

with respect to the basis $e_1, \ldots, e_n, e_{n+m}, \ldots, e_{2n+m-1}$, where C is the matrix of A with respect to $e_1, \ldots, e_n$. Then $A_m \to A$ and $A_m \in U(\mathcal{H})$. ∎

We define the relation $\leq$ on $S(\mathcal{H})$ by

$$A \leq B \quad \text{if} \quad \langle x, Ax \rangle \leq \langle x, Bx \rangle \quad \forall x \in \mathcal{H}.$$

Lemma 9.6. *The following assertions hold:*
  (i) $C(\mathcal{H}) = \{A \in B(\mathcal{H}) : A^* A \leq 1\}$.
  (ii) $U(\mathcal{H})$ *is the group of units of* $C(\mathcal{H})$.
  (iii) $C(\mathcal{H}) = U(\mathcal{H}) C(\mathcal{H})_s$, *where* $C(\mathcal{H})_s$ *denotes the set of self-adjoint contractions.*

Proof. (i) This follows from $\|Ax\|^2 = \langle x, A^* Ax \rangle$ for all $x \in H$.
(ii) Let A be a unit in $C(\mathcal{H})$. Then $A^* A \leq 1$ and $(A^{-1})^* A^{-1} \leq 1$. We conclude that $A^* A = 1$ because the order $\leq$ is partial. Hence $A \in U(\mathcal{H})$.
(iii) This is the polar decomposition for operators ([We76, p.186]). For $A \in C(\mathcal{H})$ we explicitly have

$$A = \left(A(A^* A)^{-\frac{1}{2}}\right)(A^* A)^{\frac{1}{2}}.$$

∎

9.2. Holomorphic representations of half planes

In this section we deal with the abelian case of the general analytic continuation process for Ol'shanskiĭ semigroups. This means that we deal with the semigroups

$$\mathbb{C}_+ := \{z \in \mathbb{C} : \operatorname{Re} z \geq 0\} \quad \text{and} \quad \mathbb{C}^+ := \{z \in \mathbb{C} : \operatorname{Im} z \geq 0\}.$$

Since most of the general resuls will be proved by a reduction to this case, it is very instructive to see how the theory works for these simple examples.

Let M be a real analytic (complex) manifold and F a complete locally convex (complex) topological vector space. Then a mapping $f: M \to F$ is said to be *analytic* (*holomorphic*) if it is locally the convergent sum of a (complex) power series.

The following lemma is useful to deal with analyticity ([Lan75, p.411ff]). If E is a Banach space and E' its topological dual space, then a subset $\Lambda \subseteq E'$ is said to be a *norm-determining set* if

$$\|x\| = \sup\{|\langle \omega, x \rangle| : \omega \in \Lambda\}$$

holds for all $x \in E$.

Lemma 9.7.

(i) *Let E be a complex Banach space, $U \subseteq \mathbb{C}$ open, $\Lambda \subseteq E'$ a subset which is norm-determining, and $f: U \to E$ a mapping which is locally bounded and for every $\omega \in \Lambda$ the function $\omega \circ f$ is holomorphic. Then f is holomorphic.*

(ii) *Let E be a real Banach space, $U \subseteq \mathbb{R}^n$ open, H_1, H_2 real Banach spaces and a real trilinear map*

$$E \times H_1 \times H_2 \to \mathbb{R}, \quad (x, a, b) \mapsto \langle x, a, b \rangle$$

which induces an isometric embedding $E \to \operatorname{Bil}(H_1, H_2)$, and $f: U \to E$ a mapping such that

$$x \mapsto \langle f(x), a, b \rangle$$

is analytic for all $a \in H_1, b \in H_2$. Then f is analytic. ∎

Note that Lemma 9.7(ii) applies in particular when $H_1 = E'$ is the dual space of a Banach space E, $H_2 = \mathbb{C}$ or $\mathbb{R}$, $(x, a, b) = b\langle x, a \rangle$, and f is weakly analytic.

Let X be a set, $\mathfrak{B}$ a σ-algebra of subsets of X, and $\mathcal{H}$ a Hilbert space. Let $P(\mathcal{H})$ be the set of all projections on $\mathcal{H}$. A mapping $E: \mathfrak{B} \to P(\mathcal{H})$ is called a *spectral measure* on $\mathfrak{B}$ in $\mathcal{H}$ if it satisfies the following two conditions:

(1) $E(X) = 1$.

(2) If $(A_i)_{i \in I}$ is an at most countable family of disjoint sets in $\mathfrak{B}$, then

$$E(\bigcup_{i \in I} A_i) = \sum_{i \in I} E(A_i),$$

where the right hand side converges in the strong topology.

This implies in particular that for $u, v \in \mathcal{H}$ the function

$$\mu_{u,v} : \mathfrak{B} \to \mathbb{C}, \quad A \mapsto \big(E(A)u, v\big)$$

is a bounded complex measure on $\mathfrak{B}$.

Proposition 9.8.   *Let $E: \mathfrak{B} \to P(\mathcal{H})$ be a spectral measure and $f, g: X \to \mathbb{C}$ measurable functions. Then the set*

$$\mathcal{D}(f) := \{u \in \mathcal{H} : \int_X |f(x)|^2 d\mu_{u,u}(x) < \infty\}$$

is a subspace of $\mathcal{H}$ and there exists a linear operator

$$T(f) : \mathcal{D}(T(f)) := \mathcal{D}(f) \to \mathcal{H}$$

such that

(i) $(T(f)u, v) = \int_X f(x) \, d\mu_{u,v}(x)$ *for all* $u \in \mathcal{D}(f)$, $v \in \mathcal{H}$,

(ii) $\mathcal{D}(f) = \mathcal{D}(\overline{f})$ *and* $(T(f)u, v) = (u, T(\overline{f})v)$ *for all* $u, v \in \mathcal{D}(f)$, *and*

(iii)

$$(T(f)u, T(g)v) = \int_X f(x)\overline{g}(x) \, d\mu_{u,v}(x)$$

for all $u \in \mathcal{D}(f)$ *and* $v \in \mathcal{D}(g)$.

Proof.   [Sug75, III.3.12] ∎

The following proposition is the key ingredient in the analytic extension result for the semigroup $\mathbb{C}_+$.

Proposition 9.9.   *Let A be a negative self-adjoint operator on $\mathcal{H}$. Then there exists a spectral measure E on the Borel sets in $\mathbb{R}$ in $\mathcal{H}$ such that $A = E(\mathrm{id}_{\mathbb{R}})$ and the prescription $\gamma(z) := e^{zA} := T(e_z)$, where $e_z(t) = e^{zt}$ for $\mathrm{Re}\, z \geq 0$, defines a mapping*

$$\gamma : \mathbb{C}_+ := \{z \in \mathbb{C} : \mathrm{Re}\, z \geq 0\} \to B(\mathcal{H}), \qquad z \mapsto T(e_z),$$

with the following properties:

(i) $\mathcal{D}(\gamma(z)) = \mathcal{H}$ *and* $\|\gamma(z)\| \leq 1$ *for all* $z \in \mathbb{C}_+$.

(ii) $\gamma(0) = 1$, $\gamma(z + z') = \gamma(z)\gamma(z')$, *and* $\gamma(z)^* = \gamma(\overline{z})$ *for all* $z, z' \in \mathbb{C}_+$.

(iii) γ *is strongly continuous.*

(iv) γ *is holomorphic on* $\mathrm{int}\,\mathbb{C}_+ = \{z \in \mathbb{C} : \mathrm{Re}\, z > 0\}$.

Proof.   The existence of E follows from [We76, p.181]. Moreover, the fact that A is negative implies that $E(] - \infty, 0]) = 1$ ([We76, p.186])

(i) Let $u \in \mathcal{H}$. Then

$$\int_{\mathbb{R}} |e_z(x)|^2 \, d\mu_{u,u}(x) = \int_{-\infty}^{0} e^{2x\,\mathrm{Re}\,z} \, d\mu_{u,u}(x) \leq \mu_{u,u}(] - \infty, 0]) = \|u\|^2.$$

(ii) The equality $\gamma(0) = 1$ follows trivially from the definitions. Using Proposition 9.8, we also find that

$$(\gamma(z)u, v) = (T(e_z)u, v) = (u, T(\overline{e}_z)v) = (u, T(e_{\overline{z}})v) = (u, \gamma(\overline{z})v).$$

Now we use (i) and Proposition 9.8 to see that

$$
\begin{aligned}
(\gamma(z)\gamma(z')u, v) &= (\gamma(z')u, \gamma(z)^*v) = (\gamma(z')u, \gamma(\bar{z})v) \\
&= (T(e_{z'})u, T(e_{\bar{z}})v) = \int_{\mathbb{R}} e^{z'x} e^{zx} \, d\mu_{u,v}(x) \\
&= \int_{\mathbb{R}} e^{(z+z')x} \, d\mu_{u,v}(x) = (\gamma(z+z')u, v)
\end{aligned}
$$

for all $u, v \in \mathcal{H}$. Hence $\gamma(z)\gamma(z') = \gamma(z+z')$.
(iii) Let $u, v \in \mathcal{H}$. Then

$$
(\gamma(z)u, v) = \int_{-\infty}^{0} e^{zx} \, d\mu_{u,v}(x).
$$

Since $|e^{zx}| \leq 1$ for $\operatorname{Re} z \geq 0$ and $x \leq 0$, the fact that $\mu_{u,v}$ is a finite measure, together with Lebesgue's Theorem on dominated convergence, entails that γ is weakly continuous. Moreover, in view of (ii),

$$
\|\gamma(z)u\| = (\gamma(z)u, \gamma(z)u) = (\gamma(\bar{z}+z)u, u)
$$

so that the continuity of the funtion $z \mapsto \|\gamma(z)u\|$ follows from the weak continuity. Now the strong continuity is a consequence of Proposition 9.2(iii).
(iv) In view of Lemma 9.7(i), it suffices to show that the functions $z \mapsto (\gamma(z)u, v)$ are holomorphic. Let $z_0 \in \operatorname{int}\mathbb{C}_+$.

First we note that we have for a complex number $w \in \mathbb{C}$ the following estimate:

$$
e^w - 1 = \int_0^1 w e^{tw} \, dt
$$

so that

$$
\left| \frac{e^w - 1}{w} \right| \leq e^{|w|}.
$$

For $h \in \mathbb{C}$ with $|h| < \frac{1}{2}\operatorname{Re} z_0$, this leads to

$$
\left| e^{z_0 x} \frac{e^{hx} - 1}{h} \right| \leq |x| e^{x(\operatorname{Re} z_0 - |h|)} \leq |x| e^{\frac{x}{2}\operatorname{Re} z_0}
$$

for all $x \leq 0$. Now Lebesgue's Theorem on dominated convergence yields

$$
\begin{aligned}
\lim_{h \to 0} \left((\gamma(z+h) - \gamma(z))u, v \right) &= \lim_{h \to 0} \int_{-\infty}^{0} e^{zx} \frac{e^{hx} - 1}{h} \, d\mu_{u,v}(x) \\
&= \int_{-\infty}^{0} \lim_{h \to 0} \frac{1}{h} e^{zx} \frac{e^{hx} - 1}{h} \, d\mu_{u,v}(x) \\
&= \int_{-\infty}^{0} x e^{zx} \, d\mu_{u,v}(x) \\
&= (A\gamma(z)u, v)
\end{aligned}
$$

by Proposition 9.8. We conclude that the function γ is holomorphic on $\operatorname{int}\mathbb{C}_+$ with derivative $A\gamma(z)$. Note that $\mathcal{D}(A\gamma(z)) = \mathcal{H}$. ∎

Theorem 9.10.   (Stone's Theorem) *If A is a self-adjoint operator on the Hilbert space $\mathcal{H}$, and E the spectral measure of A, then*

$$\pi(t) := e^{itA} := T(e_{it})$$

defines a continuous unitary representation of $\mathbb{R}$ on $\mathcal{H}$, and, conversely, every continuous unitary representation is obtained that way, where

$$\mathcal{D}(A) := \{u \in \mathcal{H} : \lim_{t \to 0} \frac{1}{t}\big(\pi(t)u - u\big) \text{ exists}\}$$

and

$$iAu := \lim_{t \to 0} \frac{1}{t}\big(\pi(t)u - u\big)$$

for $u \in \mathcal{D}(A)$.

Proof.   That π defines a continuous unitary representation can be proved with similar arguments as in the proof of Proposition 9.9. For the converse assertion we refer to [We76, p.208].   ∎

If $\pi: G \to U(\mathcal{H})$ is a continuous unitary representation of the Lie group G, then we write $d\pi(X)$ for the operator on $\mathcal{H}$ which satisfies $\pi(\exp tX) = e^{td\pi(X)}$ according to Theorem 9.10. We note that in this notation one supresses the reference to the point 1 where one takes the derivative in order to have $d\pi$ denote a Lie algebra representation.

We have to prepare the main theorem of this section with the following lemma which will also be a key ingredient in Ol'shanskiĭ's Theorem on Hardy spaces in the Section 9.4.

Lemma 9.11.   *Let $\mathcal{H}$ be a Hilbert space, and $F: \text{int}\,\mathbb{C}_+ \to \mathcal{H}$ a holomorphic function such that*

$$F(z + it) = U(t)F(z) \qquad \forall z \in \mathbb{C}_+, t \in \mathbb{R},$$

where $U(t) = e^{itA}$ is a unitary one-parameter group with self-adjoint generator A, and with

$$\|F\|_\infty := \sup_{\text{Re}\, z > 0} \|F(z)\| < \infty.$$

Then there exists a vector $\xi \in \mathcal{H}$ such that $F(z) = e^{zA_-}\xi$, where

$$A_- = P(]-\infty, 0])A$$

and P is the spectral measure of A. Furthermore

$$\|\xi\| = \sup_{\text{Re}\, z > 0} \|F(z)\| \quad \text{and} \quad \lim_{z \to 0} F(z) = \xi.$$

Proof.   Let $a = [\alpha_1, \alpha_2]$, $A(a)$ denote the bounded operator $P(a)A$, and $F_a(z) = P(a)F(z)$. Then the function $z \mapsto e^{zA(a)}$ is holomorphic on $\mathbb{C}$, so by analytic continuation in z_2, we obtain

(†) $$F_a(z_1 + z_2) = e^{z_2 A(a)}F_a(z_1)$$

for $\mathrm{Re}\, z_1 > 0$ and $\mathrm{Re}\, z_2 \geq 0$. But we may then use ($\dagger$) to continue the left-hand side analytically in z_2. Now for $\alpha_1 > 0$, $e^{z_2 A(a)}$ grows rapidly as $\mathrm{Re}\, z_2 \to \infty$, so by the bound for F, it follows that $F_a(z) = 0$ holds for all z in this case. Thus $F(z) \in P(]-\infty, 0])\mathcal{H}$ for $\mathrm{Re}\, z > 0$. When $\alpha_2 \leq 0$, setting $z_1 = 0$, we get from ($\dagger$) that

$$\xi_a := F_a(0) = e^{-zA(a)} F_a(z)$$

with

$$\|\xi_a\| = \sup_{\mathrm{Re}\, z > 0} \|F_a(z)\| \leq \sup_{\mathrm{Re}\, z > 0} \|F(z)\|.$$

Note that $\|F_a(z)\|$ gets bigger as $\mathrm{Re}\, z$ gets smaller since $A(a)$ is a negative operator. By the properties of the spectral measure, the vectors ξ_a form a projective system: $\xi_a = P(a)\xi_b$ when $a \subseteq b \subseteq (-\infty, 0]$. Since this is a bounded system, the limit ξ exists for $a = (-\alpha, 0]$, $\alpha \to \infty$; but this means that

$$F(z) = e^{zA} - \xi \qquad \forall z \in \mathrm{int}\,\mathbb{C}_+$$

which proves the assertions in the lemma. $\blacksquare$

To guarantee uniqueness of analytic continuations, the following lemma provides a useful argument.

Lemma 9.12.

(i) Let $r > 0$, and $B_r^+ := \{z \in \mathbb{C} : \mathrm{Re}\, z \geq 0, |z| \leq r\}$. Suppose that $f: B_r^+ \to \mathbb{C}$ is a function which is continuous on $B_r(z_0)^+$, holomorphic on the interior, and which vanishes on $\mathbb{R} \cap B_r^+$. Then $f = 0$.

(ii) Let V be a real vector space, $V_{\mathbb{C}}$ its complexification, and $W \subseteq V_{\mathbb{C}}$ a generating wedge such that $H(W) = V$. Suppose that U is an open convex neighborhood of $X \in V$ in $V_{\mathbb{C}}$ and $f: U \cap W \to \mathbb{C}$ a continuous function which is holomorphic on $U \cap \mathrm{int}\, W$ and identically 0 on $U \cap V$. Then f vanishes on $U \cap W$.

Proof.   (i) First we use the Schwarz Reflection Principle ([FL85, p.70]) to see that f has a holomorphic extension $\tilde{f}$ to the interior of $B_r := \{z \in \mathbb{C} : |z| < r\}$. Now $\tilde{f}$ vanishes on a line segment, hence $\tilde{f} = 0$.

(ii) Let $Y \in U \cap \mathrm{int}\, W$. It suffices to show that $f(Y) = 0$ for all such Y. We consider the mapping $\gamma: \mathbb{C} \to V, z \mapsto X + z(Y - X)$. Then γ is holomorphic, $\gamma^{-1}(U)$ is an open convex subset containing 0 and 1, and $\gamma^{-1}(U \cap H(W))$ is a line segment in $\gamma^{-1}(U)$. Therefore the function $f \circ \gamma$ is holomorphic on $\gamma^{-1}(U \cap \mathrm{int}\, W)$ and vanishes on $\gamma^{-1}(U \cap H(W))$. Since $\gamma^{-1}(U \cap W)$ is convex, we conclude with (i) that $f \circ \gamma$ vanishes on $\gamma^{-1}(U \cap W)$. It follows in particular that $f(Y) = f(X) = 0$. $\blacksquare$

Theorem 9.13.   (Main Theorem on Representations of Half Planes) *The following assertions hold:*

(i) *Let $\pi: \mathbb{R} \to U(\mathcal{H}), t \mapsto e^{itA}$ be a continuous unitary representation such that the self-adjoint operator $A = id\pi(1)$ is negative. Then π can be uniquely extended to a representation $\hat{\pi}: \mathbb{C}^+ := \mathbb{R} + i\mathbb{R}^+ \to C(\mathcal{H})$ such that*

   (a) *$\hat{\pi}$ is a strongly continuous semigroup representation.*

   (b) *$\hat{\pi}$ is holomorphic on $\mathrm{int}\,\mathbb{C}^+$.*

   (c) *$\hat{\pi}(z)^* = \hat{\pi}(-z)$.*

(ii) *Suppose that* $\widehat{\pi}:\mathbb{C}^+ \to C(\mathcal{H})$ *is a representation of* $\mathbb{C}^+$ *satisfying* (a) - (c). *Then* $\pi := \widehat{\pi}|_{\mathbb{R}}$ *is a continuous unitary representation such that the operator* $A := id\pi(1)$ *is negative and*

$$\widehat{\pi}(z) = e^{-izA}$$

holds for all $z \in \mathbb{C}^+$.

Proof.   (i) If follows from Proposition 9.9 that we have a representation

$$\gamma : \mathbb{C}_+ := \{z \in \mathbb{C} : \operatorname{Re} z \geq 0\} \to C(\mathcal{H})$$

with $\gamma(z) = e^{zA} = e^{izd\pi(1)}$. Setting $\widehat{\pi}(z) := \gamma(-iz)$ we obtain a representation of $\mathbb{C}^+$ with the desired properties which extends π because

$$\gamma(-ix) = e^{xd\pi(1)} = \pi(\exp x)$$

holds by Stone's Theorem. The uniqueness follows from Lemma 9.12 applied to the functions $z \mapsto (\widehat{\pi}(z)u, v)$ which are continuous on $\mathbb{C}^+$ and holomorphic on $\operatorname{int}\mathbb{C}^+$.

(ii) As a restriction of a strongly continuous representation of $\mathbb{C}^+$, the representation π is continuous. Let $u \in \mathcal{H}$. We apply Lemma 9.11 to the function $F(z) = \widehat{\pi}(iz)u$ and $U(t) = \pi(-t) = e^{itA}$ to find that

$$F(z) = e^{zA_-}u = \widehat{\pi}(iz)u \qquad \forall z \in \operatorname{int}\mathbb{C}_+,$$

where $iA = -d\pi(1)$. Since the representation

$$\alpha:\mathbb{C}_+ \to C(\mathcal{H}), \qquad z \mapsto e^{zA_-}$$

is holomorphic on $\operatorname{int}\mathbb{C}_+$ and strongly continuous (Proposition 9.9), we conclude that $\alpha(z) = \widehat{\pi}(iz)$ holds on $\mathbb{C}_+$. So $e^{tA_-} = e^{itA}$ holds for all $t \in \mathbb{R}$ and the uniqueness assertion of Stone's Theorem (Theorem 9.10) shows that $A = A_-$, i.e., A is negative.   ∎

9.3. Invariant cones and unitary representations

Let G be a Lie group with $\mathfrak{g} = L(G)$, $\pi:G \to U(\mathcal{H})$ a unitary representation of G on the Hilbert space $\mathcal{H}$, $\mathcal{H}^\infty$ ($\mathcal{H}^\omega$) the corresponding space of smooth (analytic) vectors. We write $d\pi$ for the *derived representation* of $\mathfrak{g}$ on $\mathcal{H}^\infty$. Note that this is consistent with the notation $d\pi(X)$ for the skew-adjoint infinitesimal generator of the unitary one-parameter group $t \mapsto \pi(\exp tX)$ (Theorem 9.10) because $\mathcal{H}^\infty \subseteq \mathcal{D}(d\pi(X))$. We extend the derived representation to a representation of the complexified Lie algebra $\mathfrak{g}_\mathbb{C}$ and set

$$W(\pi) := \{X \in \mathfrak{g} : id\pi(X) \leq 0\}.$$

In this section we will describe an analytic continuation precess to extend a unitary representation π of G to a representation of a certain semigroup into the semigroup $C(\mathcal{H})$ of contractions on $\mathcal{H}$ (cf. Section 9.1).

We collect some fundamental properties of $W(\pi)$ in the following lemma.

Lemma 9.14. *The following assertions hold:*

(i) $W(\pi)$ *is an invariant wedge in the Lie algebra* $\mathfrak{g}$.

(ii) $H\big(W(\pi)\big) = \mathbf{L}(\ker \pi)$.

(iii) $W(\pi)$ *is pointed if and only if* π *has discrete kernel.*

Proof.   (i) Let $X \in \mathfrak{g}$ and $g = \exp(X)$. Then $Y \in W(\pi)$ implies that

$$d\pi(e^{\operatorname{ad} X} Y) = d\pi\big(\operatorname{Ad}(g)Y\big) = \pi(g)d\pi(Y)\pi(g)^{-1}$$

and therefore $e^{\operatorname{ad} X} Y \in W(\pi)$ since conjugation does not change the spectrum of a self-adjoint operator.

(ii) Let $X \in H\big(W(\pi)\big)$. Then $id\pi(X)$ is positive and negative semidefinite, hence zero. Therefore $d\pi(X) = 0$ and Stone's Theorem (Theorem 9.10) implies that $\pi(\exp tX) = \mathbf{1}$ for all $t \in \mathbb{R}$. The converse is trivial.

(iii) This is an immediate consequence of (ii). ∎

Now we describe the semigroups which are the domains for the extended representation.

Let $\mathfrak{g}_{\mathbb{C}}$ denote the complexification of $\mathfrak{g}$ and $\sigma: X + iY \mapsto X - iY$ the conjugation. We recall that $(\mathfrak{g}_{\mathbb{C}}, \sigma)$ is a symmetric Lie algebra. We write $G_{\mathbb{C}}$ for the simply connected complex Lie group associated with $\mathfrak{g}_{\mathbb{C}}$.

Suppose that W is a pointed invariant cone in the Lie algebra $\mathfrak{g}$, $W' := \mathfrak{g} + iW$ is the associated symmetric Lie wedge in $\mathfrak{g}_{\mathbb{C}}$, and G a connected Lie group with $\mathbf{L}(G) = \mathfrak{g}$. Let $G_1 := \langle \exp_{G_{\mathbb{C}}} \mathfrak{g} \rangle$. Then we write $\Gamma_{G_1}(W)$ for the associated Lie semigroup $S_{W'}$ in $G_{\mathbb{C}}$. In view of Corollary 7.35, we know that W' is global in $G_{\mathbb{C}}$, and that $\Gamma_{G_1}(W) = G_1 \exp(iW)$. Now the universal covering semigroup $\Gamma_{\widetilde{G}}(W) := \Gamma_{G_1}(W)\widetilde{}$ is a product of $\widetilde{G}$ and $\operatorname{Exp}(iW)$ (cf. Section 3.4). Write $D := \pi_1(G) \subseteq \widetilde{G}$ for the covering kernel with respect to G. Then $D \subseteq \Gamma_{\widetilde{G}}(W)$ is a discrete central subgroup so that the semigroup

$$\Gamma_G(W) := \Gamma_{\widetilde{G}}(W)/D$$

is semigroup that will be the domain of extension for our representations. We collect its basic properties in the following theorem.

Theorem 9.15.   *Let* G *be a connected Lie group and* $W \subseteq \mathfrak{g} = \mathbf{L}(G)$ *a pointed generating invariant cone. Then there exists a locally compact semigroup* $\Gamma_G(W)$ *with the following properties:*

(i) $H\big(\Gamma_G(W)\big) \cong G$.

(ii) $\Gamma_G(W) = G \cdot \operatorname{Exp}(iW)$, *where* $\operatorname{Exp}: \mathbf{L}\big(\Gamma_G(W)\big) = \mathfrak{g} + iW \to \Gamma_G(W)$ *denotes the exponential function of* $\widetilde{\Gamma}(C)$.

(iii) *The product mapping*

$$G \times W \to \Gamma_G(W), \quad (g, X) \mapsto g \operatorname{Exp}(iX)$$

is a homemorphism.

(iv) *The subsemigroup* $\operatorname{int}\Gamma_G(W) := G \operatorname{Exp}(i \operatorname{int} W)$ *is a complex manifold with a holomorphic semigroup multiplication.*

Proof.   All these assertions are proved at the end of Section 3.4 for the case where G is simply connected. In the general case they follow from the fact that the covering kernel D is a discrete subgroup of the group of units of $\Gamma_{\widetilde{G}}(W)$ which is central, so that the covering $\Gamma_{\widetilde{G}}(W) \to \Gamma_G(W)$ is via polar decomposition topologically equivalent to the covering

$$\widetilde{G} \times W \to G \times W, \qquad (g, X) \mapsto (gD, X).$$

We want to show that every unitary representation π of G yields a representation of the semigroup $\Gamma_G\big(W(\pi)\big)$ by contractions on the Hilbert space $\mathcal{H}$ which is holomorphic on the interior of the semigroup.

Lemma 9.16.   (The Equianalyticity Lemma) *There exists a 0-neighborhood U in $\mathfrak{g}$ and a dense subspace $\mathcal{H}_1 \subseteq \mathcal{H}^\omega$ such that the series $\sum_{m=0}^\infty \frac{1}{m!} d\pi(X)^m v$ converges for every $v \in \mathcal{H}_1$ and $X \in U + iU \subseteq \mathfrak{g}_\mathbb{C}$, and that*

$$\pi(\exp X)v = \sum_{m=0}^\infty \frac{1}{m!} d\pi(X)^m v$$

holds for all $X \in U$, $v \in \mathcal{H}_1$.

Proof.   The proof is a collection of some facts in [Wa72]. Let $X_1, \ldots, X_n$ denote a basis of the Lie algebra $\mathfrak{g}$ and $\Delta := \sum_{i=1}^n X_i^2$. We consider the representation $d\pi$ of $\mathcal{U}(\mathfrak{g}_\mathbb{C})$ on $\mathcal{H}^\infty$. Let $\mathcal{H}_1 \subseteq \mathcal{H}$ denote the set of all $d\pi(\Delta)$-bounded vectors, i.e., the Paley Wiener space of the self-adjoint operator $\overline{d\pi(\Delta)}$. Then, according to Goodman's Theorem ([Wa72, 4.4.6.1]), we have that $\mathcal{H}_1 \subseteq \mathcal{H}^\omega$ because $\mathcal{H}_1$ consists of analytic vectors for the operator $\sqrt{1 - d\pi(\Delta)}$. Next the proof of Theorem 4.4.6.6 in [Wa72] yields the existence of a convex 0-neighborhood $U \subseteq \mathfrak{g}_\mathbb{C}$ such that

$$\sum_{m=0}^\infty \frac{1}{m!} d\pi(X)^m v$$

converges for every $v \in \mathcal{H}_1$ and $X \in U + iU \subseteq \mathfrak{g}_\mathbb{C}$.

Fix $X \in U$ and $v \in \mathcal{H}_1$. Since $\mathcal{H}_1 \subseteq \mathcal{H}^\omega$, Harish Chandra's Theorem ([Wa72, 4.4.5.4]) yields that

$$\pi(\exp tX).v = \sum_{m=0}^\infty \frac{1}{m!} d\pi(X)^m t^m v$$

holds for sufficiently small $t \in \mathbb{R}$. Now the remaining equality follows from the fact that the mapping

$$z \mapsto \sum_{m=0}^\infty \frac{1}{m!} d\pi(X)^m z^m v$$

is holomorphic on an open neighborhood of 0 containing $[0, 1]$.   ∎

Lemma 9.17. *Let* $f : \Gamma_G(W) \to \mathbb{C}$ *be continuous, holomorphic on* $\operatorname{int} \Gamma_G(W)$, *and zero on* G. *Then* f *vanishes.*

Proof. We apply Lemma 9.12(ii) to the function

$$\mathfrak{g} + iC \to \mathbb{C}, \quad X + iY \mapsto f\big(\operatorname{Exp}(X)\operatorname{Exp}(iY)\big)$$

to see that there exists an open subset of $\Gamma_G(W)$ where f vanishes. Since f is holomorphic on the connected interior of $\Gamma_G(W)$, we conclude that f vanishes. ∎

The following proposition contains the main result of this section for the simply connected case.

In the following we say that a semigroup morphism $\widetilde{\Gamma}(C) \to C(\mathcal{H})$ which is strongly continuous, holomorphic on the interior, and which preserves the involutions, is a *holomorphic contraction representation* of $\widetilde{\Gamma}(C)$.

Proposition 9.18. *Suppose that* G *is simply connected and that* $\pi : G \to U(\mathcal{H})$ *is a continuous unitary representation such that* $C := W(\pi)$ *is pointed and generating. Then there exists a holomorphic contraction representation*

$$\widehat{\pi} : \widetilde{\Gamma}(C) \to C(\mathcal{H})$$

such that $\widehat{\pi}|_G = \pi$.

Proof. Let $S := \Gamma(C)$ and $\widetilde{S} = \widetilde{\Gamma}(C)$.

Step 1: We define the mapping $\widehat{\pi}$ via the direct product decomposition of $\widetilde{S}$ by

$$\widehat{\pi}(g \operatorname{Exp}(X)) = \pi(g) e^{d\pi(X)}.$$

Since

$$\pi(g) e^{d\pi(X)} \pi(g)^{-1} = e^{\pi(g) d\pi(X) \pi(g)^{-1}} = e^{d\pi\big(\operatorname{Ad}(g)X\big)},$$

we have that

$$\begin{aligned}
\widehat{\pi}\big((g \operatorname{Exp} X)^*\big) &= \widehat{\pi}(\operatorname{Exp} X g^{-1}) \\
&= \widehat{\pi}\big(g^{-1} \operatorname{Exp} \operatorname{Ad}(g)X\big) = \pi(g)^{-1} e^{d\pi\big(\operatorname{Ad}(g)X\big)} \\
&= e^{d\pi(X)} \pi(g)^{-1} = \widehat{\pi}(g \operatorname{Exp} X)^*,
\end{aligned}$$

and

$$\widehat{\pi}(g \operatorname{Exp} X)^* \widehat{\pi}(g \operatorname{Exp} X) = e^{d\pi(X)} e^{d\pi(X)} = e^{2d\pi(X)} \leq 1.$$

This shows that $\widehat{\pi}(\widetilde{S}) \subseteq C(\mathcal{H})$ and that $\widehat{\pi}$ respects the involutions.

Step 2: Using Lemma 9.16, we get a neighborhood $U_{\mathbb{C}} \subseteq \mathfrak{g}_{\mathbb{C}}$ such that the mappings

$$\gamma_v : X \mapsto e^{d\pi(X)} v, \quad v \in \mathcal{H}_1$$

are complex analytic, i.e., holomorphic, and $\gamma_v(X) = \pi(\exp X) v$ holds for all $X \in \mathfrak{g} \cap U_{\mathbb{C}}$. We conclude that there exists an open convex 1-neighborhood B

in $\mathfrak{g}$ such that $V := \mathrm{Exp}(B)\,\mathrm{Exp}\left(i(B \cap C)\right)$ is an open 1-neighborhood in $\widetilde{S}$ and there exists a set of continuous mappings $\alpha_v \colon V \to \mathcal{H}$ which satisfies

$$\alpha_v(g) = \pi(g)v \qquad \forall g \in H(\widetilde{S}) \cap V$$

and which are holomorphic on $V^0 := V \cap \mathrm{int}\,\widetilde{S}$.

Step 3: Strong continuity: It follows in particular that the mappings

$$\gamma'_v : \mathrm{Exp}(U_{\mathbb{C}} \cap iC) \to \mathcal{H}, \quad \mathrm{Exp}(X) \mapsto \widehat{\pi}(\mathrm{Exp}\,X).v = e^{d\pi(X)}v$$

are continuous for $v \in \mathcal{H}_1$.

If $v_n \to v$, $v_n \in \mathcal{H}_1$ and $v \in \mathcal{H}$, then the fact that $\widehat{\pi}(\widetilde{S}) \subseteq C(\mathcal{H})$ implies that γ'_{v_n} converges uniformly to γ'_v. Since $\mathcal{H}_1$ is dense in $\mathcal{H}$, we conclude that $\widehat{\pi}$ is strongly continuous on $\mathrm{Exp}(U_{\mathbb{C}} \cap iC)$.

Let $v \in \mathcal{H}$, $X \in iC$, and $X_n \in iC$ with $X_n \to X$. We choose $m \in \mathbb{N}$ such that $\frac{1}{m}X_n \in U_{\mathbb{C}} \cap iC$ holds for all $n \in \mathbb{N}$. Then, in view of Proposition 9.2(ii) and Step 1, we find that

$$\widehat{\pi}\big(\,\mathrm{Exp}(X_n)\big)v = \widehat{\pi}\big(\,\mathrm{Exp}(\frac{1}{m}X_n)\big)^m v \to \widehat{\pi}\big(\,\mathrm{Exp}(\frac{1}{m}X)\big)^m v = \widehat{\pi}(\mathrm{Exp}\,X)v.$$

Thus $\widehat{\pi}$ is strongly continuous on $\mathrm{Exp}(iC)$. Since it is also strongly continuous on G, Proposition 9.2 and Lawson's Theorem (Theorem 7.34) imply that $\widehat{\pi}$ is strongly continuous on $\widetilde{S}$.

Step 4: $\widehat{\pi}$ is holomorphic on V^0:

Let $v \in \mathcal{H}_1$. We claim that $\widehat{\pi}(s)v = \alpha_v(s)$ for all $s \in V$. Fix $s = g\,\mathrm{Exp}(X) \in V$. Then the construction of V and Lawson's Theorem show that $g\,\mathrm{Exp}([0,1]X) \subseteq V$. For $z \in \mathbb{C}$ with $\mathrm{Re}\,z \geq 0$ we set $\gamma(z) := g\,\mathrm{Exp}(zX)$. Let $U := \gamma^{-1}(V)$. Then

$$z \mapsto \widehat{\pi}(\gamma(z))v \quad \text{and} \quad \alpha_v \circ \gamma$$

are holomorphic on $\gamma^{-1}(V^0)$ (Proposition 9.9), continuous on U, and they are equal on $U \cap i\mathbb{R}$. Now Lemma 9.12(ii) entails that these mappings are equal on the connected component of 0 in U, in particular

$$\alpha_v(g\,\mathrm{Exp}\,X) = \widehat{\pi}(g\,\mathrm{Exp}\,X)v.$$

If $v_n \to v$, $v_n \in \mathcal{H}_1$ and $v \in \mathcal{H}$, then the fact that $\widehat{\pi}(\widetilde{S}) \subseteq C(\mathcal{H})$ and $\widehat{\pi}(s)v_n = \alpha_{v_n}(s)$ implies that α_{v_n} converges uniformly to α_v. Since $\mathcal{H}_1$ is dense in $\mathcal{H}$, we conclude that $\widehat{\pi}$ is strongly holomorphic on V^0. Finally Lemma 9.7(i) implies that $\widehat{\pi}$ is a holomorphic operator valued mapping on V^0.

Step 5: Analyticity on $\mathrm{Exp}(i\,\mathrm{int}\,C)$: Let $X \in i\,\mathrm{int}\,C$ and $W \subseteq i\,\mathrm{int}\,C$ a compact neighborhood of X. Then there exists $m \in \mathbb{N}$ such that $\mathrm{Exp}(\frac{1}{m}W) \subseteq V^0$. Since the m-times multiplication mapping $B(\mathcal{H})^m \to B(\mathcal{H})$ is analytic, we conclude that the mapping

$$W \to B(\mathcal{H}), \quad Y \mapsto e^{d\pi(Y)} = (e^{\frac{1}{m}d\pi(Y)})^m$$

is analytic.

Step 6: $\hat{\pi}$ is holomorphic on $\widetilde{G}\exp(i\operatorname{int}C)$: The analyticity of $s \mapsto \langle v, \hat{\pi}(s)w\rangle$ with $w \in \mathcal{H}^\omega$ follows from Lawson's Theorem (Theorem 7.34) and Step 5. Consequently this mapping is also holomorphic on $\widetilde{G}\operatorname{Exp}(i\operatorname{int}C)$. This follows from the fact that, in view of the analyticity, the domain where the Cauchy-Riemann equations are satisfied is extendable via analytic continuation. Now Lemma 9.7(i) implies operator valued holomorphy.

Step 7: $\hat{\pi}$ is a homomorphism: If $s_2 \in \widetilde{S}$, then $s_1 \mapsto \hat{\pi}(s_1 s_2)$ is holomorphic on $\operatorname{int}\widetilde{S}$ and agrees on $\widetilde{G}$ with $s_1 \mapsto \hat{\pi}(s_1)\hat{\pi}(s_2)$. An application of Lemma 9.17 shows that $\hat{\pi}(s_1 s_2) = \hat{\pi}(s_1)\hat{\pi}(s_2)$ also holds for arbitrary s_1. ∎

Theorem 9.19. (The Extension Theorem) *Let G be a connected Lie group and $\pi: G \to U(\mathcal{H})$ a continuous unitary representation. Suppose that $W := W(\pi)$ is pointed and generating. Then there exists a covering $S = \Gamma_G(W)$ of a Lie semigroup such that $H(S) \cong G$, $S \cong G\operatorname{Exp}(iW)$, $\operatorname{int}S$ is a complex manifold, and a holomorphic contraction representation*

$$\hat{\pi} : S \to C(\mathcal{H})$$

such that $\hat{\pi}|_G = \pi$.

Proof. Let $\widetilde{G}$ denote the universal covering group of G and $p: \widetilde{G} \to G$ the covering morphism. Then $\pi' := \pi \circ p$ defines a continuous unitary representation of $\widetilde{G}$ such that $W(\pi') = W(\pi) = W$. Using Proposition 9.18 we find a strongly continuous involutive representation

$$\hat{\pi}' : \widetilde{S} \to C(\mathcal{H})$$

which extends π' and which is holomorphic on the interior. Since $\hat{\pi}'$ extends π' and $\pi'(\ker p) = \{1\}$, we conclude that $\ker p \subseteq \ker \hat{\pi}'$. Therefore $\hat{\pi}'$ factors to a continuous involutive representation $\hat{\pi}$ of the semigroup $S := \widetilde{S}/\ker p$. ∎

Using Corollary 7.36, it is also possible to obtain a generalization of the extension theorem to the cases where $W(\pi)$ is neither pointed nor generating.

Now we are going to prove a converse of the Extension Theorem. Then these two results can be interpreted as the statement that for a pointed generating cone $W \subseteq \mathfrak{g}$ the category of those continuous unitary representations π of G satisfying $W \subseteq W(\pi)$ is equivalent to the category of holomorphic contraction representations of $\widetilde{\Gamma}(W)$.

Theorem 9.20. *Let G be a connected Lie group, $W \subseteq \mathfrak{g}$ a pointed generating invariant cone, and $S = \Gamma_G(W)$. Suppose that*

$$\hat{\pi} : S \to C(\mathcal{H})$$

is a holomorphic contraction representation. Then $\pi := \hat{\pi}|_G$ is a continuous unitary representation with $W(\pi) \supseteq W$.

Proof. Since $U(\mathcal{H}) \subseteq C(\mathcal{H})$ is the unit group and carries the induced topology (cf. Section 9.1), it follows that π is a continuous unitary representation of G. Let $X \in W$. Then

$$\gamma : \mathbb{C}_+ \to C(\mathcal{H}), \quad z \mapsto \hat{\pi}(\operatorname{Exp}ziX)$$

is a holomorphic contraction representation of $\mathbb{C}_+$. So Theorem 9.13 implies that $\gamma(z) = e^{zid\pi(X)}$ and that $id\pi(X)$ is a negative self-adjoint operator, i.e., $X \in W(\pi)$. ∎

Some properties of holomorphic contraction representations

So far we have seen how to pass from representations of G to holomorphic contraction representations of the semigroups $\Gamma_G(W)$. The main advantage of this analytic extension process is that the representation obtained on the interior of $\Gamma_G(W)$ has particular nice properties. Since we do not go into the theory of irreducible representations here we only mention that the major achievements of the theory lie in its application to the irreducible representations, where it turns out that the operators coming from the interior of $\Gamma_G(W)$ are of trace class which makes it possible to develop a character theory for these representations.

In this subsection we record some properties which remain true for general representations. In the following W always denotes a pointed generating invariant cone in $\mathfrak{g}$ and G a connected Lie group with $\mathbf{L}(G) = \mathfrak{g}$.

Lemma 9.21. *Let π denote a holomorphic contraction representation of $S :=$ $\Gamma_G(W)$ on $\mathcal{H}$, and $\mathcal{H}^\infty$, $\mathcal{H}^\omega$ the corresponding spaces of smooth and analytic vectors for G. Then the following assertions hold:*

(i) $\pi(s)\mathcal{H} \subseteq \mathcal{H}^\omega$ *for all $s \in \operatorname{int} S$.*

(ii) *The spaces $\mathcal{H}^\infty$ and $\mathcal{H}^\omega$ are invariant under $\operatorname{int} S$.*

Proof.   (i) Let $v \in \mathcal{H}$ and $s \in \operatorname{int} S$. Then the holomorphy of the mapping

$$\operatorname{int} S \to \mathcal{H} \qquad t \mapsto \pi(t).v$$

shows that the mapping

$$G \to \mathcal{H} \qquad g \mapsto \pi(gs).v = \pi(g)\big(\pi(s)v\big)$$

is analytic, hence $\pi(s)v \in \mathcal{H}^\omega$.

(ii) Since $\mathcal{H}^\omega \subseteq \mathcal{H}^\infty$, this is an immediate consequence of (i). ∎

The preceding lemma was a trivial consequence of the definitions. It shows that we have a representation π^∞ of the semigroup $\operatorname{int} S$ on the space $\mathcal{H}^\infty$ of smooth vectors in $\mathcal{H}$. As we will see now, the holomorphy of π has very strong consequences for the regulariy of the representation π^∞ of $\operatorname{int} S$.

Let $A \in \mathcal{U}(\mathfrak{g}_{\mathbb{C}})$. Then

$$p_A(v) := \|d\pi(A)v\|$$

defines a seminorm on $\mathcal{H}^\infty$. These seminorms determine a locally convex topolgy on $\mathcal{H}^\infty$ turning $\mathcal{H}^\infty$ into a complex Fréchet space and the representation of G on $\mathcal{H}^\infty$ is continuous (cf. [Mag92, p.56]).

In the following we say that a representation of $\operatorname{int} S = \operatorname{int} \Gamma_G(W)$ on a Fréchet space is *holomorphic* if all the orbit mappings are holomorphic.

Proposition 9.22. *The representation π^∞ of $\operatorname{int} S$ on $\mathcal{H}^\infty$ is a holomorphic representation of $\operatorname{int} S$ on $\mathcal{H}^\infty$ with respect to the Fréchet topology.*

Proof. Let $f: \operatorname{int} S \to \mathcal{H}$ be a holomorphic mapping. For $X \in \mathfrak{g}_{\mathbb{C}}$ we define

$$(R(X)f)(s) := \left.\frac{d}{dt}\right|_{t=0} f(\exp(tX)s).$$

Let us call a vector field on a complex manifold *holomorphic* if it generates a holomorphic local flow. Then this operator corresponds to a holomorphic right invariant vector field $\mathcal{X}_X$ on $\operatorname{int} S$ such that

$$(R(X)f)(1) = df(s)\mathcal{X}_X(s) \qquad \forall s \in S.$$

Since the mapping

$$df : \operatorname{int} S \times \mathfrak{g}_{\mathbb{C}} \to \mathcal{H}$$

is holomorphic by Hartog's Theorem ([RK82, p.65]), it follows that the function $R(X)f : \operatorname{int} S \to \mathcal{H}$ is holomorphic. Inductively we find that $R(A)f$ is holomorphic for every element $A \in \mathcal{U}(\mathfrak{g}_{\mathbb{C}})$, where $\mathcal{U}(\mathfrak{g}_{\mathbb{C}})$ acts on $C^\infty(\operatorname{int} S, \mathbb{C})$ as an algebra of right invariant differential operators.

Let $v \in \mathcal{H}^\infty$ and $f_v(s) := \pi(s)v$. Then the preceding remarks show that the functions $R(A)f_v$ are holomorphic. Explicitly, we find for $X \in \mathfrak{g}_{\mathbb{C}}$ that

$$(R(X)f_v)(s) = \left.\frac{d}{dt}\right|_{t=0} f_v(\exp(tX)s) = \left.\frac{d}{dt}\right|_{t=0} \pi(\exp tX)\pi(s)v = d\pi(X)f_v(s).$$

For $A \in \mathcal{U}(\mathfrak{g}_{\mathbb{C}})$ we therefore find that

$$(R(A)f_v)(s) = d\pi(A)\big(f_v(s)\big) \qquad \forall s \in \operatorname{int} S.$$

This shows that locally the power series expansion of f_v still converges after multiplication with $d\pi(A)$. This means that $f_v: \operatorname{int} S \to \mathcal{H}^\infty$ is a holomorphic mapping. ∎

9.4. Holomorphic discrete series representations

In this section we assume that $\mathfrak{g}$ is a simple Hermitean Lie algebra and G a Lie group with Lie algebra $\mathfrak{g}$. Moreover, we assume that G is contained in a complex Lie group $G_{\mathbb{C}}$ with Lie algebra $\mathfrak{g}_{\mathbb{C}}$. Then the group K corresponding to $\mathfrak{k}$ is compact, where $\mathfrak{k}$ is a maximal compactly embedded subalgebra of $\mathfrak{g}$. Fix a Cartan algebra $\mathfrak{t}$ of $\mathfrak{k}$.

We use the notation from Chapter 7 and set

$$(9.1) \qquad (L(X)f)(g) = \left.\frac{d}{dt}\right|_{t=0} f(g \exp tX)$$

for $X \in \mathfrak{g}$ and $f \in C^\infty(G, V)$, where V is a complex finite dimensional vector space, and we extend (9.1) to $\mathfrak{g}_{\mathbb{C}}$ via complex linearity. The resulting operator will be denoted $L_{\mathbb{C}}(X)$.

Let $\Lambda = \Lambda(\mathfrak{g}_{\mathbb{C}}, \mathfrak{t}_{\mathbb{C}})$ denote the set of roots of $\mathfrak{g}_{\mathbb{C}}$ with respect to $\mathfrak{t}_{\mathbb{C}}$. As in Section 7.1 we have the subsystem Λ_k of compact roots and the set of non-compact roots. We fix a $\mathfrak{t}$-adapted positive system Λ^+. Recall the abelian subalgebras

$$\mathfrak{p}_{\mathbb{C}}^{\pm} = \sum_{\lambda \in \Lambda_p^{\pm}} \mathfrak{g}_{\mathbb{C}}^{\lambda}$$

(Lemma 7.7) and let $P^{\pm}$ be the corresponding subgroups of $G_{\mathbb{C}}$. It is well known ([Hel78, p.388]) that the map

$$P^+ \times K_{\mathbb{C}} \times P^- \to G_{\mathbb{C}}$$

given by the group multiplication is a diffeomorphism onto an open domain in $G_{\mathbb{C}}$ containing G. For $g \in G$, or more generally in $P^+ K_{\mathbb{C}} P^-$, we write $\kappa(g)$ for the $K_{\mathbb{C}}$-component in the above decomposition. We set $B := K_{\mathbb{C}} P^+$ and $\overline{B} := K_{\mathbb{C}} P^-$.

The irreducible unitary representations of the universal covering group $\widetilde{K}$ of K are parametrized by the set

$$\mathcal{P}_{\widetilde{K}} := \{\mu \in \mathfrak{t}_{\mathbb{C}}^* : (\forall \alpha \in \Delta_k^+)\langle \mu, \alpha \rangle \in \mathbb{N}_0; \mu(\mathfrak{t}) \subseteq i\mathbb{R}\}$$

(cf. the section on highest weight modules). Let $T := \exp \mathfrak{t} \subseteq K$ denote the maximal torus. We write $\widehat{T}$ for the group of unitary characters of T. Then the irreduzible unitary representations of K are parametrized by

$$\mathcal{P}_K := \{\mu \in \mathcal{P}_{\widetilde{K}} : (\exists \chi \in \widehat{T}) d\chi(\mathbf{1}) = i\mu\}.$$

The representation τ_λ associated to $\lambda \in \mathcal{P}_K$ is the representation of highest weight λ with respect to Λ_k^+. Let V_λ be the corresponding representation space.

Next we construct two homogeneous vector bundles. For the representation τ_λ of K on V_λ we first build the bundle

$$E_\lambda := G \times_K V_\lambda = \{[(g,v)] := K.(g,v) : g \in G, v \in V_\lambda\},$$

where K acts on $G \times V_\lambda$ by $k.(g,v) = (gk^{-1}, \tau_\lambda(k)v)$. Now we extend the representation τ_λ to a representation $\tau_\lambda : K_{\mathbb{C}} \to \mathrm{Gl}(V_\lambda)$ which in turn extends uniquely to a holomorphic representation

$$\sigma_\lambda : \overline{B} \cong P^- \rtimes K_{\mathbb{C}} \to \mathrm{Gl}(V_\lambda)$$

such that $P^- \subseteq \ker \sigma_\lambda$. So we obtain a holomorphic action of $\overline{B}$ on $G_{\mathbb{C}} \times V_\lambda$ by $\overline{b}.(g,v) = (g\overline{b}^{-1}, \sigma(\overline{b}).v)$ and we form the corresponding homogeneous holomorphic vector bundle

$$F_\lambda := G_{\mathbb{C}} \times_{\overline{B}} V_\lambda$$

consisting of $\overline{B}$-orbits in $G_{\mathbb{C}} \times V_\lambda$. Now the embedding $G \to G_{\mathbb{C}}$, in view of the fact that $\overline{B} \cap G = K$, induces an embedding $E_\lambda \to F_\lambda$ by $(g,v)K \mapsto (g,v)\overline{B}$. On the level of base spaces we have an open embedding $G/K \to G_{\mathbb{C}}/\overline{B}$. Hence we may identify E_λ with $F_\lambda|_{G/K}$, which defines the structure of a holomorphic vector bundle on E_λ.

The following theorem explains how to deal with holomorphic sections of holomorphic vector bundles as functions on groups.

Theorem 9.23. *Let $E = G \times_K V$ denote a homogeneous vector bundle of G over G/K defined by the representation τ of K in V which inherits the structure of a holomorphic vector bundle by an embedding $E \to F := G_{\mathbb{C}} \times_{\overline{B}} V$, where $G \to G_{\mathbb{C}}$ is an injective complexification, $\overline{B} \subseteq G_{\mathbb{C}}$ is a complex subgroup satisfying $\overline{B} \cap G = K$, and σ a holomorphic representation of $\overline{B}$ on V extending τ on K. Let $\overline{\mathfrak{b}} := \mathbf{L}(\overline{B})$. We write $\Gamma(E)$ for the space of smooth sections of E, and $\Gamma_{\mathrm{hol}}(E)$ for the space of holomorphic sections corresponding to the induced holomorphic structure. Then the following assertions hold:*

(i) *Let $\Gamma_G(E) := \{ f \in C^{\infty}(G, V) : (\forall g \in G, k \in K) f(gk) = \tau(k)^{-1}.f(g) \}$. Then the mapping*

$$\Gamma_G(E) \to \Gamma(E), \qquad f \mapsto (gK \mapsto [(g, f(g))])$$

is a well defined bijection.

(ii) *Let $\Gamma_{G_{\mathbb{C}}}(F) := \{ f \in \mathrm{Hol}(G_{\mathbb{C}}, V) : (\forall g \in G, k \in K) f(g\overline{b}) = \sigma(\overline{b})^{-1}.f(g) \}$. Then the mapping*

$$\Gamma_{G_{\mathbb{C}}}(F) \to \Gamma(F), \qquad f \mapsto (g\overline{B} \mapsto [(g, f(g))])$$

is a well defined bijection.

(iii) *A function $f \in \Gamma_G(E)$ corresponds to a holomorphic section of E if and only if in addition*

$$\mathbf{L}_{\mathbb{C}}(X).f = -d\sigma(1)(X).f \qquad \forall X \in \overline{\mathfrak{b}}.$$

Proof. (i) This is a routine verification which can, more or less, be drawn from [Wa73, Section 5.2].

(ii) This is just the holomorphic version of (i). One only has to replace smooth mappings by holomorphic mappings.

(iii) This is the crucial ingredient. For a proof we refer to [Ki76, p.203]. ∎

Given $\lambda \in \mathcal{P}_K$ we set

$$\mathcal{O}(\lambda) = \{ f \in C^{\infty}(G, V_{\lambda}) : (\forall X \in \mathfrak{p}_{\mathbb{C}}^{-}) \, \mathbf{L}_{\mathbb{C}}(X).f = 0 \, ;$$
$$(\forall g \in G, k \in K) \, f(gk) = \tau_{\lambda}(k)^{-1} f(g) \}.$$

In view of Theorem 9.23, we know that $\mathcal{O}(\lambda)$ corresponds to the space of holomorphic sections of E_{λ}.

The group G operates on $\mathcal{O}(\lambda)$ via

$$\pi(h)f(g) := h.f(g) := f(h^{-1}g).$$

On $\mathcal{O}(\lambda)$ we consider the norm

$$\|f\| := \left(\int_G \|f(g)\|^2 \, dm(g) \right)^{\frac{1}{2}}$$

and the space

$$\mathcal{H}_{\lambda} := \{ f \in \mathcal{O}(\lambda) : \|f\| < \infty \}.$$

Then $\mathcal{H}_\lambda$ is a Hilbert space and G acts unitarily and irreducibly on $\mathcal{H}_\lambda$ (cf. [Kn86, Th. 6.6]).

Let v_λ denote a normalized highest weight vector in V_λ. We define the function
$$f_\lambda : G \to V_\lambda, \qquad g \mapsto \tau_\lambda(\kappa(g)^{-1})v_\lambda.$$
Note that this function has an obviuous holomorphic extension to the open set $P^+ K_{\mathbb{C}} P^-$ in $G_{\mathbb{C}}$ which is P^--invariant on the right and P^+-invariant on the left.

One can show that $\mathcal{H}_\lambda$ is non-zero if the function $f_\lambda \in \mathcal{O}(\lambda)$ has finite norm ([Kn86, Ch. VI]).

This happens if
$$\langle \lambda + \delta, \alpha \rangle < 0$$
for all $\alpha \in \Lambda_p^+$ (cf. loc.cit.), where δ is the halfsum of the positive roots. The representations of the form π_λ are called the *holomorphic discrete series*. It is easy to check that f_λ is a common eigenvector for all the $\pi(k)$ with $k \in T$, the maximal torus of K corresponding to $\mathfrak{t}$. The holomorphic extension of f_λ shows that this function is a highest weight vector for the derived representation $d\pi_\lambda$ of π_λ.

If π is any unitary representation of G on $\mathcal{H}$, then $f \in \mathcal{H}$ is called K-*finite* if $\pi(K)f$ spans a finite dimensional vector space. We denote the vector space of K-finite vectors by $\mathcal{H}^K$. The discussion of the last paragraph shows that for $\lambda \in \mathcal{P}_K$ the space $\mathcal{H}_\lambda^K$ is a highest weight module for $\mathfrak{g}$ (cf. [Kn86, VIII.3]).

We can write
$$(9.2) \qquad \mathcal{H}_\lambda^K = \bigoplus_{\mu \in \mathcal{M}} \mathcal{H}_\lambda^{K,\mu},$$
where $\mathcal{M}$ is the set of $\mu \in \mathcal{P}_K$ occuring as weights of irreducible $\mathfrak{k}_{\mathbb{C}}$-submodules of $\mathcal{H}_\lambda^K$, and $\mathcal{H}_\lambda^{K,\mu}$ is the corresponding isotypic $\mathfrak{k}_{\mathbb{C}}$-submodule. Note that the decomposition (9.2) is orthogonal w.r.t. the inner product from $\mathcal{H}_\lambda$.

For the following proposition we recall that
$$(9.3) \qquad W(\pi_\lambda) = \{X \in \mathfrak{g} : (\forall \xi \in \mathcal{H}^\infty)\ (id\pi(X)\xi, \xi) \le 0\}.$$

Proposition 9.24. (Ol'shanskiĭ) *Let $\mathcal{H}_\lambda^K$ be a highest weight module with highest weight $\lambda : \mathfrak{t}_{\mathbb{C}} \to \mathbb{C}$ corresponding to the positive system Λ^+. Then*
$$(9.4) \qquad -i\lambda|_{\mathfrak{t}} \in C_{\min}(\Lambda^+)^* \subseteq \mathfrak{t}^*.$$

Proof. Note that $(d\pi(X)\xi, \eta) = -(\xi, d\pi(X)\eta)$ for all $X \in \mathfrak{g}$ and $\xi, \eta \in \mathcal{H}_\lambda^K$, $\mathcal{H}^K \subseteq \mathcal{H}^\omega$ ([Wa72, 4.4.5.17]), and hence
$$(d\pi(X)\xi, \eta) = -(\xi, d\pi(\overline{X})\eta)$$
for all $X \in \mathfrak{g}_{\mathbb{C}}$, where $\overline{X}$ is the complex conjugate of X with respect to the real form $\mathfrak{g}$. Now let $\alpha \in \Lambda_p^+$ and $X_\alpha \in \mathfrak{g}_{\mathbb{C}}^\alpha$. Then $\overline{X}_\alpha \in \mathfrak{g}_{\mathbb{C}}^{-\alpha}$ and $H_\alpha := i[\overline{X}_\alpha, X_\alpha] \in C_{\min} \subseteq \mathfrak{t}$ (cf. Section 7.2). If f_λ is a highest weight vector of $\mathcal{H}_\lambda^K$, then we calculate
$$\begin{aligned} 0 \le (d\pi(\overline{X}_\alpha)f_\lambda, d\pi(\overline{X}_\alpha)f_\lambda) &= -(d\pi(X_\alpha)d\pi(\overline{X}_\alpha)f_\lambda, f_\lambda) \\ &= -(d\pi([X_\alpha, \overline{X}_\alpha])f_\lambda, f_\lambda) \\ &= -i\lambda(H_\alpha)\|f_\lambda\|^2. \end{aligned}$$
Thus $-i\lambda \in C_{\min}^*$ because $C_{\min}$ is generated by the elements H_α, $\alpha \in \Lambda_p^+$. ∎

Lemma 9.25. *If β is a weight of $\mathcal{H}_\lambda^K$, then there exists a weight α of V_λ such that*

$$\beta \in \alpha - \sum_{\gamma \in \Delta_p^+} \mathbb{N}_0 \gamma.$$

Proof. In view of the Theorem of Poincaré-Birkhoff-Witt, we have that

$$\mathcal{U}(\mathfrak{g}_\mathbb{C}).f_\lambda = \mathcal{U}(\mathfrak{p}_\mathbb{C}^-)\mathcal{U}(\mathfrak{k}_\mathbb{C})\mathcal{U}(\mathfrak{p}_\mathbb{C}^+).f_\lambda = \mathcal{U}(\mathfrak{p}_\mathbb{C}^-)\mathcal{U}(\mathfrak{k}_\mathbb{C}).f_\lambda = \mathcal{U}(\mathfrak{p}^-).V_\lambda$$

because $\mathcal{U}(\mathfrak{k}_\mathbb{C}).f_\lambda$ is a $\mathfrak{k}_\mathbb{C}$-submodule of $\mathcal{H}_\lambda^K$ isomorphic to V_λ. ∎

The preceding observation makes it very easy to determine the cone $W(\pi_\lambda)$.

Theorem 9.26. *Let λ be the highest weight of of the holomorphic discrete series representation π_λ in $\mathcal{H}_\lambda$ and $\mathcal{W}_\mathfrak{k}$ the Weyl group of $\mathfrak{k}$. Then*

$$C(\pi_\lambda) := W(\pi_\lambda) \cap \mathfrak{t} = \{X \in C_{\max} : (\forall w \in \mathcal{W}_\mathfrak{k})(i\lambda, w.X) \le 0\}.$$

Proof. Let $X \in C_{\max}$. Then $X \in C(\pi_\lambda)$ is equivalent to $\beta(iX) \le 0$ for all weights of $\mathcal{H}_\lambda$. In view of Lemma 9.25, this condition reduces to $\beta(iX) \le 0$ for all weights of V_λ. Since V_λ is an irreducible $\mathfrak{k}_\mathbb{C}$-module with highest weight λ, the set of all weights is contained in the convex hull of the $\mathcal{W}_\mathfrak{k}$-orbit of λ. Hence $X \in C(\pi_\lambda)$ is equivalent to

$$\langle \mathcal{W}_\mathfrak{k}.\lambda, iX \rangle \subseteq -\mathbb{R}^+.$$

∎

Theorem 9.26, together with the Extension-Theorem (Theorem 9.15), shows that we can analytically extend the representations of the holomorphic discrete series. To have a better understanding how this continuation looks like, we give another realization of $\mathcal{H}_\lambda^K$.

Consider the Harish-Chandra realization of G/K:

$$G \subseteq G_\mathbb{C}, \quad K \subseteq K_\mathbb{C} \subseteq G_\mathbb{C}, \quad GK_\mathbb{C}P^- = \Omega K_\mathbb{C}P^-,$$

where $\Omega \cong G/K$ is a bounded domain in P^+ ([Hel78, p.391]). We write the inverse of the map $P^+ \times K_\mathbb{C} \times P^- \to P^+K_\mathbb{C}P^-$ as follows:

$$g \mapsto \big(\zeta(g), \kappa(g), \zeta'(g)\big).$$

The group action of G on Ω is given by

$$g.\omega := \zeta(g\omega).$$

Lemma 9.27. *For any $f \in \mathcal{O}(\lambda)$ the formula*

$$\tilde{f}(gkq) = \tau_\lambda(k)^{-1}f(g)$$

defines a holomorphic function $\tilde{f}: GK_\mathbb{C}P^- \to V_\lambda$.

Proof.   Let $F \in \Gamma_{\mathrm{hol}}(E_\lambda) = \Gamma_{\mathrm{hol}}(F_\lambda|_{G/K})$ denote the holomorphic section of E_λ corresponding to f via Theorem 9.23(iii). Using Theorem 9.23(ii), we find a holomorphic function

$$\widetilde{f}: GK_{\mathbb{C}}P^- \to V_\lambda$$

with

$$F(gK_{\mathbb{C}}P^-) = [(g, \widetilde{f}(g))] \qquad \forall g \in GK_{\mathbb{C}}P^-.$$

Since this function satisfies

$$\widetilde{f}(gkq) = \sigma_\lambda(\overline{k}q)^{-1}.f(g) = \tau_\lambda(\overline{k})^{-1}.f(g) \qquad \forall g \in G, k \in K_{\mathbb{C}}, q \in P^-,$$

the assertion follows.   ∎

The preceding lemma shows how to relate elements of $\mathcal{O}_\lambda$ to functions on Ω which are simpler objects. We will see that this translation process does not forget any information. We set

$$\mathcal{O}_\Omega(\lambda) = \{F: \Omega \to V_\lambda : F \text{ holomorphic}\}.$$

Lemma 9.28.   *The mapping $\Phi: \mathcal{O}(\lambda) \to \mathcal{O}_\Omega(\lambda)$ defined by $\Phi(f) = \widetilde{f}|_\Omega$ is a linear bijection.*

Proof.   According to Lemma 9.27 the mapping Φ is well defined. Next we simply give the inverse mapping $\Psi: \mathcal{O}_\Omega(\lambda) \to \mathcal{O}(\lambda)$:

$$(9.5) \qquad\qquad \Psi(F)(g) = \tau_\lambda(\kappa(g))^{-1} F(\zeta(g)).$$

In order to show that this makes sense, note first that $g = \zeta(g)\kappa(g)\zeta'(g)$ implies

$$gk = \zeta(g)\kappa(g)\zeta'(g)k = \zeta(g)\kappa(g)k(k^{-1}\zeta'(g)k)$$

which shows that $\kappa(gk) = \kappa(g)k$ and $\zeta(gk) = \zeta(g)$. Therefore we can calculate

$$\Psi(F)(gk) = \tau_\lambda(\kappa(g)k)^{-1}.F(\zeta(g)) = \tau_\lambda(k)^{-1}.\Psi(F)(g).$$

Moreover, the fact that (9.5) even defines a holomorphic mapping on $P^+K_{\mathbb{C}}P^-$, shows that we can compute $L_{\mathbb{C}}(X)\Psi(F)$ for $X \in \mathfrak{p}_{\mathbb{C}}^-$ as follows:

$$L_{\mathbb{C}}(X)\Psi(F)(g) = \frac{d}{dt}\bigg|_{t=0} \Psi(F)(g \exp tX) = \frac{d}{dt}\bigg|_{t=0} \Psi(F)(g) = 0$$

because the holomorphic extension of $\Psi(F)$ is right P^--invariant.

To see that Ψ and Φ are inverse to each other, let first $f \in \mathcal{O}_\lambda$. Then

$$\Psi(\widetilde{f}|_\Omega)(g) = \tau_\lambda(\kappa(g))^{-1}.\widetilde{f}(\zeta(g)) = \widetilde{f}(\zeta(g)\kappa(g)) = \widetilde{f}(g) = f(g).$$

If, conversely, $F \in \mathcal{O}_\Omega$, then

$$\begin{aligned}
\Phi(\Psi(F))(\zeta(g)) &= \Psi(F)\widetilde{\;}(\zeta(g)) \\
&= \tau_\lambda(\kappa(g)).\Psi(F)(g) \\
&= F(\zeta(g)).
\end{aligned}$$

∎

Now we use Φ to transport the G-action on $\mathcal{O}(\lambda)$ to $\mathcal{O}_\Omega(\lambda)$.

Lemma 9.29.   *The action of G on $\mathcal{O}_\Omega(\lambda)$ is given by the following formula*

$$(9.6) \qquad (\widetilde{\pi}_\lambda(g)F)(\omega) = \tau_\lambda(\kappa(g^{-1}\omega))^{-1}.F(g^{-1}.\omega) \qquad \forall g \in G.$$

Proof.   A simple calculation using

$$\Phi \circ \pi_\lambda(g) = \widetilde{\pi}_\lambda(g) \circ \Phi$$

shows

$$
\begin{aligned}
(\widetilde{\pi}_\lambda(g)\Phi(f))(\zeta(x)) &= (\Phi \circ \pi_\lambda(g))(f)(\zeta(x)) \\
&= (f \circ \lambda_{g^{-1}})\check{\ }(\zeta(x)) \\
&= \widetilde{f}(g^{-1}\zeta(x)) \\
&= \tau_\lambda(\kappa(g^{-1}\zeta(x)))^{-1}.\widetilde{f}(\zeta(g^{-1}\zeta(x))) \\
&= \tau_\lambda(\kappa(g^{-1}\zeta(x)))^{-1}.\Phi(f)(g^{-1}.\zeta(x)).
\end{aligned}
$$

In view of Lemma 9.28, this proves the lemma.   ∎

Now we apply the results of Section 8.4 (Theorem 8.49) to the parabolic subgroup $K_{\mathbb{C}}P^-$ of $G_{\mathbb{C}}$ and find that the semigroup $\Gamma(-W_{\max})$ maps Ω, which may be viewed as an open G-orbit in $G_{\mathbb{C}}/K_{\mathbb{C}}P^-$, into itself. Thus we have a holomorphic action of $\Gamma(W_{\max})$ on $\mathcal{O}_\Omega(\lambda)$ given by

$$(s.F)(\omega) = \tau_\lambda(\kappa(s^{-1}\omega))^{-1}F(s^{-1}.\omega)$$

which is an extension of the G-action.

If we restrict this action to the semigroup $\Gamma(W(\pi_\lambda))$ and the image of $\mathcal{H}_\lambda^K$ under Φ, then we obtain the desired extension $\widehat{\pi}_\lambda$ whose existence was guaranteed by Theorem 9.19.

In order to illustrate the situation, we discuss a simple example. Let $G = SU(1,1)$, so that

$$\mathfrak{k} = \mathfrak{t} = \left\{ \begin{pmatrix} ir & 0 \\ 0 & -ir \end{pmatrix} : r \in \mathbb{R} \right\}.$$

Then there is only one positive root given by

$$\lambda \begin{pmatrix} ir & 0 \\ 0 & -ir \end{pmatrix} = 2ir$$

and the corresponding root space is

$$\mathfrak{p}_{\mathbb{C}}^+ = \left\{ \begin{pmatrix} 0 & z \\ 0 & 0 \end{pmatrix} : z \in \mathbb{C} \right\}.$$

The domain Ω then consists of those matrices in $\mathfrak{p}_{\mathbb{C}}^+$ for which $|z| < 1$. Since

$$\begin{pmatrix} a & b \\ \overline{b} & \overline{a} \end{pmatrix}\begin{pmatrix} 1 & \omega \\ 0 & 1 \end{pmatrix} = \begin{pmatrix} a & a\omega + b \\ \overline{b} & \overline{b}\omega + \overline{a} \end{pmatrix} = \begin{pmatrix} 1 & \frac{a\omega+b}{\overline{b}\omega+\overline{a}} \\ 0 & 1 \end{pmatrix}\begin{pmatrix} \gamma & 0 \\ 0 & \overline{\gamma} \end{pmatrix}\begin{pmatrix} 1 & 0 \\ * & 1 \end{pmatrix},$$

the group action is given by

$$\begin{pmatrix} a & b \\ \overline{b} & \overline{a} \end{pmatrix}.\omega = \frac{a\omega + b}{\overline{b}\omega + \overline{a}}.$$

The highest weight λ is given via

$$\lambda \begin{pmatrix} ir & 0 \\ 0 & -ir \end{pmatrix} = irm_0$$

for some $m_0 \in \mathbb{N}$. Thus the representation $\widetilde{\pi}_\lambda$ can be written

$$\widetilde{\pi}_\lambda \left(\begin{pmatrix} a & b \\ \overline{b} & \overline{a} \end{pmatrix} \right) F(\omega) = (a - \overline{b}\omega)^{-m_0} F \left(\frac{\overline{a}\omega - b}{-\overline{b}\omega + a} \right)$$

(cf. also [Lan75, Sect. IX.3]). The functions $\omega \mapsto \omega^k$ are eigenfunctions under the action of $T_\mathbb{C}$:

$$\begin{pmatrix} c & 0 \\ 0 & c^{-1} \end{pmatrix}.\omega^k = c^{-m_0} \left(\frac{c^{-1}\omega}{c} \right)^k = c^{-(m_0+2k)}\omega^k.$$

Note that

$$C_{\max} = \left\{ \begin{pmatrix} ir & 0 \\ 0 & -ir \end{pmatrix} : r \le 0 \right\}.$$

Thus $\begin{pmatrix} c & 0 \\ 0 & c^{-1} \end{pmatrix} \in \Gamma(W_{\max})$ iff $|c| \ge 1$ and these are exactly the elements that will induce contractions on the Hilbert space since $|c^{-(m_0+2k)}| \le 1$ is equivalent to $|c| \ge 1$.

9.5. Hardy spaces

Let G be a connected Lie group and $W \subseteq \mathfrak{g}$ a pointed generating invariant cone. In this section we will construct Hilbert spaces on which the semigroups

$$\Gamma(W) := \Gamma_G(W) = G \operatorname{Exp} iW$$

whose existence is guaranteed by Theorem 9.15 naturally act by contractions. These Hilbert spaces consist of holomorphic functions on the interior $\Gamma^0(W)$ of $\Gamma(W)$. The action is simply the right-regular action

(9.7) $(\widetilde{\pi}(\gamma)f)(z) = f(z\gamma) \quad \forall \gamma \in \Gamma(W),\ z \in \Gamma^0(W).$

Consider $L = L^2(G, \mu_G)$, where μ_G is a Haar measure which is biinvariant according to Proposition 7.3(v) because $\mathfrak{g}$ contains a compactly embedded Cartan algebra (Proposition 7.10). Therefore the right-regular representation of G is simply given by $(R(g)f)(x) = f(xg) = (g.f)(x)$. Note that for $f: \Gamma^0(W) \to \mathbb{C}$ and $\gamma \in \Gamma^0(W)$ we may define $\gamma.f: G \to \mathbb{C}$ via $(\gamma.f)(x) = f(x\gamma)$ since $\Gamma^0(W)$ is a semigroup ideal in $\Gamma(W)$. In the following we write

$$\exp : \mathbf{L}(\Gamma(W)) = \mathfrak{g} + iW \to \Gamma(W)$$

for the exponential mapping of $\Gamma(W)$, even if $\Gamma(W)$ is not contained in a complex group.

Definition 9.30. By $\mathcal{H}(W)$ we denote the space of all holomorphic functions $f:\Gamma^0(W)\to\mathbb{C}$ with

$$\|f\|_H := \sup_{\gamma\in\Gamma^0}\|\gamma.f\|_2 < \infty.$$

Since $\mathcal{H}(W)$ is invariant under the right-regular representation, we obtain a representation $\hat{\pi}$ of $\Gamma(W)$ in $\mathcal{H}(W)$ as in (9.7).

A unitary representation π of G in a Hilbert space is said to be W-*admissible* if $W\subseteq W(\pi)$. According to Theorem 9.20, every holomorphic contraction representation of $\Gamma(W)$ restricts to a W-admissible representation on G, and every admissible representation can in a natural way be extended to a holomorphic representation of $\Gamma(W)$.

Theorem 9.31. (Olshanskiĭ's Theorem on Hardy spaces)

(i) $\mathcal{H}(W)$ *is a Hilbert space with norm* $\|\cdot\|_H$.

(ii) *There is an isometry* $I:\mathcal{H}(W)\to L^2(G)$ *given by* $If=\lim\gamma_j.f$ *for any sequence* $\gamma_1,\gamma_2,\ldots$ *in* Γ^0 *converging to* 1. *Here the limit is taken in* $L^2(G)$.

(iii) I *is an intertwining operator for the* G-*actions, i.e.,* $I\hat{\pi}(g)=R(g)I$ *for all* $g\in G$.

(iv) $\hat{\pi}$ *is a holomorphic contraction representation of* $\Gamma(W)$ *on* $\mathcal{H}(W)$.

Proof. The proof of Theorem 9.31 uses Lemma 9.11, which in a very abstract setting gives the key ingredient in constructing and identifying Hardy spaces.

For a fixed $f\in\mathcal{H}(W)$ the map $M_f:\Gamma^0(W)\to L^2(G)$, defined by $M_f(\gamma)=\gamma.f$, is holomorphic since for a $\varphi\in C_c^\infty(G)$ we get that $\gamma\to\int_G f(x\gamma)\varphi(x)dx$ is holomorphic (Lemma 9.7(i)). For a fixed $X\in i\operatorname{int}W$ we may therefore consider the holomorphic function $F:\operatorname{int}\mathbb{C}_+\to L^2(G)$

$$F(z)=\exp zX.f$$

and apply Lemma 9.11 to the Hilbert space $L^2(G)$ and $U(t)=R(\exp itX)$. Note that $iX\in\mathfrak{g}$. We conclude that the limit

$$I_Xf := \lim_{z\to 0}F(z)=\lim_{z\to 0}\exp(zX).f$$

exists.

We want to show that I_X does not depend on X. To this end, note first that, in the the notation from Lemma 9.11, we have

(9.8) $$\exp zX.f = e^{zA_-}.I_Xf.$$

If $\gamma\in\Gamma^0(W)$, then we have

(9.9) $$I_X(\hat{\pi}(\gamma)f)=\gamma.f$$

because the continuity of M_f shows that $(\exp zX)\gamma.f\to\gamma.f$ as $z\to 0$. Note that (9.8) shows that $\|\exp zX.f\|\le\|e^{zA_-}\|\|I_Xf\|_2\le\|I_Xf\|_2$, where $\|\cdot\|$ denotes the operator norm and $\| \cdot \|_2$ the norm on $L^2(G)$. Now for $\gamma_1 = g \exp Y \in \Gamma^0(W)$ and $\gamma_2 \in \Gamma^0(W)$ we calculate

$$
\begin{aligned}
\|\gamma_1 \gamma_2 . f\|_2 &= \|R(g)(\exp Y \gamma_2 . f)\|_2 \\
&= \|\exp Y.(\hat{\pi}(\gamma_2)f)\|_2 \\
&\leq \|I_Y(\hat{\pi}(\gamma_2)f)\|_2 \\
&= \|\gamma_2 . f\|_2,
\end{aligned}
$$

where the last equality is a consequence of (9.9). Thus we have shown

$$\|\gamma_1 \gamma_2 . f\|_2 \leq \|\gamma_2 . f\|_2.$$

We claim that I_X is an isometry. In fact, for $X \in i \operatorname{int} W$, $\varepsilon > 0$, and $\gamma \in \Gamma^0(W)$ with $\|\gamma . f\|_2 \geq \|f\|_H - \varepsilon$ we can find a $t_0 \in \mathbb{R}^+$ such that $\gamma' = \gamma \exp(-t_1 X) \in \Gamma^0(W)$ for any $t_1 \in [0, t_0]$. But then

$$
\begin{aligned}
\|\gamma . f\|_2 &= \|\gamma'(\exp t_1 X).f\|_2 \\
&\leq \|\exp t_1 X.f\|_2 \\
&\leq \|\exp t_2 X.f\|_2
\end{aligned}
$$

for $0 \leq t_2 \leq t_1 \leq t_0$. Thus

$$\|\gamma . f\|_2 \leq \sup_{0 \leq t \leq t_0} \|\exp tX.f\|_2 = \lim_{t \to 0} \|\exp tX.f\|_2 = \|I_X f\|_2,$$

so we have proved $\|f\|_H \leq \|I_X f\|_2$. Conversely, we clearly have

$$\|I_X f\|_2 \leq \sup_{\gamma \in \Gamma^0(W)} \|\gamma . f\|_2 = \|f\|_H,$$

so that we now have

(9.10) $\|I_X f\|_2 = \|f\|_H.$

Now that we know that I_X is an isometry, we can also show the that I_X is independent of X. Let $X, X' \in i \operatorname{int} W$ and calculate

$$
\begin{aligned}
\|I_X f - I_{X'} f\|_2 &\leq \|I_X f - I_X(\hat{\pi}(\gamma_\varepsilon)f)\|_2 \\
&+ \|I_X(\hat{\pi}(\gamma_\varepsilon)f) - I_{X'}(\hat{\pi}(\gamma_\varepsilon)f)\|_2 \\
&+ \|I_{X'}(\hat{\pi}(\gamma_\varepsilon)f) - I_{X'}(\hat{\pi}(\gamma_\varepsilon')f)\|_2 \\
&+ \|I_{X'}(\hat{\pi}(\gamma_\varepsilon')f) - I_{X'} f\|_2,
\end{aligned}
$$

where $\gamma_\varepsilon = \exp \varepsilon X$ and $\gamma_\varepsilon' = \exp \varepsilon X'$. For the first and fourth term we note that because of (9.8) and (9.9), the term $I_X(\hat{\pi}(\gamma_\varepsilon)f) = \gamma_\varepsilon . f = e^{\varepsilon A} . I_X f$ converges to $I_X f$ for $\varepsilon \to 0$, i.e.,

(9.11) $I_X(\hat{\pi}(\gamma_\varepsilon)f) \to I_X f.$

Thus the first and the fourth term get arbitrarily small for appropriate ε. The second term is equal to $\|\gamma_\varepsilon.f - \gamma_\varepsilon.f\|_2 = 0$ by (9.9). Finally for the third term, using (9.10), we calculate

$$
\begin{aligned}
&\|I_{X'}(\widehat{\pi}(\gamma_\varepsilon)f) - I_{X'}(\widehat{\pi}(\gamma'_\varepsilon)f)\|_2 \\
&= \|\widehat{\pi}(\gamma_\varepsilon)f - \widehat{\pi}(\gamma'_\varepsilon)f\|_H \\
&\leq \|\widehat{\pi}(\gamma_\varepsilon)f - f\|_H + \|\widehat{\pi}(\gamma'_\varepsilon)f - f\|_H \\
&= \|I_X(\widehat{\pi}(\gamma_\varepsilon)f) - I_X f\|_2 + \|I_{X'}(\widehat{\pi}(\gamma'_\varepsilon)f) - I_{X'} f\|_2,
\end{aligned}
$$

where the last expression converges to zero as $\varepsilon \to 0$ because of (9.11). Thus we have shown that I_X is independent of X. Our next goal is to prove the completeness of $\mathcal{H}(W)$. We shall make use of the fact that the norm convergence implies local uniform convergence. Thus we first show that $\mathcal{H}(W)$ satisfies a Bergman-type condition. To be more precise, we claim that for any compact subset M of $\Gamma^0(W)$ we find a constant c_M depending only on M such that

(9.12)
$$
\sup_{z \in M} |f(z)| \leq c_M \|f\|_H
$$

for all $f \in \mathcal{H}(W)$.

In order to prove (9.12), we consider a relatively compact open neighborhood U of 0 in $\mathfrak{g}$ and a relatively compact open neighborhood V of 0 in $i\mathfrak{g}$. Then $\exp V m \exp U$ is a neighborhood of $m \in \Gamma^0(W)$ in $\Gamma^0(W)$ if V is small enough. Conversely each compact neighborhood of $m \in \Gamma^0(W)$ in $\Gamma^0(W)$ can be covered by finitely many sets of this type. Thus in order to prove (9.12), we may assume that $M \subseteq \exp V m \exp U$ with $\exp: U \to \exp U$ and $\exp: V \to \exp V$ diffeomorphisms. Consider the map

$$
\begin{aligned}
\psi: \exp V m \exp U &\to \mathfrak{g}_{\mathbb{C}} \\
(\exp v) m (\exp u) &\mapsto u + v
\end{aligned}
$$

which yields holomorphic coordinates on M. Using a lower bound c for the Jacobian of this coordinate map on M and an upper bound d for the measure of $V \subseteq i\mathfrak{g}$ we find

$$
\int_{\psi(M)} |f \circ \psi^{-1}(u+v)|^2 \, du\, dv \leq \frac{1}{c} \int_V \int_{\exp U} |f((\exp v)mg)|^2 \, d\mu_G(g) dv
$$

$$
\leq \frac{1}{c} \int_V \|f\|_H^2 \, dv \leq \frac{d}{c} \|f\|_H^2.
$$

On the other hand, using the compactness of M, we see that there exists an $\varepsilon > 0$ such that for any $z \in M$ the set $\psi(\exp V m \exp U)$ contains the ε-ball B_ε around $\psi(z)$ in $\mathfrak{g}_{\mathbb{C}}$. Now, according to [Hel78, p.364], there exist a constant $e > 0$ with

$$
|f(z)|^2 \leq e \int_{B_\varepsilon} |f \circ \psi^{-1}(\zeta)|^2 d\zeta
$$

$$
\leq e \int_{\psi(\exp V m \exp U)} |f \circ \psi^{-1}(\zeta)|^2 d\zeta.
$$

Taken together, the two estimates yield (9.12) with $c_M^2 = e\frac{d}{c}$.

Now we assume that $f_1, f_2, \ldots$ is a Cauchy sequence in $\mathcal{H}(W)$. Then (9.12) shows that the sequence $(f_j)_{j \in \mathbb{N}}$ converges uniformly on compact sets to a holomorphic function $f: \Gamma^0(W) \to \mathbb{C}$. Thus for any $\gamma \in \Gamma^0(W)$ the sequence $\{\gamma.f_j\}$ converges uniformly on compact sets to $\gamma.f$, so that

$$(9.13) \qquad \int_B \|\gamma.f(x)\|^2 \, d\mu_G(x) = \lim_{j \to \infty} \int_B \|\gamma.f_j(x)\|^2 \, d\mu_G(x) \leq \lim_{j \to \infty} \|f_j\|_H$$

for any compact subset B of G. This implies $f \in \mathcal{H}(W)$ and

$$\|f\|_H \leq \lim_{j \to \infty} \|f_j\|_H.$$

In order to prove the completeness of $\mathcal{H}(W)$, it only remains to show that the f_j converge to f in $\mathcal{H}(W)$. Therefore (9.10) makes sure that it suffices to show $I_X f_j \to I_X f$ in $L^2(G)$ for some $X \in i \operatorname{int} W$. But note that (9.13) together with the dominated convergence theorem proves that $\gamma.f_j \to \gamma.f$ in $L^2(G)$ for any $\gamma \in \Gamma^0(W)$. In particular (9.8) shows

$$e^{tA_-}.I_X f_j = (\exp tX).f_j \to (\exp tX).f = e^{tA_-}.I_X f$$

for any $t > 0$. Thus for any $h \in L^2(G)$ we have

$$(e^{tA_-} I_X f_j, h) = (I_X f_j, e^{tA_-} h),$$

and, setting $M := \sup_{j \in \mathbb{N}} \|f_j\|_H$, for every $\varepsilon > 0$ we can find a $t_\varepsilon > 0$ and j_ε such that

$$\|e^{tA_-} h - h\|_2 \leq \frac{\varepsilon}{4M} \qquad \forall t \in [0, t_\varepsilon]$$

and

$$\|e^{t_\varepsilon A_-}.I_X f_j - e^{t_\varepsilon A_-}.I_X f\|_2 \leq \frac{\varepsilon}{2\|h\|_2} \qquad \forall j \geq j_\varepsilon.$$

Hence, for $j \geq j_\varepsilon$,

$$\begin{aligned}
&|(I_X f_j - I_X f, h)_H| \\
&\leq |(I_X f_j - I_X f, e^{t_\varepsilon A_-} h)| + |(I_X f_j - I_X f, e^{t_\varepsilon A_-}.h - h)| \\
&\leq \|e^{t_\varepsilon A_-}.I_X f_j - e^{t_\varepsilon A_-}.I_X f\|_2 \|h\|_2 + \|I_X f_j - I_X f\|_2 \|e^{t_\varepsilon A_-}.h - h\|_2 \\
&\leq \frac{\varepsilon}{2\|h\|_2} \|h\|_2 + 2M \frac{\varepsilon}{4M} = \varepsilon.
\end{aligned}$$

Therefore $I_X f_j$ converges weakly to $I_X f$. On the other hand $I_X f_j$ is a Cauchy sequence in the complete space $L^2(G)$ so that it converges to a function $f' \in L^2(G)$. Since it converges weakly to $I_X f$, we conclude that $f' = I_X f$, hence $I_X f_j \to I_X f$ holds in the norm topology.

By now we have proved that $\mathcal{H}(W)$ is a Hilbert space with respect to the norm $\| \cdot \|_H$ and constructed the isometry $I: \mathcal{H}(W) \to L^2(G)$. Next we complete the proof of (ii). Therefore let $(\gamma_j)_{j \in \mathbb{N}}$ be a sequence in $\Gamma^0(W)$ converging to 1 in $G_{\mathbb{C}}$. The operators $B_j: \mathcal{H}(W) \to L^2(G)$ defined by $f \mapsto \gamma_j.f$ clearly satisfy $\|B_j\| \leq 1$. Moreover, the continuity of the map M_f shows that $B_j.f \to If$ in

$L^2(G)$ for all $f \in \widetilde{\mathcal{H}} := \{\widehat{\pi}(\gamma)h : h \in \mathcal{H}(W), \gamma \in \Gamma^0(W)\}$. In fact, $B_j(\widehat{\pi}(\gamma)h) = \gamma_j.(\widehat{\pi}(\gamma)h) = \gamma_j\gamma.h \to \gamma.h$ and $\gamma.h = I(\widehat{\pi}(\gamma)h)$ by (9.9). But (9.11) shows that $I(\widetilde{\mathcal{H}})$ is dense in $I(\mathcal{H}(W))$, so that $\widetilde{\mathcal{H}}$ is dense in $\mathcal{H}(W)$, and hence $B_j.f \to If$ for all $f \in \mathcal{H}(W)$.

To prove (iii), we let $X \in i \operatorname{int} W$ and compute

$$
\begin{aligned}
R(g)If &= R(g)I_X f \\
&= R(g)\lim_{t \to 0} \exp(tX).f \\
&= \lim_{t \to 0} R(g)\big(\exp(tX).f\big) \\
&= \lim_{t \to 0} R(g)\big(\exp(tX)\widehat{\pi}(g)^{-1}\widehat{\pi}(g).f\big) \\
&= \lim_{t \to 0} \exp\big(t \operatorname{Ad}(g)X\big).(\widehat{\pi}(g)f) \\
&= I_{\operatorname{Ad}(g)X}\big(\widehat{\pi}(g).f\big) = I\big(\widehat{\pi}(g).f\big)
\end{aligned}
$$

since $\operatorname{Ad}(g)X \in i \operatorname{int} W$.

(iv) It is clear from the definitions that $\widehat{\pi}$ is a semigroup representation with $\|\widehat{\pi}(\gamma)\| \le 1$ for $\gamma \in \Gamma(W)$. To show the holomorphy of $\widehat{\pi}$, we note that the Bergman condition (9.12) implies the continuity of the evaluation functional $f \mapsto f(z)$ on $\mathcal{H}(W)$. Therefore, given $z \in \Gamma^0(W)$, there exists a vector $f_z \in \mathcal{H}(W)$ such that $(f, f_z)_H = f(z)$ for all $f \in \mathcal{H}(W)$. Then clearly the set $\{f_z \in \mathcal{H}(W) : z \in \Gamma^0(W)\}$ is dense in $\mathcal{H}(W)$. If now $z \in \Gamma^0(W)$ and $f \in \mathcal{H}(W)$, then the function

$$
\gamma \mapsto \big(\widehat{\pi}(\gamma)f, f_z\big)_H = \big(\widehat{\pi}(\gamma)f\big)(z) = f(z\gamma)
$$

is holomorphic. But this proves the holomorphy of the function $\gamma \mapsto \widehat{\pi}(\gamma).f$ since the (strong) holomorphy is a consequence of the weak holomorphy on a dense subset (Lemma 9.7(i)). Thus it only remains to show that $\widehat{\pi}(\gamma^*) = \widehat{\pi}(\gamma)^*$. To do this, recall first that $g^* = g^{-1}$ and $\widehat{\pi}(g^{-1}) = \widehat{\pi}(g)^*$ by (iii). Moreover $\gamma \mapsto \gamma^*$ is antiholomorphic so that the above shows that both, $\gamma \mapsto \widehat{\pi}(\gamma^*)$ and $\gamma \mapsto \widehat{\pi}(\gamma)^*$ are antiholomorphic. But G is a set of uniqueness for $\Gamma(W)$ (cf. Lemma 9.17) so that $\widehat{\pi}(\gamma^*) = \widehat{\pi}(\gamma)^*$ for all $\gamma \in \Gamma(W)$. ∎

The following proposition gives an intrinsic characterization of the subspace $I(\mathcal{H}(W)) \subseteq L^2(G)$ (cf. [HiOl90, 10.2], [Ols82c]).

Proposition 9.32. *The subspace $I(\mathcal{H}(W))$ is the largest subspace $\mathcal{F} \subseteq L^2(G)$ such that all the self-adjoint operators $idR(X)$, $X \in W$ are negative on $\mathcal{F}$.*

Proof. For $X \in W$ let $\mathcal{F}_X \subseteq L^2(G)$ denote the negative subspace of the self-adjoint operator $idR(X)$, where R as before denotes the right regular representation. If P_X denotes a spectral measure of $idR(X)$, then $\mathcal{F}_X$ is the image of $P_X(] - \infty, 0])$. From $R(g)dR(X)R(g^{-1}) = dR(\operatorname{Ad}(g)X)$ it follows that $R(g)\mathcal{F}_X = \mathcal{F}_{\operatorname{Ad}(g)X}$. Hence

$$
\mathcal{F} := \bigcap_{X \in W} \mathcal{F}_X
$$

is an invariant subspace of $L^2(G)$. In view of Theorems 9.20 and 9.31, it is clear that $I(\mathcal{H}(W)) \subseteq \mathcal{F}$. So we have to show the converse inclusion.

Using the Extension Theorem (Theorem 9.19), we find a holomorphic contraction representation $\pi : \Gamma_G(W) \to C(\mathcal{F})$ such that $\pi(g) = R(g)|_{\mathcal{F}}$ holds for all $g \in G$. Since the set of smooth vectors in $\mathcal{F}$ is dense, it suffices to show that $\mathcal{F}^\infty \subseteq I(\mathcal{H}(W))$. Let $f \in \mathcal{F}^\infty$. According to Proposition 9.22, the function

$$\varphi : \text{int } \Gamma_G(W) \to \mathcal{F}^\infty, \quad s \mapsto \pi(s).f$$

is holomorphic with respect to the Fréchet topology on $\mathcal{F}^\infty \subseteq L^2(G)^\infty$. Now Proposition 3.27 in [Mag92] and its proof which consists of an application of a Sobolev Lemma for the Lie group setting shows that $L^2(G)^\infty \subseteq C^\infty(G)$ and that the point evalutations $\delta_g : f \mapsto f(g)$ are continuous complex linear functionals on $L^2(G)^\infty$. Hence the function

$$F : \Gamma_G(W) \to \mathbb{C}, \quad s \mapsto \big(\pi(s).f\big)(1)$$

is holomorphic on $\text{int } \Gamma(W)$ as a composition of two holomorphic functions.

We claim that $F \in \mathcal{H}(W)$ with $IF = f$. Let $s \in \text{int } \Gamma(W)$. Then

$$\int_G |F(gs)|^2 d\mu_G(g) = \int_G |(\pi(gs).f)(1)|^2 d\mu_G(g)$$

$$= \int_G |(\pi(s).f)(g)|^2 d\mu_G(g)$$

$$= \|\pi(s).f\|_2^2 \leq \|f\|_2^2.$$

Thus $F \in \mathcal{H}(W)$.

Now the continuity of the mapping $s \mapsto \pi(s).f = s.F$ in $L^2(G)$ yields that

$$IF = \lim_{s \to 1} s.F = \lim_{s \to 1} \pi(s).f = f.$$

∎

Note that so far we have no guarantee that the Hardy space $\mathcal{H}$ is non-trivial. After the next section we will see some examples where the Hardy spaces are non-trivial.

Cauchy-Szegö Kernels

As the proof of Theorem 9.31 showed, $\mathcal{H}(W)$ is always a *reproducing kernel Hilbert space*, i.e., for each $z \in \Gamma^0(W)$ there is a vector $f_z \in \mathcal{H}(W)$ with

(9.14) $$f(z) = (f, f_z)_H \qquad \forall f \in \mathcal{H}(W).$$

From (9.9) it follows for $\gamma \in \Gamma_G(W)$ that

$$(f, f_{z\gamma}) = f(z\gamma) = (\hat{\pi}(\gamma).f)(z)$$

$$= (\hat{\pi}(\gamma).f, f_z) = (f, \hat{\pi}(\gamma^*).f_z),$$

so that

(9.15) $$\hat{\pi}(\gamma^*).f_z = f_{z\gamma}.$$

For a fixed $z \in \Gamma^0(W)$ we consider a neighborhood of the form $z_0\Gamma^0(W)$ so that (9.15) together with the continuity of $\gamma \mapsto \gamma.f$ shows that the map

$$\Gamma^0(W) \to \mathcal{H}(W), \qquad z \mapsto f_z$$

is continuous.

We now consider the kernel function

(9.16) $$K(z,w) = f_w(z)$$

which is holomorphic in the z-variable, with z, w in $\operatorname{int}\Gamma(W)$. This function is called the *Cauchy-Szegö kernel*. We will see that it defines the projection onto the Hardy space, considered as a subspace of $L^2(G)$.

Note that each f_z is an analytic vector in $\mathcal{H}(W)$, relative to the representation $\hat{\pi}$ of G. Indeed,

$$(f, \hat{\pi}(g)f_z)_H = (\hat{\pi}(g^{-1})f, f_z)_H = f(zg^{-1})$$

is analytic in g for each $f \in \mathcal{H}(W)$. From this it follows that If_z is also an analytic vector, so If_z is an analytic function on G ([Mag92, 3.27]).

Theorem 9.33.   *Fix a pointed generating cone $W \subseteq \mathfrak{g}$ and consider the corresponding Hardy space $\mathcal{H} = \mathcal{H}(W)$ on the domain $\Gamma_G^0(W) := \operatorname{int}\Gamma_G(W)$. Then there exists a holomorphic function K on $\Gamma_G^0(W)$ with the following properties:*

  (i) *$K(z,w) = K(zw^*)$ for all $z, w \in \Gamma_G^0(W)$.*
 (ii) *$K = \lim_{z \to 1} f_z$ uniformly on compact subsets of $\Gamma_G^0(W)$.*
(iii) *$K(z^*) = \overline{K(z)}$ for all $z \in \Gamma_G^0(W)$.*
 (iv) *$(If_z)(g) = K(gz^*)$ for all $z \in \Gamma_G^0(W), g \in G$.*
  (v) *The mapping $P: L^2(G) \to \mathcal{H}(W)$ defined by*

$$(P\varphi)(z) = (\varphi, If_z)$$

  satisfies $PI = \operatorname{id}_{\mathcal{H}(W)}$ and IP is the projection onto $I\bigl(\mathcal{H}(W)\bigr)$.
 (vi) *$K(gzg^{-1}) = K(z)$ for all $z \in \Gamma^0(W), g \in G$.*

Proof.   (i),(ii) We go back to the proof of the reproducing kernel property of the Hardy space, specifically formula (9.12). Here we have an estimate by the Hardy norm of the pointwise values of the functions in $\mathcal{H}(W)$.

To profit from this estimate we need some results from Section 3.4 which are trivial if the semigroup $\Gamma_G(W)$ sits in a complex group. Let $A \subseteq \Gamma^0(W)$ be a compact subset and K a compact neighborhood of A. Then, according to Theorem 3.20, there exists a 1-neighborhood U in $\Gamma_G(W)$ such that

$$A \subseteq \Gamma_H(W)\gamma = \rho_\gamma(\Gamma_G(W)) \quad \text{and} \quad \rho_\gamma^{-1}(A) \subseteq K$$

holds for each $\gamma \in U$. Hence we find for every $w \in A$ and $\gamma \in U$ an element $v_\gamma \in K$ such that $w = v_\gamma \gamma$. Then

$$|f(w)| = |f(v_\gamma \gamma)| = |(\hat{\pi}(\gamma).f)(v_\gamma)| \leq c_K \|\hat{\pi}(\gamma).f\|_H$$

for all $f \in \mathcal{H}(W)$.

Using (9.16), we conclude that

$$|f_z(w) - f_{z'}(w)| \leq c_K \|f_{z\gamma^*} - f_{z'\gamma^*}\|_H \qquad \forall w \in A, \gamma \in U.$$

But this formula allows us, since $z \mapsto f_z, \Gamma^0(W) \to \mathcal{H}(W)$ is continuous, to assert the existence of the limit

$$K(w) := f_1(w) = \lim_{z \to 1} f_z(w)$$

as a holomorphic function on $\Gamma^0(W)$ since the limit exists in the sense of uniform convergence on compact subsets of $\Gamma^0(W)$.

Using

(9.17) $$K(z, w) = (f_w, f_z)_H = \overline{(f_z, f_w)_H} = \overline{K(w, z)},$$

and (9.15) we now find for $w, z \in \Gamma^0(W)$ that

$$\begin{aligned} K(wz^*) = f_1(wz^*) &= \lim_{z' \to 1} f_{z'}(wz^*) \\ &= \lim_{z' \to 1} K(wz^*, z') = \lim_{z' \to 1} \overline{K(z', wz^*)} \\ &= \lim_{z' \to 1} \overline{f_{wz^*}(z')} = \lim_{z' \to 1} \overline{f_w(z'z)} \\ &= \overline{f_w(z)} = \overline{K(z, w)} = K(w, z) \end{aligned}$$

This proves (i) and (ii).

(iii) Let $x \in \Gamma^0(W)$. Then there exist $z, w \in \Gamma^0(W)$ such that $x = zw^*$ (cf. Theorem 3.20). Now the assertion follows from

$$K(x^*) = K(wz^*) = K(w, z) = \overline{K(z, w)} = \overline{K(zw^*)} = \overline{K(x)}.$$

(iv) Recall that If_z is given as an L^2-limit of

$$(\gamma.f_z)(g) = f_z(g\gamma) = K(g\gamma, z) = K(g\gamma z^*).$$

Locally we have

$$\lim_{\gamma \to 1} \gamma.f_z(g) = K(gz^*),$$

where the limit exists in the sense of uniform convergence on compact sets, so that $If_z(g) = K(gz^*)$ holds everywhere since If_z is an analytic function.

(v) First we note that $K(z^*) = \overline{K(z)}$ and (iv) show that

$$\int_G \varphi(g)\overline{K(gz^*)}d\mu_G(g) = (\varphi, If_z).$$

So $P(\varphi) = 0$ for $\varphi \in I(\mathcal{H}(W))^{\perp}$ and

$$P(If)(z) = (If, If_z) = (f, f_z)_H = f(z).$$

This proves that $PI = \mathrm{id}_{\mathcal{H}(W)}$ and that IP is the orthogonal projection onto $I(\mathcal{H}(W))$ since $IPI = I$ and $PI(\mathcal{H}(W)^{\perp}) = \{0\}$.

(vi) For $\gamma \in \Gamma_G^0(W)$ let

$$\pi(\gamma) := I \circ \hat{\pi}(\gamma) \circ P$$

denote the corresponding operator on $L^2(G)$. Calculating $\hat{\pi}(\gamma)(P\varphi)(z)$ for $z \in \Gamma_G^0(W)$ and $\varphi \in C_c(G)$, and letting then z tend to $x \in G$, we find that

$$\big(\pi(\gamma)\varphi\big)(x) = \lim_{z \to x} (\varphi, If_{z\gamma})$$

$$= \lim_{z \to x} \int_G \varphi(g)\overline{K(g\gamma^*z^*)}\, d\mu_G(g)$$

$$= \lim_{z \to x} \int_G \varphi(g)K(z\gamma g^{-1})\, d\mu_G(g)$$

$$= \int_G \varphi(g)K(x\gamma g^{-1})\, d\mu_G(g)$$

(cf. [Rud86, 3.12]). Note that the space $I(\mathcal{H}(W)) \subseteq L^2(G)$ is also invariant under the left regular representation because all the operators $idR(X)$, $X \in \mathfrak{g}$ commute with the left regular representation (cf. Proposition 9.32). Hence P commutes with the left regular representation and also the operators $\pi(\gamma)$.

Pick $g \in G$. Let $\varphi \in C_c(G)$. Then the above observation leads to

$$\big(\pi(\gamma)\varphi\big)(g^{-1}x) = \int_G \varphi(y)K(g^{-1}x\gamma y^{-1})d\mu_G(y)$$

$$\big(\pi(\gamma)L(g)\varphi\big)(x) = \int_G \varphi(g^{-1}y)K(x\gamma y^{-1})d\mu_G(y)$$

$$= \int_G \varphi(y)K(x\gamma y^{-1}g^{-1})d\mu_G(y).$$

Since φ was arbitrary, we conclude that $K(g^{-1}x\gamma y^{-1}) = K(x\gamma y^{-1}g^{-1})$ holds for all $x, y \in G$. For $x = y = 1$ we obtain

$$K(\gamma g^{-1}) = K(g^{-1}\gamma)$$

which proves (vi).   ∎

The invariance of K under conjugation with elements of G has an interesting consequence for the operators $\pi(\gamma)$ representing $\Gamma_G(W)$ on the subspace $I(\mathcal{H}(W))$ of $L^2(G)$. For $\varphi \in C_c(G)$ we have that

$$\big(\pi(\gamma)\varphi\big)(x) = \int_G \varphi(y)K(x\gamma y^{-1})d\mu_G(y)$$

$$= \int_G \varphi(y)K(y^{-1}x\gamma)d\mu_G(y)$$

$$= \int_G \varphi(y)(\gamma.K)(y^{-1}x)d\mu_G(y)$$

$$= (\varphi * (\gamma.K))(x).$$

So the representation of $\operatorname{int}\Gamma_G(W)$ on $I(\mathcal{H}(W))$ can be described by the convolution operators

$$R_\gamma(\varphi) := \varphi * (\gamma.K) = \varphi * (If_{\gamma^*}).$$

Examples: Cones in euclidean space

We start with the abelian case, i.e., $G = \mathbb{R}^n$ and $W \subseteq \mathbb{R}^n$ is a pointed generating cone. For the proofs of the following results we refer to Section IX.4 in [FK92]. We recall from Section I.3 the definition of the characteristic funtion

$$\varphi_W(x) := \int_{W^*} e^{-\langle \omega, x \rangle} d\mu(\omega).$$

Then the function K on $\Gamma^0(W) = \mathbb{R}^n + i\operatorname{int} W$ is given by

$$K(z) = \frac{1}{(2\pi)^n}\varphi_W(-iz).$$

So

$$K(z) = \frac{1}{(2\pi)^n} \int_{W^*} e^{\langle \omega, iz \rangle} d\mu(\omega) = \int_{W^*} e^{2\pi\langle \omega, iz \rangle} d\mu(\omega) = \int_{-W^*} e^{-2\pi i\langle \omega, z \rangle} d\mu(\omega).$$

This means that, in the sense of tempered distributions, K is the Fourier transform of the characteristic function of $-W^*$.

For the coherent states we find with (9.15) the formula

$$f_z(w) = ((-\bar{z}).f_0)(w) = K(w - \bar{z}) = \int_{W^*} e^{2\pi i\langle \omega, (w - \bar{z}) \rangle} d\mu(\omega).$$

For the mapping $P : L^2(\mathbb{R}^n) \to \mathcal{H}(W)$ we have the explicit formula

$$\begin{aligned}
(P\varphi)(z) &= (\varphi, If_z) = (\varphi, -\bar{z}.K) \\
&= \int_{\mathbb{R}^n} \int_{W^*} \varphi(x)\overline{e^{2\pi i\langle \omega, x - \bar{z} \rangle}} d\mu(\omega) d\mu(x) \\
&= \int_{W^*} \int_{\mathbb{R}^n} \varphi(x)e^{-2\pi i\langle \omega, x - z \rangle} d\mu(x) d\mu(\omega) \\
&= \int_{W^*} \mathcal{F}\varphi(\omega)e^{2\pi i\langle \omega, z \rangle} d\mu(\omega) \\
&= \mathcal{F}^{-1}(\chi_{W^*}\mathcal{F}\varphi),
\end{aligned}$$

where $\mathcal{F}f(\omega) = \int_{\mathbb{R}^n} \varphi(x)e^{-2\pi i\langle \omega, x \rangle} d\mu(x)$ denotes the Fourier transform of f and χ_{W^*} denotes the characteristic function of W^*. Thus the Fourier transform intertwines the projection onto $I(\mathcal{H}(W))$ with the multiplication with the characteristic function of W^*.

For a generalization of these results to the solvable case we refer to [HiO190].

Examples: The polydisc

Let $D := \{z = (z_1, \ldots, z_n) \in \mathbb{C}^n : |z_k| \leq 1\}$ and G the torus ∂D. Then $D \setminus \{0\}$ can be viewed as the Olshanskiĭ semigroup in $G_{\mathbb{C}} := (\mathbb{C}^*)^n$ with unit group G corresponding to the cone $W := i(\mathbb{R}^+)^n \subseteq i\mathbb{R}^n = \mathbf{L}(G)$. The Hardy space $\mathcal{H}(W)$ consists of holomorphic functions $f : D \setminus \{0\} \to \mathbb{C}$ such that the integrals

$$\int_G |f(g.z)|^2 \, d\mu_G(g) = \int_{[0,1]^n} |f(e^{2\pi i t_1} z_1, \ldots, e^{2\pi i t_n} z_n)|^2 \, dt_1 \ldots dt_n$$

are bounded by a constant that does not depend on $z = (z_1, \ldots, z_n)$. This implies in particular that the singularity at 0 can be removed so that $\mathcal{H}(W)$ actually is the classical Hardy space for the polydisc.

The reproducing kernel on this space is provided by the Poisson kernel

$$P(z, e^{2\pi i t}) = \frac{1}{1 - e^{-2\pi i t} z}$$

(cf. [Ru86, Ch. 17]). More precisely, the functions f_z are given by

$$f_z(w) = \frac{1}{1 - w\overline{z}}$$

and since in our example the involution on $G \exp(iW)$ is given by $w \mapsto w^* = \overline{w}$, the function K is given by

$$K(z) = \frac{1}{1 - z}.$$

Examples: The holomorphic discrete series

Let G be a connected simple Hermitean Lie group which is contained in a complex Lie group $G_{\mathbb{C}}$ with $\mathbf{L}(G_{\mathbb{C}}) = \mathfrak{g}_{\mathbb{C}}$ and $W \subseteq \mathfrak{g}$ a pointed generating invariant cone. In this case the Hardy space $\mathcal{H}(W)$ has been studied by Ol'shanskiĭ in [Ols82c]. We state the main result without proof.

Theorem 9.34. (Ol'shanskiĭ) *The Hardy space $\mathcal{H}(W)$ is always non-trivial and the subspace $I(\mathcal{H}(W)) \subseteq L^2(G)$ is a discrete direct sum of those holomorphic discrete series representations in $L^2(G)$ whose highest weights $\lambda \in \mathfrak{t}_{\mathbb{C}}^*$ satisfy $i\lambda \in (W \cap \mathfrak{t})^*$.*

Proof. This is Theorem 5.1 in [Ols82c]. ∎

In addition to this result, Ol'shanskiĭ's paper includes the following statements: Let π_λ denote the holomorphic contraction representation of $\Gamma_G(W)$ corresponding to a unitary highest weight representation of G with highest weight $\lambda \in -i(W \cap \mathfrak{t})^*$. Then, for every $s \in \operatorname{int} \Gamma_G(W)$ the operator $\pi_\lambda(s)$ is a *trace class operator* and the function K can be written as

$$K(s) = \sum_{\lambda \in -i(W \cap \mathfrak{t})^*} d_\lambda \operatorname{tr}(\pi_\lambda(s)),$$

where d_λ is the *formal dimension* of the representation π_λ. The series for K converges uniformly on compact subsets of $\operatorname{int} \Gamma_G(W)$.

9.6. Howe's oscillator semigroup

In this section we briefly describe a semigroup of contraction operators which can also be viewed as an analytic continuation of a unitary representation. The representation in question is the so-called metaplectic representation which is the direct sum of two unitary highest weight modules for the double cover of the symplectic group. The semigroup will be realized by integral operators with Gaussian functions as kernels.

We call a function on $\mathbb{R}^n$ a function of Gaussian type if it is of the form $\xi \mapsto e^{\pi i \xi^\top X \xi}$, where X is a symmetric complex matrix. We call it a *Gaussian function* if it is square integrable or, equivalently, if the imaginary part of X is positive definite. The set of such X is called the *Siegel upper halfplane* and denoted by S_n. Similarly we call a function on $\mathbb{C}^n$ of *Gaussian type* if it is of the form $\zeta \mapsto e^{\frac{\pi}{2}\zeta^\top X \zeta}$, where X is a complex symmetric matrix. We will call a function of Gaussian type on $\mathbb{C}^n$ a *Gaussian function* if it belongs to the *Bargmann-Fock Hilbert space* $\mathcal{F}_n$ of entire functions on $\mathbb{C}^n$ with the L^2-norm given by the measure $d\mu(\zeta) = e^{-\pi|\zeta|^2}d\zeta$. Using an explicit isomorphism $L^2(\mathbb{R}^n) \to \mathcal{F}_n$ one can show that $\zeta \mapsto e^{\frac{\pi}{2}\zeta^\top X \zeta}$ is a Gaussian function for $X \in \Omega_n$, the *Siegel domain* consisting of complex symmetric $n \times n$-matrices with $1 - X^*X$ positive definite (cf. [Fo89, 4.5]). A function $F: \mathbb{R}^{2n} \to \mathbb{C}$ is called a *real Gauss kernel* if it is of the form $F(v) = ce^{\pi i v^\top X v}$ for some $X \in S_{2n}$ and $c \in \mathbb{C}^*$. The set of real Gauss kernels will be denoted by $GK_{\mathbb{R}}$. Similarly, a function $F: \mathbb{C}^n \times \overline{\mathbb{C}}^n \to \mathbb{C}$ is called a *complex Gauss kernel* if it is of the form $F(v) = ce^{\frac{\pi}{2}v^\top X v}$ for some $X \in \Omega_{2n}$ and $c \in \mathbb{C}^*$ with $v^\top = (\zeta^\top, \overline{\omega}^\top)$. The set of complex Gauss kernels will be denoted by $GK_{\mathbb{C}}$.

We associate integral operators with real and complex Gaussian kernels as follows: For

$$X = \begin{pmatrix} A & B \\ B^\top & D \end{pmatrix} \in S_{2n}$$

we set

$$K_X(\xi, \eta) = e^{\pi i (\xi^\top A \xi + 2\xi^\top B \eta + \eta^\top D \eta)} = e^{\pi i v^\top X v},$$

where $v^\top = (\xi^\top, \eta^\top)$. The corresponding kernel operator is

$$T_X f(\xi) = \int_{\mathbb{R}^n} K_X(\xi, \eta) f(\eta) \, d\mu(\eta).$$

Similarly, for

$$X = \begin{pmatrix} A & B \\ B^\top & D \end{pmatrix} \in \Omega_{2n}$$

we set

$$K_X(\zeta, \omega) = e^{\frac{\pi}{2}(\zeta^\top A \zeta + 2\zeta^\top B \overline{\omega} + \overline{\omega}^\top D \overline{\omega})} = e^{\frac{\pi}{2}v^\top X v},$$

where $v^\top = (\zeta^\top, \overline{\omega}^\top)$. The corresponding kernel operator is

$$T_X f(\zeta) = \int_{\mathbb{C}^n} K_X(\zeta, \omega) f(\omega) d\mu(\omega).$$

The key observation is now that the integral operators associated to Gaussian kernels form a semigroup under composition.

The proof of Proposition 9.36 below consists of a straight forward calculation based on the following lemma:

Lemma 9.35.

(i) *Let* $A \in \mathbb{M}(n,\mathbb{C})$ *be symmetric and* $\operatorname{Re} A$ *positive definite. Then*

$$\int_{\mathbb{R}^n} e^{-\pi z^\top A z - 2\pi i z^\top x} dx = \det(A)^{-\frac{1}{2}} e^{-\pi z^\top A^{-1} z} \qquad \forall z \in \mathbb{C}^n,$$

where the branch of the square root is determined by the requirement that

$$\det(A)^{-\frac{1}{2}} > 0$$

when A *is real and positive definite.*

(ii) *Let* A *and* D *be symmetric in* $M(n,\mathbb{C})$ *such that*

$$\|A\| \le 1, \quad \|D\| \le 1, \quad and \quad \|A\| \|D\| < 1.$$

Then for any $z, w \in \mathbb{C}^n$

$$\int_{\mathbb{C}^n} e^{\frac{\pi}{2}(w^\top A w + \overline{w}^\top D \overline{w} + 2u^\top w + 2v^\top \overline{w})} e^{-\pi \|w\|^2} dw$$

$$= \det(1 - AD)^{-\frac{1}{2}} e^{\frac{\pi}{2}(u^\top D(1-AD)^{-1}u + 2v^\top(1-AD)^{-1}u + v^\top A(1-DA)^{-1}v)}.$$

Proof. (i) [Fo89, p.256], (ii) [Fo89, p.258]. ∎

Proposition 9.36.

(i) *Let* $X_1, X_2 \in S_{2n}$ *and*

$$X_1 = \begin{pmatrix} A_1 & B_1 \\ B_1^\top & D_1 \end{pmatrix}, \quad X_2 = \begin{pmatrix} A_2 & B_2 \\ B_2^\top & D_2 \end{pmatrix},$$

Then we have

$$T_{X_1} \circ T_{X_2} = \frac{1}{\det\left(-i(D_1 + A_2)\right)^{\frac{1}{2}}} T_Y,$$

where

$$Y = \begin{pmatrix} A_1 - B_1(D_1 + A_2)^{-1} B_1^\top & -B_1(D_1 + A_2)^{-1} B_2 \\ -B_2^\top(D_1 + A_2)^{-1} B_1^\top & D_2 - B_2^\top(D_1 + A_2)^{-1} B_2 \end{pmatrix}.$$

(ii) *Let* $X_1, X_2 \in \Omega_{2n}$ *and*

$$X_1 = \begin{pmatrix} A_1 & B_1 \\ B_1^\top & D_1 \end{pmatrix}, X_2 = \begin{pmatrix} A_2 & B_2 \\ B_2^\top & D_2 \end{pmatrix},$$

Then we have

$$T_{X_1} \circ T_{X_2} = \frac{1}{\left(\det(1 - A_2 D_1)\right)^{\frac{1}{2}}} T_Y,$$

where

$$Y = \begin{pmatrix} A_1 + (B_1 A_2(1 - D_1 A_2)^{-1} B_1^\top) & B_1(1 - A_2 D_1)^{-1} B_2 \\ -B_2^\top(1 - D_1 A_2)^{-1} B_1^\top & D_2 + B_2^\top D_1(1 - A_2 D_1)^{-1} B_2 \end{pmatrix}.$$

Proof. [Fo89, 5.6] and [Fo89, 5.50]. ∎

One can, again using the isomorphism $L^2(\mathbb{R}^n) \to \mathcal{F}_n$, show that the semi-groups $(GK_\mathbb{R}, \circ)$ and $(GK_\mathbb{C}, \circ)$ are isomorphic (cf. [Fo89, 5.51]).

The semigroup of contraction operators we are looking for is a subsemigroup of $(GK_\mathbb{R}, \circ)$. In order to define it, we consider the following dense open subsets of S_{2n} and Ω_{2n}:

$$\mathcal{D}_\mathbb{R} := \left\{ X = \begin{pmatrix} A & B \\ B^\top & D \end{pmatrix} \in S_{2n} : \det B \neq 0 \right\},$$

$$\mathcal{D}_\mathbb{C} := \left\{ X = \begin{pmatrix} A & B \\ B^\top & D \end{pmatrix} \in \Omega_{2n} : \det B \neq 0 \right\}.$$

Proposition 9.37.

(i) *The set*
$$S_\mathbb{R} := \left\{ cK_X \in GK_\mathbb{R} : X \in \mathcal{D}_\mathbb{R}, c^2 = \det(iB) \right\}$$
is a subsemigroup of $GK_\mathbb{R}$.

(ii) *The set*
$$S_\mathbb{C} := \left\{ cK_X \in GK_\mathbb{C} : X \in \mathcal{D}_\mathbb{C}, c^2 = \det(B) \right\}$$
is a subsemigroup of $GK_\mathbb{C}$.

Proof. This follows from the formulas in Proposition 9.36 (cf. [Fo89, p.245] and [Fo89, 5.51]). ∎

Again one can show that $S_\mathbb{R}$ and $S_\mathbb{C}$ are isomorphic (cf. [Fo89, 5.51]). In order to see that this semigroup is an analytic continuation of a representation for the metaplectic group we note the following lemma (cf. [Hi89, 5.1]).

Lemma 9.38. *Consider the hermitean form on $\mathbb{C}^{2n}$ given by the matrix*

$$L = \begin{pmatrix} -1_n & 0 \\ 0 & 1_n \end{pmatrix}$$

and $S_L \subseteq \mathrm{Sp}(n, \mathbb{C})$ the subsemigroup of contractions of this form. Then

$$\varphi \begin{pmatrix} A & B \\ B^\top & D \end{pmatrix} = \begin{pmatrix} (B^\top)^{-1} & -(B^\top)^{-1} D \\ A(B^\top)^{-1} & B - A(B^\top)^{-1} D \end{pmatrix}$$

defines a diffeomorphism $\varphi : \mathcal{D}_\mathbb{C} \to \mathrm{int}\, S_L$. Its inverse is given by

$$\psi \begin{pmatrix} A & B \\ C & D \end{pmatrix} = \begin{pmatrix} CA^{-1} & (A^\top)^{-1} \\ A^{-1} & -A^{-1} B \end{pmatrix}.$$

∎

Now one can show that the map

$$S_\mathbb{C} \to \mathrm{int}\, S_L$$
$$cK_X \mapsto \varphi(X)$$

is a double covering and a semigroup homomorphism (cf. [BK80,(3.8)], which, how-ever uses a different normalisation). But S_L is the, up to sign unique, Olshanskiĭ semigroup with unit group $U(n,n) \cap Sp(n,\mathbb{C})$ in $Sp(n,\mathbb{C})$ and $U(n,n) \cap Sp(n,\mathbb{C})$ is canonically isomorphic to $Sp(n,\mathbb{R})$ under a Cayley transform. Thus the homomorphism property of $cK_X \mapsto \varphi(X)$ follows also by analytic continuation once one has seen that the map $X \mapsto K_{\psi(X)}$ gives rise to the metaplectic representation. To this end, recall from [Fo89, 4.31] that the metaplectic representation of the double cover $Sp_c(n)\widetilde{}$ of the group

$$Sp(n)_c := U(n,n) \cap Sp(n,\mathbb{C}) \cong Sp(n,\mathbb{R})$$

on Fock space is given by the formula

$$\left(\pi\begin{pmatrix} A & B \\ B & A \end{pmatrix} f\right)(\zeta) = \frac{1}{\sqrt{\det A}} \int_{\mathbb{C}^n} e^{\frac{\pi}{2}(\zeta^\top \overline{B} A^{-1}\zeta + 2\overline{\omega}^\top A^{-1}\zeta - \overline{\omega} A^{-1} B\overline{\omega})} f(\omega) \, d\mu(\omega)$$

$$= \frac{1}{\sqrt{\det A}} \int_{\mathbb{C}^n} K_X(\zeta,\omega) f(\omega) \, d\mu(\omega)$$

with

$$X = \begin{pmatrix} \overline{B}A^{-1} & (A^\top)^{-1} \\ A^{-1} & -A^{-1}B \end{pmatrix} = \psi\left(\begin{pmatrix} A & B \\ B & A \end{pmatrix}\right).$$

The analytic extension to S_L is given by

$$\left(\pi\begin{pmatrix} A & B \\ C & D \end{pmatrix} f\right)(\zeta) = \frac{1}{\sqrt{\det A}} \int_{\mathbb{C}^n} e^{\frac{\pi}{2}(\zeta^\top C A^{-1}\zeta + 2\overline{\omega}^\top A^{-1}\zeta - \overline{\omega} A^{-1} B\overline{\omega})} f(\omega) \, d\mu(\omega).$$

9.7. The Lüscher-Mack Theorem

Let (G,τ) be a symmetric Lie group (cf. Section 7.3) and $H = G_0^\tau$ the identity component of the fixed point set for τ. We set

$$\mathfrak{h} := \{X \in \mathfrak{g} : d\tau(1)(X) = X\} = \mathbf{L}(H)$$

and

$$\mathfrak{q} := \{X \in \mathfrak{g} : d\tau(1)(X) = -X\}.$$

We write $\mathfrak{g}^c := \mathfrak{h} + i\mathfrak{q} \subseteq \mathfrak{g}_\mathbb{C}$ for the dual symmetric Lie algebra. Further let and $W \subseteq \mathfrak{q}$ be an $\mathrm{Ad}(H)$-invariant pointed generating regular cone. Then, according to Lawson's Theorem (Theorem 7.34), the set

$$\Gamma(W) := H \exp(W)$$

is a closed subsemigroup of G.

We note that the main assertion of the following theorem is the result that the representation of H and of the infinitesimal generators of the one-parameter semigroups generated by the elements in W fit nicely together, so that one obtains a unitary representation of G^c. This is a remarkable fact because it is not alwas true that a representation of a Lie algebra $\mathfrak{g}^c$ by essentially skew-adjoint operators on a dense subspace of a Hilbert space integrates to a representation of the corresponding simply connected group (cf. [Wa72, p.296]).

Theorem 9.39. (Lüscher and Mack) . *Let* $\rho: \Gamma(W) \to C(\mathcal{H})$ *be a strongly continuous involutive and contractive representation and* G^c *the simply connected Lie group with* $\mathbf{L}(G^c) = \mathfrak{g}^c$. *Then there exists a continuous unitary representation* $\pi: G^c \to U(\mathcal{H})$ *such that*

$$d\pi(X) = d\rho(X) \quad \forall X \in \mathfrak{h}.$$

For $Y \in W$ *we write* $d\rho(Y)$ *for the self-adjoint generator of the semigroup* $t \mapsto \rho(\exp tY)$ *of contractions. Then*

$$d\pi(iX) = id\rho(X) \quad \forall X \in W.$$

Proof. Step 1: We start with the observation that the involutivity of the representation ρ yields that $\alpha := \rho \,|_H$ is a continuous unitary representation of H. As we have already seen in Section 9.2, for the purpose of analytic continuation, the Equianalyticity-Lemma (Lemma 9.10) is a very useful tool. Here it asserts the existence of a dense subspace $\mathcal{H}_1$ of $\rho \,|_H$-analytic vectors in $\mathcal{H}$, and of a 0-neighborhood $U \subseteq \mathfrak{h}$ such that the power series $e^{d\alpha(X)}.v := \sum_{m=0}^{\infty} \frac{1}{m!} d\alpha(X)^m v$ converges for every $v \in \mathcal{H}_1$ and $X \in U_{\mathbb{C}} := U + iU \subseteq \mathfrak{h}_{\mathbb{C}}$. Moreover we have

$$\rho(\exp X)v = \alpha(\exp X)v = \sum_{m=0}^{\infty} \frac{1}{m!} d\alpha(X)^m v$$

for all $X \in U$, $v \in \mathcal{H}_1$.

Step 2: Let $X \in W$. Then

$$\gamma_X : \mathbb{R}^+ \to C(\mathcal{H}), \quad z \mapsto \rho\big(\exp(zX)\big)$$

is a strongly continuous one-parameter semigroup of self-adjoint contractions. Hence Theorem 9.13 yields the existence of a strongly continuous extension

$$\gamma_X : \mathbb{C}_+ := \{z \in \mathbb{C} : \operatorname{Re} z \geq 0\} \to C(\mathcal{H})$$

which is holomorphic on the open right half plane $\operatorname{int}\mathbb{C}_+$.

These two steps describe the information we have, and it remains to combine these pieces to a obtain a representation of G^c.

Step 3: Let $\mathcal{B} := \{X_1, \ldots, X_k\}$ be a basis of $\mathfrak{q}$ contained in W and $v \in \mathcal{H}_1$. Then the mapping

$$\beta_v' : (\mathbb{C}_+)^k \times U_{\mathbb{C}} \to \mathcal{H},$$

$$(z_1, \ldots, z_k, Y) \mapsto \gamma_{X_1}(z_1) \cdot \ldots \cdot \gamma_{X_k}(2z_k)\gamma_{X_{k-1}}(z_{k-1}) \cdot \ldots \cdot \gamma_{X_1}(z_1) e^{d\alpha(Y)}.v$$

is continuous and holomorphic on $(\operatorname{int}\mathbb{C}_+)^k \times U_{\mathbb{C}}$ by Steps 1 and 2 because it is holomorphic in each argument separately (Hartog's Theorem, [RK82, p.65]). Note that

$$\beta_v'(t_1, \ldots, t_n, Y)$$
$$= \rho\big(\exp(t_1 X_1) \exp(t_2 X_2) \cdot \ldots \cdot \exp(2t_k X_k) \cdot \ldots \cdot \exp(t_1 X_1) \exp(Y)\big).v$$

holds on $(\mathbb{R}^+)^n \times U$.

Let $V \subseteq \mathfrak{g}_\mathbb{C}$ denote a circled Baker-Campbell-Hausdorff neighborhood such that $(2k + 2)$-fold products are defined. We pick $\varepsilon > 0$ and set

$$U_\mathfrak{q} := \{z_1 X_1 * \ldots * z_k X_k * z_k X_k * \ldots * z_1 X_1 : |z_i| < \varepsilon\}$$

such that the mapping

$$\varphi : (\mathbb{C} X_1 \times \mathbb{C} X_2 \times \ldots \times \mathbb{C} X_k \times U_\mathbb{C}) \cap V^{k+1} \to \mathfrak{g}_\mathbb{C},$$
$$(z_1 X_1, \ldots, z_k X_k, Y) \mapsto z_1 X_1 * \ldots * 2 z_k X_k * \ldots * z_1 X_1 * Y$$

has a holomorphic local inverse on $U_0 := U_\mathfrak{q} * U_\mathfrak{h}$, where $U_\mathfrak{h} \subseteq U_\mathbb{C} \subseteq \mathfrak{h}_\mathbb{C}$ is sufficiently small (the existence of such an ε follows from the holomorphic version of the inverse function theorem). We set $U_1 := U_0 \cap -d\tau(\mathbf{1}) U_0$.

It follows that the mappings

$$\beta_v : U_1 \to \mathcal{H}, \quad (z_1 X_1 * \ldots * 2 z_k X_k * \ldots * z_1 X_1 * Y) \mapsto \beta_v'(z_1, \ldots, z_k, Y)$$

are holomorphic. We set

$$U_{1,+} := \{t_1 X_1 * \ldots 2 t_k X_k * \ldots * t_1 X_1 * Y : t_i > 0, Y \in U_1 \cap \mathfrak{h}\} \cap U_1.$$

Then $Z = t_1 X_1 * \ldots * 2 t_k X_k * \ldots * t_1 X_1 * Y \in U_{1,+}$ implies that

$$\begin{aligned}
&\beta_v(Z) \\
=&\beta_v(t_1 X_1 * \ldots 2 t_k X_k * \ldots * t_1 X_1 * Y) \\
=&\rho\big(\exp(t_1 X_1)\big) \cdot \ldots \cdot \rho\big(\exp(2 t_k X_k)\big) \cdot \ldots \cdot \rho\big(\exp(t_1 X_1)\big) \rho\big(\exp(Y)\big).v \\
=&\rho\big(\exp(t_1 X_1) \exp(t_2 X_2) \cdot \ldots \cdot \exp(2 t_k X_k) \cdot \ldots \cdot \exp(t_1 X_1) \exp(Y)\big).v \\
=&\rho\big(\exp(t_1 X_1 * t_2 X_2 * \ldots * 2 t_k X_k * \ldots * t_1 X_1 * Y)\big).v \\
=&\rho\big(\exp(Z)\big).v.
\end{aligned}$$

The main problem is that the open set $U_{1,+}$ is not a neighborhood of 0. So we have to modify $\mathcal{H}_1$ in such a way that we can obtain mappings which are holomorphic on a 0-neighborhood in $\mathfrak{g}_\mathbb{C}$.

Step 4: Pick $X_0 \in \operatorname{int} W$. The key idea is to use the fact that the operators in $\rho(\exp \operatorname{int} W)$ are mollifying. So we set

$$\mathcal{H}_2 := \rho(\exp X_0) \mathcal{H}_1.$$

We claim that $\mathcal{H}_2$ is a dense subspace of $\mathcal{H}$. In fact, it is clear that it is a linear subspace and for any $v' \in \mathcal{H}_2^\perp$ we have

$$0 = \big(v', \rho(\exp X_0) v\big) = \big(\rho(\exp X_0) v', v\big) \quad \forall v \in \mathcal{H}_1.$$

Since $\mathcal{H}_1$ is dense, this shows $\rho(\exp X_0) v' \in \mathcal{H}_1^\perp = \{0\}$. But then the same is true for X_0 replaced by $t X_0$ with $t > 1$ because of the semigroup property and hence also for zX_0 with $\operatorname{Re} z > 0$ by the holomorphy (Step 2). Thus finally continuity at zero shows that $v' = 0$.

Step 5: Since the set $U_{\mathfrak{q},+} := U_{1,+} \cap \mathfrak{q}$ is an open subset of $W \subseteq \mathfrak{q}$, it contains an element $X_0 \in \operatorname{int} W$. Next we modify ε above and U_1, such that $\rho_{X_0}^* : X \mapsto X * X_0$ maps U_1 diffeomorphically onto $U_1 * X_0$. Then we choose an open 0-neighborhood $U' \subseteq \mathfrak{g}_{\mathbb{C}}$ which is connected, symmetric, invariant under $-d\tau(1)$, and satisfies $U' * X_0 \subseteq U_1$.

Let $v \in \mathcal{H}_2$. Then there exists $v' \in \mathcal{H}_1$ with $v = \rho(\exp X_0).v'$. We define the mapping

$$\pi_v : U' \to \mathcal{H}, \quad Y \mapsto \beta_{v'}(Y * X_0).$$

This mapping is holomorphic because $\beta_{v'}$ is holomorphic on U_1. If $Y * X_0 \in U_{1,+} \cap U'$, then Step 3 yields

$$\pi_v(Y) = \rho\big(\exp(Y * X_0)\big).v' = \rho\big(\exp(Y)\exp(X_0)\big).v' = \rho(\exp Y).v,$$

so that we have a holomorphic continuation of the mapping $Y \mapsto \rho(\exp Y).v$ to a 0-neighborhood in $\mathfrak{g}_{\mathbb{C}}$.

Finally we want to show that these π_v may be used to construct a unitary representation of G^c with the desired properties. Since G^c is simply connected, it suffices to find a neighborhood U'' of zero in $\mathfrak{g}_{\mathbb{C}}$ and a continuous mapping $\pi : U'' \to U(\mathcal{H})$ such that

$$\pi(0) = 1 \quad \text{and} \quad \pi(X * Y) = \pi(X)\pi(Y)$$

for all $X, Y \in U'' \cap \mathfrak{g}^c$ (cf. [Mag92, 5.2]).

We choose $U'' \subseteq \mathfrak{g}_{\mathbb{C}}$ open and convex such that the following assertions hold:

(i) $U'' * U'' \subseteq U'$.

(ii) $U'' = -U'' = d\tau(1)(U'')$.

Let X, X_1, and $X_2 \in U_{1,+} \cap U'$ with $X_1 * X_2 \in U_{1,+} \cap U'$, and $v, w \in \mathcal{H}_2$. Then

$$(w, \pi_v(X)) = (w, \rho(\exp X).v)$$
$$= \big(\rho(\exp -d\tau(1)X).w, v\big)$$
$$= \big(\pi_w(-d\tau(1)X), v\big)$$

and

$$\big(\pi_w(-d\tau(1)(X_1)), \pi_v(X_2)\big) = \big(\rho(\exp -d\tau(1)(X_1)).w, \rho(\exp X_2).v\big)$$
$$= (w, \rho(\exp X_1)\rho(\exp X_2).v)$$
$$= (w, \rho(\exp X_1 * X_2).v)$$
$$= (w, \pi_v(X_1 * X_2)).$$

Note that there are open non-empty subsets V_1 and V_2 of $U_{1,+}$ such that $V_1 * V_2 \subseteq U_{1,+} \cap U'$. In fact, keeping $X_1 \in U_{1,+} \cap U'$ fixed and choosing X_2 sufficiently close to 1, the openess of $U_{1,+} \cap U'$ implies $X_1 * X_2 \in U_{1,+} \cap U'$ so that our claim follows by continuity.

If $\Theta(X) = -d\tau(\mathbf{1})(\overline{X})$, where $\overline{X}$ is the complex conjugate of X with respect to the real form $\mathfrak{g}$ of $\mathfrak{g}_{\mathbb{C}}$, analytic continuation yields

$$(9.18) \qquad \big(w, \pi_v(X)\big) = \Big(\pi_w(\Theta(X)), v\Big),$$

and

$$(9.19) \qquad \big(\pi_w(\Theta(X_1)), \pi_v(X_2)\big) = (w, \pi_v(X_1 * X_2))$$

for all $X, X_1, X_2 \in \mathcal{U}''$ and $v, w \in \mathcal{H}_2$.

An element $X \in \mathfrak{g}_{\mathbb{C}}$ is in $\mathfrak{g}^c$ iff $\Theta(X) = -X$, so that (9.18) and (9.19) together with $\pi_v(0) = v$ implies that the linear operator

$$v \mapsto \pi_v(X), \quad \mathcal{H}_2 \to \mathcal{H}$$

for $X \in \mathfrak{g}^c \cap \mathcal{U}''$ extends to a unitary operator

$$\pi'(X) \colon \mathcal{H} \to \mathcal{H}.$$

Moreover (9.19) implies now that $\pi'(X * Y) = \pi'(X)\pi'(Y)$ for $X, Y \in \mathcal{U}'' \cap \mathfrak{g}^c$. As already mentioned above, these properties permit to find a homomorphism

$$\pi : G^c \to U(\mathcal{H})$$

such that $\pi(\exp X) = \pi'(X)$ for all $X \in \mathcal{U}'' \cap \mathfrak{g}^c$ (cf. [Mag92, 5.2]).

Since the space $\mathcal{H}_2$ obviously consists of analytic vectors for this representation, it follows in particular that the space of continuous vectors is dense. Whence π is continuous by a standard 3ε-argument.

Let $X \in \mathfrak{h}$. Then $id\pi(X)$ and $id\rho(X)$ coincide on the space $\mathcal{H}_2$ which consists of analytic vectors for both self-adjoint operators. Hence they coincide (cf. Nelson's Theorem, [We76, 8.31]).

For $X \in W$ the operator $d\rho(X)$ is closed by the Hille-Yosida Theorem because it is the generator of a semigroup of contractions. Now the same argument as in the preceding paragraph yields that

$$d\rho(X) = -id\pi(iX).$$

This completes the proof.   ∎

For further results concerning analytic continuation in a similar setting, but for representations on dense subspaces of Hilbert spaces, we refer to [Jø86] and [Jø87].

Remark 9.40.   Let ρ and π as in Theorem 9.39. Then $W(\pi)$ is an invariant cone in $\mathfrak{g}^c$, and the fact that the operators $-id\pi(X)$, $X \in iW$ are negative shows that $-iW \subseteq W(\pi)$. Now the Extension-Theorem (Theorem 9.15) provides an extension of π to a holomorphic contraction representation

$$\widehat{\pi} : S := G^c \operatorname{Exp}\big(iW(\pi)\big) \to C(\mathcal{H}).$$

Let $H^c := \langle \exp_{G^c} \mathfrak{h} \rangle$ and $\Gamma(W)^c := H^c \operatorname{Exp}(W) \subseteq S$. Then $\Gamma(W)^c$ is a semigroup locally isomorphic to $\Gamma(W)$, but these two semigroups need not be coverings of each other because ρ need not be injective.

If ρ is injective, then the mapping $\pi \mid_{\Gamma(W)^c}$ yields a is a semigroup covering onto $\rho(\Gamma(W))$. Let $D \subseteq H^c$ denote its kernel. This is a subgroup which is central in $\Gamma(W)^c$. Hence $\operatorname{Ad}(d) \mid_W = \operatorname{id}_W$ for all $d \in D$. Since W was supposed to be generating, we find that $\operatorname{Ad}_{G^c}(d) = \operatorname{id}$, i.e., $D \subseteq Z(G^c)$. So we can consider the quotient group $G_1^c := G^c/D$. The subgroup $D \subseteq H^c$ is obviuously contained in the kernel of π so that π factors to a unitary representation

$$\pi_1 : G_1^c \to U(\mathcal{H})$$

which in turn has an extension

$$\hat{\pi}_1 : S_1 := S/D \to C(\mathcal{H}).$$

Now H is isomorphic to the subgroup of G_1^c corresponding to the Lie algebra $\mathfrak{h}$, so that $\Gamma(W)$ is isomorphic to a subgroup of S_1. This means that we have found an extension of ρ to a *complex* Ol'shanskiĭ semigroup S_1 containing $\Gamma(W)$.   ∎

Notes

Sections 9.1 and 9.2 contain standard material from functional analysis and the the theory of one-paramter semigroups of operators. The ideas for most of the material in the remainder of this chapter are essentially due to Olshanskiĭ. We have nevertheless expanded his treatment (cf. [Ols82], [Ols89]) quite a bit in order to make the presentation more accessible, and removed the hypothesis that the group in question has to be simple and that the semigroup in question needs to be a subsemigroup of a group. For the case where the groups are solvable many of the results appeared in [HiOl90].

For a variant of the Hardy space theory on symmetric spaces see [HOØ91]. The oscillator semigroup has been found independently by Howe (cf. [How88]) and Brunet-Kramer (cf. [BK80]). For a comprehensive treatment see [Fo89]. The Theorem of Lüscher-Mack appeared in [LM75].

10. The theory for Sl(2)

In this chapter we collect the key results on the theory of Lie semigroups described in this book for the case of Sl(2). This means we will deal with $Sl(2, \mathbb{R})$, $Sl(2, \mathbb{C})$ and their related homogeneous spaces. We adopt the notation from Section 2.2.

Lie wedges and globality

It is clear that all pointed cones in $\mathfrak{sl}(2, \mathbb{R})$ are Lie wedges. If a Lie wedge W contains a line (but is not a line itself), by conjugating we can restrict our attention to three cases:

Case 1): W contains $\mathbb{R}(X_+ - X_-)$.

Case 2): W contains $\mathbb{R}X_0$.

Case 3): W contains $\mathbb{R}X_+$.

In Case 1) W has to be all of $\mathfrak{sl}(2, \mathbb{R})$ since the inner automorphisms coming from $\mathbb{R}(X_+ - X_-)$ are the rotations around this axis (cf. formula (1.2)). In Case 2), up to sign W is either $\mathbb{R}X_0 + \mathbb{R}^+ X_+ + \mathbb{R}^+ X_-$ or $\mathbb{R}X_0 + \mathbb{R}^+ X_+ - \mathbb{R}^+ X_-$ as one easily sees by considering the hyperbolas obtained as orbits under the action of $\exp(\mathbb{R}X_0)$. In the third case we have two possibilities. If W contains a point with non-zero X_--component, then the projection of the $\exp(\mathbb{R}X_+)$-orbit onto $\mathbb{R}X_0 + \mathbb{R}X_-$ along $\mathbb{R}X_+$ contains a line parallel to $\mathbb{R}X_0$ so that W has to contain $\mathbb{R}X_+ + \mathbb{R}X_0$. Thus W is either a half space containing that plane or all of W. If, on the other hand, W is completely contained in the almost abelian algebra $\mathbb{R}X_+ + \mathbb{R}X_0$, then it is a halfplane there or all of that algebra.

The up to sign unique invariant cone in $\mathfrak{sl}(2, \mathbb{R})$ is obtained by rotating $\mathbb{R}^+ X_+$ around the axis $\mathbb{R}(X_+ - X_-)$.

It is no longer so easy to classify all the Lie wedges in $\mathfrak{sl}(2, \mathbb{C})$ and we content ourselves to recall the most important class, namely the ones conjugate to $\mathfrak{sl}(2, \mathbb{R}) + iW$, where W is an invariant cone in $\mathfrak{sl}(2, \mathbb{R})$.

If we want to determine whether a given Lie wedge is global, we have to fix the group first. If G is either $PSl(2, \mathbb{R})$ or any finite covering group thereof, then the analytic subgroup corresponding to $\mathbb{R}(X_+ - X_-)$ is compact and so are all its conjugates. Thus Proposition 1.39 says that a Lie wedge can't be global if it intersects an invariant wedge in the interior. On the other hand we have seen in Section 2.2 that the wedge $\mathbb{R}X_0 + \mathbb{R}^+ X_+ + \mathbb{R}^+ X_-$ is global in $PSl(2, \mathbb{R})$. It therefore is also global in all other groups with Lie algebra $\mathfrak{sl}(2, \mathbb{R})$ (cf. Proposition

1.41). For $G = \mathrm{Sl}(2, \mathbb{R})\widetilde{\ }$, the universal covering group of $\mathrm{Sl}(2, \mathbb{R})$, Proposition 1.40 shows that the half spaces bounded by $\mathbb{R}X_0 + \mathbb{R}X_+$ are global there. Then Proposition 1.37 shows globality for all Lie wedges contained in that half space such that the edge is contained in $\mathbb{R}X_0 + \mathbb{R}X_+$. Proposition 1.37 is not applicable to the invariant cones, but there globality follows from Lemma 6.23. Moreover, Theorem 6.22 shows that rotationally symmetric cones (about the $(X_+ - X_-)$-axis) generate all of $\mathrm{Sl}(2, \mathbb{R})\widetilde{\ }$ as semigroups if they contain an invariant one.

It should be noted here that, because of Corollary 5.5, the globality of any Lie wedge in $\mathfrak{sl}(2, \mathbb{R})$ can be tested by just considering the groups $\mathrm{Sl}(2, \mathbb{R})$ and its universal covering group.

The theory of Olshanskiĭ semigroups shows that wedge $\mathfrak{sl}(2, \mathbb{R}) + iW$ with W invariant in $\mathfrak{sl}(2, \mathbb{R})$ is global in $\mathrm{Sl}(2, \mathbb{C})$ and $\mathrm{PSl}(2, \mathbb{C})$. On the other hand Corollary 3.33 shows that, apart from these two, none of the covering semigroups can be embedded in a group. This of course corresponds to the fact that $\mathrm{Sl}(2, \mathbb{R})$ and $\mathrm{PSl}(2, \mathbb{R})$ are the only Lie groups with Lie algebra $\mathfrak{sl}(2, \mathbb{R})$ which can be embedded in a complex Lie group.

Global hyperbolicity

We have seen in Example 4.28 that $\mathrm{Sl}(2, \mathbb{R})\widetilde{\ }$, ordered by the invariant cone in $\mathfrak{sl}(2, \mathbb{R})$, is not globally hyperbolic, i.e., does not admit Cauchy surfaces (cf. Theorem 4.29). By contrast $\mathrm{Sl}(2, \mathbb{C})/\mathrm{Sl}(2, \mathbb{R})$ is globally hyperbolic and hence a Lorentzian manifold in which two timelike points may be connected by a (Lorentzian)-distance maximizing conal (i.e., timelike) curve (cf. Theorem 4.42). Similar statements are true about the one sheeted hyperboloid in $\mathfrak{sl}(2, \mathbb{R})$ ordered by the cone $\mathbb{R}^+X_+ + \mathbb{R}^-X_-$.

Maximal semigroups with interior points

The maximal semigroups containing generating Lie semigroups in $\mathrm{Sl}(2, \mathbb{R})\widetilde{\ }$ are the ones generated by halfspaces which are bounded by a subalgebra, i.e., a tangent plane to the invariant cone (cf. [Ne91a]). In $\mathrm{PSl}(2, \mathbb{R})$ all the maximal subsemigroups with non-empty interior are conjugate to $\mathrm{PSl}(2, \mathbb{R})^+$ (cf. Theorem 2.5). In $\mathrm{Sl}(2, \mathbb{C})$ the Olshanskiĭ semigroups provide examples of maximal semigroups (cf. Theorem 8.53).

The holomorphic discrete series for $\mathrm{SU}(1,1)$

Recall that

$$\mathrm{Sl}(2, \mathbb{R}) \cong \mathrm{SU}(1,1) = \left\{ \begin{pmatrix} a & b \\ \bar{b} & \bar{a} \end{pmatrix} : a, b \in \mathbb{C}^2, |a|^2 - |b|^2 = 1 \right\}.$$

Let $2 \leq m \in \mathbb{N}$. We consider the unit disc $D := \{z \in \mathbb{C} : |z| < 1\}$ with the measure

$$d\nu_m(re^{i\theta}) = 4^{1-m}(1-r^2)^{m-2} r \, dr d\theta.$$

Then

$$\mathcal{H}_m := L^2(D, \nu_m) \cap \mathrm{Hol}(D)$$

is a Hilbert space and the functions $f_n(z) = z^n$, $n \in \mathbb{N}_0$ form an orthogonal basis (cf. [Lan75, IX.3]).

As in Section 9.4 we define an action of $\mathrm{SU}(1,1)$ on $\mathcal{H}_m$ via

$$\pi_m \begin{pmatrix} a & b \\ \bar{b} & \bar{a} \end{pmatrix} f(z) = (a - \bar{b}\omega)^{-m} f\left(\frac{\bar{a}z - b}{-\bar{b}z + a}\right).$$

The maximal compact subgroup K of $G = \mathrm{SU}(1,1)$ is

$$K = \left\{ \begin{pmatrix} a & 0 \\ 0 & \bar{a} \end{pmatrix} : |a|^2 = 1 \right\} \cong \mathrm{U}(1).$$

Note that $\mathfrak{k} = \mathbf{L}(K)$ is at the same time a compactly embedded Cartan algebra and a maximal compactly embedded subalgebra in $\mathfrak{g} = \mathfrak{su}(1,1)$.

The orthogonal system $\{z^n : n \in \mathbb{N}_0\}$ diagonalizes the action of K, in fact

$$\pi_m \begin{pmatrix} e^{ir} & 0 \\ 0 & e^{-ir} \end{pmatrix} f_n(z) = e^{-imr} f_n\left(\frac{e^{-ir}z}{e^{ir}}\right) = e^{-(m+2n)ir} f_n(z),$$

i.e., $\pi_m(k)f_n = \chi_{m,n}(k)f_n$, where

$$\chi_{m,n} \begin{pmatrix} e^{ir} & 0 \\ 0 & e^{-ir} \end{pmatrix} = e^{-(m+2n)ir}.$$

The functions f_n are analytic vectors for π_m (cf. [Lan75, loc. cit.]), and the derived representation is given by

$$d\pi_m \begin{pmatrix} i & 0 \\ 0 & -i \end{pmatrix} f_n = -i(2n + m)f_n.$$

The corresponding weight is

$$\lambda_{m,n} \begin{pmatrix} ir & 0 \\ 0 & -ir \end{pmatrix} = -ir(2n + m).$$

This shows that the weights of π_m are precisely the functionals

$$\lambda_{0,m}, \lambda_{1,m}, \ldots$$

The root system $\Lambda = \Lambda(\mathfrak{g}_\mathbb{C}, \mathfrak{k}_\mathbb{C})$ consists of two elements $\{\pm\alpha\}$, where

$$\alpha \begin{pmatrix} z & 0 \\ 0 & -z \end{pmatrix} = 2z.$$

Then $\Lambda^+ := \{\alpha\}$ is a positive system. Now

$$\lambda_{n,m} = \lambda_{0,m} - n\alpha,$$

so that $\mathcal{H}_m^K := \mathrm{span}\{f_n : n \in \mathbb{N}_0\}$ is a highest weight module with highest weight $\lambda_m := \lambda_{0,m}$. More precisely, we have

$$\mathfrak{g}_{\mathbb{C}}^{\alpha} = \left\{ \begin{pmatrix} 0 & b \\ 0 & 0 \end{pmatrix} : b \in \mathbb{C} \right\} \quad \text{and} \quad \mathfrak{g}_{\mathbb{C}}^{-\alpha} = \left\{ \begin{pmatrix} 0 & 0 \\ c & 0 \end{pmatrix} : c \in \mathbb{C} \right\}$$

and a simple but tedious calculation shows that

$$d\pi_m \begin{pmatrix} 0 & 1 \\ 1 & 0 \end{pmatrix} f_n = \begin{cases} (n+m)f_{n+1} - nf_{n-1} & \text{for } n \geq 1 \\ mf_1 & \text{for } n = 0 \end{cases}$$

and

$$d\pi_m \begin{pmatrix} 0 & i \\ -i & 0 \end{pmatrix} f_n = \begin{cases} -i(n+m)f_{n+1} - inf_{n-1} & \text{for } n \geq 1 \\ -imf_1 & \text{for } n = 0. \end{cases}$$

Thus

$$d\pi_m \begin{pmatrix} 0 & 1 \\ 0 & 0 \end{pmatrix} f_n = \begin{cases} -nf_{n-1} & \text{for } n \geq 1 \\ 0 & \text{for } n = 0 \end{cases},$$

$$d\pi_m \begin{pmatrix} 0 & 0 \\ 1 & 0 \end{pmatrix} f_n = (n+m)f_{n+1}$$

and

$$d\pi_m \begin{pmatrix} 1 & 0 \\ 0 & -1 \end{pmatrix} f_n = -2(n+m)f_n.$$

Note that the operator $id\pi_m \begin{pmatrix} ri & 0 \\ 0 & -ir \end{pmatrix}$ is negative if and only if $r \leq 0$. Thus the cone $W(\pi_m)$ coincides with the unique invariant cone satisfying

$$C(\pi_m) := W(\pi_m) \cap \mathfrak{k} = \mathbb{R}^+ \begin{pmatrix} -i & 0 \\ 0 & i \end{pmatrix}.$$

Next we establish the link between $\mathcal{H}_m$ and $\mathcal{O}(\lambda_m)$. For $f \in \mathcal{H}_m$ we set

$$F \begin{pmatrix} a & b \\ \bar{b} & \bar{a} \end{pmatrix} := f\left(\frac{b}{a}\right) a^m$$

and it is easy to check that

$$F(gk) = \chi_{0,m}(k)^{-1} F(g) \qquad \forall g \in G, k \in K.$$

Moreover we find that

$$L \begin{pmatrix} 0 & 1 \\ 1 & 0 \end{pmatrix} . F \begin{pmatrix} a & b \\ \bar{b} & \bar{a} \end{pmatrix} = df\left(\frac{b}{a}\right) \frac{1}{\bar{a}^2}$$

and

$$L \begin{pmatrix} 0 & i \\ -i & 0 \end{pmatrix} . F \begin{pmatrix} a & b \\ \bar{b} & \bar{a} \end{pmatrix} = df\left(\frac{b}{a}\right) \frac{i}{\bar{a}^2},$$

so that

$$L_{\mathbb{C}} \begin{pmatrix} 0 & 1 \\ 0 & 0 \end{pmatrix} .F = 0.$$

Thus $F \in \mathcal{O}(\lambda_{0,m})$ and we see that the π_m are precisely the holomorphic discrete series representations.

We want to use the Paley-Wiener Theorem (Theorem 9.32) to determine the image of the Hardy space $\mathcal{H}(W)$ in $L^2(G)$ for $W = W(\pi_m)$. The antiholomorphic discrete series whose representations are the contragredient ones to the holomorphic discrete series satisfy $-W = W(\bar{\pi}_m)$. The other representations that occur in the Plancherel formula (cf. [Lan75, Ch. VII]), that is to say the principle series representations have $\mathfrak{k}$-weights which are symmetric with respect to 0, so for such a representation π we certainly have $W(\pi) = \{0\}$. Thus we see that, in the notation of Section 9.5,

$$I\mathcal{H}(W) = \bigoplus_{m \geq 2} \mathcal{H}_m \otimes \overline{\mathcal{H}_m},$$

because the isotypic component in $L^2(G)$ corresponding to $\mathcal{H}_m$ is isomorphic to $\mathcal{H}_m \otimes \overline{\mathcal{H}_m}$, where the left regular representations acts on the left factor and the right regualr representation acts on the right factor.

We calculate the function K on the semigroup int $\Gamma_G(W)$. Since this holomorphic function is determined by its values on $\exp(iW)$ and since it is invariant under conjugation with G, it suffices to compute it on the elements $a_t := \exp \begin{pmatrix} e^t & 0 \\ 0 & e^{-t} \end{pmatrix}$. Note that

$$K(\gamma^2) = f_\gamma(\gamma) = \|f_\gamma\|_H^2 = \|If_\gamma\|_2^2 \qquad \forall \gamma \in \mathrm{Exp}(W^0),$$

so that

$$R_\gamma(\varphi) = \varphi * (If_{\gamma^*})$$

together with the Plancheral formula for $Sl(2, \mathbb{R})$ yields that

$$K(\gamma^2) = \|If_\gamma\|_2^2 = \sum_{m \geq 2} (m-1) \|\pi_m (If_\gamma)\|_{HS}^2 = \sum_{m \geq 2} (m-1) \|\pi_m(\gamma)\|_{HS}^2,$$

where $\|A\|_{HS} = \mathrm{tr}(AA^*)$ denotes the Hilbert-Schmidt norm of an operator on a Hilbert space.

It remains to calculate the Hilbert-Schmidt norm of the operators $\pi_m(a_t)$ on $\mathcal{H}_m$. First we note that

$$\|\pi_m(a_t)\|_{HS}^2 = \mathrm{tr}\left(\pi_m(a_t)\pi_m(a_t)^*\right) = \mathrm{tr}\left(\pi_m(a_t)^2\right) = \mathrm{tr}\left(\pi_m(a_{2t})\right).$$

So the fact that $\pi_m(a_t)f_n = e^{-t(2n+m)}f_n$ shows that

$$\mathrm{tr}\,\pi_m(a_{2t}) = e^{-2tm} \sum_{n \geq 0} (e^{-4t})^n = \frac{e^{-2tm}}{1 - e^{-4t}}.$$

The calculation of $K(a_{2t})$ is now an easy exercise:

$$
\begin{aligned}
K(a_{2t}) &= \frac{1}{1 - e^{-4t}} \sum_{m \geq 2} (m-1) e^{-2tm} \\
&= \frac{1}{1 - e^{-4t}} \sum_{m \geq 1} m e^{-2t(m+1)} \\
&= \frac{-e^{-2t}}{2(1 - e^{-4t})} \sum_{m \geq 1} (-2m) e^{-2tm} \\
&= \frac{-e^{-2t}}{2(1 - e^{-4t})} \frac{d}{dt} \sum_{m \geq 1} e^{-2tm} \\
&= \frac{-e^{-2t}}{2(1 - e^{-4t})} \frac{d}{dt} \frac{e^{-2t}}{1 - e^{-2t}} \\
&= \frac{-e^{-2t}}{2(1 - e^{-4t})} \frac{-2e^{-2t}(1 - e^{-2t}) - e^{-2t}(-2e^{-2t})}{(1 - e^{-2t})^2} \\
&= \frac{e^{-4t}(1 - 2e^{-2t})}{(1 - e^{-4t})(1 - e^{-2t})^2} \\
&= \frac{1 - 2e^{-2t}}{4 \sinh(2t) \sinh(t)^2}.
\end{aligned}
$$

Finally we obtain

$$
K(a_t) = \frac{1 - 2e^{-t}}{4 \sinh(t) \sinh(\frac{t}{2})^2}.
$$

References

[Alt87]   Alt, H. W., "Lineare Funktionalanalysis", Springer Verlag, Heidelberg, 1987.

[AL91]   Arnal, D., and J. Ludwig, *La convexité de l'application moment d'un groupe de Lie*, J. Funct. Anal. **105**(1992), 256–300.

[BE81]   Beem, J. K., and P. E. Ehrlich, "Global Lorentzian Geometry", Marcel Dekker, New York, Basel, 1981.

[Bou67]   Bourbaki, N., "Espaces vectoriels topologiques, Chap. III-V", Hermann, Paris, 1967.

[Bou68]   —, "Groupes et algèbres de Lie, Chap. IV-VI", Hermann, Paris, 1968.

[Bou71a]   —, "Groupes et algèbres de Lie, Chap. I-III", Hermann, Paris, 1971.

[Bou71b]   —, "Topologie Générale", Chap. I – X, Hermann, Paris, 1971.

[Bou75]   —, "Groupes et algèbres de Lie, Chap. VII, VIII", Hermann, Paris, 1975.

[BK79]   Brunet, M., and P. Kramer, *Semigroups of length increasing transformations*, Reports on Math. Physics **15**(1979), 287–304.

[BK80]   —, *Complex extensions of the representation of the symplectic group associated with the canonical commutation relations*, Reports on Math. Physics **17**(1980), 205–215.

[CHK83]   Carruth, J. N., Hildebrand, J. A., and R. J. Koch, "The Theory of Topological Semigroups", Marcel Dekker, New York, 1983.

[CHK86]   —, "The Theory of Topological Semigroups, Vol. II", Marcel Dekker, New York, 1986.

[CB67]   Choquet-Bruhat, Y., *Hyperbolic partial differential equations on a manifold*, Battelle Rencontres, Benjamin 1967, 84–106.

[CL91]   Chon, I., and J. D. Lawson, *Attainable sets and one-parameter semigroup of sets*, Glasgow Math. Journal **33**(1991), 187–201.

[CP61]   Clifford, A. H., and G. B. Preston, "The algebraic theory of semigroups", Americam Math. Soc., Mathematical Surveys No. 7, Providence, Rhode Island, 1961.

[tD91]   tom Dieck, T., "Topologie", de Gruyter, Berlin, New York, 1991.

[Dö90a]   Dörr, N., *On irreducible $\mathfrak{sl}(2,\mathbb{R})$-modules and $\mathfrak{sl}_2$-triples*, Semigroup Forum **40**(1990), 239–245.

[Dö90b]   —, *On Ol'shanskiĭ's semigroup*, Mathematische Annalen **288** (1990), 21–33.

[Dö91a]   —, *Cartan algebras in symmetric Lie algebras*, Preprint 1370, Technische Hochschule Darmstadt, 1991, submitted.

[Dö91b]   —, *Ol'shanskiĭ wedges in symmetric Lie algebras*, Preprint 1371, Technische Hochschule Darmstadt, 1991, submitted.

[Dö91c]   —, *Symmetric spaces and convex cones*, Seminar Sophus Lie **1** (1991), 65–72.

[DN92]   Dörr, N., and K.-H. Neeb, *On Lie triple wedges and ordered symmetric spaces*, Geometriae Dedicata, to appear.

[Egg90]   Eggert, A., *Lie semialgebras in reductive Lie algebras*, Semigroup Forum 41(1990), 115–121.

[Egg91a]   —, *Über Liesche Semialgebren*, Dissertation, Technische Hochschule Darmstadt, 1991.

[Egg91b]   —, *Lie semialgebras*, Seminar Sophus Lie 1(1991), 41–46.

[Fa87]   Faraut, J., *Algèbres de Volterra et Transformation de Laplace Sphérique sur certains espaces symétriques ordonnés*, Symp. Math. **29** (1987), 183–196.

[Fa91]   —, *Espaces symétriques ordonnés et algèbre de Volterra*, J. Math. Soc. Japan **43**:1(1991), 133–147.

[FHO91]   Faraut, J., Hilgert, J., and G. Olafsson, *Analysis on ordered symmetric spaces*, in preparation.

[FK92]   Faraut, J., and A. Koranyi, "Analysis on symmetric cones and symmetric domains", book in preparation.

[FL85]   Fischer, W., and I. Lieb, "Funktionentheorie", 4. Auflage, Vieweg, Wiesbaden, 1985.

[Fo89]   Folland, G. B., "Harmonic Analysis in Phase Space", Princeton University Press, Princeton, New Jersey, 1989.

[Fü65]   Fürstenberg, H., *Translation invariant cones of functions on semisimple Lie groups*, Bull. of the AMS **71**(1965), 271–326.

[GHL87]   Gallot, S., D. Hulin, and J. Lafontaine, "Riemannian Geometry", Springer Verlag, Heidelberg, 1987.

[Ga60]   Garding, L., *Vecteurs analytiques dans les représentations des groupes de Lie*, Bull. Soc. math. France **88**(1960),73–93.

[GG77]   Gel'fand, M., and S.G. Gindikin, *Complex Manifolds whose Skeletons are real Lie Groups, and Analytic Discrete Series of Representations*, Funct. Anal. and Appl. **11**(1977), 19–27.

[Gi89]   Gichev, V. M., *Invariant Orderings in Solvable Lie Groups*, Sib. Mat. Zhurnal 30(1989), 57–69.

[Gö49]   Gödel, K., *An example of a new type of cosmological solutions of Einstein's field equations of gravitation*, Reviews of modern physics **21**(1949), 447–450.

[Gr83]   Graham, G., *Differentiable semigroups*, in Lecture Notes Math. **998** (1983), 57–127.

[Gu76]   Guts, A. K., *Invariant Orders on Three-Dimensional Lie Groups*, Sib. Mat. Zhurnal 17(1976), 986–992.

[GL84]   Guts, A. K., and A. V. Levichev, *On the Foundations of Relativity Theory*, Sov. Math. Dokl. **30**, 253–257(1984).

[HE73]   Hawking, S. W., and G. F. R. Ellis, "The Large Scale Structure of Space-time", Cambride University Press, Cambridge, 1973.

[Hel78]   Helgason, S., "Differential Geometry, Lie Groups, and Symmetric Spaces", Acad. Press, London, 1978.

[He77]   Heyer, H., "Probability measures on locally compact groups", Springer, New York, Heidelberg, 1977.

[Hi86]   Hilgert, J., *Invariant Lorentzian orders on simply connected Lie groups*, Arkiv för Mat., to appear.

[Hi89]   —, *A note on Howe's oscillator semigroup*, Annales de l'institut Fourier **39**(1989), 663–688.

[Hi92a]   —, *Convexity properties of Graßmannians*, Seminar Sophus Lie **2**:1 (1991), 13-20.

[Hi92b]   —, *A convexity theorem for boundaries of ordered symmetric spaces*, submitted.

[Hi92c]   —, *Controllability on real reductive Lie groups*, Math. Z. **209** (1992), 463–466.

[HiHo85]   Hilgert, J., and K. H. Hofmann, *Old and new on* Sl(2), Manuscripta Math. **54**(1985), 17–52.

[HiHo89]   —, *Compactly embedded Cartan Algebras and Invariant Cones in Lie Algebras*, Advances in Math. **75**(1989), 168–201.

[HiHo90]   —, *On the causal structure of homogeneous manifolds*, Math. Scand. **67**(1990), 119-144.

[HHL89]   Hilgert, J., K. H. Hofmann, and J. D. Lawson, "Lie groups, Convex cones, and Semigroups", Oxford University Press, 1989.

[HiNe91]   Hilgert, J., and K.-H. Neeb, "Lie-Gruppen und Lie-Algebren", Vieweg Verlag, Wiesbaden, 1991.

[HiNe91b] —, *Wiener Hopf operators on ordered homogeneous spaces I*, submitted.

[HiNe92]   —, *Gruppoid C*-algebras of order compactified symmetric spaces*, submitted.

[HiOl90]   Hilgert, J., and G. 'Olafsson, *Analytic continuations of representations, the solvable case*, Jap. Journal of Math., to appear.

[HOØ91]   Hilgert, J., Olafsson, G., and B. Ørsted, *Hardy Spaces on Affine Symmetric Spaces*, J. reine angew. Math. **415**(1991), 189–218.

[HP74]   Hille, E., and R. S. Phillips, "Functional analysis and semigroups", American Math. Soc., Colloquium publications Vol. XXXI, Third printing of revised ed., 1974.

[Ho65]   Hochschild, G., "The Structure of Lie Groups", Holden Day, San Francisco, 1965.

[Ho81]   Hochschild, G. P., "Basic Theory of Algebraic Groups and Lie Algebras", Springer, Graduate Texts in Mathematics **75**, 1981.

[Ho90a]   Hofmann, K. H., *Hyperplane subalgebras of real Lie algebras*, Geometriae Dedicata **36**(1990), 207–224.

[Ho90b]   Hofmann, K. H., *A memo on the exponential function and regular points*, Arch. Math. **59**(1992), 24–37.

[Ho91a]   —, Ed., Seminar Sophus Lie 1(1991), Heldermann Verlag, Berlin, 1991.

[Ho91b]   —, *Einige Ideen Sophus Lies - Hundert Jahre danach*, Jahrbuch Überblicke Mathematik 1991, Vieweg, 93–125.

[Ho91c]   —, *Zur Geschichte des Halbgruppenbegriffs*, Historia Mathematica 19(1992), 40–59.

[Ho91d]   —, *The fundamental theorems*, Seminar Sophus Lie 1(1991), 33–39.

[HoLa81]   Hofmann, K. H., and J. D. Lawson, *The local theory of semigroups in nilpotent Lie groups*, Semigroup Forum 23(1981), 343–357.

[HoLa83]   —, *Foundations of Lie Semigroups*, Springer Verlag, Lecture Notes Math. 998(1983), 128–201.

[HLP90]   Hofmann, K. H., Lawson, J. D., and J. S. Pym, Eds., "The Analytic and Topological Theory of Semigroups – Trends and Developments", de Gruyter, 1990.

[HoRu88]   Hofmann, K. H., and W. A. F. Ruppert, *On the Interior of Subsemigroup of Lie Groups*, Transaction of the AMS 324(1991), 169–179.

[HoRu91a]   —, *The structure of Lie groups which support closed divisible subsemigroups*, Preprint Technische Hochschule Darmstadt, Nr. 1431, November 1991.

[HoRu91b]   —, *Divisible subsemigroups of Lie groups*, Seminar Sophus Lie 1:2 (1991), 205–213.

[Ho88]   Howe, R., *The Oscillator semigroup*, in "The Mathematical Heritage of Hermann Weyl", Proc. Symp. Pure Math. 48, R. O. Wells Ed., AMS Providence, 1988.

[I86]   Ihringer, S., *Keile und Halbgruppen*, Dissertation, Technische Hochschule Darmstadt, 1986.

[Jac57]   Jacoby, R., *Some theorems on the structure of locally compact local groups*, Ann. Math. 66(1957), 36–69.

[Jø86]   Jørgensen, P. E. T., *Analytic continuation of local representations of Lie groups*, Pac. Journal of Math. 125:2(1986), 397–408.

[Jø87]   —, *Analytic continuation of local representations of symmetric spaces*, J. of Funct. Anal. 70(1987), 304–322.

[JK81a]   Jurdjevicz, V., and I. Kupka, *Control systems on semisimple Lie groups and their homogeneous spaces*, Ann. Inst. Fourier 31(1981), 151–179.

[JK81b]   —, *Control systems subordinated to a group action: Accessibility*, J. Diff. Eq. 39(1981), 180–211.

[JS72]   Jurdjevicz, V., and H. Sussmann, *Control systems on Lie groups*, J. Diff. Eq. 12(1972), 313–329.

[Ka70]   Kahn, H. D., *Covering semigroups*, Pacific Journal of Mathematics 34(1970), 427–439.

[Kan91]   Kaneyuki, S., *On the causal structure of the Shilov boundaries of symmetric bounded domains*, Preprint, Lecture Notes Math., to appear.

[Ki76]   Kirillov, A. A., "Elements of the Theory of Representations", Sprin-
           ger, New York, 1976.

[Kn86]   Knapp, A. W., "Representation theory of semisimple Lie groups",
           Princeton University Press, Princeton, New Jersey, 1986.

[KR82]   Kumaresan, S., and A. Ranjan, *On invariant convex cones in simple
           Lie algebras*, Proc. Ind. Acad. Sci. Math. **91**(1982), 167–182.

[Lan75]   Lang, S., "SL(2)", Addisson Wessley, 1975.

[La87]   Lawson, J. D., *Maximal subsemigroups of Lie groups that are total*,
           Proceedings of the Edinburgh Math. Soc. **87**(1987), 497–501.

[La87b]   —, *Embedding local semigroups into groups*, Proc. of the New York
           Acad. Sci., to appear.

[La89]   —, *Ordered Manifolds, Invariant Cone Fields, and Semigroups*, Fo-
           rum Math. **1**(1989), 273–308.

[La91]   —, *Polar and Ol'shanskiĭ decompositions*, Seminar Sophus Lie **1**:2
           (1991), 163–173.

[Le85]   Levichev, A., *Sufficient conditions for the nonexistence of closed
           causal curves in homogeneous space-times*, Izvestia Phys. **10**(1985),
           118–119.

[LM87]   Libermann, P., and C. – M. Marle, "Symplectic geometry and analyt-
           ical mechanics", Mathematics and its applications, D. Reidel Publ.
           Comp., 1987.

[Loe88]   Loewner, Ch., "Collected Papers", Contemporary Math., Birkhäuser,
           Berlin, 1988.

[Lo69]   Loos, O., "Symmetric Spaces I : General Theory", Benjamin, New
           York, Amsterdam, 1969.

[LM75]   Lüscher, M., and G. Mack, *Global conformal invariance and quantum
           field theory*, Comm. Math. Phys. **41**(1975), 203–234.

[Mag92]   Magyar, M., "Continuous linear representations", North-Holland,
           Mathematical Studies 168, 1992.

[MN92a]   Mittenhuber, D., and K.-H. Neeb, *On the exponential function of
           ordered manifolds with affine connection*, submitted.

[MN92b]   —, *On the exponential function of an invariant Lie semigroup*, Sem-
           inar Sophus Lie **2**:1(1992), 21-30.

[Ne90a]   Neeb, K.-H., *Globality in Semisimple Lie Groups*, Annales de l'In-
           stitut Fourier **40** (1990), 493–536.

[Ne90b]   —, *The Duality between subsemigroups of Lie groups and monotone
           function*, Trans. of the Amer. Math. Soc. **329**(1992), 653–677.

[Ne91a]   —, *Semigroups in the Universal Covering Group of Sl(2)*, Semigroup
           Forum **43**(1991), 33-43.

[Ne91b]   —, *Conal orders on homogeneous spaces*, Inventiones math. **104**
           (1991), 467–496.

[Ne91c]   —, *Monotone functions on symmetric spaces*, Mathematische An-
           nalen **291**(1991), 261–273.

308 References

[Ne91d] —, *A short course on the Lie theory of semigroups III - Globality of invariant wedges*, Seminar Sophus Lie 1(1991), 47–54.

[Ne91e] —, *Objects dual to subsemigroups of Lie groups*, Monatshefte für Mathematik 112(1991), 303–321.

[Ne91f] —, *On the fundamental group of a Lie semigroup*, Glasgow Math. J. 34(1992), 379–394.

[Nc91g] , *On the foundations of Lie Semigroups*, J. reine angew. Math. 431(1992), 165–189.

[Ne92a] —, *Invariant orders on Lie groups and coverings of ordered homogeneous spaces*, submitted.

[Ne92b] —, *Invariant subsemigroups of Lie groups*, Memoirs of the AMS, to appear.

[Ne92c] —, *A convexity theorem for semisimple symmetric spaces*, Pacific Journal of Math., to appear.

[Ne92d] —, *Contraction semigroups and representations*, Forum Math., to appear.

[Ne93a] —, "Holomorphic representation theory and coadjoint orbits of convexity type", Habilitationsschrift, Technische Hochschule Darmstadt, Januar 1993.

[Ne93b] —, *On closedness and simple connectedness of adjoint and coadjoint orbits*, submitted.

[Ne93c] —, *Holomorphic representation theory I*, Preprint Nr. 1536, Technische Hochschule Darmstadt, Feb. 1993.

[Ne93d] —, *Holomorphic representation theory II*, Preprint Nr. 1537, Technische Hochschule Darmstadt, Feb. 1993.

[Ne93e] —, *Holomorphic representation theory III*, submitted.

[Ne93f] —, *Kähler structures and convexity properties of coadjoint orbits*, submitted.

[Nel70] Nelson, E., *Analytic vectors*, Ann. of Math. 70(1959), 572–615.

[Ol90] G. Olafsson, *Causal symmetric spaces*, Mathematica Gottingensis 15(1990).

[Ols82a] Ol'shanskiĭ, G. I., *Invariant cones in Lie algebras, Lie semigroups, and the holomorphic discrete series*, Funct. Anal. and Appl. 15, 275–285 (1982).

[Ols82b] —, *Invariant orderings in simple Lie groups. The solution to E. B. Vinberg's problem*, Funct. Anal. and Appl. 16(1982), 311–313.

[Ols89] —, *Complex Lie semigroups, Hardy spaces and the Gelfand Gindikin program*, Preprint, 1982.

[Ols91] —, *On semigroups related to infinite dimensional groups*, Topics in Representation Theory, Adv. in Sov. Math 2, AMS 1991, 67–101.

[OV90] Onishchick, A. L. and E. B. Vinberg, "Lie Groups and Algebraic Groups", Springer, 1990.

[OM80]   Oshima, T., and T. Matsuki, *Orbits on affine symmetric spaces under the action of the isotropy subgroups*, J. Math. Soc. Japan **32**(1980), 399-414.

[Pal57]   Palais, R.S., "A global formulation of the Lie theory of Transformation groups", Mem. of the AMS, 22m, 1957.

[Pa81]   Paneitz, S., *Invariant convex cones and causality in semisimple Lie algebras and groups*, J. Funct. Anal. **43**(1981), 313–359.

[Pa84]   —, *Determination of invariant convec cones in simple Lie algebras*, Arkiv för Mat. **21**(1984), 217–228.

[Paz83]   Pazy, A., "Semigroups of Linear Operators and Applications to Partial Differential Equations", Springer, 1983.

[Pe72]   Penrose, R., "Techniques of differential topology in relativity", Regional Conference Series in Applied Math. SIAM **7**, 1972.

[Po90]   Poguntke, D., *Invariant Cones in Solvable Lie Algebras*, Mathematische Zeitschrift **210**(1992), 661–674.

[Po62]   Pontrjagin, L. S., "The Mathematical Theory of Optimal Processes", Interscience Publishers, New York, London, 1962.

[Ro72]   Rothschild, L. P., *Orbits in a real reductive Lie algebra*, Transactions of the AMS **168**(1972), 403–421.

[RK82]   Rothstein, W., and K. Kopfermann, "Funktionentheorie mehrerer komplexer Veränderlicher", B. I. Wissenschaftsverlag, Mannheim, Wien, Zürich, 1982.

[Rud86]   Rudin, W., "Real and Complex Analysis", McGraw-Hill, New York, 1986.

[Ru88a]   Ruppert, W. A. F., *A Geometric Approach to the Bohr Compactification of Cones*, Math. Z. **199**(1988), 209–232.

[Ru88b]   —, *Bohr Compactifications of Non-abelian Lie groups*, Semigroup Forum (1988), 325–342.

[Ru89]   —, *On open subsemigroups of connected groups*, Semigroup Forum **39**(1989), 347–362.

[SM91]   San Martin, L., *Invariant control sets on flag manifolds*, to appear.

[SM92]   —, *Nonreversibility of Subsemigroups of Semi-Simple Lie Groups*, Semigroup Forum **44**(1992), 476- 487.

[Sch75]   Schubert, H., "Topologie", Teubner Verlag, Stuttgart, 1975.

[Se76]   Segal, I. E., "Mathematical Cosmology and Extragalactical Astronomy", Academic Press, New York, San Francisco, London, 1976.

[Sp88]   Spindler, K., "Invariante Kegel in Liealgebren", Mitt. aus dem mathematischen Sem. Gießen **188**, 1988.

[Sp89]   —, *Some remarks on Levi complements and roots in Lie algebras with cone potential*, Proceedings Edinb. Math. Soc., II. Ser. **35**(1992), 71–88.

[St86]   Stanton, R. J., *Analytic Extension of the holomorphic discrete series*, Amer. J. Math. **108**(1986), 1411–1424.

[Sug75] Sugiura, M., "Unitary representations and Harmonic Analysis – An introduction –", J. Wiley and Sons, New York, London, Sydney, Toronto, 1975.

[Su72] Sussmann, H. J., *The "Bang-bang" Problem for Certain Control Systems in* Gl(n, IR), SIAM J. Control **10**(1972), 470–476.

[SJ72] Sussmann, H. J., and V. Jurdjevic, *Controllability of nonlinear systems*, J. Diff. Eq. **12**(1972), 95–166.

[Ti83] Tits, J., "Liesche Gruppen und Algebren", Springer. New York, Heidelberg, 1983.

[Vin63] Vinberg, E. B., *The theory of convex homogeneous cones*, Transactions of the Mosc. Math. Soc **12**(1963), 303–358.

[Vin80] —, *Invariant cones and orderings in Lie groups*, Funct. Anal. and Appl. **14**(1980), 1–13.

[Wa73] Wallach, N., "Harmonic analysis on homogeneous spaces", Pure and applied mathematics **19**, Marcel Dekker, New York, 1973.

[We76] Weidmann, J., "Lineare Operatoren in Hilberträumen", Teubner, Stuttgart, 1976.

[W91a] Weiss, W., *Local Lie semigroups and open embeddings into global topological semigroups*, Indagationes Math., to appear.

[W91b] Weiss, W., *Sophus Lie's fundamental theorems - Categorial aspects*, in: H. Herrlich and H.-E. Porst, Eds., "Category Theory and Applications", Heldermann Verlag, Berlin, 1991.

[Wi89] Wildberger, N., *The moment map of a Lie group representation*, Transact. AMS **330**(1989), 257–268.

[Wo69] Wolf, J., *The action of a real semisimple Lie group on a complex flag manifold, I: Orbit structure and holomorphic arc components*, Bull. of the AMS **75**(1969), 1121–1237.

List of Symbols

Index

Printing: Weihert-Druck GmbH, Darmstadt
Binding: Buchbinderei Schäffer, Grünstadt